Springer Undergraduate Texts in Mathematics and Technology

Series Editors

Helge Holden, Department of Mathematical Sciences, Norwegian University of Science and Technology, Trondheim, Norway

Keri A. Kornelson, Department of Mathematics, University of Oklahoma, Norman, OK, USA

Editorial Board

Lisa Goldberg, Department of Statistics, University of California, Berkeley, Berkeley, CA, USA

Armin Iske, Department of Mathematics, University of Hamburg, Hamburg, Germany

Palle E. T. Jorgensen, Department of Mathematics, University of Iowa, Iowa City, IA, USA

Springer Undergraduate Texts in Mathematics and Technology (SUMAT) publishes textbooks aimed primarily at the undergraduate. Each text is designed principally for students who are considering careers either in the mathematical sciences or in technology-based areas such as engineering, finance, information technology and computer science, bioscience and medicine, optimization or industry. Texts aim to be accessible introductions to a wide range of core mathematical disciplines and their practical, real-world applications; and are fashioned both for course use and for independent study.

Saber N. Elaydi · Jim M. Cushing

Discrete Mathematical Models in Population Biology

Ecological, Epidemic, and Evolutionary Dynamics

Saber N. Elaydi
Department of Mathematics
Trinity University
San Antonio, TX, USA

Jim M. Cushing
Department of Mathematics
The University of Arizona
Tucson, AZ, USA

ISSN 1867-5506   ISSN 1867-5514 (electronic)
Springer Undergraduate Texts in Mathematics and Technology
ISBN 978-3-031-64794-9   ISBN 978-3-031-64795-6 (eBook)
https://doi.org/10.1007/978-3-031-64795-6

Mathematics Subject Classification: 92-10, 97-01

© Springer Nature Switzerland AG 2024

This work is subject to copyright. All rights are solely and exclusively licensed by the Publisher, whether the whole or part of the material is concerned, specifically the rights of translation, reprinting, reuse of illustrations, recitation, broadcasting, reproduction on microfilms or in any other physical way, and transmission or information storage and retrieval, electronic adaptation, computer software, or by similar or dissimilar methodology now known or hereafter developed.
The use of general descriptive names, registered names, trademarks, service marks, etc. in this publication does not imply, even in the absence of a specific statement, that such names are exempt from the relevant protective laws and regulations and therefore free for general use.
The publisher, the authors and the editors are safe to assume that the advice and information in this book are believed to be true and accurate at the date of publication. Neither the publisher nor the authors or the editors give a warranty, expressed or implied, with respect to the material contained herein or for any errors or omissions that may have been made. The publisher remains neutral with regard to jurisdictional claims in published maps and institutional affiliations.

This Springer imprint is published by the registered company Springer Nature Switzerland AG
The registered company address is: Gewerbestrasse 11, 6330 Cham, Switzerland

If disposing of this product, please recycle the paper.

Preface

Welcome to the captivating realm of discrete-time mathematical biology! Within these pages, we embark on an illuminating journey through the rich landscapes of various discrete-time mathematical models, each offering profound insights into the dynamics of biological systems.

This book grew out of lecture notes the first author used to teach at Trinity University. It was greatly expanded when Jim Cushing joined as an author and significant contributions were made by the late Abdul-Aziz Yakubu.

The book delves into a comprehensive exploration of discrete-time competition models, predator–prey models, evolutionary models, infectious disease models, structured population models, and periodically forced biological models. The fusion of mathematics and biology has yielded extraordinary revelations about living organisms' intricate workings and complex interactions with their environment. Mathematical models have become indispensable tools for studying ecological and evolutionary processes, paving the way for groundbreaking discoveries and practical applications in a multitude of fields.

In this book, we lay the foundation for understanding the beauty and power of discrete-time models. By discretizing time, we acknowledge the importance of the discrete nature of biological events and its profound implications for understanding real-world biological systems. The discrete-time framework provides a versatile platform to explore the profound effects of time intervals, periodicity, and stage-structured dynamics, which are instrumental in capturing the essence of various biological phenomena.

We begin our exploration with single-species models, where we establish a framework for discrete-time modeling. Then we investigate competition models, where multiple species vie for limited resources. Discrete-time competition models shed light on the coexistence of species, the intricacies of competitive exclusion, and the role of environmental factors in shaping ecological communities. As we navigate through these models, we unravel the delicate balance that governs the dynamics of competing species and the consequences it holds for ecosystem stability.

Next, we turn our gaze to predator–prey interactions. Through discrete-time models, we decipher the oscillatory dynamics of predator–prey populations, uncovering the pivotal role of feedback mechanisms and nonlinear interactions in shaping the trajectories of these intricate relationships.

The journey continues with evolutionary models, where the winds of adaptation and natural selection guide the path of genetic variation. In the discrete-time evolutionary landscape, we explore the emergence and spread of advantageous traits, the origin of phenotypic diversity, and the coevolutionary arms race that shapes the evolution of biological populations.

Structured population models emerge as a crucial tool for understanding the life cycles of organisms, where individuals pass through distinct life stages with varying characteristics. Through discrete-time structured population models, we gain insights into population dynamics, life history strategies, and the consequences of life stage-specific mortality and reproduction.

Other examples of discrete-time of structured population models are infectious disease models, with populations structured by disease status. These models step into the spotlight as we unravel the secrets of disease transmission and control. These models allow us to dissect the dynamics of epidemics, investigate the effects of intervention strategies, and illuminate the role of contact networks and immunity in determining the fate of infectious diseases.

Finally, we delve into the captivating world of periodically forced biological models, where the rhythms of the environment influence the fate of living systems. In this realm, we explore the consequences of periodic variations, seasonal changes, and cyclic environmental factors on the population dynamics and ecological interactions.

Throughout this book, we strive to strike a harmonious balance between theoretical principles, mathematical rigor, and practical applications. We provide illustrative examples, numerical simulations, and empirical case studies to enhance understanding and facilitate the translation of discrete-time mathematical biology to real-world challenges.

The core of the book is chapters 1–4. The general outline of the interconnections among the chapters in the book is outlined in the diagram below. For an undergraduate course, the instructor may consider chapters 1, 3, 4, 8; 1, 2, 3, 4; or 1, 3, 4, 5; depending on his or her interest. Of course, the instructor is free to use other variations.

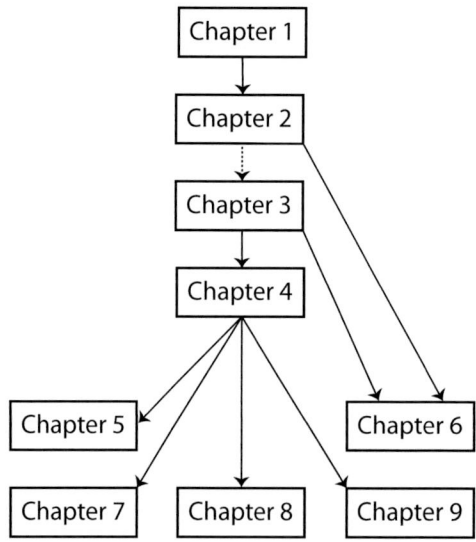

The diagram delineates the interconnections and suggested sequences among the chapters. A solid arrow leading from Chapter A to Chapter B denotes a prerequisite relationship, indicating that comprehension of Chapter A is essential before delving into Chapter B. To lay a solid foundation, we strongly advise reading the initial four chapters, which form the book's foundational segment, before exploring the more advanced chapters that follow.

The mathematical foundation required for this book includes:

- **Linear Algebra**—Understanding of matrices, vectors, and their operations. Essential for modeling interactions in complex systems.
- **Advanced Calculus**—Proficiency in multivariable calculus, and basic elements of real analysis.

Biological Prerequisites

From a biological standpoint, the students should have a basic understanding of:

- **Ecology**—Fundamental concepts such as ecosystems, population dynamics, and interspecies interactions.
- **Epidemiology**—Basic principles of disease spread, disease dynamics, and public health implications.

Course Applications

This book can be used in various educational settings:

- **First Course in Biomathematics**—An introductory course that provides students with the basics of mathematical techniques used in biological research.
- **Modeling Course**—Focuses on developing and analyzing mathematical models that represent biological phenomena.
- **Advanced Mathematical Biology Course**—A deeper dive into complex models and sophisticated mathematical frameworks to address cutting-edge problems in biology.

The book may also be used in graduate courses focusing on discrete-time modeling. Moreover, the book will be a good reference for researchers working on discrete-time biological models. It will also help researchers by providing them with challenging questions and open problems. Note that the starred problems are rather challenging for students. There are also research projects at different levels of difficulty that are suited for graduate students and researchers.

We hope this book serves as an invaluable resource for students, researchers, and practitioners, transcending disciplinary boundaries and inspiring novel collaborations. By immersing ourselves in the enthralling world of discrete-time mathematical biology, we illuminate the beauty of mathematical abstraction and its profound impact on the study of life.

San Antonio TX, USA	Saber N. Elaydi
Tucson, AZ, USA	Jim M. Cushing
2024	

Acknowledgements

It is with deep sadness that we acknowledge the significant contributions of the late Abdul-Aziz Yakubu to various sections of this book, with particular emphasis on his invaluable input in the chapters dedicated to epidemic models. Aziz's exceptional expertise and dedication have left an indelible mark on this work. His deep understanding of epidemic models and their intricate dynamics has significantly influenced the content and quality of the chapters he worked on. Throughout the writing process, Aziz demonstrated his brilliance as a researcher and collaborator. His analytical thinking, meticulous attention to detail, and ability to present complex concepts with clarity and accessibility greatly enhanced the depth and coherence of the chapters to which he contributed.

While we deeply mourn Aziz's loss, we are profoundly grateful for his significant contributions to this book. His passion for research, his unwavering dedication to advancing scientific knowledge, and his exceptional abilities as a scholar will forever be remembered and appreciated.

Contents

1 **Scalar Population Models** 1
   1.1 Introduction 1
   1.2 Linear (Density-Independent) Models 3
      1.2.1 Models Without Migration 3
      1.2.2 Models With Migration/Immigration 6
   1.3 Nonlinear Autonomous (Density-Dependent) Models 9
      1.3.1 Modeling Density Effects 10
      1.3.2 Scramble versus Contest Models 15
      1.3.3 Stability of Equilibrium Points 19
      1.3.4 Cobweb Analysis 20
   1.4 Stability by Linearization 25
      1.4.1 Global Asymptotic Stability of Equilibrium Points 29
      1.4.2 Bifurcation Analysis 37
      1.4.3 Population Cycles: Periodic Orbits 40
      1.4.4 Persistence of Species 44
      1.4.5 Period-Doubling Bifurcations Route to Chaos 45
      1.4.6 Chaos: Sensitive Dependence 49

2 **Linear Structured Population Models** 55
   2.1 Introduction 55
   2.2 Structured Population Models 56
   2.3 Eigenvalues and Eigenvectors 62
   2.4 The Perron–Frobenius Theorem 66
   2.5 Analysis of Linear Matrix Equations 75
      2.5.1 The Fundamental Theorem of Demography 75
      2.5.2 The Reproduction Number $\mathcal{R}_0$ 76
      2.5.3 Sensitivity Analysis 87

3 **Linear and Nonlinear Systems** 101
   3.1 Introduction 101
   3.2 Linear Systems 101
      3.2.1 Basic Theory 101
      3.2.2 Variation of Constants Formula 104
      3.2.3 Phase Space Analysis 105
   3.3 Nonlinear Competition Models 115
      3.3.1 Examples of Competition Models 118
      3.3.2 Stability Analysis 120
   3.4 Local Stability of Competition Systems 122

	3.5	Predator–Prey Models.	129
		3.5.1 Nicholson–Bailey Model	129
		3.5.2 Beddington et al. Model	131
		3.5.3 May Model	135
	3.6	Population Cycles: Periodic Orbits	137
	3.7	The Stable Manifold Theorem	144
	3.8	Global Stability	149
		3.8.1 Liapunov Functions	150
		3.8.2 Hierarchical Models	155
		3.8.3 Competitive Systems	158
		3.8.4 Comparison Methods	161
		3.8.5 Cooperative Systems	164
	3.9	Persistence	169
		3.9.1 Single species	169
		3.9.2 Stage-structured Models	170
	3.10	Bifurcation	174
		3.10.1 Period-Doubling Bifurcation	175
		3.10.2 Neimark–Sacker Bifurcation	177
		3.10.3 Saddle-Node Bifurcation	182
	3.11	Appendix	187
4	**Infectious Disease Models: Part I**		**189**
	4.1	Introduction	189
	4.2	Discrete-Time Kermack–McKendrick Type SIR Epidemic Model.	189
		4.2.1 Discrete-Time SIR Model	190
		4.2.2 Some Properties of the SIR Model	193
		4.2.3 Calculation of the Mean Infectious Period	199
	4.3	A Case Study: The 1978 English Boarding School Influenza Outbreak	200
	4.4	Discrete-Time SIRS Model	201
	4.5	Herd Immunity	205
	4.6	SIS Discrete-Time Epidemic Models With Birth and Death	209
	4.7	The Reproduction Number $\mathcal{R}_0$	212
	4.8	Global Stability of Equilibria	220
		4.8.1 Liapunov Functions	220
		4.8.2 Models with Constant or Asymptotically Constant Population Size	222
	4.9	Disease Acquired Herd Immunity	227
5	**Models with Multiple Attractors**		**235**
	5.1	Introduction	235
	5.2	Models with Multiple Attractors with no Allee Effects	236
	5.3	Models with the Allee Effects	242
	5.4	Model Derivation for Single Species	243
	5.5	Allee Effects and Hysteresis	248
	5.6	Competition Models with the Allee Effect	253
	5.7	Predator–Prey Models with Allee Effects	262
	5.8	Global Dynamics of Population Models with the Allee Effect	268
	5.9	Appendix	278

CONTENTS

6 Nonlinear Structured Population Models **281**
- 6.1 Introduction 281
- 6.2 Equilibria: The Fundamental Bifurcation Theorem 286
- 6.3 Stable Bifurcation: The Primitive Case 293
- 6.4 Stable Bifurcation: The Imprimitive Case 300
  - 6.4.1 The Case $k = 2$ 302
  - 6.4.2 The Case $k = 3$ 312
  - 6.4.3 The Case $k \geq 4$ 316
- 6.5 Secondary Bifurcations 316
- 6.6 A Case Study: The LPA Model 319
  - 6.6.1 Model Derivation 321
  - 6.6.2 Basic Analysis of the LPA Model 322
  - 6.6.3 Secondary Bifurcations in the LPA Model 323
  - 6.6.4 Discussion 327

7 Infectious Disease Models Part: II **335**
- 7.1 Introduction 335
- 7.2 Endemic Equilibria 336
  - 7.2.1 The Bifurcation of Endemic Equilibria 338
  - 7.2.2 Stability of Endemic Equilibria 340
- 7.3 Models with Multiple Infected Classes 345
  - 7.3.1 A Basic SI Model 349
  - 7.3.2 SIR and SEIR Models 355
- 7.4 Models with Disease-Free Population Cycles 359
- 7.5 Two Case Studies 367
  - 7.5.1 A Malaria Model 367
  - 7.5.2 Isolation and Quarantine: SARS Model 373

8 Evolutionary Models **379**
- 8.1 Introduction 379
  - 8.1.1 Darwinian Dynamic Models 379
  - 8.1.2 An Example: A Darwinian Logistic (Beverton–Holt) Model 381
  - 8.1.3 Evolutionary Stable Strategies 384
- 8.2 The Fundamental Bifurcation Theorem for Darwinian Models 385
- 8.3 Examples 391
  - 8.3.1 A Darwinian Juvenile–Adult Model 391
  - 8.3.2 A Darwinian LPA Model 396
  - 8.3.3 A Multiple Trait Discrete Logistic (Beverton–Holt) Model 397
  - 8.3.4 Non-equilibrium Dynamics: A Darwinian Ricker Model 398
  - 8.3.5 A Darwinian Leslie–Gower Competition Model 400
- 8.4 Global Stability 407
- 8.5 Secondary Bifurcations 411

9 Nonautonomous Models **417**
- 9.1 Introduction 417
- 9.2 Examples of Nonautonomous Models 417
- 9.3 Linear Periodic Difference Equations 418
  - 9.3.1 Attenuance Versus Resonance in Linear Periodic Models 419

9.4	Periodically Forced Discrete Logistic (Beverton–Holt) Model..............		424
	9.4.1	The Cushing–Henson Conjecture: Discrete Logistic (Beverton–Holt) Recruitment...................................	424
	9.4.2	Deleterious Effects of Periodic Forcing.........................	426
	9.4.3	Discrete Logistic (Beverton–Holt) Recruitment with $p=2$...........	428
9.5	Nonautonomous Ricker Model.....................................		430
	9.5.1	Periodically Forced Ricker Model.............................	433
	9.5.2	Neither Attenuance nor Resonance when $r(t) = K(t)$..............	435
9.6	Nonlinear Periodic Systems.......................................		438
	9.6.1	Monotone Periodic Systems.................................	438
	9.6.2	Mixed Monotone Periodic Systems............................	440
	9.6.3	Asymptotically Periodic Systems.............................	442
	9.6.4	Small Amplitude Perturbations..............................	445
	9.6.5	Examples..	447
	9.6.6	A Case Study...	453

Bibliography **459**

Index **477**

Chapter 1
Scalar Population Models

The main focus of this chapter is on the introduction of single-species population models and their analysis. Of fundamental interest in population dynamics is how population size changes in time. Is the population increasing, decreasing, fluctuating, or even chaotic? Is it in danger of extinction or will it persist? What are the characteristics of the population's trajectory? Will it stabilize at an equilibrium state or will it be subject to crash-and-boom sustained oscillations? The answers to these questions require an understanding of the basic demographic processes of birth and death and possibly other processes by which individuals might arrive or depart from the population, such as immigration and/or emigration. In this chapter, we consider models in which a single, aggregate state variable is used to characterize the population. These models, which in effect treat all individuals in the population as identical, are called *unstructured* population models. In Chapter 2, we will consider structured population models in which individuals can differ according to some model-designated structuring variable such as age, size, weight, etc.

1.1 Introduction

By population size, which we denote by x, we can mean many things: the number of individuals, the density of individuals, total biomass or dry weight, and so on. We are concerned with the problem of predicting or projecting the population size from one census to the next, when the census data are collected at discrete times $t_0 < t_1 < t_2 < \cdots$. The modeling methodology is based on simple bookkeeping procedures that account for the survival of existing individuals, the birth of new individuals, and the arrival or departure of other individuals through other means (such as immigration/harvesting or emigration/stocking).

Generally, it is assumed that census times are equally spaced. This time interval is at the discretion of the modeler. It could be, for example, simply a convenient time for a follow-up census or it could be a time unit chosen with regard to some significant physiological or life history event (such as a maturation period or generation time). We can assume without loss in generality (by choice of time units) that the time interval is 1. Our goal is to formulate a model that predicts the population size $x(t+1)$ at time $t+1$ from a knowledge of the population size $x(t)$ at time t.

A general population model is given by the difference equation

$$x(t+1) = f(x(t)), \qquad (1.1)$$

where f is the function that gives $x(t+1)$ knowing $x(t)$. For instance, if we are given the population size $x_0 = x(0)$ at time zero, then we can generate all future population sizes,

$$x(1) = f(x_0),$$
$$x(2) = f(x(1)) = f^2(x_0),$$
$$\vdots$$
$$x(t) = f(x(t-1)) = f^t(x_0)).$$

Here f^t is the composition of f taken t times. One of the main objectives of modeling is to find the appropriate function f.

Consider first the case of a closed population, that is to say, a population subject only to birth and death processes and is free of other means of arrivals and departures of individuals. An individual present at time $t+1$ was either present at time t or not. In the latter case, the individual is a newborn. In the former case, the individual is a survivor from the previously censused population. We have then

$$x(t+1) = \text{ newborns} + \text{survivors} \tag{1.2}$$

and the model is formulated by making assumptions and writing mathematical expressions for the number of newborns and survivors, as they depend on the previous population size $x(t)$. The most common assumption is that the number of newborns present at the next census time $t+1$ is proportional to the population's size at time t, that is

$$\text{newborns} = bx(t), \tag{1.3}$$

where $b \geq 0$ is called a (per capita) *fertility rate*. (b is often called a birth rate, but any individual born during the interval between census times must survive to the next census in order to be counted as a newborn. Thus, b also has a survival component.) If the probability (or the fraction of) individuals alive at time t that survive to time $t+1$ is denoted by s, $0 \leq s \leq 1$, then the number of

$$\text{survivors} = sx(t). \tag{1.4}$$

The fraction s is the (per capita) *probability of survival* (or the *survival rate*) over one time unit. We then arrive at the equation

$$x(t+1) = bx(t) + sx(t) \tag{1.5}$$

for the predicted population size $x(t+1)$ at time $t+1$.

If the vital rates b and s remain constant in time, then Equation (1.5) is an autonomous linear difference equation (or recursion formula). There are, however, many reasons why b and/or s might not remain constant in time. For example, they might change over time because of biological/physiological changes in the population and/or changes in the population's physical environment. At different population densities, fertility and survival rates might differ, so that b and s become functions of $x(t)$. In this case, the population dynamics are said to be *density-dependent* and Equation (1.5) becomes a *nonlinear* difference equation. Vital rates could also change over time due to a changing physical environment. For example, randomly occurring changes in the environment could be modeled by letting b and s be random variables, in which case Equation (1.5) becomes a *stochastic* difference equation. Or, if environmental changes were instead regular, say approximately periodic (due to daily, monthly, seasonal, or annual fluctuations), then there would be reason to choose b and s as periodic sequences b_t and s_t, which leads to a *nonautonomous (periodically forced)* difference equation. Another source of temporal change in b and s could be natural selection and evolutionary adaptation.

If $s = 0$, then no individual survives a time unit. If the time unit is such that an individual has at most one reproductive event during the interval, then $s = 0$ models a *semelparous (monocarpic) population*, whereas $s > 0$ models an *iteroparous (polycarpic) population*. We will use these biological terms to describe these two cases.

If, in addition to reproduction and survival processes, the population is subject to other means of arrival or departure from the population, such as by immigration/stocking or emigration/harvesting, then the bookkeeping (or balance) Equation (1.2) becomes

$$x(t+1) = \text{ newborns + survivors + migrations}, \qquad (1.6)$$

where, by migration, we mean other arrivals or departures from the population not due to births or deaths. The modeler now has the additional task of formulating hypotheses about such migrations, how they are to be formulated mathematically, and how they might or might not depend on t and/or $x(t)$. Migration could include, for example, immigration/stocking and emigration/harvesting. It can be decomposed into the difference between arrivals and departures

$$\text{migrations = arrivals - departures}$$

and could be either positive or negative. In the latter case, attention needs to be given in the model formulation so that the right side of the Equation (1.6) does not produce negative population predictions or, if it does, how the equation is then to be used and interpreted. In other words, we cannot have more individuals departing the population than there are present in the population.

If none of the processes represented on the right side of the balance Equation (1.6) depends explicitly on time t, that is to say, the dependence on t occurs only through a dependence on $x(t)$, then the equation is an autonomous difference equation. If the appearance of $x(t)$ on the right side is linear, then the equation is a linear autonomous difference equation. We consider such equations in Section 1.2. We consider nonlinear equations in Section 1.3.

1.2 Linear (Density-Independent) Models

In this section, we consider the balance Equation (1.6) under the assumption that each of the processes represented on the right side is either constant or linearly dependent on $x(t)$. The result is a linear autonomous difference equation. We consider the case when the newborns and survivors are given by (1.3) and (1.4).

1.2.1 Models Without Migration

We can write Equation (1.5) as

$$x(t+1) = rx(t), \qquad (1.7)$$

where the constant

$$r := b + s$$

denotes the *population growth rate*.[1] For each initial population size $x_0 \geq 0$, this equation produces a unique solution $x(t)$ for $t = 0, 1, 2, 3, \cdots$ of predicted future population sizes. By induction, we can calculate a formula for this sequence, namely[2]

$$x(t) = r^t x_0. \tag{1.8}$$

If $r = 0$, then $x(t) = 0$ for all $t = 1, 2, 3, \cdots$ and all $x_0 > 0$, a case of little interest. We assume from now on that

$$r > 0$$

in which case we deduce from (1.8) that

$$x(t) = x_0 e^{\alpha t} \text{ where } \alpha = \ln r.$$

If $x_0 = 0$, then $x(t) = x_0$ for all t and we refer to $x^* = 0$ as *an equilibrium solution* or simply *an equilibrium* (or *fixed point*). On the other hand, if $x_0 > 0$, then

$$\lim_{t \to \infty} x(t) = \begin{cases} 0 & \text{if } 0 < r < 1 \\ x_0 & \text{if } r = 1 \\ +\infty & \text{if } 1 < r. \end{cases} \tag{1.9}$$

For $0 < r < 1$, we conclude that the equilibrium $x^* = 0$ is *attracting*, in which case the model predicts that the population will go extinct at an exponential rate. However, if $r > 1$, then we say the equilibrium $x^* = 0$ is *repelling*, in which case the population grows without bound at an exponential rate $\ln r$. If $r = 1$, then the population remains unchanged for all time. In this case, every initial population size $x^* = x_0$ is an equilibrium (fixed point) and the population model has infinitely many positive (non-extinction) equilibria. Because of the radically different, long-term outcomes for $r < 1$ and $r > 1$, we call $r = 1$ a *bifurcation point* (see Figure 1.1).

Since $r = b + s$, it is not difficult to see algebraically that

$$\begin{aligned} r > 1 & \quad \text{if and only if} \quad b(1-s)^{-1} > 1 \\ 0 < r < 1 & \quad \text{if and only if} \quad 0 < b(1-s)^{-1} < 1. \end{aligned}$$

The threshold parameter,

$$\mathcal{R}_0 := \frac{b}{1-s},$$

is called the *reproduction number*. To interpret $\mathcal{R}_0$ biologically, we write it in the form of a geometric series

$$\mathcal{R}_0 = \sum_{i=0}^{\infty} b s^i$$

and observe that s^i is the probability that an individual survives i time units. Consequently, bs^i is the number of newborns produced by an individual conditioned on surviving i time steps. It follows that $\mathcal{R}_0$ is the expected number of newborns produced by an individual over its entire life span. It is, therefore, no surprise that, in our model of a "closed" population, in which no immigration and emigration occur, $\mathcal{R}_0 < 1$ implies population extinction, while $\mathcal{R}_0 > 1$ implies unbounded

---

[1] Some authors define the population growth rate to be $\ln r$ instead of r.
[2] Note that $x(t) \equiv r^t x_0$ implies $x(t+1) = rx(t) = r^{t+1} x_0$.

1.2. LINEAR (DENSITY-INDEPENDENT) MODELS

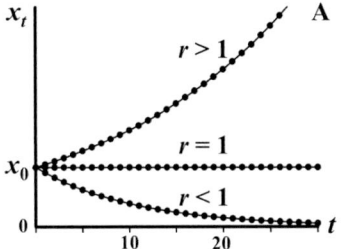 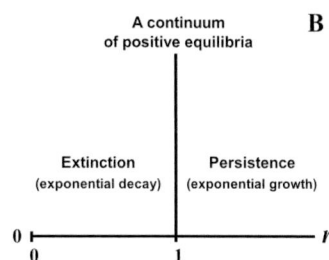

Figure 1.1: (a) Time-series plots of $x(t)$ versus t depict the three cases (1.9) for the asymptotic dynamics of Equation (1.7). If $r > 1$, the population experiences exponential growth, if $r = 1$ the population size stays constant, and if $r > 1$ the population decays exponentially and goes to extinction. (b) The asymptotic dynamics of Equation (1.7) depicted in a bifurcation diagram. In these plots, $\mathcal{R}_0$ can replace r.

population growth. Thus, $\mathcal{R}_0 = 1$ is also a bifurcation point. Figure 1.1 illustrates the asymptotic possibilities (1.9) for Equation (1.7) by means of times series plots and by means of a bifurcation diagram which depicts the asymptotic outcome as a function of r. Since exponential growth cannot be sustained indefinitely into the future, it is clear from these alternatives that Equation (1.7) cannot serve as a model for long-term, sustained population survival, with the sole exception of the isolated value of $r = 1$. We will see in Section 1.3 how density dependence of vital rates can change this conclusion by provided parameter intervals over which sustained population survival can occur (which it does by bending or warping the vertical continuum of positive equilibria).

Remark 1 *In Equation (1.7), it is assumed that both the survivors parameter s and the newborns parameter b are both time-independent and thus r is constant. In this case, Equation (1.5) is called an autonomous equation. On the other hand, if either both or one of the parameters r and s is time-dependent, then Equation (1.7) is called nonautonomous and is written as*

$$x(t+1) = r(t)x(t), \tag{1.10}$$

where the constant

$$r(t) := b(t) + s(t).$$

There are many reasons why a model's parameters might change over time, such as changes in the population's physical environment, life history cycles, metabolic processes, etc. More details on nonautonomous difference equations may be found in Chapter 8.

By mathematical induction, one may show that the solution of Equation (1.10) is given by

$$x(t) = x_0 \prod_{i=0}^{t-1} r(i). \tag{1.11}$$

1.2.2 Models With Migration/Immigration

To allow migrations (for example, harvesting and/or stocking) in Equation (1.6), we denote the net migration rate per time unit by a function m (which may or may not depend on the time or population size) and obtain the difference equation

$$x(t+1) = rx(t) + m, \tag{1.12}$$

where

$$m = \text{constant or } m = \gamma x(t),$$

which describe constant and proportional migration (or stocking), respectively. A positive constant value of $m > 0$ denotes the total immigration (or stocking) rate and a negative value $m < 0$ denotes emigration (or harvesting). Similarly, the constant γ describes the per capita immigration (or stocking) rate when $\gamma > 0$ or the per capita emigration (or harvesting) rate when $\gamma < 0$.

Equation (1.12) with constant migration rate m and initial condition $x_0 \geq 0$ gives rise to the following initial value population problem:

$$x(t+1) = rx(t) + m, \quad x(0) = x_0. \tag{1.13}$$

This equation has the equilibrium point

$$x^* = \frac{m}{1-r},$$

which is positive if either $m < 0, r > 1$ or $m > 0, r < 1$. Since

$$\begin{aligned}
x(1) &= rx_0 + m \\
x(2) &= rx(1) + m \\
&= r^2 x_0 + (1+r)m \\
x(3) &= r^3 x_0 + (1+r+r^2)m \\
&\vdots \\
x(t) &= r^t x_0 + (1+r+\cdots+r^{t-2}+r^{t-1})m \\
&= r^t x_0 + \sum_{i=0}^{t-1} r^{t-i-1} m
\end{aligned}$$

Now, using the formula for the sum of a geometric series[3], the solution of Equation (1.13) is given by

$$x(t) = \begin{cases} r^t \left(x_0 - \frac{m}{1-r}\right) + \frac{m}{1-r} & \text{if } r \neq 1, \\ x_0 + mt & \text{if } r = 1. \end{cases} \tag{1.14}$$

This mathematical formula for a solution of Equation (1.13) only makes sense in a population model if it produces nonnegative values of $x(t)$ for $t \geq 1$.

---

[3] $\quad 1 + r + r^2 + \cdots + r^{t-1} = \frac{1-r^t}{1-r}$

1.2. LINEAR (DENSITY-INDEPENDENT) MODELS

If $m > 0$ (immigration/stocking), then formula (1.14) shows that $x(t) > 0$ for all $t > 0$. This is true for all values of $r > 0$ and all initial conditions x_0. Thus, in particular, if $r < 1$ and the population will (asymptotically) die out in the absence of immigration/stocking, we see that the model predicts that any level of immigration/stocking $m > 0$ will prevent a population extinction, no matter how small the population is initially. In fact, if $0 < r < 1$,

$$\lim_{t \to \infty} x(t) = x^* = \frac{m}{1-r}$$

and x^* is attracting (see Figure 1.1).

Consider the case $m < 0$ (emigration/harvesting). Then it is possible that $x(t)$ assumes a negative value at a time $t^* \geq 1$ (in which case it remains negative for all subsequent times $t \geq t^*$). We interpret this to mean that the population is extinct at time $t = t^*$ (and the model is biologically meaningless for $t > t^*$).

For example, if $r = 1$, then, from formula (1.14), we see that t^* is the smallest integer greater than $-x_0/m$, and that extinction occurs for any population at a time depending on its initial size x_0.

Suppose $m < 0$ and $r > 1$, that is to say, the population grows exponentially without bound in the absence of migration. Since $m/(1-r) > 0$, we see from formula (1.14) that $x(t)$ is positive for all $t > 0$ and grows exponentially without bound for initial conditions satisfying $x_0 > m/(1-r)$. For (and only for) initial conditions satisfying $x_0 < m/(1-r)$, we can solve the equation

$$r^t \left(x_0 - \frac{m}{1-r} \right) + \frac{m}{1-r} = 0$$

for a positive

$$t = \frac{1}{\ln(r)} \ln \left(\frac{m}{m + (r-1)x_0} \right) > 0.$$

The population is then extinct at the smallest integer t^* greater than this value of t. We conclude that an exponentially growing population can avoid extinction due to emigration/harvesting if (and only if) its initial population size is sufficiently large.

We leave it as an exercise to show the population goes extinct for every initial condition x_0 and any value of $m < 0$ when $r < 1$ (see Figure 1.2).

Equation (1.12) with constant *per capita* migration (or variable stocking) rate γ and initial condition x_0 gives rise to the initial value population problem (1.5) where $\lambda = b + s + \gamma$ replaces r in (1.9), that is,

$$x(t+1) = \lambda x(t). \tag{1.15}$$

Example 1 *Consider the following linear difference equation, which models a population with constant immigration*

$$x(t+1) = rx(t) + m, \quad x_0 = 10$$

(a) Find $x(1), x(2), x(3), x(r)$ when $r = \frac{1}{4}$, $m = 2$.

(b) Solve the equation by finding a formula for $x(t)$ in terms of x_0, when $r = \frac{1}{4}$, $m = 2$.

(c) Solve the equation by finding a formula for $x(t)$ in terms of x_0, when $r = 1.5$, $m = 2$.

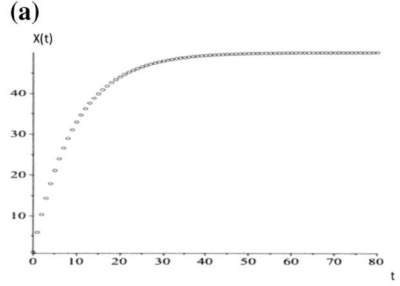
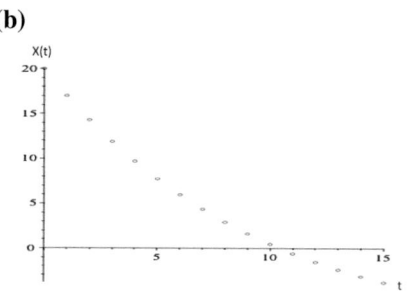

Figure 1.2: (a) The population with migration (stocking) grows to its carrying capacity (equilibrium), when $r < 1$ and $m > 0$. (b) The population with immigration goes to extinction when $r < 1$ and $m < 0$.

Solution:

(a) $x(1) = \frac{1}{4}10 + 2 = \frac{9}{2}$, $x(2) = \frac{1}{4}\frac{9}{2} + 2 = 258$, $x(3) = \frac{89}{32}$, $x(4) = \frac{345}{128}$.

(b) From Equation (1.14), we have

$$x(t) = \left(\frac{1}{4}\right)^t \cdot 10 + \left(\frac{1 - \left(\frac{1}{4}\right)^t}{1 - \frac{1}{4}}\right) \cdot 2$$

$$= 10 \cdot (4^{-t}) + \frac{8}{3}(1 - 4^{-t})$$

$$\lim_{t \to \infty} x(t) = 0 + \frac{8}{3} = \frac{8}{3}.$$

(c)

$$x(t) = \left(\frac{3}{2}\right)^t \cdot 10 + \left(\frac{\left(\frac{3}{2}\right)^t - 1}{\frac{3}{2} - 1}\right) \cdot 2$$

$$= (1.5)^t \cdot 10 + 4((1.5)^t - 1)$$

$$\lim_{t \to \infty} x(t) = \infty.$$

Let us consider the situation when both parameters r and m are time-dependent. This leads to the nonautonomous difference equation

$$x(t+1) = r(t)x(t) + m(t). \tag{1.16}$$

Now
$$x(1) = r(0)x_0 + m(0)$$
$$x(2) = r(1)x(1) + m(1)$$
$$= r(1)(r(0)x_0 + r(1)m(0) + m(1)$$
$$x(3) = r(2)r(1)r(0)x_0 + r(2)r(1)m(0) + r(2)m(1) + m(2)$$
$$\vdots$$
$$x(t) = x_0 \prod_{i=0}^{t-1} r(i) + \prod_{j=1}^{t-1} r(j)m(0) + \prod_{j=2}^{t-2} r(j)m(1) + \cdots + m(t-1).$$

This leads to the one-dimensional variation of constants formula

$$x(t) = x_0 \prod_{i=0}^{t-1} r(i) + \sum_{i=0}^{t-1}\left(\prod_{j=i+1}^{t-1} r(j))m(i)\right), \tag{1.17}$$

where

$$\prod_{j=t+1}^{t} r(j) = 1. \tag{1.18}$$

1.3 Nonlinear Autonomous (Density-Dependent) Models

If the fertility and survival rates b and s appearing in the model difference equation (1.5) in Section 1.2 are constant in time, then the only alternative to extinction ($r < 1$), is unlimited exponential growth ($r > 1$), which of course is not sustainable in the long run. Thus, fertility b and/or survival s must change in time if a model population is to avoid the alternatives of extinction or unlimited growth.

As pointed out in Section 1.1, there are numerous reasons why the fertility rates and survival probabilities might change over time. In this section, we consider one possibility, namely, that of a population's self-regulation of its fertility and survival. This we do mathematically by assuming that fertility and survival are functions of population density x. In this way, unlimited population growth can be avoided if fertility and survival are sufficiently decreased as population density becomes high. A replacement of b and s by functions of x in the balance equation (1.5) will result in a nonlinear difference equation. The dynamics predicted by such a model equation depends significantly on the detailed properties of the nonlinearities and can vary considerably from one model equation to another.

There are, however, some quite general dynamic properties of nonlinear population models, valid under minimal mathematical assumptions, which can serve as a starting point for the analysis of specific model equations.

After a brief look in Section 1.3.1 at some issues involved in the construction of nonlinear models, we study these general properties of population models in Section 1.3.3.

1.3.1 Modeling Density Effects

If we assume the fertility and survival rates in the balance equation (1.5) are dependent on population density x, we have the nonlinear difference equation

$$x(t+1) = b(x(t)) x(t) + s(x(t)) x(t). \tag{1.19}$$

Let $\mathbb{R}$ denote the real numbers, $\mathbb{R}_+$ denote the nonnegative real numbers, and Ω an open interval containing $\mathbb{R}_+$ We assume throughout that the functions $b(x)$ and $s(x)$ satisfy the following conditions.

A1: *$b(x)$ and $s(x)$ are twice continuously differentiable, real value functions on Ω that satisfy $b(x) > 0$ and $0 \leq s(x) < 1$ for all $x \in \mathbb{R}_+$.*

We can also write Equation (1.19) as

$$x(t+1) = r(x(t)) x(t), \tag{1.20}$$

where we call the factor

$$r(x) := b(x) + s(x)$$

the *population growth rate*. The population growth rate $r(x)$ predicts the ratio of successive population densities

$$r(x(t)) = \frac{x(t+1)}{x(t)}. \tag{1.21}$$

The dynamics described by Equation (1.19) depends crucially on the properties of the functions $b(x)$ and $s(x)$. One general property of these functions that is important, for both the biological interpretation of the model equation and for its dynamics, is their monotonicity.

If $b(x)$ (or $s(x)$) is a decreasing function on some interval of x values, then an increase in population density (in this interval) is deleterious in the sense that they reduce fertility (or survival). Such an assumption on a model component $b(x)$ or $s(x)$ is widely referred to as a *negative density effect* (or a negative feedback). Such component negative density effects are generally assumed to occur at least at high densities and, in many models, at all density levels $x > 0$. In this section, we consider only the latter case, that is to say $b(x)$ and $s(x)$ are decreasing (or non-increasing) functions for all values of $x > 0$.

The cases when $b(x)$ or $s(x)$ is increasing for some range of $x > 0$ values do arise when modeling some important biological phenomena (particularly when they increase on some interval near $x = 0$). We postpone this case of *positive density effects* (or positive feedbacks) to Chapter 5.

An issue to keep in mind when deriving and interpreting a discrete-time model (1.19) is how reproduction occurs and when the population census is taken [45]. For example, suppose reproduction is pulsed, i.e., occurs regularly during a very narrow time interval (such as during a short breeding season). If the census is taken just before reproduction occurs, then to appear in the next census the newborns will be exposed to the survival probability over one time unit. In this case, the fertility function $b(x)$ records the number of births times the probability of survival to the next census. On the other hand, if the census is taken just after a reproductive pulse, then the newborns recorded will be the births that occur. For populations with non-pulsed reproduction, careful consideration must be given to how fertility is to be modeled between census times. For a discussion of this issue and how it effects, the resulting difference equation see [45].

1.3. NONLINEAR AUTONOMOUS (DENSITY-DEPENDENT) MODELS

Example 2 *Consider a population with pulsed reproduction that is censused just prior to the reproduction event. Suppose we mathematically model the density dependence of both the fertility and survival by decreasing rational functions of the form*

$$\frac{1}{1+cx}, \quad c>0 \tag{1.22}$$

in order to describe a negative density dependence at all population densities $x \geq 0$. Specifically, we could model the number of births by

$$b_i \frac{1}{1+c_1 x}, \quad b_i, c_1 > 0$$

(b_i is the inherent, i.e., density free, birth rate) and the survival probabilities of existing individuals and newborns, respectively, by

$$s\frac{1}{1+c_2 x}, \quad s_b \frac{1}{1+c_3 x}, \quad 0 \leq s, s_b < 1, \quad c_i \geq 0$$

(s_b is the density-free newborn survival rate and s is the density-free adult survival rate). Then

$$b(x) = b_i s_b \frac{1}{1+c_1 x}\frac{1}{1+c_3 x}, \quad s(x) = s\frac{1}{1+c_2 x}$$

in Equation (1.19), which becomes

$$x(t+1) = b\frac{1}{1+c_1 x(t)}\frac{1}{1+c_3 x(t)}x(t) + s\frac{1}{1+c_2 x(t)}x(t), \tag{1.23}$$

where we have defined $b := b_i s_b$. For this model equation, the density-dependent growth rate is

$$r(x) = b\frac{1}{1+c_1 x}\frac{1}{1+c_3 x}x + s\frac{1}{1+c_2 x}.$$

If, on the other hand, the census is taken just after the reproductive pulse, then

$$b(x) = b_i \frac{1}{1+c_1 x}$$

(because newborns are counted before they suffer any survival probability) and we obtain the difference equation

$$x(t+1) = b\frac{1}{1+c_1 x(t)}x(t) + s\frac{1}{1+c_2 x(t)}x(t), \tag{1.24}$$

where now $b = b_i$. For this model equation,

$$r(x) = \frac{b}{1+c_1 x}x + \frac{s}{1+c_2 x}.$$

*A special case is the **discrete logistic (Beverton–Holt) equation***

$$x(t+1) = \frac{b}{1+cx(t)}x(t), \quad b, c > 0 \tag{1.25}$$

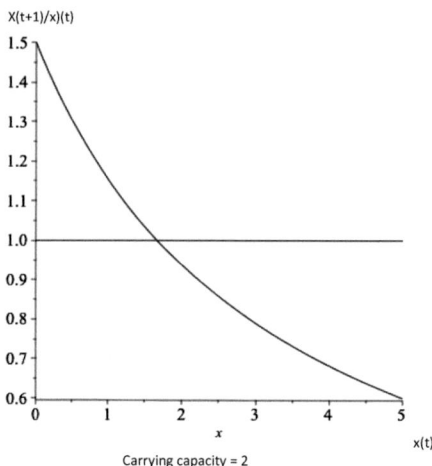

Figure 1.3: The population growth rate function of the discrete logistic (Beverton–Holt) model.

with
$$r(x) = \frac{b}{1+cx},$$
which arises from Equation (1.24) if $s = 0$, i.e., the population is semelparous, or if $s > 0$ and $c = c_1 = c_2$, i.e., density affects fertility and survival in the same way [210] (Figure 1.3).

If we let $r_0 := r(0) = b$ denote the density-free (or inherent or intrinsic) population growth rate and if $r_0 > 1$, then the positive equilibrium (fixed point)
$$K := \frac{r_0 - 1}{c}$$
of the discrete logistic (Beverton–Holt) equation (1.25) is called the environmental carrying capacity. In the special case when $r_0 = b > 1$, we can re-write the discrete logistic (Beverton–Holt) equation as
$$x(t+1) = \frac{r_0 K}{K + (r_0 - 1)x(t)} x(t), \quad r_0 > 1, \ K > 0. \tag{1.26}$$
(Note: the subscript "0" is usually dropped and r_0 replaced by r.)

Example 3 *Difference equations different from those in Example 2 result if negative density effects on vital rates are described by exponential functions of the form*
$$\exp(-cx), \quad c > 0$$
in place of rational functions of the form (1.22).

For example, in this case instead of (1.23), we have
$$x(t+1) = bx(t)\exp(-c_1 x(t))\exp(-c_3 x(t)) + sx(t)\exp(-c_2 x(t)), \quad c_i > 0 \tag{1.27}$$

1.3. NONLINEAR AUTONOMOUS (DENSITY-DEPENDENT) MODELS

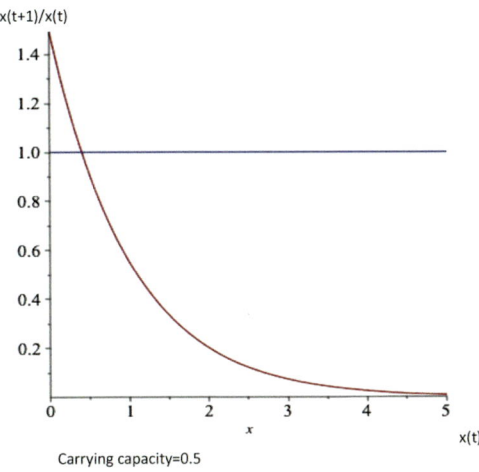

Figure 1.4: The population growth rate function of the Ricker model.

with
$$r(x) = b\exp(-c_1 x)\exp(-c_3 x) + s\exp(-c_2 x)$$
for a population censused just before the reproductive pulse. Or, if censused just after the pulse, we have the difference equation
$$x(t+1) = bx(t)\exp(-c_1 x(t)) + sx(t)\exp(-c_2 x(t)), \quad c_i > 0 \tag{1.28}$$
with
$$r(x) = b\exp(-c_1 x) + s\exp(-c_2 x)$$
instead of (1.24). A special case of this equation is the **exponential model equation**
$$x(t+1) = bx(t)\exp(-cx(t)), \quad b, c > 0 \tag{1.29}$$
with
$$r(x) = b\exp(-cx)$$
which arises from Equation (1.24) if $s = 0$, i.e., the population is semelparous, or if $s > 0$ and $c = c_1 = c_2$, i.e., density affects fertility and survival in the same way [210] (Figure 1.4).

If $b > 1$, then
$$K := \frac{\ln b}{c} > 0$$
is a positive equilibrium (fixed point) of the exponential model equation (1.29). In the special case when $b > 1$, we let $\alpha := \ln b$ and re-write the exponential model equation as
$$x(t+1) = x(t)\exp(\alpha - cx), \quad \alpha, c > 0 \tag{1.30}$$
which is called the **Ricker equation** (Figure 1.5).

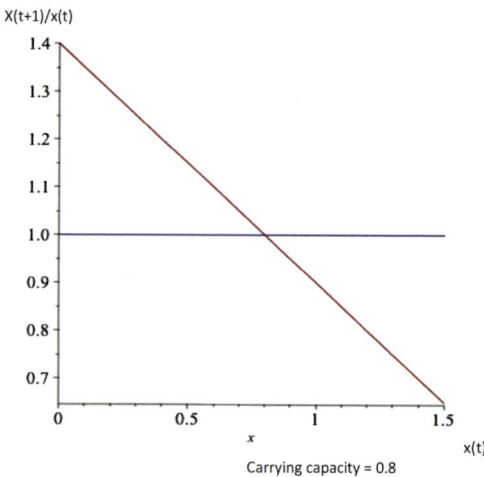

Figure 1.5: The population growth rate function of the quadratic model.

Example 4 *Another example, popularized in mathematical biology by May [233], is the* **quadratic model equation**

$$x(t+1) = b(1 - cx(t))x(t), \quad b, c > 0 \tag{1.31}$$

(sometimes also called the logistic model equation). This model is a feasible biological model only if the right side of the equation produces nonnegative values. Therefore, we assume $b < 4$ so that it maps the square $[0, 1/c] \times [0, 1/c]$ into itself. If $b > 1$, then

$$K = \frac{1}{bc}(b-1)$$

is a positive equilibrium (fixed point). In the special case when $b > 1$, we can re-write the equation as

$$x(t+1) = b\left(1 - \frac{1}{bK}(b-1)x(t)\right)x(t), \quad 1 < b < 4, \ K > 0. \tag{1.32}$$

We can simplify this equation by re-scaling the population density by setting

$$y(t) := \frac{a-1}{aK}x(t)$$

to obtain the equation

$$y(t+1) = ay(t)(1 - y(t)).$$

Yet another version that appears in the literature (e.g., see [37]) arises from the re-scaling

$$y(t) = \frac{a-1}{a}x(t)$$

to obtain the equation

$$y(t+1) = ay(t)\left(1 - \frac{y(t)}{K}\right).$$

Some analysis of this model is left to the reader as Problem 13 in Exercises 1.3.1–1.3.2.

1.3. NONLINEAR AUTONOMOUS (DENSITY-DEPENDENT) MODELS

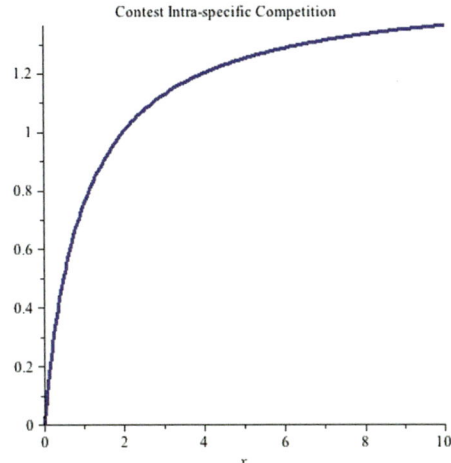

Figure 1.6: A contest competition model.

In Examples 2 and 3, we chose mathematical expressions for the effects that population density has on vital rates only on the basis of their monotonicity properties in order to capture negative density effects, to which we restrict our attention in this chapter. More sophisticated models for $b(x)$ and $s(x)$ are derived from more detailed biological/ecological/life history underpinnings (and might involve positive density effects). For example, the rational and exponential type nonlinearities used in the examples above have been derived from binomial processes due to interactions or "collisions" among individuals [88], probabilistic models of mate finding [104], [236], predation by predators [104], and spatial dispersal [293].

1.3.2 Scramble versus Contest Models

There are two main kinds of competition that are based on the way that resources are distributed among individuals of a species. If resources are allocated equally, then we have a scramble competition, which is also called overcompensatory competition. The exponential model (or the Ricker model) is an example of a scramble (overcompensatory) competition among individuals of a species. On the other hand, if resources are allocated unequally with monopolization, then we have a contest competition or compensatory competition. The discrete logistic (Beverton–Holt) model is an example of contest (compensatory) competition among individuals of a species.

For instance, assume that there are S available resources, and there are N individuals. Then in a scramble competition model, each individual gets S/N units from the available resources. Now if l is the maximum number of units an individual may use, then S/l is the maximum number of individuals that may have the maximum share of the resources, while $N - S/l$ individuals will share the rest of the resources in a contest competition model (see Figure 1.6).

It turns out that some species do not always follow these two paradigms. For instance, the southern pine beetle *Dendroctonus frontalis* shifts from contest competition early in its life cycle, when adults attack a tree to lay eggs in galleries, to scramble competition later on [257]. Another example is the Mediterranean flour moth *Ephestia kuehniella* who shifts from scramble competition prior to parasitization to contest competition [206] (Figure 1.7).

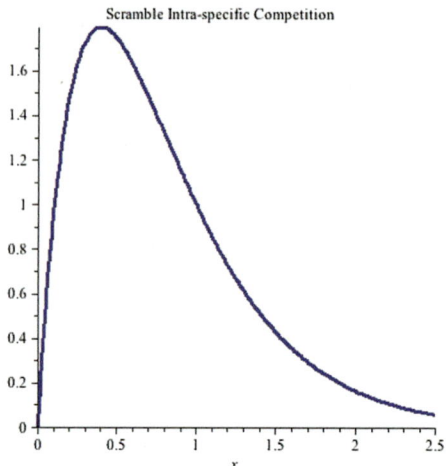

Figure 1.7: A scramble competition model.

These examples show the need for more general models. One of these general models are attributed to Hassell [156] and is given by the difference equation

$$x(t+1) = b \frac{1}{\left(1 + \frac{c}{k} x(t)\right)^k} x(t).$$

This model allows a wide spectrum of competition types, depending on the value of k. At the extreme cases, $k = 1$, we get the discrete logistic (Beverton–Holt) model (contest competition) and, at the other extreme $k \to \infty$, we get the exponential model (scramble competition), since

$$\lim_{k \to \infty} \left(1 + \frac{c}{k} x\right)^{-k} = e^{-cx}.$$

Exercises 1.3.1–1.3.2

In problems 1–4, we assume density-independent growth.

1. Suppose a population of cells divides synchronously, with each member producing r daughter cells. Denote the number of cells in each generation as $N(0), N(1), N(2), \ldots$.

   (a) Write a difference equation that relates the number of cells in generations $t+1$ and the number of cells in generation t.

   (b) If the number of daughter cells $r = 2$ and $N(0) = 10^3$, find $N(5)$.

2. A moth has a generation time of 10 days and a population per generation growth rate of $r = 1.4$. If there are 20,000 moths now, how many moths will there be in 60 days?

3. Mountain goats were introduced in the Olympic Peninsula in Washington State. The population increased from 12 individuals in 1929 to 1,175 individuals in 1983.

(a) Calculate the population growth rate r for this time frame.

(b) What is the doubling time, i.e., the time when the population reaches the size of 24, of this population?

(c) Assuming that r remains constant in the future, in what year will the mountain goat population reach 10,000?

4. One hundred rabbits are introduced into a small island at the beginning of 1995. If this population has a constant population growth rate $r = 1.3$ per year on the island, what will be the population size at the beginning of 2000? At the beginning of 2005?

5. Adult female aphids produce galls on the leaves of poplars. All the progeny of a single aphid are contained in one gall. Some fraction of these will emerge and survive to adulthood. Let

$$a(t) = \text{number of adult female aphids in generation } t,$$
$$p(t) = \text{number of progeny in generation } t,$$
$$m = \text{fractional mortality of the young aphids},$$
$$s = \text{number of progeny per female aphid},$$
$$r = \text{ratio of female aphids to total adult aphids}.$$

(a) Write the difference equation relating $p(t+1)$ and $a(t)$.

(b) Write the difference equation describing $a(t)$.

(c) Find $a(t)$ in terms of $a(0)$.

6. A species has an annual constant population growth rate $r = 0.5$. Suppose that the population size now is 20,000 and the annual emigration is 200 individuals.

(a) Write the difference equation that models the growth of the population.

(b) What is the size of the population after 10 years?

(c) Does the population survive or go extinct? If it goes extinct, at what time? If it survives, what is the population size as $t \to \infty$?

7. A species grows according to the discrete logistic (Beverton–Holt) model $x(t+1) = \frac{bKx(t)}{K+(b-1)x(t)}$. If $x_0 = 2$, $b = 1.5$, $K = 4$, find $x(1)$, $x(2)$, $x(3)$, $x(4)$, $x(5)$. Draw the time-series graph for these parameter values.

8. A variation of the discrete logistic (Beverton–Holt) model is Hassell's model

$$x(t+1) = \frac{bx(t)}{(1+x(t))^l}. \tag{1.33}$$

9. A species grows according to the Hassell's model (1.33). If $l = 2$, $r = 2.25$, $K = 4$, $x(0) = 10$.

(a) Draw the population growth rate function $r(x)$ versus x.

(b) Find $x(1)$, $x(2)$, $x(3)$, $x(4)$, $x(5)$.

(c) Draw the time-series graph.

(d) Does the population survive or go extinct? If it survives, what is the upper bound of the size of the population?

10. A plot of $\ln x(t+1)/x(t)$ versus $x(t)$ reveals a linear relationship described by
$$\ln \frac{x(t+1)}{x(t)} = 1 - 0.01x(t),$$
where $x(t)$ is the size of a population at time t.

   (a) Write the difference equation that models the population growth.
   (b) Does the population survive or go extinct? If it survives, what is the upper bound of the size of the population?

11. A species grows according to a population growth rate function of the form
$$r(x) = \frac{ax}{1+x^2}.$$
The population growth rate function passes through $(0,0)$, $(A,1)$, $(K,1)$, with $0 < A < K$. If $K = 2$, find a, A, and K.

   (a) Draw the graph of the population growth rate function $r(x)$ versus x.
   (b) Draw the time-series graph of the model
   $$r(x) = \frac{ax}{1+x^2}$$
   for $x_0 = 0.2$, $x_0 = 1$. What can you conclude from the graph about the size of the population at future generations?

12. A species grows according to a linear population growth rate function of the form $r(x) = -mx + b$, where $m, b > 0$. Show that the resulting model has the form of the the quadratic model
$$x(t+1) = rx(t)\left(1 - \frac{x(t)}{K}\right). \tag{1.34}$$

   (a) Assume $r < 4$. Draw the graph of the population growth rate function $r(x)$.
   (b) Draw the time-series graph of the model
   $$x(t+1) = rx(t)\left(1 - \frac{x(t)}{K}\right)$$
   for $x_0 = 0.2$, $x_0 = 1$. What can you conclude from the graph about the size of the population at future generations?

13. Consider the discrete logistic (Beverton–Holt) equation
$$x(t+1) = \frac{bKx(t)}{K + (b-1)x(t)}.$$

(a) Use the substitution $y(t) = \frac{1}{x(t)}$, to convert the equation into a linear difference equation.

(b) Solve the resulting linear difference equation.

(c) Show that the solution of the discrete logistic (Beverton–Holt) equation is given by
$$x(t) = \frac{Kx_0 b^t}{K + (b^t - 1)x_0},$$
where $x_0 = x(0)$.

1.3.3 Stability of Equilibrium Points

A basic question concerning population dynamics is the existence and stability of equilibria. An *equilibrium (fixed point)* of the equation solution $x(t)$ of the equation $x(t+1) = r(x(t))x(t)$ is a constant solution, i.e., a solution that satisfies $x(t) = x^*$, for some real number x^*, for all t. One obvious equilibrium is $x^* = 0$, which we call the *extinction equilibrium*. A *positive equilibrium* $x^* > 0$ is a positive root of the algebraic equation
$$r(x) = 1. \tag{1.35}$$

In addition to the existence of equilibria, we are also interested in their stability or instability, by which we mean the following. Consider a general difference equation
$$x(t+1) = f(x(t)) \tag{1.36}$$
under the assumption that $f(x)$ is continuously differentiable in a neighborhood of a fixed point $x^* = f(x^*)$.

Definition 1 *Let x^* be an equilibrium point of Equation (1.36). Then*

(i) *The equilibrium point x^* is stable if, for any $\varepsilon > 0$, there exists $\delta > 0$ such that $|f^t(x_0) - x^*| < \varepsilon$, whenever $|x_0 - x^*| < \delta$. In other words $|x(t) - x^*| < \varepsilon$, whenever $|x_0 - x^*| < \delta$, where $x(0) = x_0$.*

(ii) *The equilibrium point x^* is attracting if there exists $\mu > 0$ such that $\lim_{t \to \infty} f^t(x_0) = x^*$, i.e., $\lim_{t \to \infty} x(t) = x^*$, whenever $|x_0 - x^*| < \mu$.*

(iii) *The equilibrium point x^* is locally asymptotically stable (LAS) if it is stable and attracting.*

(iv) *Let D be a set contained in the domain of f. The equilibrium point x^* is globally asymptotically stable on D if it is locally asymptotically stable and $\lim_{t \to \infty} f^t(x_0) = x^*$ for all $x_0 \in D$, i.e., $\lim_{t \to \infty} x(t) = x^*$ for all $x_0 \in D$.*

Local asymptotic stability concerns the fate of initial conditions near to an equilibrium. Initial conditions sufficiently close to the equilibrium yield solution sequences that remain close for all later time and, in fact, asymptotically approach the equilibrium as $t \to +\infty$.

Definition 2 *The basin of attraction of an equilibrium of Equation (1.36) is the entire set of initial conditions whose solution sequences asymptotically approach the equilibrium as $t \to +\infty$.*

A locally asymptotically stable equilibrium x^* is globally asymptotically stable on D if D is contained in the basin of attraction. For a population model $x(t+1) = r(x(t)) x(t)$, an equilibrium point $x^* > 0$ is globally asymptotically stable if its basin of attraction is $D = R_+ \setminus 0$. There are two basic methods to determine the stability properties of an equilibrium point, one graphic and another analytic. We will start first with two graphic methods, namely, the cobweb diagram and the time-series plots. The second method, stability by linearization will be investigated in a separate section, Section 3.2.2.

At the risk of some confusion, we will often in this book, as is commonly done in the literature, use the word "stable" to mean "locally asymptotically stable".

1.3.4 Cobweb Analysis

When $f(x) \geq 0$ for $x \geq 0$, the general one-dimensional difference equation

$$x(t+1) = f(x(t)) \tag{1.37}$$

can be viewed as a map

$$f : [0, \infty) \to [0, \infty),$$

where $[0, \infty)$ denotes the interval of nonnegative real numbers $x \geq 0$. This map recursively generates a sequence starting from a nonnegative initial value x_0. We call the set

$$O^+(x_0) = \{x_0, f(x_0), f^2(x_0), \cdots, f^t(x_0), \cdots\}$$

the *forward orbit* of $x_0 \geq 0$. Here f^t denotes the tth iterate of the map f, i.e., the map obtain by composing f with itself t times:

$$f^2(x) := f(f(x)), \quad f^3(x) := f(f^2(x)), \quad \cdots, \quad f^t(x) := f(f^{t-1}(x)).$$

In population dynamic applications, $O^+(x_0)$ lists the future population densities starting from $x_0 \geq 0$.

If the function f is bijective (one to one), then it has an inverse. In this case, we can define the *backward orbit* of x_0 (the sequence of past population densities prior to x_0) as

$$O^-(x_0) = \{x_0, f^{-1}(x_0), f^{-2}(x_0), \cdots, f^{-t}(x_0), \cdots\}.$$

The collection of orbits for all possible initial conditions often has a complicated set structure. One method for the study of orbits is based on the graph of $f(x)$. This graphical method, called *cobweb*, is carried out by simultaneously graphing the function f of Equation (1.37) and the diagonal line $y = x$ on the same coordinate $x - y$ plane. Recall that a point x^* in the domain of f is an *equilibrium point (fixed point)* of (1.37) if $f(x^*) = x^*$. Graphically, x^* in the domain of f is a fixed point of f if and only if the graph of f intersects the diagonal line $y = x$ at the point (x^*, x^*).

To trace the orbit from an initial population density x_0, we begin by locating x_0 on the x-axis. The point $(x_0, f(x_0))$ lies on the graph of f and on the vertical line through the point $(x_0, 0)$. The horizontal line through the point $(x_0, f(x_0))$ crosses the diagonal line $y = x$ at the point $(f(x_0), f(x_0))$. Repeating the same process with $f(x_0)$ replacing x_0 leads to the point $(f^2(x_0), f^2(x_0))$ on the diagonal line. Repeating this process, we locate a sequence of points, $(f^3(x_0), f^3(x_0))$, $(f^4(x_0), f^4(x_0))$, $\cdots$, on the line $y = x$ and hence, every point in the forward orbit of x_0.

We illustrate the cobweb method using the discrete logistic (Beverton–Holt) and Ricker models.

1.3. NONLINEAR AUTONOMOUS (DENSITY-DEPENDENT) MODELS

Example 5 *The discrete logistic (Beverton–Holt) model* (1.25). *For this model equation, we have*

$$f(x) = \frac{bx}{1+cx} \text{ for } x \geq 0.$$

The fixed points of f (equilibrium points of the discrete logistic (Beverton–Holt) model) is the set of the solutions of $f(x^*) = x^*$, which are

$$x^* = 0, \ x^* = \frac{b-1}{c}. \tag{1.38}$$

The fixed point x^* is biologically relevant only if $b \geq 1$. When $b \leq 1$, $x^* = 0$ is the only fixed point. When $b > 1$, both fixed points are biologically relevant. The two fixed points coalesce at the bifurcation point $b = 1$.

Two differentiations with respect to x, yield

$$f'(x) = \frac{b}{(1+cx)^2} \text{ and } f''(x) = -\frac{2bc}{(1+cx)^3}.$$

Hence, $f'(x) > 0$ and $f''(x) < 0$ for all $x \geq 0$ and the graph of f is increasing and concave down.

Case 1. $b \leq 1$. *In this case, f has only one biologically relevant fixed point $x^* = 0$, where $f'(0) \leq 1$. The fact that $f(x) < x$ when $x > 0$ together with a cobweb analysis, as shown in Figure 1.8, imply $\{f^t(x_0)\}_{t\geq 0}$ is a bounded, strictly decreasing sequence for all positive initial population densities x_0. Thus, the population dynamic sequence $\{f^t(x_0)\}_{t\geq 0}$ converges to the only fixed point $x^* = 0$ for all initial population densities. Hence, this model equation implies the ultimate fate of this population is extinction when $b \leq 1$.*

Case 2. $b > 1$: *In this case, f has two biologically relevant fixed points (1.38). Notice that these two fixed points coalesce when $b = 1$, where the transcritical bifurcation described in Section 1.3.3 occurs. Graphical analysis shows that $f(x)$ remains in the open interval $(0, (b-1)/c)$ and that $x < f(x)$ as long as x is in the interval $(0, \frac{b-1}{c})$. Therefore, $\{f^t(x_0)\}_{t\geq 0}$ is a bounded, strictly increasing sequence for all initial population sizes x_0 in the interval $(0, (b-1)/c)$ and*

$$\lim_{t \to \infty} f^t(x_0) = \frac{b-1}{c}.$$

A similar analysis shows that $f(x)$ remains in the interval $(\frac{b-1}{c}, \infty)$ and that $x > f(x)$ whenever x is in the interval $(\frac{b-1}{c}, \infty)$. Therefore, $\{f^t(x_0)\}_{t\geq 0}$ is a bounded, strictly decreasing sequence for all population sizes x_0 in the interval $((b-1)/c, \infty)$ and

$$\lim_{t \to \infty} f^t(x_0) = \frac{b-1}{c}.$$

All positive population sizes x less than the positive fixed point x^* increase (monotonically) and ultimately converge to x^*, while all population sizes larger than the positive fixed point x^* decrease (monotonically), and ultimately converge to x^* as long as $x(0) > 0$ (Figure 1.9).

In conclusion, when a population is governed by the discrete logistic (Beverton–Holt) model no matter what the size of the initial population density, the population will persist as x^* is a globally attracting fixed point whenever $f'(0) = b > 1$ (Figure 1.10).

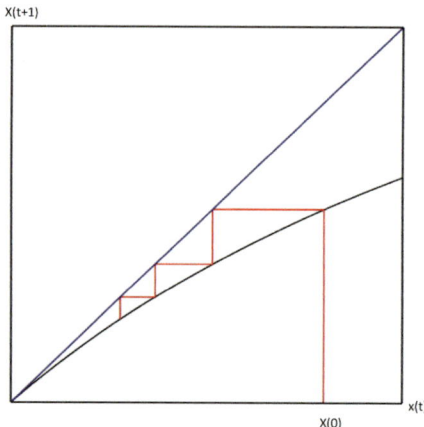

Figure 1.8: The cobweb diagram of the discrete logistic map where $b \leq 1$. The population goes extinct.

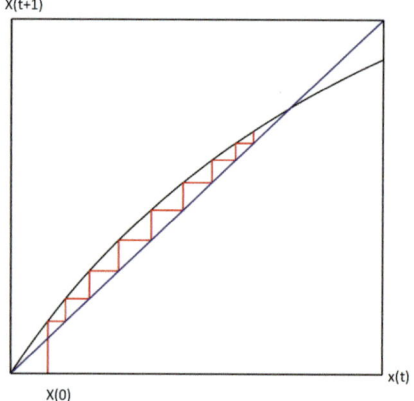

Figure 1.9: The cobweb diagram of the discrete logistic map, where $b > 1$. The population size approaches the equilibrium point.

1.3. NONLINEAR AUTONOMOUS (DENSITY-DEPENDENT) MODELS

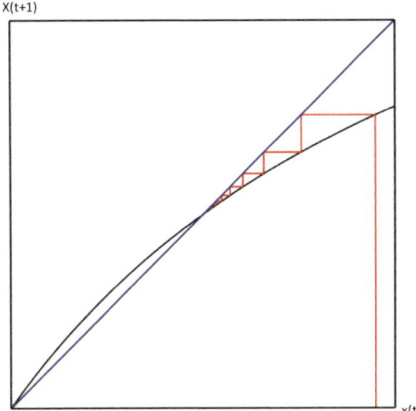

Figure 1.10: The cobweb diagram of the discrete logistic map, where $b > 1$. The population size approaches the equilibrium point.

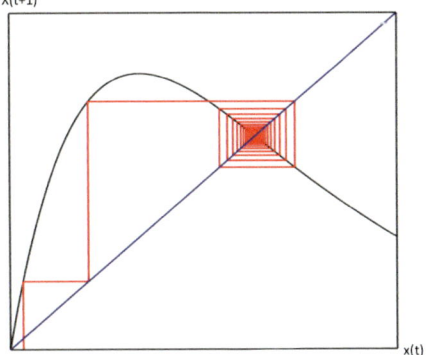

Figure 1.11: The cobweb diagram of the Ricker map where $\alpha = 1.9$. The population approaches the positive equilibrium point $x_2^* = \alpha$. The extinction equilibrium $x_1^* = 0$ is unstable, while x_2^* is attracting.

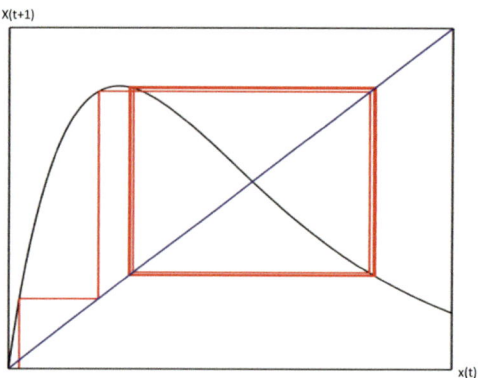

Figure 1.12: The cobweb diagram of the Ricker map where $\alpha = 2.2$. The positive equilibrium point $x_2^* = \alpha$ is unstable. The population fluctuates around two values.

Example 6 *Consider the Ricker model $x(t+1) = x(t)e^{\alpha - cx(t)}$, with $x_0 = 0.5$, $\alpha = 1.9, 2.2$, and $c = 1$. The cobweb diagram shows that the equilibrium $x^* = \alpha$ is asymptotically stable for $\alpha = 1.9$ and unstable for $\alpha = 2.2$. The analysis of this model is more complicated than the discrete logistic map and requires a general theorem on stability, which we will introduce in the next section. However, one may infer from cobweb diagrams (see Figures 1.11, 1.12) that the extinction equilibrium point $x_1^* = 0$ is always unstable, while the positive equilibrium point $x_2^* = \alpha$ is attracting if $0 < \alpha < 2$.*

Exercises 1.3.4

1. Consider the linear growth model $x(t+1) = rx(t)$.
   (a) Generate a table of values for $x(t)$, $t = 0 \cdots 10$ for $r = 0.5, 1.5$ and $x_0 = 0.1$.
   (b) Draw the time-series $(t - x(t))$ graph.
   (c) Draw the cobweb diagram.
   (d) Determine whether the equilibrium points are attracting.

2. Consider the equation $x(t+1) = \frac{1}{2}x(t) + 2$, $x_0 = 0.2$.
   (a) Generate a table of values for $x(t)$, $t = 0 \cdots 10$ for $a = 0.5, 1.5$.
   (b) Find the equilibrium points.
   (c) Use the time-series graph to determine whether the equilibrium points are attracting.
   (d) Use the cobweb-diagram to determine whether the equilibrium points are attracting.

In problems 3–5:

(a) Find the equilibrium points.

(b) Generate a table of values $x(t)$, $t = 1, 2, \ldots, 20$.

(c) Plot the time-series graph.

(d) Determine whether the equilibrium is attracting or not.

3. $x(t+1) = x(t)\left[r + r\left(1 - \frac{x(t)}{K}\right)\right]$, $x_0 = 0.1$ with $r = 0.5, 2.5, 3.5$ and $K = 1$.

4. $x(t+1) = \frac{bKx(t)}{K+(r-1)x(t)}$, $x_0 = 0.2$, with $K = 2$ and $b = 1.5, 2.5$.

5. (Ricker model) $x(t+1) = x(t)e^{\alpha - cx(t)}$ with $x_0 = 0.2$, $c = 1$, $\alpha = 0.5, 1.5, 2.5$.

6. Consider the Hassell's model
$$x(t+1) = \frac{bx(t)}{(1+cx(t))^l} \quad l, b, s > 0.$$
Find the equilibrium points and determine whether they are attracting or not.

7. In population dynamics, a frequently encountered model for fish populations is the exponential model given by
$$x(t+1) = bx(t)e^{-cx(t)},$$
where α represents the maximal growth rate of the organism and β is the inhibition of growth caused by overpopulation.

   (a) Show that this equation has an equilibrium point $x^* = \ln \frac{b}{c}$.

   (b) Show that this equilibrium point is attracting for some values of α and is not attracting for other values α.

8. Use the cobweb diagram to determine the stability of the equilibrium points of the θ-Ricker model:
$$x(t+1) = x(t)e^{\alpha - cx(t)^\theta}, \alpha > 0, \theta > 0, c > 0.$$

9. Use the cobweb diagram to determine whether the equilibrium points of the θ-quadratic model are attracting or not:
$$x(t+1) = bx(t)\left(1 - \left(\frac{x(t)}{K}\right)^\theta\right), b > 0, \theta > 0, K > 0.$$

10. Use the cobweb diagram to determine whether the equilibrium points of the equation below are attracting or not
$$x(t+1) = x(t)\left(\frac{b}{1+cx(t)} - m\right), b, c, m > 0.$$

1.4 Stability by Linearization

A basic tool for determining the local asymptotic stability of an equilibrium is the linearization principle.

Theorem 1 *[119]* **The Linearization Principle**. *A fixed point x^* of the difference equation (1.36) $x(t+1) = f(x(t))$ is locally asymptotically stable (LAS) if $|f'(x^*)| < 1$ and is unstable if $|f'(x^*)| > 1$.*

Note that Theorem 1 fails to apply if $|f'(x^*)| = 1$. The local stability properties of such an equilibrium cannot be determined from $f'(x^*)$ alone. An equilibrium of (1.36) is called *hyperbolic* if $|f'(x^*)| \neq 1$ and *nonhyperbolic* if $|f'(x^*)| = 1$. The stability properties of a nonhyperbolic equilibrium are model-dependent; a nonhyperbolic equilibrium can be stable or unstable depending on the (higher order) properties of the nonlinearity. Thus, $|f'(x^*)| < 1$ is a sufficient, but not necessary, condition for local asymptotic stability of x^*. The stability of nonhyperbolic fixed points will be treated in the next section.

It is straightforward to apply Theorem 1 to the extinction equilibrium of the equation

$$x(t+1) = b(x(t))x(t) + s(x(t))x(t) \tag{1.39}$$

for which $f(x) = r(x)x$ to find that it is stable if $r(0) < 1$ and unstable if $r(0) > 1$.

With regard to the positive equilibria, we restrict attention in this chapter to equations with only negative density effects at all population levels. Specifically, we assume

A1. $b'(x) < 0$ and $s'(x) \leq 0$ for all $x > 0$ and $b(0) > 0$,

where

$$b'(x) = \frac{db(x)}{dx}, \quad s'(x) = \frac{ds(x)}{dx},$$

denote derivatives with respect to x. Motivated by the bifurcation diagram in Figure 1.1(b) for the linear Equation (1.7) we introduce the inherent population growth rate $r(0) = b(0) + s(0)$, which for notational simplicity we re-label as

$$r_0 = r(0),$$

explicitly into Equation (1.19) as follows. We assume, using $b(0) > 0$ in A1, that $b(x)$ is written

$$b(x) = b_0 \beta(x),$$

where $\beta(x)$ satisfies the smoothness requirements in A1 in Section 1.3.1, namely:

A2. $b(x)$ and $s(x)$ are twice continuously differentiable, real-valued functions on Ω that satisfy $b(x) > 0$ and $0 \leq s(x) < 1$ for all $x \in \Omega$

and

$$\beta(0) = 1.$$

This implies $b_0 = b(0) > 0$ is the *inherent fertility rate*. We re-write Equation (1.39) as

$$x(t+1) = (r_0 - s(0))\beta(x)x + s(x)x. \tag{1.40}$$

The equilibrium Equation (1.35) for positive equilibria is

$$1 = (r_0 - s(0))\beta(x) + s(x). \tag{1.41}$$

From this, see that, for any $x^* > 0$, there is a value of $r_0 = r_0(x^*)$ given by

$$r_0(x^*) = s(0) + \frac{1 - s(x^*)}{\beta(x^*)} > 0$$

1.4. STABILITY BY LINEARIZATION

for which x^* is an equilibrium. By A1, $r_0(x^*)$ is an increasing function of $x^* > 0$ with $r_0(0) = 1$. It follows that the inverse function $x^* = x^*(r_0)$ is an increasing function of $r_0 > 1$ with $x^*(1) = 0$. The graph of $x^*(r_0)$ plotted in the (r_0, x)-plane, together with the plot of the extinction equilibria for all r_0, constitute the equilibrium bifurcation diagram for Equation (1.40). These graphs intersect at the bifurcation point $(r_0, x) = (1, 0)$. This bifurcation graph is the analog of the vertical bifurcation diagram of the linear Equation (1.7) in Figure 1.1(b). The nonlinear bifurcation diagram, however, is not vertical since

$$r_0'(0) = -s'(0) - (1 - s(0))\beta'(0) > 0$$

by A1. This implies that a (unique) positive equilibrium exists each for r_0 on some interval $1 < r_0 < r_m \leq +\infty$.

We know that the extinction equilibrium is unstable for $r_0 > 1$. What can we say about the stability of the positive equilibria $x^*(r_0)$ for $r_0 > 1$? To apply the linearization principle, we consider

$$f'(x^*(r_0)) = r(x^*(r_0)) + r'(x^*(r_0))x^*(r_0)$$
$$= 1 + r'(x^*(r_0))x^*(r_0)$$
$$= 1 + r'(0)\left.\frac{dx^*(r_0)}{dr_0}\right|_{r_0=1}(r_0 - 1) + O\left((r_0 - 1)^2\right).$$

By an implicit differentiation of

$$1 = (r_0 - s(0))\beta(x^*(r_0)) + s(x^*(r_0))$$

with respect to r_0, we have

$$r'(0)\left.\frac{dx^*(r_0)}{dr_0}\right|_{r_0=1} = [(1 - s(0))\beta'(0) + s'(0)]\left[-\frac{1}{(1 - s(0))\beta'(0) + s'(0)}\right] = -1$$

and hence

$$f'(x^*(r_0)) = 1 - (r_0 - 1) + O\left((r_0 - 1)^2\right).$$

and $|f'(x^*(r_0))| < 1$ for $r_0 \gtrsim 1$ (by which is meant r_0 is greater than, but close to 1).

In summary, we have

Theorem 2 *The extinction equilibrium of Equation (1.39) is locally asymptotically stable if $r_0 = r(0) < 1$ and unstable if $r_0 > 1$. There exists a unique positive equilibrium for each r_0 on an interval $1 < r_0 < r_m \leq +\infty$ which is locally asymptotically stable for $r_0 \gtrsim 1$.*

Graphically, we have, under Assumptions A1 and A2, a bifurcation diagram similar to that in Figure 1.1(b) for linear equations except that the graph of positive equilibria is no longer vertical. The negative density effects bend the graph to the right so that the range of r_0 values for which there exist positive equilibria is no longer an isolated point, but an interval $1 < r_0 < r_m \leq +\infty$. In most population models found in the literature, $r_m = +\infty$, i.e., there is a unique positive equilibrium for each $r_0 > 1$. This follows when the range of the function $r_0(x^*)$ for $x^* > 1$ is the half line $r_0 > 1$, which occurs if

$$\lim_{x \to +\infty} \beta(x) = 0. \tag{1.42}$$

This limit property holds for most models used in population dynamics. It embodies the assumption that the per capita birth rate vanishes as population density increases without bounds.

Corollary 1 *If (1.42) holds, then $r_m = +\infty$ in Theorem 2.*

Example 7 *Theorem 2 and Corollary 1 apply to all of the model equations in both Examples 2 and 3. In particular, they apply to the special cases of the discrete logistic (Beverton–Holt) and exponential models*

$$x(t+1) = bx(t)\frac{1}{1+cx(t)} \tag{1.43}$$

and

$$x(t+1) = bx(t)\exp\left(-cx(t)\right) \tag{1.44}$$

with $c > 0$. In these cases, $r_0 = b$ and explicit formulas for the positive equilibria are, for all $r_0 > 1$,

$$x^*(r_0) = \frac{r_0 - 1}{c} \text{ in Equation (1.43)}$$

$$x^*(r_0) = \frac{\ln r_0}{c} \text{ in Equation (1.44)}$$

respectively. By Theorem 2, these equilibria are stable for $r_0 \gtrsim 1$, but more can be said for these special cases by applying the linearization principle directly.

For the discrete logistic equation (Beverton–Holt), we have

$$|f'(x^*(r_0))| = \frac{1}{r_0} < 1$$

for all $r_0 > 1$, and therefore all positive equilibria are stable. For the the exponential equation, we have

$$|f'(x^*(r_0))| = |1 - \ln r_0|$$

and find that the positive equilibria are stable if $1 < r_0 < e^2$ and are unstable if $r_0 > e^2$ (see Figure 1.13).

The stability analysis in this section (e.g., Theorems 1 and 2) concerns the local stability of equilibria, that is to say, it concerns initial conditions near an equilibrium. This local analysis does not (necessarily) imply anything about the fate of initial conditions further from the equilibrium.

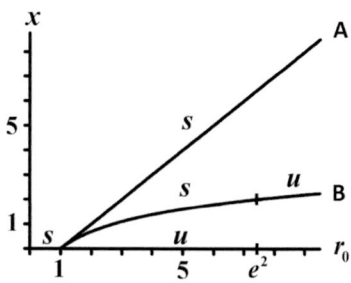

Figure 1.13: The bifurcation diagrams for A. the discrete logistic (Beverton–Holt) equation (1.25) (drawn with $c = 1$) and B. the exponential equation (1.29) (drawn with $c = 1$) in which the local stability properties of equilibria are indicated by s (locally asymptotically stable) and u (unstable).

1.4. STABILITY BY LINEARIZATION

Also of interest is the nature of the dynamics when the positive equilibria destabilize (as in the exponential model for $r_0 > e^2$). We will take up these questions in Sections 1.3.4.

The cobweb analysis of the discrete logistic (Beverton–Holt) and exponential models in Example 5, shows that the locally asymptotically stable positive fixed point x^* (as determined by the Linearization Principle) is in fact globally attracting, i.e., orbits at all initial $x_0 > 0$ tend to the positive fixed point x^* (not just those near x^*). That is to say, x^* is globally asymptotically stable; see Section 1.4.1.

1.4.1 Global Asymptotic Stability of Equilibrium Points

In the previous section, we studied the local stability of equilibrium points. In this section, we study global asymptotic stability. Recall the definition

Definition 3 *A fixed point x^* of the difference equation $x(t+1) = r(x(t))x(t)$ is **globally asymptotically stable** if x^* is (locally asymptotically) stable and $\lim_{t\to\infty} x(t) = x^*$ for all $x(0) > 0$, i.e., x^* is stable and globally attracting.*

For many single-species population models such as the discrete logistic (Beverton–Holt), and exponential models, the positive equilibria are not only locally asymptotically stable but also globally asymptotically stable or, as we might say, in those examples "local stability implies global stability". A natural question to ask is: under what circumstances will the local stability of an equilibrium point imply its global stability? To prove the global stability of an equilibrium point analytically, there are two main methods used in the literature. The oldest and most celebrated method is the method of Liapunov functions [226], whose use and utility have been developed by many authors, including LaSalle [207] and Elaydi [119] [120].

One drawback of the Liapunov method is the difficulty in constructing an appropriate Liapunov function for a model equation of interest. The method, however, will be explored in Chapters 3 and 4.

The second method for investigating global asymptotic stability, one that is being less explored, is based on a theorem of Coppel [56], Cull [63], Elaydi and Sacker [126]. This theorem states, roughly, that global stability is guaranteed when there are no 2-cycles. This criterion of ruling out the existence of 2-cycles is often not difficult to verify for specific model equations. Of course, in some cases, one might have to result to a numerical investigation (although this is not rigorously conclusive, of course).

To state the first main results in this section, we apply the next theorem to the general single-species population model, Equation (1.37) and its defining function f.

Theorem 3 *[119] Assume that $f: I \to I$ is continuous on an interval I such that all orbits of the equation $x(t+1) = f(x(t))$ are bounded. Then every orbit converges to an equilibrium point if and only if there are no 2- cycles (that is, the only solutions to the equation $f(f(x)) = x$ are the equilibrium points of f).*

Remark 2 *It should be noted that the above theorem states that under the stated conditions if there is only one equilibrium point, then it is globally attracting and there is nothing mentioned about its stability. The good news is that one might show in the case of one-dimensional maps, attractivity implies local stability, as shown in the next result.*

Theorem 4 *[119], Let x^* be an attracting fixed point of a continuous map $f : I \to I$, where I is an interval in $\mathbb{R}$. Then x^* is (locally asymptotically) stable.*

Proof. For a proof, see [120] or [119].

The discrete logistic (Beverton–Holt) and exponential models have two equilibria, the equilibrium point $x_1^* = 0$, and a positive point $x_2^* > 0$. Furthermore, due to the nature of the density dependence modeled in those equations, all orbits are bounded. This is because, in both cases $f(x)$ has a bound for all $x \geq 0$:

$$0 \leq b\frac{x}{1+cx} = \frac{b}{c}\frac{cx}{(1+cx)} \leq \frac{b}{c}$$

$$0 \leq bxe^{-cx} \leq \frac{b}{c}e^{-1}.$$

These are the characteristics of many density-dependent single-species population models found in the literature. By Theorem 3, if $x^* > 0$ is locally asymptotically stable, then it is globally stable provided there are no 2-cycles.

One of the challenges in using Theorem 3 is to show that the difference equation has no periodic orbits of minimal period 2. The following result may help in this situation. A map is monotone increasing if whenever $x_1 < x_2$, then $f(x_1) < f((x_2)$.

Lemma 1 *Let $g(x) = x + f(x)$, If $g(x)$ is monotone, then f has no points of minimal period 2. Equivalently, if $1 + f'(x) \neq 0$ for all x, then f has no periodic orbits of period 2.*

Proof. Suppose $g(x)$ is monotone and let $\{x_1, x_2\}$ be a periodic orbit of period 2. Then $g(x_1) = x_1 + f(x_1) = x_1 + x_2$ and $g(x_2) = x_2 + f(x_2) = x_2 + x_1$, which contradicts the assumption of monotonicity of $g(x)$. ∎

It should be noted that this result gives sufficient conditions for the nonexistence of periodic orbits of period 2 but not necessary conditions. There is a class of maps, namely, monotone maps, where we can show global stability with very little effort. We may use the monotone convergence theorem from real analysis (every bounded monotone sequence must converge) to establish global stability. However, we need the following critical lemma to facilitate the proof of the main stability results for maps.

Lemma 2 *Suppose that the map $f : D \longrightarrow \mathbb{R}_+$ is continuous. If for $x_0 \in D$, $\lim_{t\to\infty} f^t(x_0) = z$, where $z \in D$, then z is a fixed point of the map f.*

Proof. Suppose that for $x_0 \in D$, $\lim_{t\to\infty} f^t(x_0) = z$, where $z \in D$. Since f is continuous on D, $f(\lim_{t\to\infty} f^t(x_0)) = f(z)$. But

$$f(\lim_{t\to\infty} f^t(x_0)) = \lim_{t\to\infty} f(f^t(x_0)) = \lim_{t\to\infty} f^{t+1}(x_0)) = z.$$

Since the limit of a sequence is unique, it follows that $f(z) = z$, and thus z is a fixed point of the map f. ∎

Theorem 5 *Suppose that the map $f : D \longrightarrow D$ is continuous and monotone on a subset D of $\mathbb{R}_+$ and x^* is a fixed point of f in D. Assume for x_0 in D that: (i) if $x_0 < x^*$, then $f(x_0) > x_0$ and (ii) if $x_0 > x^*$, then $f(x_0) < x_0$. Then x^* is globally asymptotically stable relative to D.*

1.4. STABILITY BY LINEARIZATION

Proof. Let $x_0 < x^*$. Then
$$\ldots f^t(x_0) < f^{t-1} < \ldots < f^2(x_0) < f(x_0) < f(x^*) = x^*.$$

By the monotone convergence theorem, the orbit $\{f^t(x_0)\}$ converges to a point $z \leq x^*$ and by Lemma 2, z is a fixed point of the map f. But x^* is the only fixed point in its neighborhood and thus $z = x^*$. ∎

One may also use Theorem 3 to prove global stability by showing there are no cycles of period 2. The following lemma establishes that for monotone maps.

Lemma 3 *If $f : D \to D$ is monotone on its domain D, then the map f has no points in D of minimal period 2.*

Proof. Suppose $f(x)$ is monotone and let $\{x_1, x_2\}$ be a periodic orbit of period 2, and assume, without loss of generality, that $x_1 < x_2$, and f is increasing. Then $f(x_1) = x_2 > x_1 = f(x_2)$, which contradicts the assumption. ∎

Example 8 *The discrete logistic (Beverton–Holt) model* (1.25). For this model equation, we have
$$f(x) = b\frac{x}{1+cx} \text{ for } x \geq 0.$$

We know from Example 5 that, if $0 < b \leq 1$, then we have only the extinction equilibrium $x_1^* = 0$, and if $b > 1$, then we have both the extinction equilibrium point and the positive equilibrium point
$$x_2^* = \frac{b-1}{c}.$$

Note that
$$f'(x) = \frac{b}{(1+cx)^2} > 0.$$

Hence, the map
$$f(x) = \frac{bx}{1+cx}$$

is monotone. It follows that (i) if $0 < b < 1$, then the extinction equilibrium is globally asymptotically stable, (ii) if $b > 1$, then the positive equilibrium is globally asymptotically stable.

Example 9 *Exponential model* (1.29). For this model equation, we have
$$f(x) = bxe^{-cx} \text{ for } x \geq 0.$$

Note that there is a positive equilibrium $x^* = c^{-1} \ln b$ if $b > 1$ and that it is locally asymptotically stable provided $1 < r < e^2$. To apply Theorem 3, we consider the equation $f(f(x)) = x$, which leads to
$$b^2 x \exp\left(-bcxe^{-cx} - cx\right) = x.$$

We are interested in nonzero solutions so we cancel x from both sides. After some algebra, the resulting equation reduces to b
$$2\ln b = bcxe^{-cx} + cx.$$

The equilibrium $x^* = c^{-1} \ln b$ is a solution of this equation. There is no other positive solution if $b < e^2$. This is because, under this assumption, the right side

$$g(x) = bcxe^{-cx} + cx$$

is a strictly increasing function for $x > 0$. To see this, calculate

$$g'(x) = c\left(-e^{-cx}b(cx-1) + 1\right)$$

and use calculus to show that the minimum of $h(x) = -e^{-cx}b(cx-1)+1$ (which occurs at $x = 2c^{-1}$) is positive.

We conclude that, when $1 < b < e^2$, the exponential model (1.29) has no 2-cycles and Theorem 3 implies $x^* = c^{-1} \ln b$ is globally asymptotically stable.

An alternative method to show global stability is to use the Allwright–Singer Theorem attributed to Singer in the literature, although Allwright published the result a few months before Singer. But before stating the theorem, it is necessary to introduce the following definition.

Definition 4 *A map $f : [a, b] \to b$, b may be ∞, is an S-map if*

(i) *f is C^3-map and f' vanishes at most one point d that is a relative extremum of f,*

(ii) *there exists $x^* \in (a, b)$ such that $f(x) > x$ if $x < x^*$, and $f(x) < x$, if $x > x^*$,*

(iii) *the Schwarzian derivative [119], [120] $Sh(x) < 0$ for all $x \in [a, b]$, except at the critical point d, where*

$$Sh(x) = \frac{f'''(x)}{f'(x)} - \frac{3}{2}\left(\frac{f''(x)}{f'(x)}\right)^2.$$

Note that a map f is C^3, if $f'''(x)$ exists and continuous.

Theorem 6 (Allwright–Singer *[12], [278])* Let f be an S-map and x^* is a fixed point of f. If $|f'(x^*)| \leq 1$, then x^* is globally asymptotically stable relative to $[a, b]$.

Example 10 *Consider the Ricker model $x(t+1) = x(t)e^{\alpha - cx(t)}$ on $[0, \infty)$ with the equilibrium points $x_1^* = 0$ and $x_2^* = \frac{\alpha}{c}$. Now $|f'(x_2^*)| \leq 1$ if and only if $0 < \alpha \leq 2$. The Schwarzian derivative is given by*

$$Sf(x) = -\left(\frac{c^4x^2 - 4c^3x + 6c^2}{2(1-cx)^2}\right).$$

Note that the quantity

$$p(x) = c^4x^2 - 4c^3x + 6c^2 > 0$$

holds for all $x \in \mathbb{R}$, and $c > 0$, since it has complex roots only and is positive at $x = 0$. Hence, $Sf(x) < 0$ for all $x \in [0, \infty)$. This shows that the map f is an S-map. Now $|f'(x_2^*)| = |1 - \alpha| \leq 1$ if $0 < \alpha \leq 2$. By Allwright–Singer Theorem 6, x_2^* is globally asymptotically stable for $x(0) > 0$ if $0 < \alpha \leq 2$.

Of course, one may have used Theorem 3 by showing that the Ricker map has no periodic orbits of minimal period 2, though, it is probably easier to use Allwright–Singer's Theorem here as shown above.

1.4. STABILITY BY LINEARIZATION

It should be noted that, in case an equilibrium point is locally but not globally asymptotically stable, one may want to know its basin of attraction (see Definition 2). This is particularly of interest when the difference equation possesses multiple attractors. The following two examples concern difference equations with multiple attractors. They involve the equations with multiple locally asymptotically stable equilibria. Biologically, the equations involve a so-called Allee effect. A (demographic) Allee effect is present in a model equation

$$x(t+1) = r(x(t)) x(t) \tag{1.45}$$

when an increase in low-level population density causes an *increase* in the population growth rate $r(x)$, i.e., when $r'(0) > 0$. This phenomenon was extensively studied by the sociologist A. C. Allee [6], [7]. As we will see, an Allee effect can sometimes be the cause of multiple attractors in population models. An important case of multiple attractors caused by an Allee effect is when one of the attractors is the extinction equilibrium $x = 0$, i.e., when $r'(0) < 1$ and a survival attractor is also present (which could be a stable equilibrium or any other attractor).

Definition 5 *A demographic Allee effect (or simply an Allee effect) is present in Equation* (1.45) *if $r(x)$ is increasing for small $x > 0$. A strong Allee effect occurs if there exist two attractors, one of which is $x = 0$ and the other of which is a positive (survival) attractor.*

Clearly, a sufficient condition for an Allee effect is $r'(0) > 0$. A necessary condition for an Allee effect to be associated with a strong Allee effect is that $0 < r'(0) < 1$, so that $x = 0$ is stable (by the linearization principle). However, for a strong Allee effect to be present, there must also exist a stable positive attractor, such as a stable positive equilibrium (i.e., an $x^* > 0$ such that $r(x^*) = 1$ and $|r'(x^*)| < 1$). This is illustrated by the following example.

Example 11 *[8] Consider the following single-species Ricker model with the strong Allee effect*

$$x(t+1) = x(t) \exp((\alpha - x(t))) \exp\left(-\frac{m}{1+sx(t)}\right) = f(x(t)), \tag{1.46}$$

where the factor

$$\exp\left(-\frac{m}{1+sx(t)}\right)$$

represents the effect of the presence of a strong Allee effect due to predator saturation. Equation (1.46) *has three fixed points $x_1^* = 0$, $x_2^* = A$, and $x_3^* = K$, where*

$$A = \frac{(\alpha s - 1) - \sqrt{(\alpha s - 1)^2 - 4s(m - \alpha)}}{2s}$$

$$K = \frac{(\alpha s - 1) + \sqrt{(\alpha s - 1)^2 - 4s(m - \alpha)}}{2s}.$$

To ensure that both A and K exist and are positive, one must assume that

$$\alpha s > 1, \quad m > \alpha, \quad (\alpha s + 1)^2 > 4ms.$$

For maps with multi-stable fixed points, Theorem 3 is the only option. Now for $0 < K\left(1 - \frac{s}{m}(\alpha - K)^2\right) < 2$, Equation (1.46) has no periodic orbits of minimal period 2. Hence, every orbit must converge to one of the three fixed points. Under this condition, one may show that $x_1^* = 0$ is attracting, $x_2^* = A$ is a repeller, and $x_3^* = K$ is attracting. The basin of attractions of $x_1^* = 0$

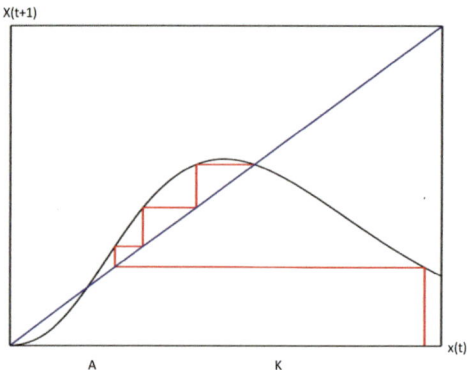

Figure 1.14: The cobweb diagram of the Ricker model with the strong Allee effect. There are three equilibrium point $x_1^* = 0$, which is asymptotically stable, the Allee equilibrium point A, which is unstable, and the carrying capacity K, which is asymptotically stable.

and $x_3^* = K$ are given by the sets $\mathcal{B}(0) = (0, A) \cup (f^{-1}(A), \infty)$, $\mathcal{B}(K) = (A, f^{-1}(A))$, as shown in Figure 1.14.

An Allee effect does not always produce a strong Allee effect. Even so, it might still result in multiple positive attractors. The next example illustrates this.

Example 12 *Consider a model of a single species that possesses a cooperative (Allee) low-density effect on per capita growth rate given by*

$$r_0 + uv\frac{x^m}{v+x^m}.$$

Here u is a measure of the intrinsic effectivity of cooperation and v is the level of cooperation. The high-density negative effect is modeled by the factor $1/(1+cx)$. Hence model equation is

$$x(t+1) = x(t)\left(r_0 + uv\frac{x^m(t)}{v+x^m(t)}\right)\frac{1}{1+cx(t)}. \tag{1.47}$$

The equation for a positive equilibrium is

$$1 = \left(r_0 + uv\frac{x^m}{v+x^m}\right)\frac{1}{1+cx}.$$

To get a bifurcation diagram that plots x as a function of r_0, we would need to solve this equation for $x > 0$. But $m \geq 2$ is intractable so instead we solve for $r_0 = r_0(x)$ to get

$$r_0(x) = 1 + cx - uv\left(\frac{x^m}{v+x^m}\right) \tag{1.48}$$

and plot r_0 as a function of x. The desired bifurcation of x plotted as a function of r_0 is obtained by re-orienting axes. For $m \geq 2$, this procedure will show that the difference equation (1.47) can have multiple positive equilibria.

For example, the bifurcation diagram for a case with $m = 4$ is shown in Figure 1.15. The s-shaped bifurcation seen in Figure 1.15 shows that, for some values of r_0 (such as $r_0 = 1.75$), there

1.4. STABILITY BY LINEARIZATION

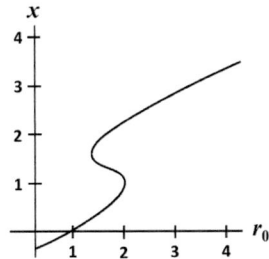

Figure 1.15: The bifurcation diagram associated with Equation (1.47) with parameter values $v = 4$, $u = 1.5$, $c = 2$, and $m = 4$ shows a forward bifurcation at $r_0 = 1$ and a hysteresis shape with multiple positive equilibria for r_0 between approximately 1.4 and 2.0.

are three positive equilibria. (Such an s-shaped bifurcation diagram is called a hysteresis diagram.) Using a computer algebra program to solve the equilibrium equation with the parameter values

$$v = 2.5, \quad u = 1.5, \quad c = 2$$

used in Figure 1.15, we obtain the three positive equilibria indicated in the bifurcation diagram at $r_0 = 1.75$ to be (rounded to three significant digits)

$$x_1^* = 0.393, \quad x_2^* = 1.16, \quad x_3^* = 1.99.$$

Applying the linearization principle to these equilibria and the extinction equilibrium $x_0^* = 0$ with

$$f(x) = x\left(r_0 + uv\frac{x^4}{v + x^4}\right)\frac{1}{1 + cx},$$

we find from (rounded to three significant digits)

$$f'(0) = 1.75, \quad f'(0.393) = 0.638, \quad f'(1.15) = 1.40, \quad f'(1.99) - 0.558$$

that there are two stable positive equilibria x_1^* and x_3^* and two unstable equilibria $x_0^* = 0$ and x_2^*.

For more on Allee effects, see Chapter 5 and Chapter 6.

Exercises 1.4.1

In problems 1–5:

(a) Find the equilibrium points and determine their local stability properties.

(b) Find the conditions under which a locally stable equilibrium point is global asymptotically stable. If an equilibrium point is locally but not globally asymptotically stable, find its basin of attraction.

1. $x(t+1) = rx(t)(1 - x(t))$, $0 < r < 4$
2. $x(t+1) = \frac{bx(t)}{1+cx(t)}$

3. $x(t+1) = \dfrac{bx^2(t)}{1+x^2(t)}$, $b > 0$

4. $x(t+1) = \dfrac{bx(t)}{(1+x(t))^{\frac{1}{2}}}$, $b > 0$.

5. The equation below has been used to model the dynamics of bobwhite quail populations [139, 239]
$$p(t+1) = p(t)\left[r + \dfrac{s}{1+p(t)^\ell}\right] \quad r, s, \ell > 0.$$

   (a) Find conditions under which there exists a positive equilibrium point.

   (b) If $\ell = 3, 4$, find conditions under which the positive equilibrium point is globally asymptotically stable.

   (c) Find conditions under which the positive equilibrium point is globally asymptotically stable, for $\ell > 0$

6. Consider Hassell's model
$$p(t+1) = \dfrac{ap(t)}{[1+bp(t))^k]} \quad a, b, k > 0.$$

   (a) Prove that if $a < 1$, the extinction equilibrium is globally asymptotically stable.

   (b) Show that if $z > 0$, the positive equilibrium point is globally asymptotically stable.

7. Consider the model below
$$p(t+1) = m + (1-m)\dfrac{(1+s)p(t)}{1+sp(t)}, \quad m, s > 0.$$

   Determine the global stability of the equilibrium points.

8. Consider the discrete logistic (Beverton–Holt) model with constant immigration
$$x(t+1) = \dfrac{bx(t)}{1+cx(t)} + m, \quad b, c > 0.$$

   Show that the positive equilibrium point is globally asymptotically stable.

9. Use the monotone convergence theorem to show that the positive equilibrium point of the discrete logistic (Beverton–Holt) model
$$x(t+1) = \dfrac{rKx(t)}{K + (r-1)x(t)}$$
is globally asymptotically stable if $r > 1$.

10. The following difference equation is the discrete logistic (Beverton–Holt) model with a strong Allee effect. Find conditions under which the model exhibits multiple attractors and then find their basins of attraction.
$$x(t+1) = \dfrac{rKx(t)}{K + (r-1)x(t)} \dfrac{m}{1+sx(t)}. \tag{1.49}$$

11. The following equation is the Ricker model with a strong Allee effect. Find conditions under which the model exhibits multiple attractors and then find their basins of attraction.

$$x(t+1) = x(t) \exp\left(r\left(1 - \frac{x(t)}{K}\right)\frac{m}{1+sx(t)}\right) \tag{1.50}$$

12. The following equation is the Hassell's model with a strong Allee effect. Find conditions under which the model exhibits multiple attractors and then find their basins of attraction

$$x(t+1) = \frac{rx(t)}{(1+kx(t))^\theta}\frac{m}{1+sx(t)}. \tag{1.51}$$

13. Using a cobweb analysis, find the basins of attraction for the two stable positive equilibria in Example 12.

1.4.2 Bifurcation Analysis

Consider the one-parameter difference equation

$$x(t+1) = f(x(t), \mu), \tag{1.52}$$

where $\mu \in \mathbb{R}$ is the parameter. Let x^* be the equilibrium point at the parameter value μ^*, that is $f(x^*, \mu^*) = x^*$. Then, by the linearization principle (Theorem 1), x^* is locally asymptotically stable if $|f'(x^*, \mu^*)| < 1$ and x^* is unstable if $|f'(x^*, \mu^*)| > 1$.

The equilibrium points x^* for which $|f'(x^*)| \neq 1$ are called hyperbolic equilibrium (fixed) points and those for which $|f'(x^*)| = 1$ are called nonhyperbolic equilibrium points. Next, we are going to investigate the stability of equilibria x near x^* and μ near μ^* where x^* is a nonhyperbolic fixed point, i.e. where $f'(x^*, \mu^*) = 1$ or -1. In these cases, μ^* is called a bifurcation point. We consider here two types of bifurcations:

1. Period-doubling bifurcation that occurs when $f'(x^*, \mu^*) = -1$. In this case, the fixed point loses its stability and a 2-cycle appears as μ increases (or decreases) through μ^*.

2. Transcritical or a tangent (blue-sky) bifurcation that occurs when $f'(x^*, \mu^*) = 1$. In this case, x^* loses its stability and new asymptotically stable fixed points appear as μ increases (or decreases) through μ^*.

The two types of bifurcation are illustrated in the following example.

Example 13 *Consider the exponential model*

$$x(t+1) = bx(t)\exp\left(-cx(t)\right).$$

We have seen that $x_1^* = 0$ is asymptotically stable when $b < 1$ and unstable when $b > 1$. Now $f'(x) = (1 - cx)x\exp(-cx)$, $f'(0) = 1$, when $b = 1$. Hence, we have a transcritical bifurcation, when $b = 1$.

For $b \geq 1$, the bifurcation curve is given by $x = \frac{1}{c}\ln b$. The equilibrium $x_2^* = \frac{1}{c}\ln b$ is asymptotically stable if $1 < b < e^2$. Now

$$f'(x_2^*) = \left(b - c\frac{b\ln b}{c}\right) \cdot \frac{1}{b} = 1 - \ln b$$

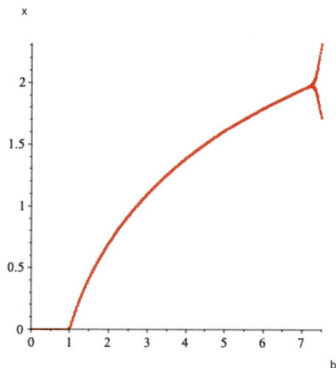

Figure 1.16: The bifurcation diagram of the exponential map, at $b = 1$ we have a transcritical bifurcation, and at $b = e^2$, we have period-doubling bifurcation.

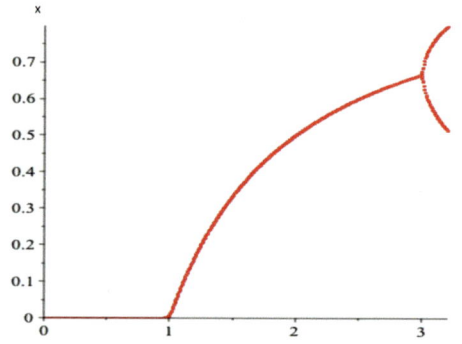

Figure 1.17: The bifurcation diagram of the quadratic map, at $r = 1$ we have a transcritical bifurcation, and at $r = 3$, we have period-doubling bifurcation.

and when $b = e^2$, we have $f'(x_2^*) = 1 - 2 = -1$. Hence, there is a period-doubling bifurcation at $b = e^2$ (see Figure 1.16).

We still have not addressed the question of stability at these bifurcation points. The following two theorems provide the definitive answer [119, 120].

Theorem 7 *Suppose that $f'(x^*) = -1$, for a fixed point x^* of f. Then x^* is asymptotically stable if the Schwarzian derivative $Sf(x^*) < 0$, where $Sf(x^*) = -f'''(x^*) - \frac{3}{2}(f''(x^*))^2$.*

Now if $f'(x^*) = 1$, we have a new phenomenon in which the fixed point is asymptotically stable relative to points on its right (left) but unstable to the points on its left (right). Next, a formal definition of this phenomenon is called semi-asymptotic stability.

Definition 6 *Let x^* be an equilibrium point of Equation* (1.36). *Then*

1.4. STABILITY BY LINEARIZATION

(i) The equilibrium point x^* is semi-stable from the right (left) if for any $\varepsilon > 0$, there exists $\delta > 0$ such that $|x(t) - x^*| < \varepsilon$, whenever $0 < (x(0) - x^*) < \delta$ $(0 < x^* - x(0) < \delta)$.

(ii) The equilibrium point x^* is semi-attracting from the right (left) if there exists $\mu > 0$ such that $\lim_{t \to \infty} x(t) = x^*$, whenever $0 < (x(0) - x^*) < \mu$ $(0 < (x^* - x(0)) < \mu)$.

(iii) The equilibrium point x^* is locally semi-asymptotically stable (SLAS) if it is semi-stable and semi-attracting.

(iv) The equilibrium point x^* is globally semi-asymptotically stable if it is semi-asymptotically stable and $\lim_{t \to \infty} x(t) = x^*$, for either all $x(0) > x^*$ or all $x(0) < x^*$, where $x(0)$ is in the domain of the map f.

Theorem 8 *Suppose that $f'(x^*) = 1$, for a fixed point x^* of f. The following statements hold true:*

(i) *If $f''(x^*) \neq 0$, x^* is semi-asymptotically stable from the right if $f''(x^*) < 0$ and semi-asymptotically stable from the left if $f''(x^*) > 0$.*

(ii) *If $f''(x^*) = 0$, then x^* is asymptotically stable if $f'''(x^*) < 0$ and x^* is unstable if $f''' > 0$.*

Now let us apply these theorems to the exponential map. First, for $x_1^* = 0$, we have $f'(x_1^*) = 1$, $f''(x) = (-2cb + c^2 bx)e^{-cx}$. When $b = 1$ and $f''(0) = -2c \neq 0$. Hence, $x_1^* = 0$ is semi-asymptotically stable from the right, which means asymptotically stable, since we are not interested in negative x_0.

The Schwarzian derivative $Sf(x_2^*) = c^2 > 0$ and thus x_2^* is unstable at the bifurcation point $b = e^2$.

Example 14 *Consider the quadratic model*

$$x(t+1) = rx(t)(1 - x(t)) = f(x(t))$$

where $0 \, r < 4$ so that $f(x)$ maps $[0,1] \times [0,1]$ into itself. We have seen that $x_1^* = 0$ is asymptotically stable when $0 < r < 1$ and unstable when $r > 1$. Now $f'(x) = r - 2rx$, $f'(0) = r = 1$, when $r = 1$. Hence, we have a transcritical bifurcation, when $r = 1$. For $r \geq 1$, the bifurcation curve is given by $x = \frac{r-1}{r}$. Now $x_2^* = \frac{r-1}{r}$ is asymptotically stable if $1 < r < 3$. Now if $r = 3$, $f'(x_2^*) = -1$. Hence, we have a period-doubling bifurcation at $r = 3$ (see Figure 1.17). Computing the Schwarzian derivative at x_2^*, we find it to be negative. Hence, x_2^* is asymptotically stable at the bifurcation point $r = 3$.

Exercises 1.4.2

1. For the following equations, find the equilibrium points and determine the stability of all equilibrium points:

    (a) The Ricker model: $x(t+1) = x(t) \exp\left(\alpha \left(1 - \frac{x(t)}{K}\right)\right)$

    (b) The discrete logistic (Beverton–Holt) model: $x(t+1) = \frac{bKx(t)}{K + (b-1)x(t)}$

    (c) Theta exponential model: $x(t+1) = bx(t) \exp\left(-cx^\theta(t)\right)$, $\theta > 0$

    (d) Hassell's model: $x(t+1) = \frac{bx(t)}{[1+cx(t)]^l}$, $b, s, l > 0$

2. Allee effect is a phenomenon in which there is a positive correlation between the growth rate of a population and its size at low population levels (x near 0). The following two problems are models that contain an Allee effect.

   (a) $x(t+1) = \frac{bx^2(t)}{1+x^2(t)}$, $b > 0$

   (b) $x(t+1) = x(t) \exp\left(\alpha \left(1 - \frac{x(t)}{K}\right)\left(\frac{x(t)}{A} - 1\right)\right)$

3. For the following equations:

   (i) Find the bifurcation points and determine their types.

   (ii) Draw the bifurcation diagrams.

   (iii) Determine the stability of the fixed points at the bifurcation points.

   (a) $x(t+1) = \frac{bKx(t)}{K+(b-1)x(t)}$ (the parameter r for a fixed K)

   (b) $x(t+1) = x(t) \exp\left(\alpha \left(1 - \frac{x(t)}{K}\right)\right)$ (the parameter α for a fixed K)

   (c) $x(t+1) = \frac{bx^2(t)}{1+x^2(t)}$

   (d) $x(t+1) = \frac{bx(t)}{\sqrt{(1+x(t))}}$

1.4.3 Population Cycles: Periodic Orbits

Due to a built-in time delay and the influence of density on the per-capita growth rate, population models are capable of generating complex patterns from equilibrium points, cyclic oscillations between two population densities, cycles with three points, four points, and so on. For example, the exponential model $x(t+1) = bx(t)e^{-cx(t)}$ exhibits cyclic oscillations between two population densities, whenever $b > e^2$ (see Figure 1.18). In this section, we introduce definitions and theoretical results that are used to study the stability of such cyclic oscillations in single-species models.

Definition 7 *Consider*
$$x(t+1) = f(x(t)).$$
Then
$$c_k = \{\overline{x}_1, \overline{x}_2, \cdots, \overline{x}_k\}$$
is a k-cycle of f if
$$f(\overline{x}_i) = \overline{x}_{i+1} \quad \text{for each } i \in \{1, 2, \cdots, (k-1)\} \quad \text{and} \quad f(\overline{x}_k) = \overline{x}_1.$$

Note that a k-cycle is the orbit of a periodic point of period k. Equivalently, k-cycles correspond to fixed points p of the k-composite map
$$f^k(p) = p.$$
Consequently, the characterization of the stability of k-cycles of the equation $x(t+1) = f(x(t))$ reduces to the characterization of the stability of fixed points of the equation $x(t+1) = g(x(t))$,

1.4. STABILITY BY LINEARIZATION

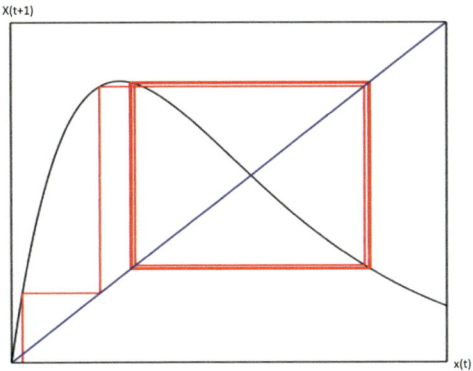

Figure 1.18: The appearance of a 2-cycle in the exponential map when $b > e^2$.

where $g = f^k$. As a consequence, one may obtain stability results for hyperbolic k-cycles. We say a periodic point x of period-k is hyperbolic if

$$\left|\frac{df^k(x)}{dx}\right| \neq 1.$$

Equivalently, the k-cycle $c_k = \{\overline{x}_1, \overline{x}_2, \cdots, \overline{x}_k\}$ is hyperbolic if

$$\prod_{i=1}^{k} f'(x_i) \neq 1.$$

Next, we give the definitions of the stability notions for cycles (Figure 1.19).

Definition 8 *Let $c_k = \{\overline{x}_1, \overline{x}_2, \cdots, \overline{x}_k\}$ be a k-cycle of the map f or of Equation (1.36). Then*

(i) *The cycle c_k is stable if for any $\varepsilon > 0$, there exists $\delta > 0$ such that $|f^{tk+i-1}(x_0) - \overline{x}_i| < \varepsilon$, whenever $|x_0 - \overline{x}_i| < \delta$. In other words $|x(tk+i-1) - \overline{x}_i| < \varepsilon$, whenever $|x_0 - \overline{x}_i| < \delta$, where $x(0) = x_0$, for $1 \leq i \leq k$.*

(ii) *The cycle c_k is attracting if there exists $\mu > 0$ such that $\lim_{t\to\infty} f^{tk+i-1}(x_0) = \overline{x}_i$, i.e., $\lim_{t\to\infty} x(tk+i-1) = \overline{x}_i$, whenever $|x_0 - \overline{x}_i| < \mu$, for $1 \leq i \leq k$.*

(iii) *The cycle c_k is locally asymptotically stable (LAS) if it is stable and attracting.*

(iv) *Let D be a subset in the domain of f. The cycle c_k is globally asymptotically stable on D if it is locally asymptotically stable and $\lim_{t\to\infty} f^{tk+i-1}(x_0) = \overline{x}_i$, i.e., $\lim_{t\to\infty} x(tk+i-1) = \overline{x}_i$, for all $x_0 \in D$ and $1 \leq i \leq k$.*

Theorem 9 *[119] Let $c_k = \{\overline{x}_1, \overline{x}_2, \cdots, \overline{x}_k\}$ be a k-cycle of f. Then*

1. *c_k is locally asymptotically stable if*

$$|f'(\overline{x}_1)f'(\overline{x}_2)\ldots f'(\overline{x}_k)| < 1.$$

2. c_k is unstable if
$$|f'(\overline{x}_1)f'(\overline{x}_2)\ldots f'(\overline{x}_k)| > 1.$$

For the proof of the theorem, we can use Theorem 1 and the chain rule.

Example 15 *Consider the mathematically tractable quadratic model,*
$$x(t+1) = x(t)\left[1 + r\left(1 - \frac{x(t)}{K}\right)\right],$$
where r and K are positive constants. To find the 2-cycle, we solve the equation $f^2(x) = x$, where
$$f(x) = x\left[1 + r\left(1 - \frac{x}{K}\right)\right].$$
Assume $r < 3$ to that $f(x)$ maps the square $[0, K(r+1)/2r] \times [0, K(r+1)/2r]$ into itself. First, we note that the fixed points of f are $x^* = 0$ and $x^* = K$. The 2-cycle of f is $\{\overline{x}_1, \overline{x}_2\}$ where
$$\begin{aligned}\overline{x}_1 &= \tfrac{K}{2r}\left(2 + r - \sqrt{r^2 - 4}\right),\\ \overline{x}_2 &= \tfrac{K}{2r}\left(2 + R + \sqrt{r^2 - 4}\right).\end{aligned}$$
Notice that both $\overline{x}_1$ and $\overline{x}_2$ are positive and not equal only when $r > 2$. The 2-cycle $\{\overline{x}_1, \overline{x}_2\}$ coalesce and $\overline{x}_1 = \overline{x}_2 = x^* = K$ when $2 = 2$, and they are complex numbers when $r < 2$.

To investigate the stability of this 2-cycle, we apply Theorem 9. Since
$$f'(\overline{x}_1) = -1 - \sqrt{r^2 - 4} \text{ and } f'(\overline{x}_2) = -1 + \sqrt{r^2 - 4},$$
we obtain that the 2-cycle $\{\overline{x}_1, \overline{x}_2\}$ is locally asymptotically stable if
$$|f'(\overline{x}_1)f'(\overline{x}_2)| = |5 - r^2| < 1$$
and unstable if
$$|f'(\overline{x}_1)f'(\overline{x}_2)| = |5 - r^2| > 1.$$
That is, the 2-cycle $\{\overline{x}_1, \overline{x}_2\}$ is locally asymptotically stable when
$$2 < r < \sqrt{6} \text{ and unstable when } r > \sqrt{6}.$$
See Figure 1.20.

In Section 1.4.1, we studied global stability of fixed points. Can a k-cycle with period $k > 1$ be globally asymptotically stable? This question was answered in 2002 by Elaydi and Yakubu [134] who showed that this is not possible in **autonomous discrete-time** models such as model (1.37). We summarize the result of Elaydi and Yakubu in the following theorem.

Theorem 10 *[134] Assume that f is continuous on an interval I. Then Equation (1.37), $x(t+1) = f(x(t))$, has no globally asymptotically stable k-cycle with period $k > 1$.*

1.4. STABILITY BY LINEARIZATION

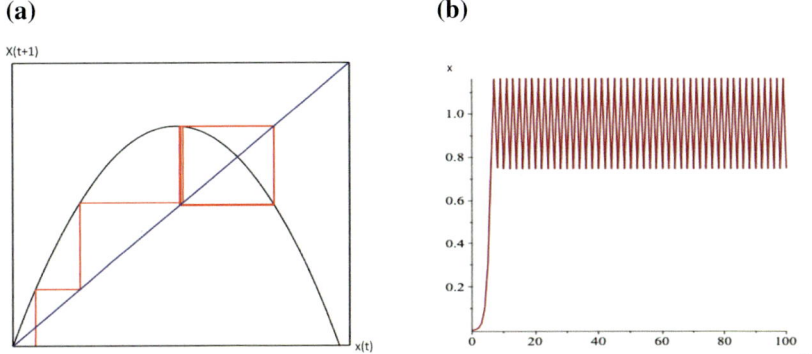

Figure 1.19: (a) Cobweb. Asymptotically stable 2-cycle, $r = 2.2$. (b) Time series. Asymptotically stable 2-cycle in (t, x_t)-plane, $r = 2.2$.

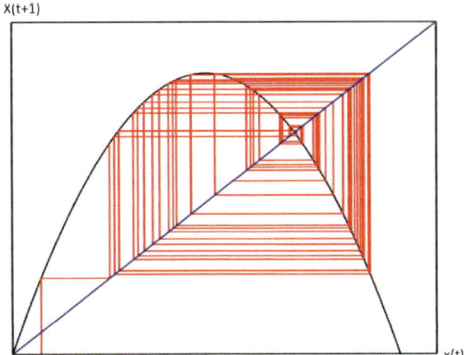

Figure 1.20: An unstable 2-cycle, $r > \sqrt{(6)}$.

Exercises 1.4.3

For Equations 1–4, find all the 2-cycles and determine their stability.

1. $x(t+1) = 1 - x^2(t)$
2. $x(t+1) = 3.2x(t)(1 - x(t))$
3. $x(t+1) = \frac{2x(t)}{1+x(t)}$
4. $x(t+1) = x(t)e^{2.5-x(t)}$
5. Let $x(t+1) = ax^2(t) + bx(t) + c = f(x(t))$ $a, b, c \in \mathbb{R}$, $a \neq 0$.

    (a) If $\{\bar{x}_1, \bar{x}_2\}$ is a 2-cycle such that $f'(\bar{x}_1) \cdot f'(\bar{x}_2) = -1$, show that this cycle is asymptotically stable.

    (b) If $\{\bar{x}_1, \bar{x}_2\}$ is a 2-cycle such that $f'(\bar{x}_1) \cdot f'(\bar{x}_2) = 1$, determine whether this cycle is asymptotically stable or unstable.

6. Let $x(t+1) = ax^3(t) - bx(t) + 1$. Find the values of a and b for which the periodic orbits $\{0, 1\}$ is asymptotically stable.

For Equations 7–9, use numerical approximation to find the periodic orbits of periods 2 and 4 and determine their stability.

7. $x(t+1) = 9x(t)e^{-x(t)}$
8. $x(t+1) = x(t)e^{2.5-x(t)}$
9. $x(t+1) = 3.5x(t)(1 - x(t))$

1.4.4 Persistence of Species

Persistence of species is considered an important concept in ecology and epidemiology. It concerns the long-time survival of a species, genotypes, strategies in the evolutionary game, etc. In other words, persistence of a species means its survival under the specified natural conditions. In epidemics, however, the focus is on persistence of a disease for a long period of time. Next, we give the formal definition of persistence.

Definition 9 ([170]) *Consider the single-species model represented by the difference equation $x(t+1) = f(x(t))$. Then we say that species x is "uniformly" persistent if there exists $\eta > 0$ such that for any initial point $x(0) > 0$, $\lim\limits_{t \to \infty} x(t) > \eta$.*

Now if in a model, there is a globally asymptotically stable positive equilibrium point, or if the extinction equilibrium $x^* = 0$ is a repeller (e.g., if $|f'(0)| > 1$), then species x is "uniformly" persistent. Similarly, if there are multiple positive equilibrium points that attract all positive points, then we say species x persist. Now if a model has a k-cycle $c_k = \{\bar{x}_1, \bar{x}_2, \ldots, \bar{x}_k\}$, then, from the Elaydi–Yakubu Theorem 10, c_k cannot be globally asymptotically stable. In this case, to show persistence, one only needs to show that the extinction equilibrium is a repeller.

An earlier Example 11,

$$x(t+1) = x(t)\exp(\alpha - x(t) - \frac{m}{1+sx(t)}) = f(x(t)) \qquad (1.53)$$

illustrates, clearly, the concept of persistence. Here we have the extinction equilibrium $x^* = 0$ is not a repeller and, in fact, it is locally asymptotically stable. There are two positive equilibria, A and K, where A is a repeller and K is locally asymptotically stable. We showed in this example that the basin of attraction of K is given by the interval $(A, f^{-1}(A))$, as shown in Figure 1.14. Thus, species x will persist if its initial size $x(0)$ lies in the interval $(A, f^{-1}(A))$.

1.4.5 Period-Doubling Bifurcations Route to Chaos

In the last section, we observed that the discrete logistic (Beverton–Holt) exhibits only non-periodic fixed point dynamics, while the exponential, and Ricker models can exhibit more complicated periodic dynamics. For example, the Ricker model with $f(x) = xe^{\alpha - cx}$ exhibits a locally asymptotically stable fixed point whenever $0 < \alpha < 2$. As α increases past 2, the fixed point $x^* = \frac{\alpha}{c}$ loses its stability and a new 2-cycle is born. As α increases further, the Ricker model exhibits the phenomenon of *period-doubling bifurcations route to chaos* and cycles of period 4, 8, 16, 32, $\cdots$ appear. The exponential and the Ricker with Allee effect also exhibit similar period-doubling bifurcations. Recall that the fixed points of the Ricker model

$$f(x) = x\exp(\alpha - cx)$$

are $x^* = 0$ and $x^* = \alpha/c$. Since $f'(0) = e^\alpha > 1$, $x^* = 0$ is always unstable (Theorem 1). However, since $f'(\alpha/c) = 1-\alpha$, the stability of $x^* = \alpha/c$ depends on the value of α (Theorem 1). The positive fixed point $x^* = \alpha/c$ is locally asymptotically stable when $0 < \alpha < 2$ and unstable when $\alpha > 2$. At $\alpha = 2$, we have $f'(\alpha/c) = -1$, a case not covered by the linearization principle (Theorem 1). In fact, using cobwebbing, one can show that $x^* = \alpha/c$ is unstable in this case.

As α increases past $\alpha^* = 2$, the positive fixed point loses its stability and a new 2-cycle is born (period-doubling bifurcation at $\alpha^* = 2$). By Theorem 9, this new 2-cycle is locally asymptotically stable when $2 < \alpha < 2.52646$ and unstable when $\alpha > 2.52646$.

As α increases past $\alpha^* = 2.52646 \cdots$, the 2-cycle loses its stability and a new 2^2-cycle is born (period-doubling bifurcation at $\alpha^* = 2.52646 \cdots$). This 2^2-cycle is locally asymptotically stable when $2.52646 \cdots < \alpha < 2.65635 \cdots$ and unstable when $\alpha > 2.65635 \cdots$. To compute the 2^2-cycle, $\{\bar{x}_1, \bar{x}_2, \bar{x}_3, \bar{x}_4\}$, we can use a software to solve $f^4(x) = x$, for $\bar{x}_1, \bar{x}_2, \bar{x}_3, \bar{x}_4 > 0$. This process of period-doubling bifurcations continues indefinitely and produces a sequence of α values, $\{\alpha_n\}_{n=1}^\infty$, where $\alpha_1 = 2$, $\alpha_2 = 2.52646\cdots$, $\alpha_3 = 2.65635 \cdots$, $\alpha_4 = 2.6846 \cdots$, $\alpha_5 = 2.6907 \cdots$, are the α values where the bifurcations occur. Table 1.1 sheds some light on some remarkable patterns (see Figure 1.21).

From Table 1.1, we make the following observations which can be verified numerically.

1. The sequence $\{\alpha_n\}_{n=1}^\infty$ seems to converge to $\alpha_\infty \approx 2.692$.

2. The window size $(\alpha_n - \alpha_{n-1})$ between successive α_{n-1} and α_n values gets smaller and smaller, and eventually approaches zero as $n \to \infty$.

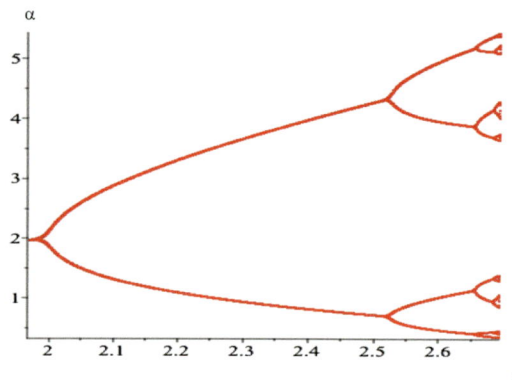

Figure 1.21: Period-doubling: The appearance of asymptotically stable cycles of period 2, 4, 8, 16, $\cdots$.

n	α_n	$\alpha_n - \alpha_{n-1}$	$\frac{\alpha_n - \alpha_{n-1}}{\alpha_{n+1} - \alpha_n}$
1	2	-	-
2	2.52646	0.52646	-
3	2.65635	0.12989	4.05312
4	2.6846	0.02825	4.59787
5	2.6907	0.0061	4.63114

Table 1.1: The bifurcation points for the Ricker equation.

3. As $n \to \infty$, the ratio of successive window sizes,
$$\frac{\alpha_n - \alpha_{n-1}}{\alpha_{n+1} - \alpha_n},$$
approaches a constant, $\delta \approx 4.669201609$, called the **Feigenbaum number**; named after its discoverer Mitchell Feigenbaum [137]. In fact
$$\delta = \lim_{n \to \infty} \frac{\alpha_n - \alpha_{n-1}}{\alpha_{n+1} - \alpha_n} \approx 4.669201609 \cdots, \tag{1.54}$$
Feigenbaum discovered that the number δ is universal and does not depend on the family of maps under discussion; it is the same for a large class of **one-hump** or **unimodal** maps such as the Ricker and quadratic maps.

Equation (1.54) may be used to generate the sequence $\{b_n\}_{n=1}^{\infty}$ with good accuracy. To illustrate this, we let $\delta = \frac{\alpha_n - \alpha_{n-1}}{\alpha_{n+1} - \alpha_n}$ and solve for α_{n+1}. Then
$$\alpha_{n+1} = \alpha_n + \frac{\alpha_n - \alpha_{n-1}}{\delta}. \tag{1.55}$$
For example, given $\alpha_1 = 2$ and $\alpha_2 = 2.52646$ (from Table 1.1), then from Equation (1.55) with $\delta \approx 4.6692$, we get $\alpha_3 \approx 2.6392$, which is a good approximation to the value of α_3 in Table 1.1. The best way to summarize the above discussion is with a bifurcation diagram.

1.4. STABILITY BY LINEARIZATION

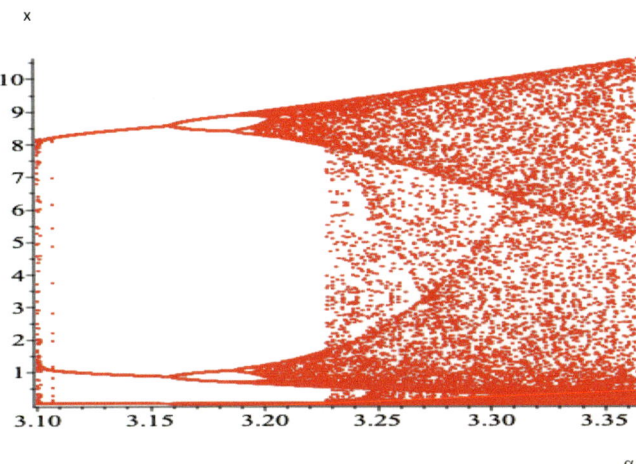

Figure 1.22: The appearance of period 3.

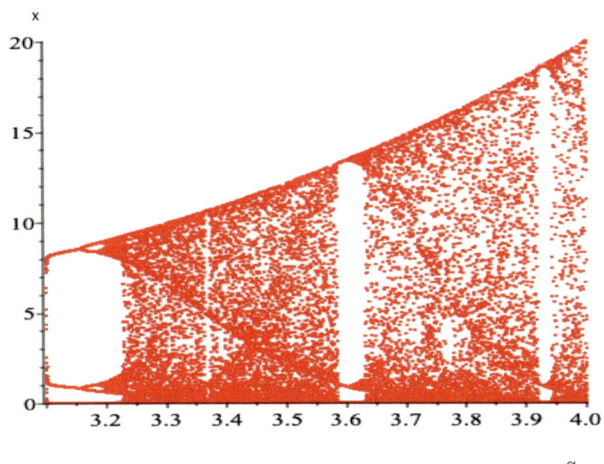

Figure 1.23: The appearance of odd periods and the "maximum" chaos at $\alpha = 4$.

1. First, fix the value of the parameter c, say $c = 1$. For an initial choice of the parameter $\alpha = \alpha_0 \in 0, 2]$, choose an initial condition $x_0 = 1$ in the closed interval $(0, \infty)$ and compute $f^t(x_0)$. For example, take $t = 1,000$ and obtain the following iterates of f:

$$x_0, f(x_0), f^2(x_0), \cdots, f^{800}(x_0), f^{801}(x_0), \cdots, f^{1,000}(x_0).$$

2. Discard the first 800 iterates,

$$x_0, f(x_0), f^2(x_0), \cdots, \text{and } f^{800}(x_0),$$

and plot the rest of the iterates

$$f^{801}(x_0), \cdots, \text{and } f^{1,000}(x_0)$$

in the bifurcation diagram. Thus, only asymptotically stable sets will be seen in the bifurcation diagram.

3. Steps (1) and (2) are done repeatedly for increasing values of $\alpha \in [\alpha_0, 4]$ in increments of $\frac{1}{100}$ with $c = 1$ and $x_0 = 1$.

4. (a) In the bifurcation diagram, we see that, for $\alpha = \alpha_0 \in (0, 2]$, the orbit of the initial condition $x_0 = 1$ converges to $x^* = \alpha/c$. For example, if $\alpha = 0.5$, the point $x = (0.5, 0.5)$ appears in the diagram. Similarly, when $\alpha = 0.51$, the point $(\alpha, x^*) = (0.51, 0)$ appears in the diagram, etc. At $\alpha = 2$, the point $(\alpha, x) = (2, 2)$ appears in the diagram. For values of b beyond $\alpha_1 = 2$, $x^* = \alpha$ loses its stability, disappears from the bifurcation diagram and a 2-cycle, denoted by $\{\bar{x}_1, \bar{x}_2\}$, is born ($x^* = \alpha$ undergoes period-doubling bifurcation at $\alpha = \alpha_1$).

   (b) For values of $\alpha \in (\alpha_1, \alpha_2]$, the orbit of $x_0 = 1$ converges to the 2-cycle. Hence, to each $\alpha \in (\alpha_1, \alpha_2]$, there corresponds two points, $\bar{x}_1$ and $\bar{x}_2$, in the diagram. For values of α beyond α_2, the 2-cycle loses its stability, disappears from the bifurcation diagram and a 4-cycle, denoted by $\{\bar{x}_1, \bar{x}_2, \bar{x}_3, \bar{x}_4\}$, is born ($\{\bar{x}_1, \bar{x}_2\}$ undergoes period-doubling bifurcation at $\alpha = \alpha_2$). These period-doubling bifurcations continue until $\alpha = \alpha_\infty$.

Now, we turn our attention to the parameter values $\alpha > \alpha_\infty$. The situation in this parameter regime is much more complicated than the region of the period-doubling bifurcation, $0 < \alpha \leq \alpha_\infty$ (where only stable cycles appear in the bifurcation diagram). The best way to explain the dynamics of the orbit of x_0 is to start from $\alpha = 4$ and march backward to α_∞. At $\alpha = 4$, we see only one band covering the whole interval $0 < \alpha < 20$ (see Figure 1.23). This band slowly narrows as α decreases.

From the bifurcation diagram (Figure 1.22), we observe that the biggest window occurs for values of α between $\alpha = 3.1 \cdots$ and $\alpha = 3.364 \cdots$. This window is called a period 3-window. An asymptotically stable 3-cycle appears first at $\alpha \approx 3.1 \cdots$, after which, the period-doubling bifurcations dominate. This 3-cycle loses its stability and gives birth to an asymptotically stable 6-cycle. The period-doubling bifurcations continue until $\alpha \approx 3.2 \cdots$ (corresponding to α_∞ in the first part of the bifurcation diagram), after which, we get into a more complicated dynamics region. Windows of all odd periods appear between $\alpha - \infty$ and $\alpha = 4$ (see Figure 1.22).

We have seen that the Ricker model with $f(x) = xe^{\alpha - cx}$, exhibits a 3-cycle when $\alpha \approx 3.1 \cdots$. In 1975 Li and Yorke [224] proved an abstract mathematical result which states that if a continuous population function generates a 3-cycle, then it necessarily generates periodic k-cycles where k is any positive integer. The significant result of Li and Yorke can be seen as a special case of a general theorem of A. N. Sharkovsky [277].

In his remarkable result, Sharkovsky [277] used the following ordering of the positive integers.

To state the theorem, Sharkovsky introduced an ordering $\triangleright$ of the positive integers which we indicate below.

$$3 \triangleright 5 \triangleright 7 \triangleright 9 \triangleright \cdots \triangleright (2n+1) \times 2^0 \triangleright \cdots \cdots$$
$$3 \times 2 \triangleright 5 \times 2 \triangleright 7 \times 2 \triangleright 9 \times 2 \triangleright \cdots \triangleright (2n+1) \times 2^1 \triangleright \cdots$$
$$\vdots$$
$$3 \times 2^n \triangleright 5 \times 2^n \triangleright 7 \times 2^n \triangleright 9 \times 2^n \triangleright \cdots \triangleright (2n+1) \times 2^n \triangleright \cdots$$
$$\vdots$$
$$\cdots \triangleright 2^n \triangleright 2^{n-1} \triangleright \cdots \triangleright 2^3 \triangleright 2^2 \triangleright 2^1 \triangleright 2^0$$

Then Sharkovsky's Theorem [277] is as follows.

Theorem 11 (Sharkovsky [277]) *Let $F : I \to I$ be a continuous map that has a periodic orbit of prime period k. Then for any positive integer l that is preceded by k in Sharkovsky's order $k \triangleright l$, there is a periodic orbit of prime period l.*

Sharkovsky's Theorem implies that if a population model has a cycle of period 2 then it has a fixed point. If it does not have a cycle of period 4, then it does not have a cycle of period 8. A population model with a cycle of period 3 has cycles of all periods (also, Li and Yorke [224]). For example, $f(x) = x \exp(\alpha - cx)$ has a 3-cycle when $\alpha \approx 3.1 \cdots$, and, consequently, has cycles of every period (infinitely many periodic points). In agreement with the Theorem of Coppel [56], by Sharkovsky's Theorem, a population model with a continuous population function f can only have equilibrium points when f has no 2-cycles.

1.4.6 Chaos: Sensitive Dependence

We saw in Section 1.4.5 how complicated the dynamics can be for a difference equation as mathematical simple as the exponential (Ricker) equation. These dynamics have come to be called "chaotic". Mathematicians have formulated several different rigorous definitions of chaos and of a chaotic attractor (not all of which are equivalent) that involve sophisticated concepts that we will not discuss here. All these definitions, however, involve a property called sensitive dependence (to initial conditions). This concept captures the phenomenon that two solutions, no matter how close initially, will diverge and ultimately become significantly different from one another. While this property is not equivalent to chaos, it is an important feature of chaos.

Definition 10 *A map of an interval I is said to possess sensitive dependence on initial conditions if there exists $\nu > 0$ such that for any $x_0 \in I$ and $\delta > 0$, there exists $y_0 \in (x_0 - \delta, x_0 + \delta)$ and a positive integer k such that*
$$\left| f^k(x_0) - f^k(y_0) \right| \geq \nu.$$

The number ν will be called the sensitivity constant of f.

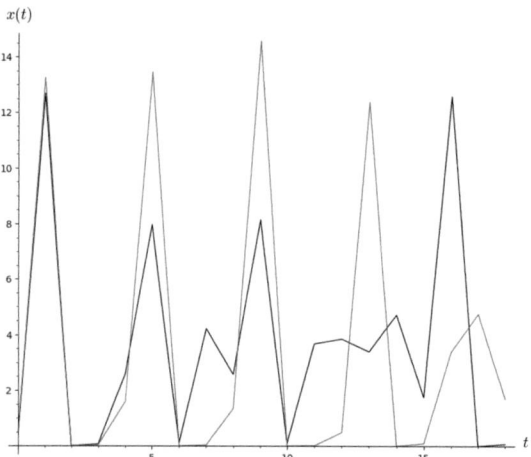

Figure 1.24: The error almost doubles after each iteration.

The simplest function with sensitive dependence is the linear map $f(x) = cx$, $c > 1$. For the initial points x_0, and $x_0 + \delta$, we have

$$f^n(x_0 + \delta) - f^n(x_0) = c^n(x_0 + \delta) - c^n x_0$$
$$= c^n \delta.$$

Hence, $|f^n(x_0 + \delta) - f^n(x_0)|$ will increase to ∞ as n goes to ∞, regardless of how small δ is. However, this linear map is not an interesting example because it does not possess any of the other properties of chaos.

A more interesting example is provided by the Ricker map $f(x) = x \exp(3.9 - x)$. In Figure 1.24, we let $x_0 = 0.25$ and $x_0 + \delta = 0.25$. After eight iterations, we observe from the time series how the error gets bigger and bigger, and after eight iterations, the difference between $f^8(0.2)$ and $f^8(0.25)$ is substantial.

This exponential stretching exhibited by the preceding map may be expressed by the **Liapunov exponent** λ. Roughly speaking, the Liapunov exponent $\lambda(x)$ at a point x measures the growth in error per iteration or the average loss of information during successive iterates of points near x.

How do we define this formally? We begin by considering a point x_0 and a neighboring point $x_0 + \delta$. Then the error e_n is defined as

$$e_n = |f^n(x_0 + \delta) - f(x_0)|,$$

and the relative error

$$\left|\frac{e_n}{\delta}\right| = \frac{|f^n(x_0 + \delta) - f^n(x_0)|}{\delta}.$$

If the map f possesses sensitive dependence on initial conditions, we expect the relative error $\frac{e_n}{\delta}$ to grow exponentially with n, and thus

$$e^{n\tilde{\lambda}} = \lim_{\delta \to 0} \frac{e_n}{\delta} = \lim_{\delta \to 0} \frac{|f^n(x_0 + \delta) - f^n(x_0)|}{\delta}, \quad \text{for some } \tilde{\lambda} > 0.$$

1.4. STABILITY BY LINEARIZATION

Hence

$$e^{n\tilde{\lambda}} = \left|\frac{d}{dx}f^n(x_0)\right| = |f'(x_0)f'(x(1))\ldots f'(x(n-1))|$$

and

$$\tilde{\lambda} = \frac{1}{n}\sum_{k=0}^{n-1} \ln f'(x(k)).$$

This motivates us to define the Liapunov exponent $\lambda(x_0)$ for a map f as

$$\lambda(x_0) = \lim_{n\to\infty} \frac{1}{n} \ln |[f^n(x_0)]'|. \qquad (1.56)$$

We can easily verify that

$$\ln |[f^n(x_0)]'| = \sum_{k=0}^{n-1} \ln |f'(x(k))|, \qquad (1.57)$$

where $x(k) = f^k(x_0)$. Thus, Equation (1.56) becomes

$$\lambda(x_0) = \lim_{n\to\infty} \frac{1}{n}\sum_{k=0}^{n-1} \ln |f'(x(k))|. \qquad (1.58)$$

The formula (1.58) tells us that the Liapunov exponent (the rate of convergence of two orbits) is the rate of change of the natural logarithm of the absolute value of the derivatives of the map evaluated at the orbit points. Note that if the application of the map to two nearby points leads to two points further apart, then the absolute value of the derivative of the map is greater than 1 when evaluated at these orbit points, and hence its logarithm is positive. If the orbit points continue to diverge, then the rate of change of the logarithm of the absolute values of the derivatives is positive, and hence the presence of sensitive dependence on initial conditions.

As we will see in the examples that follow if the Liapunov exponent λ is positive, then sensitive dependence exists. Moreover, as the Liapunov exponent becomes larger, and the magnification of error becomes greater.

A Numerical Scheme to Compute Liapunov Exponents

It is often the case that one may not be able to exactly compute Liapunov exponents. In this case, one resort to numerical schemes. We will illustrate the scheme for the Ricker map $f(x) = x\exp(\alpha - x)$. For a fixed value of α, start with an initial point say 0.5. Discard the first 400 (transient) iterates. Then compute an additional 100 iterates. The Liapunov exponent is now estimated by the formula

$$\lambda(0.5) = \frac{1}{500}\sum_{k=401}^{500} \ln |(1-x(k))\exp(\alpha - x(k)))|.$$

Starting with $\alpha = 2$ and increasing α by $1/1000$, we end up with Figure 1.25.

The negative spikes correspond to the 2^n-cycles where we have stable cycles that do not possess sensitive dependence. Note also that λ remains negative for $2 < \alpha < \alpha_\infty \approx 2.69$, and approaches zero at the period-doubling bifurcation point. As α increases toward 4, it oscillates between positive and negative values. The positive values of λ increase as we get closer and closer to $\alpha = 4$, which demonstrates that f_α is increasingly sensitive to initial conditions.

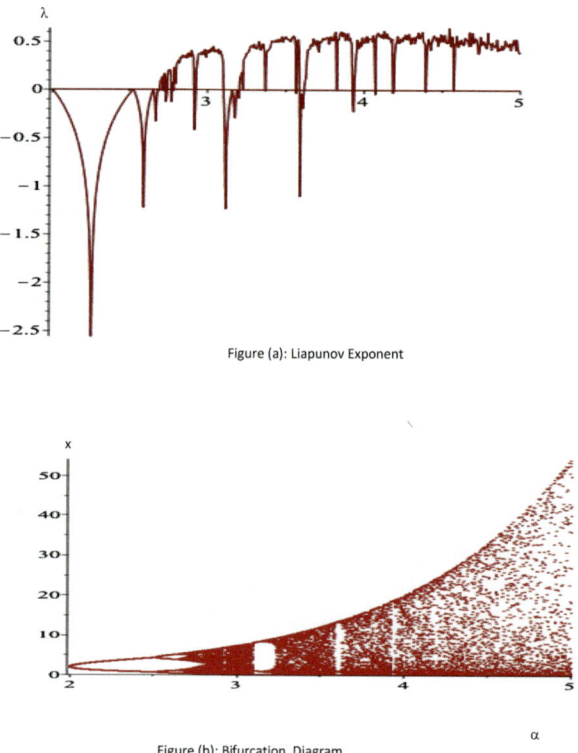

Figure 1.25: The top graph is the Liapunov exponent λ versus α graph, while the lower graph is the bifurcation diagram x versus α. Sensitivity to initial conditions occurs when $\lambda > 0$.

Exercises 1.4.6

A rough estimate of a Liapunov exponent $\lambda(x_0)$ of a map f maybe obtained by the formula

$$\lambda(x_0) \approx \frac{1}{n} \ln \left| \frac{e_n}{e_0} \right|, \qquad (1.59)$$

where e_0 is the initial error in x_0 and $e_n = |f^n(x_0 + e_0) - f^n(x_0)|$ is the error after n iterations.

In problems 1–5, use formula (1.59) to approximate the Liapunov exponent.

1. $f(x) = 4x(1-x)$ on $[0,1]$, $\quad n = 5, 6, \quad x_0 = 0.1, \quad e_0 = 0.01$.
2. $f(x) = 4x^3 - 3x$ on $[-1,1]$, $\quad n = 5, 6, \quad x_0 = 0.1, \quad e_0 = 0.01$.
3. $f(x) = \sin x$ on $[0, 2\pi]$, $\quad n = 5, 6, \quad x_0 = 0.3, \quad e_0 = 0.01$.
4. $f(x) = 8x^4 - 8x^2$ on $[-1,1]$, $\quad n = 5, 6, \quad x_0 = 0.1, \quad e_0 = 0.01$.

In problems 5–6, draw the graphs of the Liapunov exponent and the bifurcation diagram and show that chaos occurs if the Liapunov exponent is positive.

5. $f(x) = bx \exp -x$, $1 < b < 50$
6. $f(x) = rx(1-x)$, $0 < r < 4$

Chapter 2
Linear Structured Population Models

2.1 Introduction

In Chapter 1, we considered single (scalar) difference equations as models for the discrete-time dynamics of a biological population. In these models, x_t represents the population's size at discrete census times $t = 0, 1, 2, \ldots$. Population size can mean many different things: the number of individuals, the density of individuals, or some other measure such as total population biomass, dry weight, and so on. In any case, in such models, there are no distinctions made among individual members of the population, all of whom are in effect considered identical. In reality, however, individuals in biological populations do differ from one another and often do so quite significantly with regard to physiological and behavioral characteristics that determine their survival and fertility rates and other important vital rates and characteristics, such as metabolic rates, resource consumption rates, exposure to predators and pathogens, life cycle stages, and so on. Indeed, biologists know a lot about and expend considerable energy in observing and measuring the characteristics, activities, and attributes of individuals. The scalar difference equations we considered in Chapter 1 do not capture any of these important biological details and therefore they are difficult to relate to data describing the physiological and behavioral characteristics of individuals. This deficiency prevents an understanding of how the characteristics of individuals can affect and determine the dynamics of the population as a whole.

A more accurate dynamic model would incorporate, in some way or other, differences among individuals. What kinds of differences to use in a model are up to the modeler to decide. They are typically chosen for a specific population by what is considered, by biologists, to be the most significant characteristics of an individual that determine its survival and reproductive output. For example, often an individual's age is a factor that correlates with its survival and fertility rates (and other important factors, such as susceptibility to predation or diseases, resource-gathering efficiency, and so on). In other circumstances, an individual's body size or weight, or its life cycle stage, might be the most important physiological characteristic.

At one extreme, one could build a model that tracks each and every individual in a population and accounts for each individual's survival and reproduction. Obviously, except for populations consisting of only a very few individuals, such individual-based models require a large number of equations (i.e., are of very high dimension) and are mainly tractable only by means of computer simulations. Alternatively, an intermediate point of view is to place all individuals into one of a finite number of classes and to build a model that tracks the dynamics of each class (treating all individuals within a class as identical). These kinds of models are called *structured population models*. Mathematically, they consist of a finite number of difference equations that describe the discrete-time dynamics of the population size of each class. These systems are often written using vector and matrix notation and are therefore called *matrix models* (for the discrete-time dynamics of a structured population).

A common classification scheme is based on discrete (chronological) age classes and, in Section 1.1, we consider age-structured matrix models. In principle, however, any classification scheme for individuals can be used to a build structured population model [45], [67].

2.2 Structured Population Models

In the unstructured single-species models of Chapter 1, all individuals become equally reproductive after one time interval. In general, this is not always the case. Mammals have pre-reproductive ages, and human females have pre-reproductive, reproductive, and post-reproductive periods. Analytic generalization of the unstructured model that focuses on the population dynamics of females under the assumption that individuals have three options (getting older, dying, or (possibly) reproducing) was developed by Lewis, Leslie, and Bernadelli [31]. P. H. Leslie, an English biologist, first introduced a model that takes into account age structure of a population of females [209]. Earlier versions of the model were studied by E. Lewis (1941) and Bernadelli (1942) [31], [214]. However, it was Leslie who developed the model in detail and popularized its use. The Leslie matrix model is widely used by human demographers and biologists interested in the study of population dynamics.

The Leslie age-structured matrix model is used to study the population dynamics of a population structured by a discrete set of chronological age classes (age is time since birth). We begin by specifying a finite set of k age intervals, starting with age $a = 0$ at birth and all of the same length l:

$$0 \leq a < l, \quad l \leq a < 2l, \quad \cdots \quad (k-2)\,l \leq a < (k-1)\,l, \quad (k-1)\,l \leq a.$$

We enumerate these classes by $i = 1, 2, \ldots, k$ and let $x_i(t)$ denote the number (or density) of individuals in the population whose age is in the i-age interval.[1] (In common parlance, we would say that an individual in age class $i \geq 2$ is of "age $i - 1$".)

In the Leslie age-structured model, it is assumed that the population is censused at time intervals of length l. For notational simplicity, we let one unit of model time correspond to the age interval length l. The vector[2]

$$\mathbf{x}(t) = \begin{pmatrix} x_1(t) \\ x_2(t) \\ \vdots \\ x_k(t) \end{pmatrix}$$

---

[1] We use the notation $x(t)$ to denote the dependence of x on t, rather than x_t as in Chapter 1, in order to avoid cumbersome double subscripts.

[2] Bold face, lower case letters denote column vectors.

2.2. STRUCTURED POPULATION MODELS

is the population demographic vector of age class-specific densities at time t. We wish to build a model that predicts $\mathbf{x}(t+1)$ from $\mathbf{x}(t)$. To do this, we need to take into account births, deaths, and movement from one age class to another. (We ignore immigration and emigration.)

Let $s_i \leq 1$ be the probability an individual in age class i, $1 \leq i \leq k$, is alive at time $t+1$. If an individual in age class i, $1 \leq i \leq k-1$, survives, then it necessarily moves to the age class $i+1$. If an individual in age class k survives, then it remains in age class k (a class that therefore consists of all individuals of age $k-1$ or older). Thus

$$\begin{aligned} x_{i+1}(t+1) &= s_i x_i(t) \quad \text{for } 1 \leq i \leq k-2 \\ x_k(t+1) &= s_{k-1} x_{k-1}(t) + s_k x_k(t) \end{aligned} \quad (2.1)$$

We assume

$$s_i > 0 \text{ for } 1 \leq i \leq k-2$$

so that it is possible for an individual to reach the final age class k. We allow that $s_k = 0$, in which case no individual in age class k survives one unit of time, or in other words, no individual reaches age k.

The formula in (2.1) accounts for all but the first component (first age class) in the demographic vector $\mathbf{x}(t+1)$. The first component $x_1(t+1)$ is the class of newborns and the only entries into this age class are due to births. Reproductive processes can be quite complicated and varied across different species, ranging from asexual cloning to sexual mating. In a model of human populations, for example, if the population being modeled consists of both females and males, then the model would have to account in some way for the mating habits, i.e., the formation of mating pairs, etc. While there are models that do this, we will avoid this problem by assuming that only females are counted in the demographic vector. We then assume each female in age class i produces $b_i \geq 0$ newborns whose survival probability to the next census is s_{0i}. Then the number of newborns at time $t+1$ is

$$x_1(t+1) = \sum_{i=1}^{k} f_i x_i(t),$$

where $f_i = s_{0i} b_i$ denotes the age-specific, per-capita fertility rate per unit time. Combining this equation with Equations (2.1), we obtain the following system of difference equations:

$$\begin{aligned} x_1(t+1) &= f_1 x_1(t) + f_2 x_2(t) + \ldots + f_k x_k(t) \\ x_2(t+1) &= s_1 x_1(t) \\ x_3(t+1) &= s_2 x_2(t) \\ &\vdots \\ x_k(t+1) &= s_{k-1} x_{k-1}(t) + s_k x_k(t) \end{aligned}$$

for all the components of the demographic vector $\mathbf{x}(t)$, which we can write using vector/matrix notation as

$$\begin{pmatrix} x_1(t) \\ x_2(t) \\ \vdots \\ x_k(t) \end{pmatrix} = \begin{pmatrix} f_1 & f_2 & \ldots & f_{k-1} & f_k \\ s_1 & 0 & \ldots & 0 & 0 \\ 0 & s_2 & \ldots & 0 & 0 \\ \vdots & \vdots & & \vdots & \vdots \\ 0 & 0 & \ldots & s_{k-1} & s_k \end{pmatrix} \begin{pmatrix} x_1(t-1) \\ x_2(t-1) \\ \vdots \\ x_{k-1}(t-1) \\ x_k(t-1) \end{pmatrix}.$$

More succinctly, we write
$$\mathbf{x}(t+1) = A\mathbf{x}(t), \tag{2.2}$$
where
$$A = \begin{pmatrix} f_1 & f_2 & \cdots & f_{k-1} & f_k \\ s_1 & 0 & \cdots & 0 & 0 \\ 0 & s_2 & \cdots & 0 & 0 \\ \vdots & \vdots & & \vdots & \vdots \\ 0 & 0 & \cdots & s_{k-1} & s_k \end{pmatrix}. \tag{2.3}$$

The recursive equation map (2.2), often called a *matrix equation*, describes the discrete-time dynamics of the population's demographic vector $\mathbf{x}(t)$ through time once an initial demographic vector $\mathbf{x}(0) = \mathbf{x}_0$ is prescribed. The matrix A is called the (population) *projection matrix*.

Usually, it is assumed that the youngest individuals are not reproductive, so that $f_1 = 0$. Indeed, if individuals are non-reproductive until they reach age class $i = m \leq k$, then $f_1 = f_2 = \ldots = f_{m=1} = 0$ and $f_m > 0$. For example, the simplest (lowest dimensional) age structured model based on distinguishing between non-reproducing (juvenile) and reproducing (adult) age classes, has the 2×2 projection matrix
$$\begin{pmatrix} 0 & f_2 \\ s_1 & s_2 \end{pmatrix}, \quad f_2 > 0. \tag{2.4}$$
Here the time unit is the maturation time (which accounts for $f_1 = 0$), f_2 is adult fertility, s_1 is juvenile survival, and s_2 is adult survival.

If other classification schemes are used for individuals in a population, then the projection matrix A will take a different form from that in (2.3). The particular projection matrix (2.3), which arises when chronological age is used to classify individuals, is called a *Leslie matrix*. Note that we can write a Leslie matrix as
$$A = F + T, \tag{2.5}$$
where
$$F = \begin{pmatrix} f_1 & f_2 & \cdots & f_{k-1} & f_k \\ 0 & 0 & \cdots & 0 & 0 \\ 0 & 0 & \cdots & 0 & 0 \\ \vdots & \vdots & & \vdots & \vdots \\ 0 & 0 & \cdots & 0 & 0 \end{pmatrix}$$
$$T = \begin{pmatrix} 0 & 0 & \cdots & 0 & 0 \\ s_1 & 0 & \cdots & 0 & 0 \\ 0 & s_2 & \cdots & 0 & 0 \\ \vdots & \vdots & & \vdots & \vdots \\ 0 & 0 & \cdots & s_{k-1} & s_k \end{pmatrix}. \tag{2.6}$$

Population models can be based on other classification schemes for its individuals. For example, the *standard size-structured population model* [45] categorize individuals by means of increasing body size (e.g., weight, height, surface area, etc.) and assumes that after a unit of time a (surviving) individual either remains in the same size class or advances to the next larger class. Then in place of (2.1), we have
$$\begin{array}{l} x_{i+1}(t+1) = s_i p_i x_i(t) + s_{i+1}(1 - p_{i+1}) x_{i+1}(t) \quad \text{for } 1 \leq i \leq k-2 \\ x_k(t+1) = s_{k-1} p_{k-1} x_{k-1}(t) + s_k x_k(t) \end{array},$$

2.2. STRUCTURED POPULATION MODELS

where s_i is the probability an i-class individual survives and p_i is the probability that, given survival, it grows to the next size class. The size class k consists of individuals of size k and larger. Assuming all newborns lie in the smallest size class, we arrive again with a matrix equation of the form (2.2) with a projection matrix (2.5), but with

$$F = \begin{pmatrix} f_1 & f_2 & \cdots & f_{k-1} & f_k \\ 0 & 0 & \cdots & 0 & 0 \\ 0 & 0 & \cdots & 0 & 0 \\ \vdots & \vdots & & \vdots & \vdots \\ 0 & 0 & \cdots & 0 & 0 \end{pmatrix}$$

$$T = \begin{pmatrix} s_1(1-p_1) & 0 & 0 & \cdots & 0 & 0 \\ s_1 p_1 & s_2(1-p_2) & 0 & \cdots & 0 & 0 \\ 0 & s_2 p_2 & s_3(1-p_3) & \cdots & 0 & 0 \\ \vdots & \vdots & & & \vdots & \vdots \\ \vdots & \vdots & & \cdots & s_{k-1}(1-p_{k-1}) & 0 \\ 0 & 0 & & \cdots & s_{k-1} p_{k-1} & s_k \end{pmatrix}. \quad (2.7)$$

In [45], this matrix is called the standard size structured matrix. It is also sometimes called an *Usher matrix*, since it arose in renewal management studies of forests in which trees are classified by (girth) size [289], [290], [291].

In general, if a population is structured by means of a finite number of categories for its individuals, its discrete-time dynamics are described by a matrix equation (2.2) with a projection matrix (2.5) where the matrix F consists of all entries in A that account for births and the T consists of all class transition probabilities between classes. Both matrices contain nonnegative entries and are therefore called *nonnegative matrices*. In the absence of immigration/emigration, the sum of the column entries in T must be less than or equal to 1 (and unless there are no class specific deaths they must be strictly less than 1).

An alternative way to describe a matrix model (2.2) is by means of a life cycle graph, which is constructed as follows.

(i) Choose a set of classes (sometimes called stages, individual states or i-states) such that each individual in the population falls into one and only one class.

(ii) Define the unit of time to be used in the model. For example, time interval from t to $t+1$ might represent a day, a week, a year, a maturation period, or a convenient census schedule.

(iii) Create a node for each stage and number the nodes from 1 to k.

(iv) Put a directed line from node i to node j if an individual in stage i at time t can contribute an individual(s) (by birth or transition) to stage j at time $t+1$. If an i-class individual at time t can contribute to the i-class at $t+1$, put a directed line from node i to itself.

Example 17 *Suppose that an animal reaches sexual maturity after 2 years of age and remains fertile for the rest of its life. Suppose no individual survives more than 4 years of age and that each*

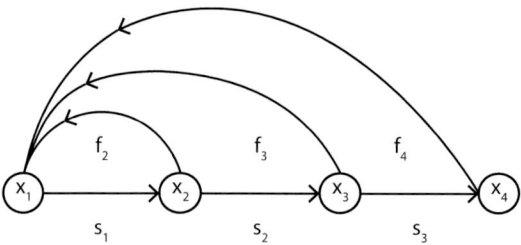

Figure 2.1: The life cycle graph for the Leslie matrix model in Example 17.

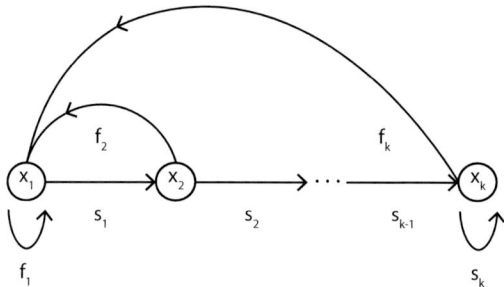

Figure 2.2: The life cycle graph for a general Leslie matrix.

age class suffers some mortality each year. Then using a time interval of 1 year, we have the Leslie matrix model

$$A = \begin{pmatrix} 0 & f_2 & f_3 & f_4 \\ s_1 & 0 & 0 & 0 \\ 0 & s_2 & 0 & 0 \\ 0 & 0 & s_3 & 0 \end{pmatrix} \quad f_i > 0, \quad 0 < s_i < 1$$

and the associated life cycle graph (Figure 2.1).

Example 18 *The life cycle graph of a general Leslie model with fertility and transition matrices (2.6) appears in Figure 2.2.*

Example 19 *The life cycle graph of a standard size-structured model with fertility and transition matrices (2.7), assuming maturity is first reached at size m, appears in Figure 2.3.*

Exercises 2.2

1. Consider the table below obtained from two consecutive censuses for an animal population.

    (a) Construct a 3 × 3 Leslie matrix for this population with three age classes under the assumption that no individual lives past the third age class.

2.2. STRUCTURED POPULATION MODELS

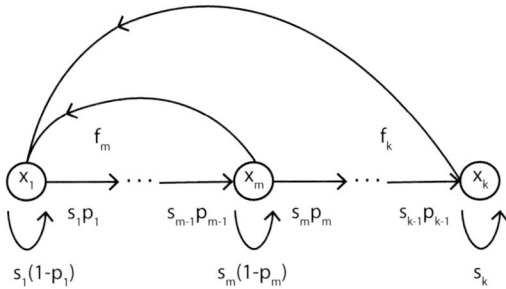

Figure 2.3: The life cycle graph for the Usher size structured matrix model (2.7).

(b) Draw the life cycle graph.

(c) Verify that your Leslie matrix is correct by seeing if it computes the year 2 numbers based on the year 1 numbers as an input. What would the population size of each age group be in year 3?

(d) Starting with the age distribution $x_1(0) = 900$, $x_2(0) = 800$, $x_3(0) = 400$, compute the population vector for six consecutive years. Round off so that you have only whole animals in the population vector.

Age class i	No. of females year 1	No. of females year 2	No. of female births by mothers in age class i in year 1
1	1000	2220	0
2	600	700	600
3	540	540	1620

2. Consider the table below obtained from two consecutive censuses for an animal population.

   (a) Construct a 4×4 Leslie matrix for this population with four age classes under the assumption that no individual lives past the fourth age class.

   (b) Draw the life cycle graph.

   (c) Starting with the age distribution $x_1(0) = 1200$, $x_2(0) = 1000$, $x_3(0) = 900$, $x_4(0) = 0$ compute the population vector for six consecutive years.

Age i	No. females year 1	No. females year 2	No. female births by mothers of age i in year 1
1	2000	3200	0
2	1400	1500	1200
3	1200	1000	700
4	900	800	1300

3. Construct the population projection matrix corresponding to the following life cycle graph in Figure 2.4. Describe the biological features of the model.

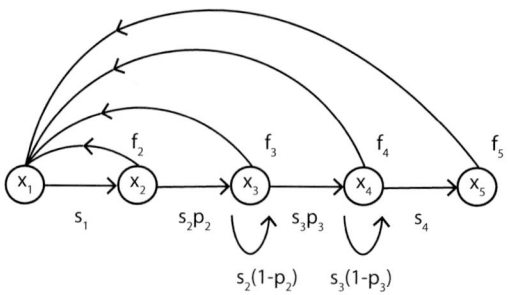

Figure 2.4: The life cycle graph for Exercise 3.

4. Draw the life cycle graph corresponding to the following population projection matrix.

$$A = \begin{pmatrix} 0 & 0 & 0 & 2 \\ 0.5 & 0 & 0 & 0 \\ 0 & 1 & 0 & 0 \\ 0 & 0 & 0.75 & 0 \end{pmatrix}.$$

If $x_1(0) = 10$, $x_2(10) = 12$, $x_3(0) = 15$, $x_4(0) = 2$, compute the population vector for six consecutive years.

5. Draw the life cycle graph corresponding to the following population projection matrix.

$$A = \begin{pmatrix} 2 & 0 & 1 & 3 & 0.2 \\ 0.75 & 0.4 & 0 & 0 & 0 \\ 0 & 0.5 & 0 & 0 & 0 \\ 0 & 0 & 1 & 0 & 0 \\ 0 & 0 & 0 & 0.6 & 0 \end{pmatrix}.$$

If $x_1(0) = 100$, $x_2(0) = 100$, $x_3(0) = 80$, $x_4(0) = 60$, $x_5(0) = 50$, compute the population vector for five consecutive years.

2.3 Eigenvalues and Eigenvectors

Consider a matrix model

$$\mathbf{x}(t+1) = A\mathbf{x}(t), \qquad (2.8)$$

where A is a general $k \times k$ matrix of the form

$$A = \begin{pmatrix} a_{11} & a_{12} & \cdots & a_{1k} \\ a_{21} & a_{22} & \cdots & a_{2k} \\ \vdots & \vdots & \vdots & \vdots \\ a_{k1} & a_{k2} & \cdots & a_{kk} \end{pmatrix}$$

consisting of nonnegative entries $a_{ij} \geq 0$. By repeated application of (2.8), we find that

$$\mathbf{x}(t) = A^t \mathbf{x}_0$$

2.3. EIGENVALUES AND EIGENVECTORS

which is a formula for the solution $\mathbf{x}(t)$ of Equation (2.8) in terms of its initial condition $\mathbf{x}_0 = \mathbf{x}(0)$. In principle, we could use this formula to study the properties of $\mathbf{x}(t)$, such as its long-term asymptotic dynamics, but a difficulty in doing this is with calculating high powers of the matrix A. Fortunately, some methods from linear algebra and matrix theory can help in this regard.

First, we need to understand the eigenvalue–eigenvector problem for matrices and be able to compute the eigenvalues and the associated eigenvectors.

A number λ is an *eigenvalue* of the matrix A if there exists a *nonzero* vector $\mathbf{v} \neq \mathbf{0}$ such that $\lambda \mathbf{v} = A\mathbf{v}$ or
$$(A - \lambda I)\mathbf{v} = \mathbf{0} \tag{2.9}$$
in which case $\mathbf{v}$ is an associated *eigenvector*. Note that if $\mathbf{v}$ is an eigenvector associated with an eigenvalue λ then so is any nonzero multiple of $\mathbf{v}$. Thus, $c \neq 0$ and $\lambda \mathbf{v} = A\mathbf{v}$ implies $c(\lambda \mathbf{v}) = c(A\mathbf{v})$ and $\lambda(c\mathbf{v}) = A(c\mathbf{v})$ so that $c\mathbf{v} \neq \mathbf{0}$ is an eigenvector. To find eigenvalues we use a fact from linear algebra that Equation (2.9) can have a nonzero vector solution $\mathbf{v}$ only if the matrix $\lambda I - A$ is singular (i.e., not invertible), or in other words,
$$\det(A - \lambda I) = 0,$$
where I is the identity matrix
$$I = \begin{pmatrix} 1 & 0 & \cdots & 0 \\ 0 & 1 & \cdots & 0 \\ \vdots & \vdots & \vdots & \vdots \\ 0 & 0 & \cdots & 1 \end{pmatrix}.$$

For a given $k \times k$ matrix A, there are k eigenvalues (some of them are repeated) $\lambda_1, \ldots, \lambda_k$, and at most k eigenvectors $\mathbf{v}_1, \mathbf{v}_2, \ldots \mathbf{v}_r$. If some eigenvalues are repeated, then r may be less than k.

Note that the eigenvalue with the largest absolute or modulus value is called the *dominant eigenvalue*.

Example 20 *Consider the 2×2 Leslie projection matrix*
$$A = \begin{pmatrix} 2 & 3 \\ 1 & 0 \end{pmatrix}.$$

The characteristic polynomial is
$$\det(A - \lambda I) = \det\left[\lambda \begin{pmatrix} 1 & 0 \\ 0 & 1 \end{pmatrix} - \begin{pmatrix} 2 & 3 \\ 1 & 0 \end{pmatrix}\right] = \det \begin{pmatrix} \lambda - 2 & -3 \\ -1 & \lambda \end{pmatrix} = \lambda^2 - 2\lambda - 3$$

and the eigenvalues are its roots $\lambda_1 = 3$ and $\lambda_2 = -1$. Clearly, $\lambda_1 = 3$ is the dominant eigenvalue. To find eigenvectors, we solve Equation (2.9) for each eigenvalue. For $\lambda_1 = 3$, the equation is
$$\begin{pmatrix} 2-3 & 3 \\ 1 & -3 \end{pmatrix} \begin{pmatrix} v_1 \\ v_2 \end{pmatrix} = \begin{pmatrix} 0 \\ 0 \end{pmatrix}, \quad \mathbf{v} = \begin{pmatrix} v_1 \\ v_2 \end{pmatrix} \neq \begin{pmatrix} 0 \\ 0 \end{pmatrix}$$

which is equivalent to the two algebra equations
$$-v_1 + 3v_2 = 0$$
$$v_1 - 3v_2 = 0.$$

for v_1 and v_2. Note that these equations are multiples of each other and hence solving one for v_1 and v_2 automatically solves the other equation. (This is generally true for 2×2 matrices A.) We can, for example, solve the first equation in any way we want, so long as not both equal 0. A simple choice is $v_1 = 3$ and $v_2 = 1$ so that

$$\mathbf{v} = \begin{pmatrix} 3 \\ 1 \end{pmatrix}$$

is an eigenvector associated with eigenvalue $\lambda_1 = 3$. As pointed out above, any nonzero scalar multiple of this vector is also an eigenvector associated with λ_1.

We repeat these steps for the second eigenvalue $\lambda_2 = -1$. Equation (2.9)

$$\begin{pmatrix} 2-(-1) & 3 \\ 1 & -(-1) \end{pmatrix} \begin{pmatrix} v_1 \\ v_2 \end{pmatrix} = \begin{pmatrix} 0 \\ 0 \end{pmatrix}$$

is equivalent to two equations for v_1 and v_2 only one of which we need to solve, say, the second equation

$$v_1 + v_2 = 0.$$

Thus, $v_1 = 1$ and $v_2 = -1$ yield an eigenvector

$$\mathbf{v}_2 = \begin{pmatrix} 1 \\ -1 \end{pmatrix}$$

associated with eigenvalue $\lambda_2 = -1$. Of course any nonzero scalar multiple of this vector can be an eigenvector of λ_2. For instance,

$$\begin{pmatrix} 2 \\ -2 \end{pmatrix}, \begin{pmatrix} 3 \\ -3 \end{pmatrix}, \begin{pmatrix} -2 \\ 2 \end{pmatrix}$$

are all eigenvectors for λ_2.

Example 21 *Consider the 3×3 Leslie matrix*

$$A = \begin{pmatrix} 0 & 2 & 1 \\ 1 & 0 & 0 \\ 0 & 1 & 0 \end{pmatrix}.$$

The characteristic polynomial is

$$\det \begin{pmatrix} \lambda & -2 & -1 \\ -1 & \lambda & 0 \\ 0 & -1 & \lambda \end{pmatrix} = 0$$

$$\lambda \det \begin{pmatrix} \lambda & 0 \\ -1 & \lambda \end{pmatrix} - (-2) \det \begin{pmatrix} -1 & 0 \\ 0 & \lambda \end{pmatrix} + (-1) \det \begin{pmatrix} -1 & \lambda \\ 0 & -1 \end{pmatrix} = 0$$

$$\lambda \left(\lambda^2 - 0 \right) + 2 \left(-\lambda - 0 \right) - \left(1 - 0 \right) = 0$$

$$\lambda^3 - 2\lambda - 1 = 0.$$

By inspection, we see that $\lambda = -1$ is a root. To find an associated eigenvector, we solve Equation (2.9)

$$\begin{pmatrix} -1 & -2 & -1 \\ -1 & -1 & 0 \\ 0 & -1 & -1 \end{pmatrix} \begin{pmatrix} v_1 \\ v_2 \\ v_3 \end{pmatrix} = \begin{pmatrix} 0 \\ 0 \\ 0 \end{pmatrix}.$$

2.3. EIGENVALUES AND EIGENVECTORS

i.e., the equations

$$-v_1 - 2v_2 - v_3 = 0$$
$$-v_1 - v_2 = 0$$
$$-v_2 - v_3 = 0$$

for v_1, v_2, and v_3 not all 0. Since the first equation is the sum of the second and third equations, we need only solve the last two equations, which we can do by choosing any nonzero value for v_1 and taking $v_2 = -v_1$ and $v_3 = -v_2$. Say, $v_1 = 1$ and $v_2 = -1$, $v_3 = 1$

$$\begin{pmatrix} v_1 \\ v_2 \\ v_3 \end{pmatrix} = \begin{pmatrix} 1 \\ -1 \\ 1 \end{pmatrix}.$$

The remaining eigenvalues are the roots of the second factor in $\lambda^3 - 2\lambda - 1 = (\lambda + 1)(\lambda^2 - \lambda - 1)$ whose eigenvectors can be calculated in this same manner. In fact, modern computer programs will find the eigenvalues and eigenvectors of matrices. Using such a program, we find three real eigenvalues and associated eigenvectors

$$\lambda_1 = \frac{1}{2} + \frac{1}{2}\sqrt{5}, \quad \mathbf{v} = \begin{pmatrix} 3 + \sqrt{5} \\ 1 + \sqrt{5} \\ 2 \end{pmatrix}$$

$$\lambda_2 = \frac{1}{2} - \frac{1}{2}\sqrt{5}, \quad \mathbf{v} = \begin{pmatrix} 3 - \sqrt{5} \\ 1 - \sqrt{5} \\ 2 \end{pmatrix}$$

$$\lambda_3 = -1, \quad \mathbf{v} = \begin{pmatrix} 1 \\ -1 \\ 1 \end{pmatrix}$$

The dominant eigenvalue is λ_1.

Example 22 *The characteristic polynomial of the Leslie matrix*

$$A = \begin{pmatrix} f_1 & f_2 & \cdots & f_{k-1} & f_k \\ s_1 & 0 & \cdots & 0 & 0 \\ 0 & s_2 & \cdots & 0 & 0 \\ \vdots & \vdots & & \vdots & \vdots \\ 0 & 0 & \cdots & s_{k-1} & 0 \end{pmatrix}$$

is

$$\lambda^k - f_1\lambda^{k-1} - f_2 s_1 \lambda^{k-2} - f_3 s_1 s_2 \lambda^{k-3} - f_4 s_1 s_2 s_3 \lambda^{k-4} - \cdots - f_k s_1 s_2 s_3 \cdots s_{k-1}$$

or using summation notation

$$\sum_{i=0}^{k-1} \left(-f_{k-i} \prod_{j=1}^{k-i-1} s_j \right) \lambda^i + \lambda^k.$$

There exists no general formula to find roots of high-degree polynomials; so, for large Leslie matrices, we are left with numerical approximations obtained from calculators or computer programs.

Exercises 2.3

1. For the matrices below, calculate the eigenvalues and find an associated eigenvector for each one.

   (a) $\begin{pmatrix} 8 & -18 \\ 3 & -7 \end{pmatrix}$
   (b) $\begin{pmatrix} 1 & 1 \\ -1 & 1 \end{pmatrix}$
   (c) $\begin{pmatrix} 4 & -1 \\ 1 & 2 \end{pmatrix}$

   (d) $\begin{pmatrix} 7 & -4 & -6 \\ 2 & 0 & -2 \\ 2 & -1 & -1 \end{pmatrix}$
   (e) $\begin{pmatrix} 5 & -9 & -4 \\ 2 & -3 & -2 \\ 2 & -4 & -1 \end{pmatrix}$
   (f) $\begin{pmatrix} -11 & 17 & 17 \\ -5 & 11 & 5 \\ -5 & 5 & 11 \end{pmatrix}$

2. (i) Find the characteristic polynomial and the dominant eigenvalue of each Leslie matrix below. Is the dominant eigenvalue strictly greater than all other eigenvalues (or is there another eigenvalue of the same absolute value)?

   (a) $\begin{pmatrix} 0 & 2 & 1 \\ 0.9 & 0 & 0 \\ 0 & 0.8 & 0 \end{pmatrix}$
   (b) $\begin{pmatrix} 0 & 3 & 2 \\ 0.1 & 0 & 0 \\ 0 & 0.9 & 0.5 \end{pmatrix}$
   (c) $\begin{pmatrix} 0 & 0 & 2 \\ 0.95 & 0 & 0 \\ 0 & 0.9 & 0 \end{pmatrix}$

   (d) $\begin{pmatrix} 0 & 0.75 & 1 & 0.25 \\ 0.8 & 0 & 0 & 0 \\ 0 & 0.8 & 0 & 0 \\ 0 & 0 & 0.8 & 0 \end{pmatrix}$
   (e) $\begin{pmatrix} 0 & 0.75 & 0 & 0.25 \\ 0.8 & 0 & 0 & 0 \\ 0 & 0.8 & 0 & 0 \\ 0 & 0 & 0.8 & 0 \end{pmatrix}$

   (ii) Show that the dominant eigenvalue of each matrix in (i) has a positive eigenvector associated with it.

3. (i) Find the characteristic polynomial and the dominant eigenvalue of each Usher matrix below. Is the dominant eigenvalue strictly greater than all other eigenvalues (or is there another eigenvalue of the same absolute value)?

   (a) $\begin{pmatrix} 0.1 & 2 & 1 \\ 0.8 & 0.1 & 0 \\ 0 & 0.8 & 0.5 \end{pmatrix}$
   (b) $\begin{pmatrix} 0 & 1 & 1 \\ 0.9 & 0.5 & 0 \\ 0 & 0.4 & 0.1 \end{pmatrix}$

   (c) $\begin{pmatrix} 0.5 & 0 & 1 & 1 \\ 0.4 & 0.5 & 0 & 0 \\ 0 & 0.4 & 0.5 & 0 \\ 0 & 0 & 0.4 & 0 \end{pmatrix}$
   (e) $\begin{pmatrix} 0.5 & 0 & 1 & 1 \\ 0.3 & 0.5 & 0 & 0 \\ 0 & 0.3 & 0.5 & 0 \\ 0 & 0 & 0.3 & 0 \end{pmatrix}$

   (ii) Show that the dominant eigenvalue of each matrix in (i) has a positive eigenvector associated with it.

4. Suppose λ is an eigenvalue of a $k \times k$ matrix A. Show that any nonzero linear combination $c_1 \mathbf{v}_1 + c_2 \mathbf{v}$ of two eigenvectors $\mathbf{v}_1$ and $\mathbf{v}_2$ associated with λ is also an eigenvector associated with λ. It follows that the set of eigenvectors associated with λ is a linear subspace of k-dimensional Euclidean space.

2.4 The Perron–Frobenius Theorem

A *nonnegative matrix* A is a matrix whose entries are all nonnegative real numbers. We denote a nonnegative matrix by writing $A \geq 0$. A *positive matrix* A is a matrix all of whose entries are

2.4. THE PERRON–FROBENIUS THEOREM

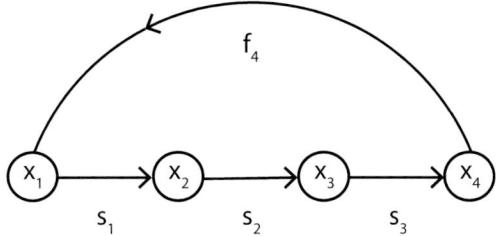

Figure 2.5: The life cycle graph of the Leslie matrix model in Example 23.

positive, in which case we write $A > 0$. We also call a vector $\mathbf{v}$ nonnegative or positive, and write $\mathbf{v} \geq 0$ and $\mathbf{v} > 0$, according to whether all of its entries are nonnegative or positive, respectively.

Definition 11 *A nonnegative matrix $A \geq 0$ is called **irreducible** if its life cycle graph contains a path (possibly with more than one step) from every node to every other node. Such a graph is called strongly connected.*

Nodes refer to stages (or classes) in a population model.

A projection matrix is not irreducible, i.e., is *reducible*, if the population's life cycle contains at least one stage that cannot contribute, by any path, to at least one other stage. Most life cycle graphs used in population models are irreducible. There are important applications where they are not, however.

Example 23 *The 4×4 Leslie matrix*

$$A = \begin{pmatrix} 0 & 0 & 0 & f_4 \\ s_1 & 0 & 0 & 0 \\ 0 & s_2 & 0 & 0 \\ 0 & 0 & s_3 & 0 \end{pmatrix}$$

with $0 < s_1, s_2, s_3 \leq 1$ and $f_4 > 0$. From its life cycle below, we see a closed loop path (of length 4) that connects all classes and hence this matrix is irreducible (Figure 2.5).

Example 24 *The 4×4 Usher matrix*

$$A = \begin{pmatrix} 0 & f_2 & f_3 & 0 \\ s_1 & s_2(1-p_2) & 0 & 0 \\ 0 & s_2 p_2 & s_3(1-p_3) & 0 \\ 0 & 0 & s_3 p_4 & s_4 \end{pmatrix},$$

with $0 < s_i, p_i \leq 1$ for $i = 1, 2, 3$ and $0 \leq s_4 \leq 1$ and with $f_2, f_3 > 0$, has the life cycle (Figure 2.6). Since the fourth class connects to no other class, this matrix is reducible.

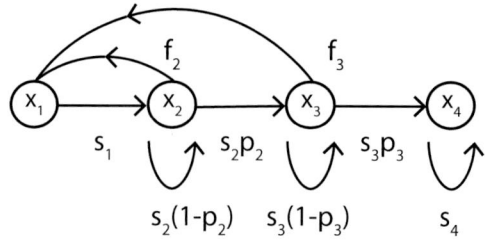

Figure 2.6: The life cycle graph of the Leslie matrix model in Example 24.

Example 25 *Consider the general Leslie matrix $A = F + T$ given by (2.6), i.e.,*

$$A = \begin{pmatrix} f_1 & f_2 & \cdots & f_{k-1} & f_k \\ s_1 & 0 & \cdots & 0 & 0 \\ 0 & s_2 & \cdots & 0 & 0 \\ \vdots & \vdots & & \vdots & \vdots \\ 0 & 0 & \cdots & s_{k-1} & s_k \end{pmatrix}$$

with $0 < s_k \leq 1$, $f_i \geq 0$ and

$$0 < s_i \leq 1 \text{ for } i = 1, 2, \cdots k-1.$$

By these latter inequalities we see, from the life cycle figure (Figure 2.2), there is a closed loop path (of length k) that connects all classes if $f_k > 0$. Thus, if $f_k > 0$, this matrix is irreducible. On the other hand, if $f_k = 0$, then the class k connects to no other class and the matrix is reducible.

By inspection of a matrix's entries, it is easy enough to determine if the matrix is nonnegative or positive, but not always so easy to determine if it is irreducible. The following theorem provides an algebraic way to determine whether a nonnegative matrix is irreducible.

Theorem 12 *A $k \times k$ matrix $A \geq 0$ is irreducible if and only if $(I + A)^{k-1} > 0$.*

Example 26 *Consider the 2×2 matrix*

$$A = \begin{pmatrix} 0 & f_2 \\ s_1 & s_2 \end{pmatrix}$$

with $f_2 > 0$, $0 < s_1 \leq 1$, and $0 \leq s_2 \leq 1$. This Leslie matrix describes a basic juvenile–adult model in which the first age class is not reproductive (hence the 0 in the upper right corner). Clearly, this matrix is nonnegative. It is also irreducible (by Theorem 13 with $k = 2$) because

$$(I+A)^{k-1} = I + A = \begin{pmatrix} 1 & f_2 \\ s_1 & 1+s_2 \end{pmatrix}$$

is positive.

2.4. THE PERRON–FROBENIUS THEOREM

Example 27 *Consider the 4×4 Leslie and Usher matrices from Examples 23 and 24:*

$$A = \begin{pmatrix} 0 & 0 & 0 & f_4 \\ s_1 & 0 & 0 & 0 \\ 0 & s_2 & 0 & 0 \\ 0 & 0 & s_3 & 0 \end{pmatrix} \quad \text{and} \quad \begin{pmatrix} 0 & f_2 & f_3 & 0 \\ s_1 & s_2(1-p_2) & 0 & 0 \\ 0 & s_2 p_2 & s_3(1-p_3) & 0 \\ 0 & 0 & s_3 p_4 & s_4 \end{pmatrix}.$$

In the first case, we calculate

$$(I+A)^{k-1} = (I+A)^3 = \begin{pmatrix} 1 & 0 & 0 & f_4 \\ s_1 & 1 & 0 & 0 \\ 0 & s_2 & 1 & 0 \\ 0 & 0 & s_3 & 1 \end{pmatrix}^3$$

$$= \begin{pmatrix} 1 & f_4 s_3 s_2 & 3 f_4 s_3 & 3 f_4 \\ 3 s_1 & 1 & s_1 s_3 f_4 & 3 s_1 f_4 \\ 3 s_1 s_2 & 3 s_2 & 1 & s_1 s_2 f_4 \\ s_1 s_2 s_3 & 3 s_2 s_3 & 3 s_3 & 1 \end{pmatrix}$$

and conclude from Theorem 12, since this matrix is positive, that A is irreducible.

For the second case, we find that

$$(I+A)^{k-1} = (I+A)^3 = \begin{pmatrix} 0 & f_2 & f_3 & 0 \\ s_1 & s_2(1-p_2) & 0 & 0 \\ 0 & s_2 p_2 & s_3(1-p_3) & 0 \\ 0 & 0 & s_3 p_4 & s_4 \end{pmatrix}^3 = \begin{pmatrix} * & * & * & 0 \\ * & * & * & * \\ * & * & * & * \\ * & * & * & * \end{pmatrix}$$

where the asterisks are (nonnegative) entries which are not needed. This is because the upper right corner of the matrix equals 0 and, as a result, $(I+A)^3$ is not a positive matrix. It follows from Theorem 12 that this matrix A is reducible.

The following famous theorem concerns the eigenvalues of a nonnegative irreducible matrix.

Theorem 13 **Perron–Frobenius**. *If a nonnegative matrix A is irreducible, then the following are true.*

(i) There exists a real, positive eigenvalue of A that is not repeated, i.e., is a simple root of the characteristic equation. This eigenvalue is dominant, i.e., is greater than or equal to the absolute values of all other eigenvalues of A (and is often called the Perron eigenvalue).

(ii) There exists a positive eigenvector associated with the dominant eigenvalue. No other eigenvalue of A has a nonnegative eigenvector.

As we will see Section 2.5, the dominant eigenvalue of a matrix A plays an important role in the asymptotic dynamics of the linear matrix equation

$$\mathbf{x}(t+1) = A\mathbf{x}(t).$$

This is especially true if the dominant eigenvalue is *strictly dominant*, i.e., it is strictly greater than the absolute value of all other eigenvalues.

Definition 12 *If the dominant eigenvalue of a nonnegative irreducible matrix is strictly dominant, then the matrix is called **primitive**. Otherwise, it is **imprimitive**.*

Example 28 *Consider the 2×2 the nonnegative irreducible Leslie matrix*
$$A = \begin{pmatrix} 0 & f_2 \\ s_1 & s_2 \end{pmatrix}$$
with $f_2 > 0$, $0 < s_1 \leq 1$, and $0 \leq s_2 \leq 1$ in Example 26. The eigenvalues are
$$\lambda_1 = \frac{1}{2}s_2 + \frac{1}{2}\sqrt{s_2^2 + 4f_2 s_1}, \quad \lambda_2 = \frac{1}{2}s_2 - \frac{1}{2}\sqrt{s_2^2 + 4f_2 s_1}$$
and a little algebra shows λ_1 is strictly dominant if $s_2 > 0$ and
$$\mathbf{v}_1 = \begin{pmatrix} \sqrt{s_2^2 + 4f_2 s_1} - s_2 \\ 2s_1 \end{pmatrix}$$
is an associated positive eigenvector.

On the other hand, $s_2 = 0$ then $\lambda_1 = |\lambda_2|$ and λ_1 is not strictly dominant. In this case, A is imprimitive. Biologically, $s_2 = 0$ means in this model that juveniles survive a unit of time (a maturation period) with probability s_1 to become adults who produce f_2 newborns per capita during the next time unit, but do not survive a second time unit. This model is therefore sometimes used as the basis for studying semelparous organisms, that is, organisms whose adults have only one reproductive episode in their lifetime (e.g., annual plants, many species of fish and insects, and some species of mollusks, squid, and octopus).

Usually, it is not easy or even possible to calculate the eigenvalues of a matrix A, as it is for 2×2 matrices such as in Example 28. The following theorem contains two methods for determining the primitivity of a matrix without needing to calculate eigenvalues.

Theorem 14 *Assume a $k \times k$ nonnegative A. Then the following statements hold true.*

(a) A is primitive if and only if A^m is positive, for some $m \geq 1$.

(b) A is primitive if and only if the greatest common divisor of the lengths of loops in its cycle graph is 1.

Note from Theorem 14(b) it follows that, if a population projection matrix is primitive, then each class is connected to any other class by a path of length no longer than $k^2 - 2k + 2$. This may be interpreted by the following theorem

Theorem 15 *A nonnegative $k \times k$ matrix A is primitive if and only if A^{k^2-2k+2} is positive.*

If, in a nonnegative, irreducible matrix model, a class is connected to itself by a loop of length 1, that is to say, if it is possible for individuals in a class to remain in that class in one time unit, then by Theorem 14(b) the matrix is primitive.

Corollary 2 *If a nonnegative irreducible matrix A has a positive diagonal element, then it is primitive.*

2.4. THE PERRON–FROBENIUS THEOREM

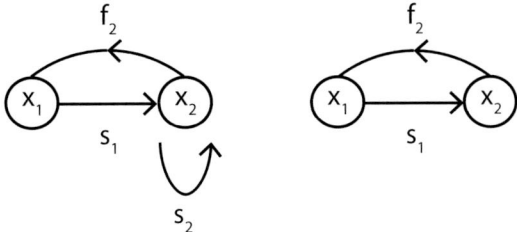

Figure 2.7: The life cycle graphs for the Leslie matrix models in Example 29.

Example 29 *Consider the 2×2 nonnegative irreducible Leslie matrix*

$$A = \begin{pmatrix} 0 & f_2 \\ s_1 & s_2 \end{pmatrix}$$

in Example 28. If $s_2 > 0$ then by Corollary 2 this matrix is primitive. This can also be seen by Theorem 14(a) from the calculation

$$A^{k^2-2k+2} = A^2 = \begin{pmatrix} f_2 s_1 & f_2 s_2 \\ s_1 s_2 & s_2^2 + f_2 s_1 \end{pmatrix}$$

which yields a positive matrix if $s_2 > 0$. Also note that the life cycle graph (on the left in Figure 2.7) has two loops, one of length 2 and one of length 1, whose greatest common divisor is 1.

However, if $s_2 = 0$, then, by Theorem 15(a), A is imprimitive since

$$A^2 = \begin{pmatrix} f_2 s_1 & 0 \\ 0 & f_2 s_1 \end{pmatrix}$$

is not positive. This can also be seen from the life cycle graph of A (on the right in Figure 2.7) and Theorem 14(b), which has only one loop of length 2.

Example 30 *The nonnegative irreducible matrix*

$$A = \begin{pmatrix} 0 & 0 & 0 & f_4 \\ s_1 & 0 & 0 & 0 \\ 0 & s_2 & 0 & 0 \\ 0 & 0 & s_3 & 0 \end{pmatrix}$$

in Example 27 has only one loop in its life cycle graph and it is of length 4. Hence, the greatest common divisor of the lengths of all loops is 4. By Theorem 14(b), this matrix is imprimitive. We can reach this same conclusion by means of Theorem 14(a) using the (computer-assisted) calculation

$$A^{k^2-2k+2} = A^{10} = \begin{pmatrix} 0 & 0 & f_4^3 s_1^2 s_2^2 s_3^3 & 0 \\ 0 & 0 & 0 & f_4^3 s_1^3 s_2^2 s_3^2 \\ f_4^2 s_1^2 s_2^3 s_3^2 & 0 & 0 & 0 \\ 0 & f_4^2 s_1^2 s_2^2 s_3^3 & 0 & 0 \end{pmatrix}.$$

Since this is not a positive matrix, A is not primitive.

The Leslie matrix in Example 30 is a special case of

$$A = \begin{pmatrix} f_1 & f_2 & \cdots & f_{k-1} & f_k \\ s_1 & 0 & \cdots & 0 & 0 \\ 0 & s_2 & \cdots & 0 & 0 \\ \vdots & \vdots & & \vdots & \vdots \\ 0 & 0 & \cdots & s_{k-1} & s_k \end{pmatrix} \quad \begin{matrix} f_i \geq 0, & 0 < s_i \leq 1 \\ f_k > 0, & 0 \leq s, < 1 \end{matrix} \tag{2.10}$$

which is primitive if either $f_1 > 0$ or $s_k > 0$ by Corollary 2. If both $f_1 = s_k = 0$ (the youngest age class is immature and no individual survives past the last age class k), then, for each $f_i > 0$, there is a closed loop in the life cycle graph (going through each age class up to the ith which, by reproduction, returns to the first age class.). From Corollary 2 and Theorem 14(b), we obtain the following results for a general Leslie matrix.

Corollary 3 *The (nonnegative irreducible) Leslie matrix (2.10) is primitive if either $f_1 > 0$ or $s_k > 0$. If both $f_1 = s_k = 0$, then it is primitive if (and only if) the greatest common divisor of the set of indices for which $f_i > 0$ equals 1. In particular, it is primitive if two consecutive classes are fertile.*

Example 31 *The Leslie matrix*

$$A = \begin{pmatrix} 0 & 1 & 2 & 1 \\ 0.5 & 0 & 0 & 0 \\ 0 & 0.5 & 0 & 0 \\ 0 & 0 & 0.75 & 0 \end{pmatrix}$$

has consecutive fertile classes and is therefore primitive by Corollary 3. We can see this specifically from the eigenvalues, which are approximately (to the nearest thousandth)

$$\lambda_1 \approx 1.065, \quad \lambda_2 \approx -0.283 + 0.522i, \quad \lambda_3 \approx -0.283 - 0.522i, \quad \lambda_4 \approx -0.500.$$

The Perron eigenvalue λ_1 is strictly dominant since $|\lambda_2| = |\lambda_3| \approx 0.594$ and $|\lambda_4| \approx 0.500$. A positive eigenvector associated with λ_1 is

$$\mathbf{v}_1 = \begin{pmatrix} 0.542 \\ 0.254 \\ 0.119 \\ 0.084 \end{pmatrix}.$$

Example 32 *Consider the Usher matrix*

$$A = \begin{pmatrix} \frac{1}{2} & \frac{1}{4} & \frac{1}{2} \\ \frac{1}{2} & \frac{1}{2} & 0 \\ 0 & \frac{1}{4} & \frac{1}{2} \end{pmatrix}.$$

Since

$$A^{k-1} = \begin{pmatrix} 1+\frac{1}{2} & \frac{1}{4} & \frac{1}{2} \\ \frac{1}{2} & 1+\frac{1}{2} & 0 \\ 0 & \frac{1}{4} & 1+\frac{1}{2} \end{pmatrix}^2 = \begin{pmatrix} \frac{19}{8} & \frac{7}{8} & \frac{3}{2} \\ \frac{3}{2} & \frac{19}{8} & \frac{1}{4} \\ \frac{1}{8} & \frac{3}{4} & \frac{9}{4} \end{pmatrix}$$

is positive, this nonnegative matrix is irreducible by Theorem 12. Since there is a positive diagonal entry (actually three), the matrix is primitive by Corollary 2.

Also we can note that the eigenvalues are

$$\lambda_1 = 1, \quad \lambda_2 = \frac{1}{4} + \frac{1}{4}i, \quad \lambda_3 = \frac{1}{4} - \frac{1}{4}i$$

and that the Perron eigenvalue $\lambda_1 = 1$ is strictly dominant because $|\lambda_2| = |\lambda_3| = \sqrt{2}/4 < 1$. A positive eigenvector associated with λ_1 is

$$\mathbf{v} = \begin{pmatrix} \frac{2}{5} \\ \frac{2}{5} \\ \frac{1}{5} \end{pmatrix}.$$

Exercises 2.4

1. Determine which one of the following matrices is (i) irreducible and (ii) primitive, by using the algebraic method and the life cycle graph method.

(a) $\begin{pmatrix} 0 & 0.5 & 0 & 0 & 0 \\ 0.5 & 0 & 0.5 & 0 & 0 \\ 0 & 0.5 & 0 & 0 & 0 \\ 0.5 & 0 & 0 & 1 & 0 \\ 0 & 0 & 0.5 & 0 & 1 \end{pmatrix}$ (b) $\begin{pmatrix} 0.5 & 0.25 & 0 \\ 0.5 & 0.5 & 0.5 \\ 0 & 0.25 & 0.5 \end{pmatrix}$

(c) $\begin{pmatrix} 0 & 1 & 3 & 0 & 0 & 0 \\ 0.5 & 0 & 0 & 0 & 0 & 0 \\ 0 & 0.5 & 0 & 0 & 0 & 0 \\ 0 & 0 & 0 & 0 & 1 & 2 \\ 0.2 & 0 & 0 & 0.5 & 0 & 0 \\ 0 & 0.2 & 0 & 0 & 0.5 & 0 \end{pmatrix}$ (d) $\begin{pmatrix} 1 & 2 & 3 & 0 \\ 0 & 1 & 2 & 0.5 \\ 0 & 1 & 2 & 0.2 \\ 0 & 0 & 1 & 0.2 \end{pmatrix}$

2. Which of the nonnegative matrices below are primitive and which are not? Justify your answers.

(a) $\begin{pmatrix} 0 & 1 & 1 \\ 1 & 0 & 1 \\ 1 & 1 & 0 \end{pmatrix}$ (b) $\begin{pmatrix} 0 & a & a \\ a & 0 & a \\ a & a & 0 \end{pmatrix}, a > 0$ (c) $\begin{pmatrix} a_{11} & a_{12} & a_{13} \\ a_{21} & a_{22} & a_{23} \\ a_{31} & a_{32} & 1 \end{pmatrix}$

(d) $\begin{pmatrix} 0 & 0 & 1 & 2 \\ 0 & 0 & 1 & 2.1 \\ 0.5 & 0.3 & 0 & 0 \\ 0.4 & 0.4 & 0 & 0 \end{pmatrix}$ (e) $\begin{pmatrix} 0 & 0 & 1 & 2 \\ 0 & 0 & 1 & 2.1 \\ 0.5 & 0.3 & 0 & 0 \\ 0.4 & 0.4 & 0 & a \end{pmatrix}, a \geq 0$

3. Consider the population model with projection matrix $A = F + T$ with

$$F = \begin{pmatrix} 0 & 0 & 1 \\ 0 & 0 & 1.1 \\ 0 & 0 & 0 \end{pmatrix}, \quad T = \begin{pmatrix} 0 & 0 & 0 \\ 0 & 0 & 0 \\ 0.9 & 0.8 & s_3 \end{pmatrix}$$

where $0 \leq s_3 < 1$. This is a population classified into two different juvenile newborn classes $i = 1$ and 2 and a single mature adult class $i = 3$. The time step is the maturation period for both juvenile classes. s_3 is the probability an adult survives one time step.

(i) Show A is irreducible by means of the associated life cycle graph and by use of Theorem 12.

(ii) For what values of s_3 is A primitive? Justify your answer.

(iii) Calculate the eigenvalues of A by solving the characteristic polynomial.

4. Repeat Problem 3 with

$$F = \begin{pmatrix} 0 & 0 & f_1 \\ 0 & 0 & f_2 \\ 0 & 0 & 0 \end{pmatrix}, \quad T = \begin{pmatrix} 0 & 0 & 0 \\ 0 & 0 & 0 \\ s_1 & s_2 & s_3 \end{pmatrix}$$

where $f_i > 0$, $0 < s_1, s_2 \leq 1$ and $0 \leq s_3 < 1$

5. Reduction Rule 1. Suppose that all ages past some age (say age 2) have zero net fecundity, then the Leslie matrix has the form

$$A = \begin{pmatrix} f_1 & f_2 & 0 & 0 \\ s_1 & 0 & 0 & 0 \\ 0 & s_2 & 0 & 0 \\ 0 & 0 & s_3 & 0 \end{pmatrix}.$$

Show that the dominant eigenvalue of A is equal to the dominant eigenvalue of the reduced matrix B formed by deleting the columns with all 0s and the corresponding rows:

$$B = \begin{pmatrix} f_1 & f_2 \\ s_1 & 0 \end{pmatrix}.$$

Reduction Rule 2. Suppose that all ages beyond age 1 have the same net fecundity f_i and the survival rate s_i.

(a) Write down the Leslie 4×4 matrix A.

(b) Show that the dominant eigenvalue of the matrix A is the same dominant eigenvalue of the reduced 2×2 matrix

$$B = \begin{pmatrix} f_1 & f_2 \\ s_1 & s_2 \end{pmatrix}.$$

2.5 Analysis of Linear Matrix Equations

2.5.1 The Fundamental Theorem of Demography

Consider a linear matrix equation
$$\mathbf{x}(t+1) = A\mathbf{x}(t). \tag{2.11}$$
with an irreducible nonnegative projection matrix A. Let $\lambda_1, \lambda_2, \ldots, \lambda_k$ be the eigenvalues of A and let $\mathbf{v}_1, \mathbf{v}_2, \ldots, \mathbf{v}_k$ be corresponding eigenvectors. Then $\varphi(t) = c_1 \lambda_1^t \mathbf{v}_1$ is a solution of Equation (2.11). To show this notice that
$$\begin{aligned}\varphi_1(t+1) &= c_1 \lambda_1^{t+1} \mathbf{v}_1 = c_1 \lambda_1^t (\lambda_1 \mathbf{v}_1) \\ &= c_1 \lambda_1^t (A\mathbf{v}_1) = A\left(c_1 \lambda_1^t \mathbf{v}_1\right) \\ &= A\varphi_1(t).\end{aligned}$$
Similarly, one may show that $\varphi_2(t) = c_2 \lambda_2^t \mathbf{v}_2, \ldots, \varphi_k(t) = c_k \lambda_k^t \mathbf{v}_k$ are also solutions of Equation (2.11). If the k eigenvalues are $\lambda_1, \lambda_2, \ldots, \lambda_k$ distinct, or more generally, if their eigenvectors are independent, then a formula for the general solution of Equation (2.11) is
$$\mathbf{x}(t) = c_1 \lambda_1^t \mathbf{v}_1 + c_2 \lambda_2^t \mathbf{v}_2 + \cdots + c_k \lambda_k^t \mathbf{v}_k = \sum_{i=1}^{k} c_i \lambda_i^t \mathbf{v}_i, \tag{2.12}$$
where the c_i are arbitrary scalars. Note that, under our assumption, the eigenvectors form a basis and we can write
$$\mathbf{x}(0) = c_1 \mathbf{v}_1 + c_2 \mathbf{v}_2 + \cdots + c_k \mathbf{v}_k, \tag{2.13}$$
where the constants c_i are the coordinates of the initial population vector $\mathbf{x}(0)$. Equation (2.13) consists of k linear equations in $c_1, c_2, \ldots, c_k$ which may be solved to compute these coordinates.

We assume that the matrix A is primitive and we label its strictly dominant eigenvalue by λ_1 (cf. Theorem 13). Then $\lambda_1 > |\lambda_2| \geq |\lambda_3| \geq \cdots \geq |\lambda_k|$ and λ_1^t dominates all the other terms λ_i^t as $t \to +\infty$ in (2.12). Moreover, $\mathbf{v}_1$ can be chosen as a positive eigenvector (cf. Theorem 13) and it can be shown that, if $\mathbf{x}(0)$ is a nonnegative vector, then [172]
$$c_1 > 0.$$
This means the leading term $c_1 \lambda_1^t \mathbf{v}_1$ in $\mathbf{x}(t)$ will, regardless of the initial population $\mathbf{x}(0)$, come to dominate the other exponential terms $c_i \lambda_i^t \mathbf{v}_i$ in (2.12). As a result, as $t \to +\infty$ the population vector $\mathbf{x}(t)$ will behave like $c_1 \lambda_1^t \mathbf{v}_1$,
$$\mathbf{x}(t) \approx c_1 \lambda_1^t \mathbf{v}_1,$$
or in other words, the population vector will decay or grow at the exponential rate λ_1 if $\lambda_1 < 1$ or $\lambda_1 > 1$, respectively. For population models, the dominant eigenvalue λ_1 of the projection matrix is often denoted by r (or r_0) and is called the *population growth rate*.

We introduce the vector norm
$$||\mathbf{x}|| = \sum_{i=1}^{k} |x_i|$$
which when applied to a nonnegative population vector $\mathbf{x}$ equals the *total population size*. The ratio
$$\frac{\mathbf{x}}{||\mathbf{x}||}$$

is a normalized population distribution whose components are less than 1 and sum to 1 (a partition of unity) and therefore equal to the proportion of each class within the total population.

If we divide both sides of (2.12) by $\|x(t)\|$, we obtain

$$\frac{\mathbf{x}(t)}{\|\mathbf{x}(t)\|} = \frac{c_1 \lambda_1^t \mathbf{v}_1 + c_2 \lambda_2^t \mathbf{v}_2 + \cdots + c_k \lambda_k^t \mathbf{v}_k}{\|c_1 \lambda_1^t \mathbf{v}_1 + c_2 \lambda_2^t \mathbf{v}_2 + \cdots + c_k \lambda_k^t \mathbf{v}_k\|}$$

and, after dividing numerator and denominator by $c_1 \lambda_1^t$,

$$\frac{\mathbf{x}(t)}{\|\mathbf{x}(t)\|} = \frac{\mathbf{v}_1 + \frac{c_2}{c_1}\left(\frac{\lambda_2}{\lambda_1}\right)^t \mathbf{v}_2 + \cdots + \frac{c_k}{c_1}\left(\frac{\lambda_k}{\lambda_1}\right)^t \mathbf{v}_k}{\left\|\mathbf{v}_1 + \frac{c_2}{c_1}\left(\frac{\lambda_2}{\lambda_1}\right)^t \mathbf{v}_2 + \cdots + \frac{c_k}{c_1}\left(\frac{\lambda_k}{\lambda_1}\right)^t \mathbf{v}_k\right\|}.$$

Since $\lambda_1 > |\lambda_i|$ for $i \geq 2$ and hence

$$\lim_{t \to \infty} \left(\frac{\lambda_i}{\lambda_1}\right)^t = 0,$$

we find that, regardless of the initial population vector $\mathbf{x}(0)$,

$$\lim_{t \to \infty} \frac{\mathbf{x}(t)}{\|\mathbf{x}(t)\|} = \frac{\mathbf{v}_1}{\|\mathbf{v}_1\|}.$$

This result is known as the *strong ergodic property*. It implies that, regardless of the initial condition $\mathbf{x}(0)$ and regardless of whether the population is dying out ($r = \lambda_1 < 1$) or growing exponentially ($r = \lambda_1 > 1$), the normalized population distribution "stabilizes" asymptotically as $t \to +\infty$ to the normalized ("unit") eigenvector $\mathbf{v}_1/\|\mathbf{v}_1\|$, which we see gives the long-term proportions of individuals within each age class. While we obtained this result under the assumption that there are k independent eigenvectors, it remains valid even if there are not k independent eigenvectors [172].

Theorem 16 *Fundamental Theorem of Demography.*

Assume the nonnegative matrix A is primitive and let r denote its strictly dominant, positive eigenvalue and $\mathbf{v}$ an associated positive eigenvector. If $r \neq 1$, the demographic vector $\mathbf{x}(t)$ with nonnegative, nonzero initial condition $\mathbf{x}(0)$, decays or grows at an exponential rate (i.e., each component decays or grows like r^t) if $r < 1$ or $r > 1$, respectively. Moreover, the normalized age distribution $\mathbf{x}/\|\mathbf{x}\|$ asymptotically approaches $\mathbf{v}_1/\|\mathbf{v}_1\|$.

2.5.2 The Reproduction Number $\mathcal{R}_0$

The bifurcation diagram in Figure 1.1 of Chapter 1, which summarized the asymptotic dynamics of a linear scalar difference equation, we now see also summarizes the asymptotic dynamics of the linear matrix equation (2.11) with r taken to be the dominant eigenvalue of the primitive projection matrix A. Thus, we say that the extinction equilibrium $\mathbf{x} = \mathbf{0}$ destabilizes as r increases through 1.

Unlike the scalar equation ($k = 1$ in Chapter 1) in which r is easily identifiable as a numerical coefficient in the equation, the dominant eigenvalue r of the projection matrix needs to be calculated from the projection matrix A. Since the degree of the characteristic polynomial equals k, for large matrices (i.e., for population structured by a large number of classes) no formula is in general

2.5. ANALYSIS OF LINEAR MATRIX EQUATIONS

available for the calculation of r from the entries in the matrix A. However, for the vast majority of projection matrices used in population dynamics, there is another quantity that can be used to determine population growth or decay and for which there is a tractable formula based on the entries of the projection matrix A. This quantity, the *reproduction number*, is a generalization to matrix equations (2.11) of the quantity

$$\mathcal{R}_0 = b\,(1-s)^{-1}$$

introduced in Chapter 1 for scalar equations.

Before seeing how to calculate the reproduction number $\mathcal{R}_0$ for a general matrix equation, we first consider the Leslie matrix equation with projection matrix

$$A = \begin{pmatrix} f_1 & f_2 & \cdots & f_{k-1} & f_k \\ s_1 & 0 & \cdots & 0 & 0 \\ 0 & s_2 & \cdots & 0 & 0 \\ \vdots & \vdots & & \vdots & \vdots \\ 0 & 0 & \cdots & s_{k-1} & s_k \end{pmatrix}. \tag{2.14}$$

Recall that

$$\begin{aligned} f_i \geq 0 \text{ and } 0 < s_i \leq 1 \text{ for } i = 1, 2, \cdots, k-1 \\ 0 \leq s_k < 1 \text{ and } f_k > 0 \end{aligned} \tag{2.15}$$

which are assumptions that ensure individuals in the last age class k are fertile and there is a positive probability of a newborn will reach the final age class k, which is a fertile age class ($f_k > 0$). As a result, this matrix is irreducible. If $s_k = 0$, then no individual in age class k survives one time unit (i.e., no individual reaches age k). If $s_k > 0$, then the requirement $s_k < 1$ ensures the expected life span of a newborn is finite.

The interpretation of $\mathcal{R}_0 = b\,(1-s)^{-1}$ for a scalar (i.e., one-dimensional matrix model) can be determined from the harmonic series

$$\mathcal{R}_0 = b + bs + bs^2 + \cdots + bs^i + \cdots,$$

which we can interpret as a probabilistic expectation. The fraction s^i is the probability that a newborn lives i time units and, if that event occurs, then the expected reproductive payoff is b. Thus, $\mathcal{R}_0$ is the expected number of offspring per newborn per lifetime. For a Leslie age-structured model,

$$p_i = s_1 s_2 \cdots s_{i-1}, \quad 2 \leq i \leq k$$

is the probability a newborn (which necessarily starts in age class $i=1$) lives i time units, and thereby becomes a member of age class i. Once reaching age class k an individual survives $j \geq 1$ more time units, remaining in age class k, with probability s_k^j. The age class specific expected reproductive outputs of a newborn, as it passes through the age classes $i=2$ up to $k-1$ (the next to last age class), are

$$f_1, \quad f_2 s_1, \quad f_3 s_1 s_2, \cdots, \quad f_{k-1}\,(s_1 s_2 \cdots s_{k-2})$$

and the expected reproductive outputs while remaining in the last age class k are

$$f_k\,(s_1 s_2 \cdots s_{k-1}), \quad f_k\,(s_1 s_2 \cdots s_{k-1})\,s_k, \quad \cdots \quad , \quad f_k\,(s_1 s_2 \cdots s_{k-1})\,s_k^j, \cdots$$

To simplify these expressions notationally, we define
$$p_i := s_0 s_1 s_2 \cdots s_{i-1}, \quad 1 \le i \le k$$
where, for convenience, we set $s_0 = 1$. The quantity p_i is the probability a newborn reaches age class i. With this notation the age class-specific reproductive outputs above are
$$f_1 p_1, \quad f_2 p_2, \quad f_3 p_3, \cdots, \quad f_{k-1} p_{k-1}$$
from age classes $i = 1$ to $k-1$ and
$$f_k p_k, \quad f_k p_k s_k, \quad f_k p_k s_k^2, \cdots, \quad f_k p_k s_k^j, \cdots$$
for age class k. The reproduction number is the total lifetime expected reproductive output of a newborn, i.e.,
$$\mathcal{R}_0 = \sum_{i=1}^{k-1} f_i p_i + f_k p_k \sum_{j=0}^{k} s_k^j$$
or
$$\mathcal{R}_0 = \sum_{i=1}^{k-1} f_i p_i + f_k p_k \frac{1}{1-s_k}. \tag{2.16}$$

It is reasonable to conjecture that if $\mathcal{R}_0 < 1$, i.e., if a newborn is not expected to replace itself by reproduction during its lifetime, then the population will die out ($r < 1$) If a newborn will more than replace itself, i.e., $\mathcal{R}_0 > 1$, then the population will grow ($r > 1$). So we conjecture that
$$\mathcal{R}_0 < 1 \Leftrightarrow r < 1, \quad \mathcal{R}_0 > 1 \Leftrightarrow r > 1, \quad \mathcal{R}_0 = 1 \Leftrightarrow r = 1. \tag{2.17}$$
This is not mathematically obvious, however, since a formula for the dominant eigenvalue r of a Leslie matrix A is generally unavailable. An exception is the case $k = 2$
$$A = \begin{pmatrix} f_1 & f_2 \\ s_1 & s_2 \end{pmatrix}$$
for which we can easily calculate
$$r = \frac{1}{2}\left(f_1 + s_2 + \sqrt{(f_1 - s_2)^2 + 4 f_2 s_1}\right).$$
It takes only a bit of algebra to show that (2.17) holds with
$$\mathcal{R}_0 = f_1 + f_2 s_1 \frac{1}{1-s_2}.$$
That (2.17) holds is known from theorems about even more general projection matrices [84], [223].

In a general matrix population model, the i-class individuals at time $t+1$ consist (in the absence of immigration/emigration) of newborns contributed from all fertile j-class individuals plus surviving j-class individuals that move into the i-class
$$x_i(t+1) = \sum_{j=1}^{k} f_{ij} x_j(t) + \sum_{j=1}^{k} \tau_{ij} x_j(t),$$

2.5. ANALYSIS OF LINEAR MATRIX EQUATIONS

where f_{ij} is the (surviving per capita) i-class offspring per j-class parent and τ_{ij} is the fraction of j-class individuals that survive and move to the i-class. Thus,

$$0 \leq f_{ij}, \quad 0 \leq \tau_{ij} \leq 1, \quad \sum_{j=1}^{k} \tau_{ij} \leq 1 \tag{2.18}$$

(the latter sum being the fraction of surviving i-class individuals). The projection matrix can be written

$$A = F + T,$$

where

$$F = (f_{ij}), \quad T = (\tau_{ij})$$

are *fertility* and *transition matrices*, respectively. Both F and T are nonnegative matrices. Thus, at any time t, the vector of newborns is $F\mathbf{x}(t)$ and $T\mathbf{x}(t)$ is the vector of survivors from time t. We will need the following theorem.

Theorem 17 *[30] A nonnegative matrix has a nonnegative, dominant eigenvalue and there exists a nonnegative eigenvector associated with it.*

We begin with an investigation of the transition matrix T. Let $\lambda_T \geq 0$ be its dominant eigenvalue. The vector of survivors from $\mathbf{x}(0)$ at time t is $T^t\mathbf{x}(0)$ and the population of survivors is $\|T^t\mathbf{x}(0)\|$. Since it is reasonable to rule out immortality, we assume $\lim_{t \to +\infty} \|T^t\mathbf{x}(0)\| = 0$ for all nonnegative initial vectors $\mathbf{x}(0)$. This is equivalent to $\lambda_T < 1$ and implies $I - T$ is invertible and

$$(I - T)^{-1} = I + T + T^2 + \cdots$$

[171], [223]. Consider the matrix

$$F(I - T)^{-1} = F + FT + FT^2 + \cdots.$$

An initial cohort $\mathbf{x}(0)$ produces $F\mathbf{x}(0)$ offspring after one time unit, $FT\mathbf{x}(0)$ offspring after two time units, and so on. Therefore, this cohort produces a total of $F(I-T)^{-1}\mathbf{x}(0)$ offspring (the next-generation) over the course of its lifetime. For this reason, $F(I-T)^{-1}$ is called the *next-generation matrix*. Being a sum of nonnegative matrices, it is also nonnegative and therefore, by Theorem 17, has a nonnegative, dominant eigenvalue.

Definition 13 *[84] Assume $\lambda_T < 1$ for a population matrix model with projection matrix $A = F+T$ satisfying (2.18). The **reproduction number** $\mathcal{R}_0$ is the dominant eigenvalue of the next-generation matrix $F(I-T)^{-1}$.*

A Leslie matrix has fertility and transition matrices

$$F = \begin{pmatrix} f_1 & f_2 & \cdots & f_{k-1} & f_k \\ 0 & 0 & \cdots & 0 & 0 \\ 0 & 0 & \cdots & 0 & 0 \\ \vdots & \vdots & & \vdots & \vdots \\ 0 & 0 & \cdots & 0 & 0 \end{pmatrix}, \quad T = \begin{pmatrix} 0 & 0 & \cdots & 0 & 0 \\ s_1 & 0 & \cdots & 0 & 0 \\ 0 & s_2 & \cdots & 0 & 0 \\ \vdots & \vdots & & \vdots & \vdots \\ 0 & 0 & \cdots & s_{k-1} & s_k \end{pmatrix}$$

which, under conditions (2.15), satisfies the requirements (2.18) and $\lambda_T = s_k < 1$ of Definition 13. To see that Definition 13 of $\mathcal{R}_0$ is consistent with the formula (2.16) for $\mathcal{R}_0$ calculated above from first principles, we calculate

$$(I-T)^{-1} = \begin{pmatrix} 1 & 0 & 0 & \cdots & 0 \\ s_1 & 1 & 0 & \cdots & 0 \\ s_1 s_2 & s_2 & 1 & \cdots & 0 \\ s_1 s_2 s_3 & s_2 s_3 & s_3 & \cdots & 0 \\ \vdots & \vdots & \vdots & \square & \vdots \\ \frac{s_1 s_2 \cdots s_{k-1}}{1-s_k} & \frac{s_2 \cdots s_{k-1}}{1-s_k} & \frac{s_3 \cdots s_{k-1}}{1-s_k} & \cdots & \frac{1}{1-s_k} \end{pmatrix}$$

and

$$F(I-T)^{-1} = \begin{pmatrix} \sum_{i=1}^{k-1} f_i p_i + f_k p_k \frac{1}{1-s_k} & * & \cdots & * & * \\ 0 & 0 & \cdots & 0 & 0 \\ 0 & 0 & \cdots & 0 & 0 \\ \vdots & \vdots & & \vdots & \vdots \\ 0 & 0 & \cdots & 0 & 0 \end{pmatrix}. \quad (2.19)$$

(The asterisks represent unneeded terms). The eigenvalues of this diagonal matrix appear along the diagonal, and hence we get for the dominant eigenvalue $\mathcal{R}_0$ the same formula (2.16).

The formula (2.16) for $\mathcal{R}_0$ for a Leslie matrix shows that $\mathcal{R}_0$ is the lifetime expected number of newborns produced by a newborn. This interpretation of $\mathcal{R}_0$ is the same for any population model in which all newborns lie in the first class. Without loss in generality, suppose newborns lie in the first class so that all rows of R are zeros except the first. Then $F(I-T)^{-1}$ has all zero rows except the first and $\mathcal{R}_0$ is the first row, first column entry, i.e., is the dot product of the first row of F with the first column of $(I-T)^{-1}$. Then j^{th} entry in the first column of $(I-T)^{-1}$ is the expected number of time units a newborn will spend in class j and, as a result, $\mathcal{R}_0$ is the expected number of newborns produced by a newborn over its lifetime. More generally, suppose there are more than one, say m, newborn classes in the population model. Then a similar interpretation can be given to $\mathcal{R}_0$ as follows. By Theorem 17 there exists a nonnegative eigenvector $\mathbf{v}_0$ of $F(I-T)^{-1}$ associated with $\mathcal{R}_0$

$$F(I-T)^{-1} \mathbf{v}_0 = \mathcal{R}_0 \mathbf{v}_0.$$

Since $\mathbf{v}_0$ is in the range of F, it is a vector consisting only of newborns. Any initial vector $\mathbf{x}(0) = c\mathbf{v}_0$ which is a positive scalar multiple $c > 0$ of $\mathbf{v}$, i.e., which consists of only newborns in the same proportions as in $\mathbf{v}/\|\mathbf{v}\|$, satisfies

$$F(I-T)^{-1} \mathbf{x}(0) = \mathcal{R}_0 \mathbf{x}(0).$$

The left side, and therefore $\mathcal{R}_0 \mathbf{x}(0)$, is the vector of newborns produced by the newborns in $\mathbf{x}(0)$ over their lifetimes. This says a j-class newborn produces $\mathcal{R}_0$ newborns in class j. A newborn cohort $c\mathbf{v}$ can be considered "typical" in the sense that if newborns were sampled randomly using the probability distribution defined by the entries in $\mathbf{v}_0/\|\mathbf{v}_0\|$ the result would be a newborn cohort $\mathbf{x}(0) = c\mathbf{v}$. In this sense, $\mathcal{R}_0$ is the lifetime expected number of j-class newborns produced by a "typical" j-class newborn. For the case of a single newborn class $j = 1$, the eigenvector $\mathbf{v}_0 = (1, 0, \cdots, 0)^T$.

2.5. ANALYSIS OF LINEAR MATRIX EQUATIONS

It should be noted that $\mathcal{R}_0$ can be used to determine population growth or decay in place of r since they are on the same side of 1. This is the assertion of the following extension of a theorem of Cushing and Zhou [84].

Theorem 18 *[223] Assume that $\lambda_T < 1$ for an irreducible projection matrix $A = F + T$ satisfying (2.18). Assume also that $\mathcal{R}_0 > 0$. Then one of the following holds*

$$r = \mathcal{R}_0 = 1 \quad or \quad 1 < r < \mathcal{R}_0 \quad or \quad \mathcal{R}_0 < r < 1. \tag{2.20}$$

Proof. We begin by noting some facts about a matrix and its transpose[3]: they have the same eigenvalues; if one is nonnegative (positive) so is the other; and if one is irreducible so is the other.

Since $F(I-T)^{-1}$ is nonnegative, its dominant eigenvalue $\mathcal{R}_0$ is also the dominant eigenvalue of its transpose. Theorem 17 implies that there exists a nonnegative eigenvector $\mathbf{w}$ of the transpose associated with $\mathcal{R}_0$

$$\left(F(I-T)^{-1}\right)^T \mathbf{w} = \mathcal{R}_0 \mathbf{w}.$$

Taking the transpose of both sides and re-arranging terms, we obtain

$$\mathbf{w}^T F(I-T)^{-1} = \mathcal{R}_0 \mathbf{w}^T.$$

Dividing by $\mathcal{R}_0$ and multiplying (on the right) by $I - T$, we see that

$$\mathbf{w}^T F/\mathcal{R}_0 = \mathbf{w}^T (I-T)$$

and hence

$$\mathbf{w}^T (T + F/\mathcal{R}_0) = \mathbf{w}^T.$$

Finally, by taking the transpose of both sides, we arrive at

$$(T + F/\mathcal{R}_0)^T \mathbf{w} = \mathbf{w}. \tag{2.21}$$

Since the matrix $T + F/\mathcal{R}_0$ is nonnegative and irreducible, so is its transpose. Since $\mathbf{w}$ is nonnegative, it follows from (2.21) and Theorem 13(ii) that the dominant eigenvalue of $(T + F/\mathcal{R}_0)^T$, and hence the dominant eigenvalue of its transpose $T + F/\mathcal{R}_0$, equals 1. It also follows from (2.21) that

$$(\mathcal{R}_0 T + F)^T \mathbf{w} = \mathcal{R}_0 \mathbf{w}$$

i.e., $\mathcal{R}_0$ is the dominant eigenvalue of the (nonnegative, irreducible) matrix $\mathcal{R}_0 T + F$.

Applying these conclusions when $\mathcal{R}_0 = 1$, we see that the dominant eigenvalue of $A = T + F$ equals 1, which is the first alternative in (2.20).

If, on the other hand, $\mathcal{R}_0 > 1$, then, by Theorem 13(iii), the dominant eigenvalue of $T + F$ is greater than that of $T + F/\mathcal{R}_0$, i.e., $r > 1$. Moreover, the dominant eigenvalue of $\mathcal{R}_0 T + F$ is greater than that of $T + F$, i.e., $\mathcal{R}_0 > r$. This is the second alternative in (2.20).

Finally, if $\mathcal{R}_0 < 1$ then by Theorem 13(iii) the dominant eigenvalue of $T + F$ is less than that of $T + F/\mathcal{R}_0$, i.e., $r < 1$. Moreover, the dominant eigenvalue of $\mathcal{R}_0 T + F$ is less than that of $T + F$, i.e., $\mathcal{R}_0 < r$. ∎

---

[3]The transpose of a matrix $A = (a_{ij})$ is $A^T = (a_{ji})$. Note that $(A\mathbf{w})^T = \mathbf{w}^T A^T$.

It follows from this theorem that solutions of a general Leslie matrix, under assumptions (2.15), grow or decay if $\mathcal{R}_0 > 1$ or $\mathcal{R}_0 < 1$, respectively. Note that no formula for the dominant eigenvalue r is available for a general Leslie matrix.

The Leslie model illustrates an important point concerning the calculation $\mathcal{R}_0$. The algebraic reason the calculation of $\mathcal{R}_0$ is tractable for a Leslie matrix is because all rows, other than the first, in the fertility matrix F consist entirely of zeros, which is therefore also true of the next-generation matrix $F(I-T)^{-1}$, exposing its eigenvalues along the diagonal. The reason F has only one nonzero row in a Leslie age-structured model is because all newborns lie in just one class, namely the first age class. This is often true in other matrix models as well and, for any such model, the calculation of $\mathcal{R}_0$ is tractable for the same reason.

An Usher (or standard size structured) matrix model is another example [67]. For a general Usher matrix (see (2.7)), F is the same as for a Leslie matrix, under assumptions and (2.15) on f_i and s_i, and

$$T = \begin{pmatrix} \tau_{11} & 0 & 0 & \cdots & 0 & 0 \\ \tau_{21} & \tau_{22} & 0 & \cdots & 0 & 0 \\ 0 & \tau_{32} & \tau_{33} & \cdots & 0 & 0 \\ \vdots & \vdots & & & \vdots & \vdots \\ \vdots & \vdots & & \cdots & \tau_{k-1,k-1} & 0 \\ 0 & 0 & & \cdots & \tau_{k,k-1} & \tau_{kk} \end{pmatrix} \tag{2.22}$$

$$\tau_{ii} = s_i(1-p_i), \quad \tau_{i+1,i} = s_i p_i, \quad 0 < p_i \leq 1 \text{ for } i = 1, 2, \cdots, k-1$$
$$\text{and } \tau_{kk} = s_k.$$

Then
$$\lambda_T = \max\{\tau_{ii}\} = \max\{s_i(1-p_i),\ s_k\} < 1.$$

An induction calculation shows [67], [70]

$$(I-T)^{-1} = \begin{pmatrix} \frac{1}{1-\tau_{11}} & 0 & \cdots & 0 \\ \frac{t_{21}}{(1-\tau_{11})(1-\tau_{22})} & \frac{1}{1-\tau_{22}} & \cdots & 0 \\ \vdots & \vdots & \cdots & 0 \\ \prod_{j=1}^{i} \frac{\tau_{j,j-1}}{1-\tau_{jj}} & \prod_{j=2}^{i} \frac{\tau_{j,j-1}}{1-\tau_{jj}} & \cdots & 0 \\ \vdots & \vdots & \ddots & \vdots \\ \prod_{j=1}^{k} \frac{\tau_{j,j-1}}{1-\tau_{jj}} & \prod_{j=2}^{k} \frac{\tau_{j,j-1}}{1-\tau_{jj}} & \cdots & \frac{1}{1-\tau_{kk}} \end{pmatrix} \tag{2.23}$$

which gives the next-generation matrix

$$F(I-T)^{-1} = \begin{pmatrix} \sum_{i=1}^{k} \tau_i \prod_{j=1}^{i} \frac{\tau_{j,j-1}}{1-t_{jj}} & \cdots & f_m \frac{1}{1-\tau_{mm}} \\ 0 & \cdots & 0 \\ \vdots & \ddots & \vdots \\ 0 & \cdots & 0 \end{pmatrix}$$

and

$$\mathcal{R}_0 = \sum_{i=1}^{k} f_i \prod_{j=1}^{i} \frac{\tau_{j,j-1}}{1-\tau_{jj}}, \tag{2.24}$$

2.5. ANALYSIS OF LINEAR MATRIX EQUATIONS

where $t_{10} = 1$. It follows from Theorem 18 that solutions of this general Usher matrix grow or decay if $\mathcal{R}_0 > 1$ or $\mathcal{R}_0 < 1$, respectively. Note that no formula for the dominant eigenvalue r is available for a general Leslie matrix.

Example 33 *Consider a $k = 3$ matrix model in which there is one fertile class (class $i = 3$) and two newborn classes (classes $i = 1$ and 2), both of which are immature. List the juvenile class as $i = 1$ and 2 and the adult class as $i = 3$. Assume adults do not survive one time unit (e.g., are semelparous). If the time step is the juvenile maturation period, then*

$$F = \begin{pmatrix} 0 & 0 & f_{13} \\ 0 & 0 & f_{23} \\ 0 & 0 & 0 \end{pmatrix}, \quad T = \begin{pmatrix} 0 & 0 & 0 \\ 0 & 0 & 0 \\ \tau_{31} & \tau_{32} & 0 \end{pmatrix},$$

where f_{i3} is the number of i-class newborns produced per adult and τ_{3i} is the fraction of i-class newborns survive to become adults. Note that $\lambda_T = 0 < 1$. Calculations show

$$F(I - T)^{-1} = \begin{pmatrix} \tau_{31} f_{13} & \tau_{32} f_{13} & f_{13} \\ \tau_{31} f_{23} & \tau_{32} f_{23} & f_{23} \\ 0 & 0 & 0 \end{pmatrix}$$

a block diagonal matrix whose eigenvalues are 0 and those of the 2×2 matrix

$$\begin{pmatrix} \tau_{31} f_{13} & \tau_{32} f_{13} \\ \tau_{31} f_{23} & \tau_{32} f_{23} \end{pmatrix},$$

which are 0 and

$$\mathcal{R}_0 = \tau_{31} f_{13} + \tau_{32} f_{23},$$

which has the associated eigenvector

$$\frac{\mathbf{v}_0}{\|\mathbf{v}_0\|} = \begin{pmatrix} \frac{f_{13}}{f_{13} + f_{23}} \\ \frac{f_{23}}{f_{13} + f_{23}} \end{pmatrix}.$$

For this example, it is also possible to calculate the dominant eigenvalue

$$r = \sqrt{\tau_{31} f_{13} + \tau_{32} f_{23}} = \sqrt{\mathcal{R}_0}$$

of the projection matrix

$$A = F + T = \begin{pmatrix} 0 & 0 & f_{13} \\ 0 & 0 & f_{23} \\ \tau_{31} & \tau_{32} & 0 \end{pmatrix}$$

from the roots of its characteristic polynomial

$$\lambda \left(\lambda^2 + (-\tau_{31} f_{13} - \tau_{32} f_{23}) \right).$$

Example 34 *The 4×4 Leslie matrix*

$$A = \begin{pmatrix} 0 & f_2 & 0 & f_4 \\ s_1 & 0 & 0 & 0 \\ 0 & s_2 & 0 & 0 \\ 0 & 0 & s_3 & 0 \end{pmatrix}$$

describes a population classified into four age classes, the first and third of which are not reproductive. For example, the first age class could consist of young (newborn) immature juveniles and the third age class could consist of adults who are in a state of dormancy and are not reproductively active again until they reach the fourth age class. By Corollary 3, this Leslie matrix is imprimitive. A calculation shows

$$F(I-T)^{-1} = \begin{pmatrix} f_2 s_1 + f_4 s_1 s_2 s_3 & f_2 + f_4 s_2 s_3 & f_4 s_3 & f_4 \\ 0 & 0 & 0 & 0 \\ 0 & 0 & 0 & 0 \\ 0 & 0 & 0 & 0 \end{pmatrix}$$

and hence

$$\mathcal{R}_0 = f_2 s_1 + f_4 s_1 s_2 s_3.$$

Since $\lambda_T = 0 < 1$, Theorem 18 implies the population grows or decays if this reproduction number $\mathcal{R}_0 > 1$ or $\mathcal{R}_0 < 1$, respectively.

Example 35 *The Leslie matrix*

$$A = \begin{pmatrix} 0 & 1 & 2 & 1 \\ 1/2 & 0 & 0 & 0 \\ 0 & 1/2 & 0 & 0 \\ 0 & 0 & 3/4 & 0 \end{pmatrix}$$

is primitive by Corollary 3. From the formula (2.16), we calculate

$$\mathcal{R}_0 = 0 + 1 \times \left(\frac{1}{2}\right) + 2 \times \left(\frac{1}{2} \times \frac{1}{2}\right) + 1 \times \left(\frac{1}{2} \times \frac{1}{2} \times \frac{1}{2}\right) = \frac{9}{8} > 1.$$

and find that the population grows exponentially.

The Leslie matrix

$$A = \begin{pmatrix} 0 & 0 & 0 & 2 & 3 \\ 1/2 & 0 & 0 & 0 & 0 \\ 0 & 1/2 & 0 & 0 & 0 \\ 0 & 0 & 1 & 0 & 0 \\ 0 & 0 & 0 & 1/2 & 0 \end{pmatrix}$$

is also primitive by Corollary 3. From formula (2.16), we calculate

$$\mathcal{R}_0 = 2 \times \left(\frac{1}{2} \times \frac{1}{2} \times 1\right) + 3 \times \left(\frac{1}{2} \times \frac{1}{2} \times 1 \times \frac{1}{2}\right) = \frac{7}{8} < 1.$$

and find that the population dies out exponentially.

Example 36 *Consider a Leslie matrix consisting of $k = 25$ age classes. Suppose only individuals in the last two age classes, 24 and 25, are fertile adults. Suppose the probability a newborn survives to age 1 is $s_1 = 0.9$ and that an individual's probability of survival decreases by 1 percent each year.*

2.5. ANALYSIS OF LINEAR MATRIX EQUATIONS

Then

$$s_1 = 0.9$$
$$s_2 = 0.99 s_1 = (0.99)(0.9)$$
$$s_3 = 0.99 s_2 = (0.99)^2 (0.9)$$
$$s_4 = 0.99 s_3 = (0.99)^3 (0.9)$$
$$\vdots$$
$$s_{24} = (0.99)^{23} (0.9)$$
$$s_{25} = (0.99)^{24} (0.9)$$

Suppose each adult in age class 24 produces 1 newborn per unit time. We ask: how fertile must the adults in age class 25 be in order to avoid population extinction?

The resulting 25×25 Leslie matrix

$$A = \begin{pmatrix} 0 & 0 & \cdots & 1 & f_{25} \\ 0.9 & 0 & \cdots & 0 & 0 \\ 0 & (0.99)(0.9) & \cdots & 0 & 0 \\ \vdots & \vdots & & \vdots & \vdots \\ 0 & 0 & \cdots & (0.99)^{24}(0.9) & 0 \end{pmatrix}$$

is primitive by Corollary 3. It is not difficult to calculate $\mathcal{R}_0$ for this large matrix from formula (2.16) (noting that all but the last two terms equal 0):

$$\mathcal{R}_0 = 1 \times (0.99)^{23} (0.9) + f_{25} (0.99)^{24} (0.9).$$

To avoid extinction, we need $\mathcal{R}_0 > 1$, i.e.,

$$(0.99)^{23} (0.9) + f_{25} (0.99)^{24} (0.9) > 1.$$

This holds if and only if the adults in age class 25 have fertility

$$f_{25} > \frac{1 - (0.99)^{23} (0.9)}{(0.99)^{24} (0.9)} \approx 0.404.$$

Example 37 *A model for insect populations with larval, pupal, and adult life cycle stages that correlate with chronological age has a projection matrix*

$$\begin{pmatrix} 0 & 0 & f_3 \\ s_1 & 0 & 0 \\ 0 & s_2 & s_3 \end{pmatrix}, \quad f_3 > 0.$$

This is called the LPA-model and it has been extensively used to model, for example, species of beetles [59], [88]. This Leslie matrix is primitive if $s_3 > 0$ (Corollary 3), i.e., if adults live more than one time unit (which is the time spent as a larva and as a pupa in this model). For this Leslie model, we have from (2.16) that

$$\mathcal{R}_0 = f_3 \frac{s_1 s_2}{1 - s_3}$$

and the population decays or grows according to whether this quantity based on the entries in the Leslie matrix is less than or greater than 1.

Example 38 *African elephants face threats from loss of habitat and from exploitation for ivory, skins, etc. In order to design measures for conservation of elephant populations, it is essential to know their vital reproductive and mortality parameters. Demographic data, based on age classes of length 5 years, for the populations of the elephant Loxodonta africana in Amboseli National Park, Kenya, are reported in [243] and appear in Table 2.1. The survival probabilities s_i and fertilities f_i obtained from these data (see [191]) define a Leslie matrix of size $k = 13$. The value of $\mathcal{R}_0$ calculated in Table 2.1 from formula (2.16) is $\mathcal{R}_0 = 5.497 > 1$, which implies population growth.*

Age classes (years)	i	s_i	f_i	p_i	$p_i f_i$
0–4	1	0.917	0.000	1	0
5–9	2	0.976	0.014	0.917	0.013
10–14	3	0.956	0.550	0.895	0.492
15–19	4	0.939	0.925	0.856	0.791
20–24	5	0.918	1.040	0.803	0.836
25–29	6	0.911	1.053	0.738	0.777
30–34	7	0.890	1.067	0.672	0.717
35–39	8	0.832	1.090	0.598	0.652
40–44	9	0.842	0.985	0.498	0.490
45–49	10	0.844	0.829	0.419	0.347
50–54	11	0.802	0.646	0.354	0.228
55–59	12	0.718	0.472	0.284	0.134
60–64	13	0	0.099	0.204	0.020
					$R_0 = 5.497$

Table 2.1: The data in columns s_i and f_i are from [243] and [191]. These data yield p_i and $p_i f_i$ shown in columns 5 and 6 (rounded to three decimals). The net reproduction number R_0 is the sum of the entries in the column $p_i f_i$ (see (2.16))

Example 39 *In [246], the authors construct a three-life cycle stage Usher matrix model for the dynamics of an invasive plant (Cestrum aurantiacum) and use data obtained from field sites in the Mount Elgon ecosystem in Kenya to assign numerical values to entries of the projection matrix. They classify plants into seed, sapling, and adult classes in an Usher matrix $P = F + T$ with*

$$F = \begin{pmatrix} 0 & 0 & f_3 \\ 0 & 0 & 0 \\ 0 & 0 & 0 \end{pmatrix}, \quad T = \begin{pmatrix} s_1(1-p_1) & 0 & 0 \\ s_1 p_1 & s_2(1-p_2) & 0 \\ 0 & s_2 p_2 & s_3 \end{pmatrix}.$$

The time interval is 1 year. In the fertility matrix, f_3 is the per-capita number of seeds produced by an adult plant per year. In the transition matrix T, s_1, and s_2 are the annual survival probabilities of a seed and a sapling, respectively, and p_1 and p_2 are the probabilities that a seed becomes a sapling and a sapling becomes and adult during a year. Adult survival probability is s_3.

The authors collected data from several experimental field locations and used it to estimate the model parameters for each site, with the goal of determining whether the invasive plant population was growing or decreasing at that location. For example, in one location (named Kiptogot) they estimated

$$f_3 = 46, \quad s_1 = 0.877, \quad s_2 = 0.951, \quad s_3 = 0.806, \quad p_1 = 0.900, \quad p_2 = 0.900$$

2.5. ANALYSIS OF LINEAR MATRIX EQUATIONS

which results in the projection matrix

$$P = \begin{pmatrix} 0.088 & 0 & 46 \\ 0.789 & 0.095 & 0 \\ 0 & 0.856 & 0.806 \end{pmatrix}.$$

From the formula (2.24), we find that $\mathcal{R}_0 = 194.03$ (the expected number of seeds produced per seed per lifetime). Since $\mathcal{R}_0 > 1$, the population is growing. Also, one can use a computer program to calculate the dominant eigenvalue of P with the result that $r = 3.4923$ which, being greater than 1, implies the same conclusion.

2.5.3 Sensitivity Analysis

The population growth rate r and net production number $\mathcal{R}_0$ are examples of demographic statistics that result from a matrix model for a structured population. These, and many other demographic statistics derivable from a model, depend on the entries in the projection matrix A, i.e., on vital rates such as fertility, survival, class transition probabilities, and so on. It many applications it is of interest to quantify this dependence and how the statistic changes with changes in the entries in A, and to compare the relative effect of the statistic that different entries have. For example, does a perturbation in a survival rate have a greater or lesser effect on r than does a perturbation in a fertility rate? This quantification of the effects of parameter perturbation can, for example, aid in making management decisions in how best to affect an increase in population growth of a managed resource or to affect a decrease r when wanting to control a pest. Or in other contexts, such quantifications can aid in the understanding of evolutionary dynamics in that they shed light on which model parameters lead to greater fitness and are thereby likely to be favored by natural selection.

Mathematically, sensitivity analysis deals essentially with derivatives. It concerns the derivative of the statistic of interest with respect to model parameters of interest. If a mathematical formula for the statistic is available, then this is simply a calculus problem. If a formula is not available, then other means are needed.

Example 40 *Consider the juvenile–adult Leslie model with projection matrix (2.4)*

$$A = \begin{pmatrix} 0 & f_2 \\ s_1 & s_2 \end{pmatrix}, \quad f_2 > 0, \quad 0 < s_1 \leq 1, \quad 0 \leq s_2 < 1$$

whose population growth rate is

$$r = \frac{1}{2}\left(s_2 + \sqrt{s_2^2 + 4f_2 s_1}\right).$$

The sensitivities of r to the survival rates s_1 and s_2 are the derivatives

$$\frac{dr}{ds_1} = \frac{f_2}{\sqrt{s_2^2 + 4f_2 s_1}}, \quad \frac{dr}{ds_2} = \frac{s_2 + \sqrt{s_2^2 + 4f_2 s_1}}{2\sqrt{s_2^2 + 4f_2 s_1}}. \tag{2.25}$$

Their ratio

$$\frac{dr/ds_1}{dr/s_2} = 2\frac{f_2}{s_2 + \sqrt{4f_2 s_1 + s_2^2}}$$

is easily shown (by differentiation) to be an increasing function of $f_2 > 0$ which vanishes at $f_2 = 0$ and tends to $+\infty$ as $f_2 \to +\infty$. Thus, there is a unique value of f_2, namely $f_2 = s_1 + s_2$, where the ratio equals 1 and we conclude

$$\frac{dr/ds_1}{dr/s_2} > 1 \text{ if } f_2 > s_1 + s_2,$$

$$\frac{dr/ds_1}{dr/s_2} < 1 \text{ if } f_2 < s_1 + s_2.$$

In other words, r is more sensitive to changes in s_1 (than to changes in s_2) if $f_2 > s_1 + s_2$, while the opposite is true if $f_2 < s_1 + s_2$. For example, if it was desired to increase population growth through an increase of a survival rate by a small amount, then one gets a larger increase in r by increasing juvenile survival if $f_2 > s_1 + s_2$ or by increasing adult survival if $f_2 < s_1 + s_2$.

Example 41 *For the juvenile–adult model in Example 40, we can also consider the sensitivities of the reproduction number*

$$\mathcal{R}_0 = f_2 \frac{s_1}{1 - s_2}.$$

Sensitivities with respect to s_1 and s_2 are

$$\frac{d\mathcal{R}_0}{ds_1} = f_2 \frac{1}{1 - s_2}, \quad \frac{d\mathcal{R}_0}{ds_2} = f_2 \frac{s_1}{(1 - s_2)^2}. \tag{2.26}$$

Note that their ratio is

$$\frac{d\mathcal{R}_0/ds_1}{d\mathcal{R}_0/ds_2} = \frac{1 - s_2}{s_1},$$

from which we can conclude that $\mathcal{R}_0$ is more sensitive to s_1 than to s_2 if $s_1 + s_2 < 1$. On the other hand, if $s_1 + s_2 > 1$, then the opposite is true.

Note that, for the juvenile–adult model considered in Examples 40 and 41, it is possible, for certain values of f_2, s_1 and s_2, that an increase in s_1 (or s_2) can lead to opposite effects on r and $\mathcal{R}_0$. For example, consider a population with fertility and survival rates given in the Leslie matrix

$$A = \begin{pmatrix} 0 & 2 \\ \frac{3}{4} & \frac{1}{2} \end{pmatrix}.$$

Then the sensitivities of r on the survival rates are, according to formulas (2.25),

$$\frac{dr}{ds_1} = \frac{4}{5} > \frac{dr}{ds_2} = \frac{3}{5},$$

and those of $\mathcal{R}_0$ are, according to formulas (2.26),

$$\frac{d\mathcal{R}_0}{ds_1} = 4 < \frac{d\mathcal{R}_0}{ds_2} = 6.$$

For large matrices A, one does not, in general, have a formula for the dominant eigenvalue r. How then to calculate sensitivities to changes in individual entries in A? Consider A as a function of one of its entries a_{ij} and write $A = A(a_{ij})$. Assuming A is nonnegative and irreducible, the dominant eigenvalue $r = r(a_{ij})$ is a function of a_{ij} (as well as all other entries, but we hold all

2.5. ANALYSIS OF LINEAR MATRIX EQUATIONS

other entries fixed). A positive eigenvector associated with $r(a_{ij})$ is also a function of a_{ij} and we write $\mathbf{v} = \mathbf{v}(a_{ij})$. We have

$$A(a_{ij})\mathbf{v}(a_{ij}) = r(a_{ij})\mathbf{v}(a_{ij}).$$

Dropping the explicit functional dependence notation (a_{ij}) for notational convenience, we have

$$(rI - A)\mathbf{v} = \mathbf{0}.$$

Differentiating with respect to a_{ij}, we have[4]

$$(rI - A)\frac{d\mathbf{v}}{da_{ij}} + \left(\frac{dr}{da_{ij}}I - \frac{dA}{da_{ij}}\right)\mathbf{v} = \mathbf{0}$$

and we find that $d\mathbf{v}/da_{ij}$ solves the linear algebraic equation

$$(rI - A)\mathbf{z} = \left(\frac{dA}{da_{ij}} - \frac{dr}{da_{ij}}I\right)\mathbf{v} \qquad (2.27)$$

for a vector $\mathbf{z}$. From linear algebra, this equation is solvable if and only if the right side is orthogonal to the kernel of the transpose of the matrix $r(I - A)$ or in other words to the left eigenvectors of r. Let $\mathbf{w}^T$ denote a left eigenvector of A (which is also a function of a_{ij})[5]

$$\mathbf{w}^T A = r\mathbf{w}^T$$

or, equivalently, where

$$A^T \mathbf{w} = r\mathbf{w}.$$

Since A^T is also nonnegative and irreducible (and r is its dominant eigenvalue, since A^T and A have the same eigenvalues), the Perron–Frobenius Theorem tells us that $\mathbf{w}$ is a positive vector. Returning to (2.27), we find that

$$\mathbf{w}^T\left(\frac{dA}{da_{ij}} - \frac{dr}{da_{ij}}I\right)\mathbf{v} = \mathbf{0}$$

and, as a result,

$$\frac{dr}{da_{ij}} = \frac{\mathbf{w}^T \frac{dA}{da_{ij}} \mathbf{v}}{\mathbf{w}^T \mathbf{v}}.$$

Since a_{ij} is an entry in A and all other entries are being held fixed, the matrix of derivatives dA/da_{ij} consists entirely of 0s except in the ij^{th} entry where there is a 1. Thus,

$$\frac{dA}{da_{ij}}\mathbf{v} = \begin{pmatrix} 0 \\ \vdots \\ 0 \\ v_j \\ 0 \\ \vdots \\ 0 \end{pmatrix},$$

---

[4] The product rule holds for matrix and vector products, so long as the order of multiplication is unchanged.
[5] The superscript T denotes the (complex) transpose of a vector or matrix.

where v_j (the jth component of $\mathbf{v}$) appears in the ith row. Then

$$\mathbf{w}^T \frac{dA}{da_{ij}} \mathbf{v} = w_i v_j$$

and

$$\frac{dr}{da_{ij}} = \frac{w_i v_j}{\mathbf{w}^T \mathbf{v}}, \qquad (2.28)$$

where w_i is the ith component of $\mathbf{w} = (w_i)$. The denominator is the inner product

$$\mathbf{w}^T \mathbf{v} = \sum_{n=1}^{k} w_n v_n$$

of the left and right eigenvectors (which could be chosen so that the inner product equals 1). Formula (2.28) allows the computation of r sensitivities to entries in A using eigenvectors.

The sensitivities of r to all entries in A can be gathered together into a *sensitivity matrix [46]* for r, namely

$$S = \left(\frac{dr}{da_{ij}}\right) = \frac{1}{\mathbf{w}^T \mathbf{v}} (w_i v_j)$$

which can be re-written as

$$S = \frac{1}{\mathbf{w}^T \mathbf{v}} \mathbf{w} \mathbf{v}^T. \qquad (2.29)$$

Example 42 *Consider the juvenile–adult Leslie matrix model in Example 40, whose coefficient matrix*

$$A = \begin{pmatrix} 0 & f_2 \\ s_1 & s_2 \end{pmatrix}, \quad f_2 > 0, \quad 0 < s_1 \leq 1, \quad 0 \leq s_2 < 1$$

has dominant eigenvalue

$$r = \frac{1}{2} s_2 + \frac{1}{2} \sqrt{s_2^2 + 4 f_2 s_1}$$

with an associated positive eigenvector

$$\mathbf{v} = \begin{pmatrix} \sqrt{s_2^2 + 4 f_2 s_1} - s_2 \\ 2 s_1 \end{pmatrix}.$$

A positive eigenvector of the transpose

$$A^T = \begin{pmatrix} 0 & s_1 \\ f_2 & s_2 \end{pmatrix}$$

associated with r is

$$\mathbf{w} = \begin{pmatrix} \sqrt{s_2^2 + 4 f_2 s_1} - s_2 \\ 2 f_2 \end{pmatrix}.$$

From formula (2.29), we obtain the sensitivity matrix

$$S = \begin{pmatrix} \dfrac{\left(s_2 - \sqrt{s_2^2 + 4 f_2 s_1}\right)^2}{\left(s_2 - \sqrt{s_2^2 + 4 f_2 s_1}\right)^2 + 4 f_2 s_1} & -2 s_1 \dfrac{s_2 - \sqrt{s_2^2 + 4 f_2 s_1}}{\left(s_2 - \sqrt{s_2^2 + 4 f_2 s_1}\right)^2 + 4 f_2 s_1} \\ -2 f_2 \dfrac{s_2 - \sqrt{s_2^2 + 4 f_2 s_1}}{\left(s_2 - \sqrt{s_2^2 + 4 f_2 s_1}\right)^2 + 4 f_2 s_1} & \dfrac{4 f_2 s_1}{\left(s_2 - \sqrt{s_2^2 + 4 f_2 s_1}\right)^2 + 4 f_2 s_1} \end{pmatrix}.$$

2.5. ANALYSIS OF LINEAR MATRIX EQUATIONS

Some algebraic simplification show

$$S = \begin{pmatrix} \frac{1}{2} \frac{\sqrt{s_2^2+4f_2s_1}-s_2}{\sqrt{s_2^2+4f_2s_1}} & \frac{s_1}{\sqrt{s_2^2+4f_2s_1}} \\ \frac{f_2}{\sqrt{s_2^2+4f_2s_1}} & \frac{1}{2}\frac{s_2+\sqrt{s_2^2+4f_2s_1}}{\sqrt{s_2^2+4f_2s_1}} \end{pmatrix}. \tag{2.30}$$

The survival rates s_1 and s_2 are located in the a_{21} and a_{22} positions in A (i.e., in the second row of A), and therefore the sensitivities of r with respect to s_1 and s_2 appear in the second row of S, which agrees with those calculated in Example 40.

Example 43 *Consider A in Example 42 with $f_2 = 2$, $s_1 = 3/4$ and $s_2 = 1/2$, i.e.,*

$$A = \begin{pmatrix} 0 & 2 \\ \frac{3}{4} & \frac{1}{2} \end{pmatrix}. \tag{2.31}$$

Linear algebraic calculations show that the dominant eigenvalue $r = 3/2$ has associated left and right eigenvectors

$$\mathbf{w}^T = \begin{pmatrix} 1 & 2 \end{pmatrix}, \quad \mathbf{v} = \begin{pmatrix} 4 \\ 3 \end{pmatrix}$$

from which we calculate

$$\mathbf{w}\mathbf{v}^T = \begin{pmatrix} 4 & 3 \\ 8 & 6 \end{pmatrix} \quad \text{and} \quad \mathbf{w}^T\mathbf{v} = 10$$

to obtain the sensitivity matrix

$$S = \begin{pmatrix} \frac{2}{5} & \frac{3}{10} \\ \frac{4}{5} & \frac{3}{5} \end{pmatrix} \tag{2.32}$$

for the dominant eigenvalue r. (This matrix could also be calculated, of course, using the general formula (2.30).) Thus, the sensitivity of r to f_2 is $3/10$ and to s_2 is $3/5$.

One might wonder the meaning of the sensitivity $dr/da_{11} - 2/5$ of r with respect to the zero entry in the a_{11} position of projection matrix (2.31), as given in the sensitivity matrix (2.32), since the model presumably allows no nonzero entry in this position, i.e., assumes that no 1-class individual is fertile. If the modeler wishes to restrict attention to the case when the time unit is such that no newborn can reproduce in one time unit (i.e., the unit of time in the model is a fixed maturation period required of all newborns), then this sensitivity would have no relevance. However, if the modeler is interested in a different scenario, one in which the question is: what would be the effect on population growth if some newborns matured earlier than one time unit, due say to a genetic mutation or some environmental changes? Then this sensitivity would be of interest, since it describes the effect on population growth of such a change in the maturation period. The same kind of considerations can be given to the zero entries in any matrix model based on any classification scheme.

Another important point about sensitivities is the following. A comparison of sensitivities to small changes in parameters that are different scales, such as fertilities and survival rates, could be misleading. For this reason, proportional changes in r to proportional changes in parameter are often used instead. The *elasticity* of r to changes in a_{ij} is defined by

$$e_{ij} = \frac{a_{ij}}{r}\frac{dr}{da_{ij}}$$

which can also be viewed as
$$e_{ij} = \frac{d\ln r}{d\ln a_{ij}}.$$

One can gather all elasticities into an *elasticity matrix*
$$E = \left(\frac{a_{ij}}{r}\frac{dr}{da_{ij}}\right).$$

This can be re-written in terms of the Hadamard product, which is defined for two matrices $P = (p_{ij})$ and $Q = (q_{ij})$ (of the same size) by entry-wise multiplication:
$$P \odot Q = (p_{ij} q_{ij}).$$

Then
$$E = \frac{1}{r} A \odot S. \tag{2.33}$$

Example 44 *For the Leslie matrix*
$$A = \begin{pmatrix} 0 & 2 \\ \frac{3}{4} & \frac{1}{2} \end{pmatrix},$$

we have $r = 3/2$ *and*
$$S = \begin{pmatrix} \frac{2}{5} & \frac{3}{10} \\ \frac{4}{5} & \frac{3}{5} \end{pmatrix}.$$

The elasticity matrix is given by (2.33):
$$E = \frac{1}{3/2} \begin{pmatrix} 0 & 2 \\ \frac{3}{4} & \frac{1}{2} \end{pmatrix} \odot \begin{pmatrix} \frac{2}{5} & \frac{3}{10} \\ \frac{4}{5} & \frac{3}{5} \end{pmatrix} = \begin{pmatrix} 0 & \frac{2}{5} \\ \frac{2}{5} & \frac{1}{5} \end{pmatrix}.$$

In terms of sensitivities, the largest effect (4/5) on r occurs in this example with a change in juvenile survival s_1, whereas, in terms of elasticities, there is an equal effect (2/5) with a change in adult fertility.

There is no universal rule as to which measures, sensitivities or elasticities, are best. As pointed by Caswell [45], [46], they are simply derivatives on different scales and, as a result, address different questions and, therefore, can give different answers. Their use depends on what is of interest to the user: perturbations on an additive or on a proportional scale.

The sensitivities or elasticities defined above are with respect to parameters residing in an entry of the projection matrix. Sometimes it is of interest to consider the effects of changes in a lower level parameter on which one or more of the entries in A depend. For example, trade-offs are fundamental to life history strategies [264]. In a matrix model, an increase in one entry might be associated with a decrease in another. Such trade-offs are often related to resource allocations, for example, among fertility, survival, growth, etc. A basic trade-off occurs between the allocation of resources to fertility at the expense of survival.

Example 45 *Consider the juvenile–adult Leslie matrix*
$$A = \begin{pmatrix} 0 & f_2 p \\ s_1 & s_2(1-p) \end{pmatrix}, \quad f_2 > 0, \quad 0 < s_1 \leq 1, \quad 0 \leq s_2 < 1$$

2.5. ANALYSIS OF LINEAR MATRIX EQUATIONS

into which we have introduced factor p that satisfies $0 \le p \le 1$ that measures a trade-off between fertility and adult survival. Here f_2 is the maximum possible fertility and s_2 is the maximum possible adult survival probability. The population growth rate r depends on p and one can consider the sensitivity of this dependence by calculating the derivative dr/dp.

For example, starting with the projection matrix (2.31), we have

$$A = \begin{pmatrix} 0 & 2p \\ \frac{3}{4} & \frac{1}{2}(1-p) \end{pmatrix} \qquad (2.34)$$

whose growth rate

$$r = \frac{(1-p) + \sqrt{(1-p)^2 + 24p}}{4}$$

has positive sensitivity

$$\frac{dr}{dp} = \frac{1}{4} \frac{p + 11 - \sqrt{(1-p)^2 + 24p}}{\sqrt{(1-p)^2 + 24p}} > 0 \qquad (2.35)$$

to the resource allocation to fertility. For example, if the allocation fraction is $p = 1/10$, then the sensitivity to a change in p is

$$\left.\frac{dr}{dp}\right|_{p=1/10} \approx 1.30,$$

while, if $p = 9/10$, the sensitivity is

$$\left.\frac{dr}{dp}\right|_{p=9/10} \approx 0.39.$$

The effect of a perturbation in the resource allocation fraction on the population growth rate is greater when $p = 1/10$ (and $r \approx 0.67$ and the population is decaying) than it is when $p = 9/10$ (and $r = 1.19$ and the population is growing).

If a projection matrix entry a_{ij} is a function of a parameter p, then the sensitivity of r to p can be calculated by the chain rule

$$\frac{dr}{dp} = \frac{dr}{da_{ij}} \frac{da_{ij}}{dp}.$$

If two entries, a_{ij} and a_{kl}, depend on p, then by multi-variable calculus the sensitivity of r to p is

$$\frac{dr}{dp} = \frac{dr}{da_{ij}} \frac{da_{ij}}{dp} + \frac{dr}{da_{kl}} \frac{da_{kl}}{dp}.$$

In Example 45, we have

$$\frac{da_{21}}{dp} = f_2, \quad \frac{da_{22}}{dp} = -s_2$$

and from the sensitivity matrix (2.30) that

$$\frac{dr}{da_{21}} = \frac{s_1}{\sqrt{(s_2(1-p))^2 + 4f_2 p s_1}}$$

$$\frac{dr}{da_{22}} = \frac{1}{2} \frac{s_2(1-p) + \sqrt{(s_2(1-p))^2 + 4f_2 p s_1}}{\sqrt{(s_2(1-p))^2 + 4f_2 p s_1}}.$$

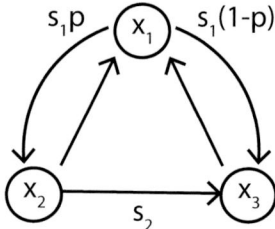

Figure 2.8: The life cycle graph for the matrix model in Example 46.

from which we calculate the sensitivity

$$\frac{dr}{dp} = \frac{dr}{da_{21}}\frac{da_{21}}{dp} + \frac{dr}{da_{22}}\frac{da_{22}}{dp}$$

$$= \frac{s_1}{\sqrt{(s_2(1-p))^2 + 4f_2ps_1}}f_2 + \frac{1}{2}\frac{s_2(1-p) + \sqrt{(s_2(1-p))^2 + 4f_2ps_1}}{\sqrt{(s_2(1-p))^2 + 4f_2ps_1}}(-s_2).$$

For the specific Example 2.34 with $f_2 = 2$, $s_1 = 3/4$, and $s_2 = /1/2$ this formula reduces to the formula (2.35) we found by calculation of the derivative of r.

In general, the sensitivity of r to a parameter p that can possibly appear in every entry in A is given by the formula

$$\frac{dr}{dp} = \sum_{i,j}\frac{dr}{da_{ij}}\frac{da_{ij}}{dp}, \tag{2.36}$$

where the sensitivities dr/da_{ij} can be calculated, in the absence of a formula for r, using eigenvectors and the formula (2.28).

Example 46 *The projection matrix*

$$A = \begin{pmatrix} 0 & f_2 & f_3 \\ s_1p & 0 & 0 \\ s_1(1-p) & s_2 & 0 \end{pmatrix}$$

describes a population structured by a juvenile class x_1 and two adults class. The class x_2 of smaller adults and the class x_3 of larger adults have fertilities f_2 and f_3. The unit of time is the juvenile maturation period. The juvenile survival rate is s_1 and p is the fraction of surviving juveniles that are smaller adults and $1-p$ is the fraction that are larger adults. Larger adults are semelparous and do not survive one unit of time while smaller adults survive with probability s_2 in which case they grow into the larger class. The matrix A is primitive if $p < 1$ (Figure 2.8).

As an example, let

$$f_2 = 0.9, \ f_3 = 1.1, \ s_1 = 0.75, \ s_2 = 0.5, \ p = 0.5.$$

In this case, the larger adults are more fertile than the smaller adults. Juveniles survive with probability $s_1 = 0.75$ and half are small adults and half are large adults at maturation. The probability

2.5. ANALYSIS OF LINEAR MATRIX EQUATIONS

a smaller adult survives and grows to a larger adult is $p = 0.5$. The dominant eigenvalue of the resulting projection matrix

$$A = \begin{pmatrix} 0 & 0.9 & 1.1 \\ 0.375 & 0 & 0 \\ 0.375 & 0.5 & 0 \end{pmatrix},$$

and corresponding left and right eigenvectors, are

$$r = 0.980\,03$$

$$\mathbf{v} = \begin{pmatrix} 0.821\,90 \\ 0.314\,49 \\ 0.474\,95 \end{pmatrix}, \qquad \mathbf{w} = \begin{pmatrix} 0.472\,30 \\ 0.704\,20 \\ 0.530\,12 \end{pmatrix}.$$

Since $r < 1$ this population is declining. Consider this question: can r be increased by a manipulation of p, the fraction of juveniles that mature as smaller adults? Would r increase by having a larger fraction of juveniles mature to larger adults (smaller p), who have the larger fertility rate? Or would it be better to have a larger fraction mature as smaller adults (larger p), in order to take advantage of a possible second reproductive event as a larger adult? We can get insight into these questions by calculating the sensitivity of r to p at these parameter values.

Formula (2.29) gives the sensitivity matrix

$$S = \begin{pmatrix} 0.427\,58 & 0.189\,19 & 0.174\,67 \\ 0.781\,40 & 0.345\,73 & 0.319\,21 \\ 0.554\,93 & 0.245\,53 & 0.226\,70 \end{pmatrix}$$

for the population grow rate r. The sensitivity of r to p at $p = 0.5$ is (cf. (2.36))

$$\left.\frac{dr}{dp}\right|_{p=0.5} = \left.\frac{dr}{da_{21}}\frac{da_{21}}{dp}\right|_{p=0.5} + \left.\frac{dr}{da_{31}}\frac{da_{31}}{dp}\right|_{p=0.5}$$
$$= 0.781\,40 \times (0.5) + 0.554\,93 \times (-0.5)$$
$$= 0.113\,24$$

is positive and hence r increases when p is increased.

We can estimate the change in r caused by a change in p by

$$\Delta r \approx \frac{dr}{dp}\Delta p$$

which, in this example, gives $\Delta r \approx 0.113\,24 \times \Delta p$. If, for example, p is increased from 0.5 to 0.8, we get $\Delta p = 0.3$ and $\Delta r \approx 0.033\,97$. This raises r from $0.980\,03$ to $0.980\,03 + 0.033\,97 = 1.014\,00$ and the population is now growing. That this approximation is fairly accurate can be seen by calculating the dominant eigenvalue $r \approx 1.014\,96$ of the projection matrix with p changed to 0.8

$$A = \begin{pmatrix} 0 & 0.9 & 1.1 \\ 0.6 & 0 & 0 \\ 0.15 & 0.5 & 0 \end{pmatrix}.$$

We have focused on the sensitivity of the population growth rate, i.e., the dominant eigenvalue, of the projection matrix A. The sensitivity of any statistic of interest derivable from A can also be of interest. For example, we looked briefly at the sensitivity of the reproduction number $\mathcal{R}_0$ in Example 41. When a formula for the statistic is available in terms of the entries of A, then sensitivity analysis is a calculus problem of finding derivatives. The books [45], [46] by Caswell contain numerous formulas for carrying out a great variety of sensitivity analyses in the absence of explicit formulas for the statistic of interest.

Exercises 2.5

1. A population starts with 100 members of age 1. Assume that each member of age 1 produces one offspring and that 2/3 survive to age 2. All members of age 2 produce three offspring and then die. Find the corresponding Leslie matrix. Find the population growth rate r and the stable age distribution of the population. Calculate the reproduction number $\mathcal{R}_0$.

2. Fefferman and Reed [136] give life-table data for red-cocked woodpeckers *(Picoides borealis)* in the southeastern United States. Based on long-term observation of marked individuals, they present the following annual values of survival and fertility s_i and f_i for females. Using this date, write the Leslie matrix A. Find the population growth rate r and the stable age distribution of the population. Calculate the reproduction number $\mathcal{R}_0$.

i	s_i	f_i	i	s_i	f_i
1	0.4010	0.0000	8	0.0255	1.5040
2	0.2943	0.1260	9	0.0170	1.5040
3	0.2829	1.1290	10	0.0114	1.5040
4	0.1290	1.5040	11	0.0076	1.5040
5	0.0860	1.5040	12	0.0051	1.5040
6	0.0574	1.5040	13	0.0000	**
7	0.0383	1.5040			

3. (a) Find the population geometric growth rate r and the reproduction number $\mathcal{R}_0$ for the projection matrix $A = F + T$ with

$$F = \begin{pmatrix} 0 & 0.25 & 0.75 \\ 0 & 0 & 0 \\ 0 & 0 & 0 \end{pmatrix}, \quad T = \begin{pmatrix} 0 & 0 & 0 \\ 0.7 & 0 & 0 \\ 0 & 0.9 & 0 \end{pmatrix}$$

Is the population growing or decaying? What is the asymptotic age distribution?

(b) Repeat (a) for the projection matrix with

$$F = \begin{pmatrix} 0 & 0.9 & 0.5 \\ 0 & 0 & 0 \\ 0 & 0 & 0 \end{pmatrix}, \quad T = \begin{pmatrix} 0.1 & 0 & 0 \\ 0.8 & 0.1 & 0 \\ 0 & 0.8 & 0.2 \end{pmatrix}$$

4. Consider Leslie model with $k = 4$ age classes and with $s_4 = 0$ (i.e., there is no survival after age class 4). Suppose the third age class is the first reproductively mature class. Suppose further that from one age class to the next survival increases by 5% and fertility drops by 10%. If newborn survival is 50%, what level of fertility f_3 at maturation will guarantee population growth?

2.5. ANALYSIS OF LINEAR MATRIX EQUATIONS

5. Consider the Usher matrix $A = F + T$ with fertilities and survival rates
$$f_1 = 0, \quad f_2 = 0, \quad f_3 = 10$$
$$s_1 = 0.5, \quad s_2 = 0.6, \quad s_3 = 0.7$$
and $p_1 = 0.5$. This represents a model with three size classes: a class of small immature newborns, a class of larger sized juveniles, and a class of large sized adults. How large should the fraction p_2 of juveniles that grow to adult size (per unit time) be in order for the population to grow?

6. Consider a $k \times k$ Usher model with T given by (2.22). Assume only two classes of newborns, taken without loss in generality to be classes $i = 1$ and 2, and assume only the largest class is fertile. Then
$$F = \begin{pmatrix} 0 & 0 & \cdots & 0 & f_{1k} \\ 0 & 0 & \cdots & 0 & f_{2k} \\ 0 & 0 & \cdots & 0 & 0 \\ \vdots & \vdots & \ddots & \vdots & \vdots \\ 0 & 0 & \cdots & 0 & 0 \end{pmatrix}, \quad f_{ik} > 0.$$
Calculate $\mathcal{R}_0$.

7. The age at which reproduction first occurs can have a major effect on whether an age-structured population will grow or decay. Consider a 3×3 Leslie matrix in which the first age class is immature
$$\begin{pmatrix} 0 & f_2 & f_3 \\ s_1 & 0 & 0 \\ 0 & s_2 & 0 \end{pmatrix}.$$
Population growth or extinction can be determined from whether $\mathcal{R}_0$ given by (2.16) is greater or less than 1. Lowering $\mathcal{R}_0$, therefore, can be viewed as a threat to population survival. By what fraction does $\mathcal{R}_0$ decrease if the second class becomes infertile (and all other parameters remain the same)? In that case, by what factor would class $i = 3$ fertility f_3 need to be increased in order to raise $\mathcal{R}_0$ back to what it was when class $i = 2$ was fertile?

8. For a standard population matrix $A = F + T$, prove that $(I - T)^{-1} = \sum_{i=0}^{\infty} T^i$. (Hint: calculate $(I - T) \sum_{i=0}^{\infty} T^i$.)

9. Calculate the sensitivity and elasticity matrices S and E for the population growth rate r of the projection matrices below.

(a) $A = \begin{pmatrix} 0.5 & 1.5 \\ 0.5 & 0.1 \end{pmatrix}$ \quad (b) $A = \begin{pmatrix} 0.2 & 0.8 \\ 0.9 & 0 \end{pmatrix}$

(c) $A = \begin{pmatrix} 0 & 0.8 & 0.5 \\ 0.9 & 0 & 0 \\ 0 & 0.8 & 0.5 \end{pmatrix}$ \quad (d) $A = \begin{pmatrix} 0.1 & 0.8 & 0.5 \\ 0.75 & 0.1 & 0 \\ 0 & 0.8 & 0.2 \end{pmatrix}$

(e) $A = \begin{pmatrix} 0 & 0 & 0.7 & 0.2 \\ 0 & 0 & 0.2 & 0.7 \\ 0.9 & 0 & 0.5 & 0 \\ 0 & 0.9 & 0 & 0.5 \end{pmatrix}$ \quad (f) $A = \begin{pmatrix} 0 & 0 & 0.5 & 0.2 \\ 0.9 & 0 & 0.2 & 0.5 \\ 0 & 0.4 & 0.5 & 0 \\ 0 & 0.4 & 0.2 & 0.1 \end{pmatrix}$

10. Consider the three size class Usher matrix (2.7) $A = F + T$ with fertility and transition matrices

$$F = \begin{pmatrix} 0 & 0 & 1 \\ 0 & 0 & 0 \\ 0 & 0 & 0 \end{pmatrix}$$

$$T = \begin{pmatrix} 0.5(1-p_1) & 0 & 0 \\ 0.5p_1 & 0.9(1-p_2) & 0 \\ 0 & 0.9p_2 & 0.95 \end{pmatrix}.$$

Here p_i is the fraction of individuals of size i that grow to the next size $i+1$ per unit time.

(a) Suppose both $p_i = 0.1$. Is the population growing or decaying? Calculate the sensitivity of r to p_1. Calculate the sensitivity of r to p_2. A change in which fraction has a greater effect on r?

(b) Suppose both $p_i = 0.25$. Is the population growing or decaying? Calculate the sensitivity of r to p_1. Calculate the sensitivity of r to p_2. A change in which fraction has a greater effect on r?

11. (a) Calculate $\mathcal{R}_0$ for the Usher matrix in Problem 10. Use your answers to calculate formulas for the sensitivities of $\mathcal{R}_0$ to p_1 and to p_2.

(b) Suppose both $p_i = 0.1$. Is the population growing or decaying? Calculate the sensitivity of $\mathcal{R}_0$ to p_1 using your formula in (a). Calculate the sensitivity of $\mathcal{R}_0$ to p_2 using your formula in (a). A change in which fraction has a greater effect on $\mathcal{R}_0$?

12. Calculate the elasticity matrix E for the projection matrix (2.4) of the juvenile–adult model considered in Example 40. Show that $e_{12} = e_{21}$, i.e., that the elasticity of r with respect to adult fertility and juvenile survival are equal.

13. (a) The simplest type of genetic inheritance of traits in animals occurs when a certain trait is determined by a specific pair of genes, each of which may be two types, G and g. An individual may have a GG combination, a Gg combination (which is the same as gG), or a gg combination. An individual with GG genes is said to be dominant; a gg recessive, a hybrid has Gg genes. In each generation, there are three possible states $a_1 = GG$, $a_2 = Gg$, and $a_3 = gg$. Let $p_i(t)$ represent the probability that state a_i occurs in the tth generation and let p_{ij} be the probability that a_i occurs in the $(t+1)^{st}$ generation given that a_j occurred in the tth generation. Let us consider a process of continued mating. We begin with an individual of genetic character GG and mate it with a hybrid Gg. Then continuing to mate the offspring with a hybrid individual. The difference equation that models this system is given by

$$p_1(t+1) = p_{11}p_1(t) + p_{12}p_2(t) + p_{13}p_3(t)$$
$$p_2(t+1) = p_{21}p_1(t) + p_{22}p_2(t) + p_{23}p_3(t)$$
$$p_3(t+1) = p_{31}p_1(t) + p_{32}p_2(t) + p_{33}p_3(t)$$

or

$$p(t+1) = Ap_t \qquad (2.37)$$

2.5. ANALYSIS OF LINEAR MATRIX EQUATIONS

with
$$A = \begin{pmatrix} p_{11} & p_{12} & p_{13} \\ p_{21} & p_{22} & p_{23} \\ p_{31} & p_{32} & p_{33} \end{pmatrix}.$$

Now p_{11} is the probability of producing an offspring GG by mating GG and Gg. Clearly, the offspring received a G gene from parent GG with probability 1 and the other G from parent Gg with probability $1/2$. By the multiplication principle, $p_{11} = 1 \times 1/2 = 1/2$.

(i) Calculate $p_{12}, p_{13}, p_{21}, p_{21}, p_{22}, p_{23}, p_{31}, p_{32}, p_{33}$

(ii) Calculate the dominant eigenvalue and $\lim_{t \to \infty} p_t$.

(b) Let us modify the model in (a) by first mating a recessive individual (gg) with a dominant individual (GG). Then, continuing to mate the offspring with a dominant individual. Find the matrix A. Find the dominant eigenvalue and calculate $\lim_{t \to \infty} p_t$.

14. In a study of the effects that the famous Deepwater Horizon Oil Spill in 2010 had on the sperm whale populations in the Gulf of Mexico, the authors in [4] and [48] utilized a matrix model with demographic vector

$$\begin{pmatrix} x_1 \\ x_2 \\ x_3 \\ x_4 \\ x_5 \end{pmatrix} = \begin{pmatrix} \text{calves} \\ \text{juveniles} \\ \text{mature females} \\ \text{mothers} \\ \text{post breeding females} \end{pmatrix}$$

and projection matrix $A = F + T$, where

$$F = \begin{pmatrix} 0 & 0 & 0.1250 & 0 & 0 \\ 0 & 0 & 0 & 0 & 0 \\ 0 & 0 & 0 & 0 & 0 \\ 0 & 0 & 0 & 0 & 0 \\ 0 & 0 & 0 & 0 & 0 \end{pmatrix}$$

$$T = \begin{pmatrix} 0.4778 & 0 & 0 & 0 & 0 \\ 0.4292 & 0.8339 & 0 & 0 & 0 \\ 0 & 0.1085 & 0.7249 & 0 & 0.4810 \\ 0 & 0 & 0.2528 & 0.4967 & 0 \\ 0 & 0 & 0 & 0.4810 & 0.4967 \end{pmatrix}.$$

The use of a computer program to carry out matrix calculations will be necessary for parts of this problem.

(a) Draw the associated life cycle graph.
(b) Show that A is irreducible.
(c) Show that A is primitive.
(d) Calculate the dominant (Perron) eigenvalue λ_1. Is the population growing or declining?
(e) Calculate the positive eigenvector $\mathbf{v}_1$ associated with λ_1 that has norm $\|\mathbf{v}_1\| = 1$. In this normalized class distribution, the highest proportion of individuals and the lowest proportion of individuals are in which classes?

(f) Calculate $\mathcal{R}_0$.

(g) Calculate the sensitivity matrix. To which entry in A is λ_1 most sensitive?

(h) Calculate the elasticity matrix. To which entry in A is λ_1 most elastic?

Chapter 3
Linear and Nonlinear Systems

3.1 Introduction

In Chapter 1, we studied both linear and nonlinear single-species population models represented by one-dimensional difference equations. In Chapter 2, we studied linear age- or size-structured population models represented by systems of linear difference equations.

In this chapter, we extend our study to nonlinear population models represented by nonlinear systems of difference equations. Examples of such models include multi-species competition models, predator–prey models, infectious disease models, and structured models. For competition models, we consider the Leslie–Gower model (an extension of the discrete logistic (Beverton–Holt) model), the Ricker competition model, and a quadratic competition model. For predator–prey or host–parasitoid models, we study the Nicholson–Bailey model and the Beddington model. For structured populations, we look at sample competition models between two-stage species and between three-stage species.

Stability analysis will be developed using analytic and graphic methods. The the graphic method utilizes the phase space diagrams, an extension of the cobweb diagrams in one dimension. The analytic methods consist of the linearization principle, comparison method, theory of monotone maps, theory of triangular maps, theory of critical curves, and Liapunov functions.

3.2 Linear Systems

3.2.1 Basic Theory

Consider the n-dimensional linear systems

$$\mathbf{x}(t+1) = A\mathbf{x}(t), \qquad (3.1)$$

where $A_{n \times n} = (a_{ij})_{n \times n}$ is an $n \times n$ matrix and $\mathbf{x} = (x_1, x_2, \ldots, x_n)^T \in \mathbb{R}^n$. Note that we have encountered a special class of these systems, the structured population models, in Chapter 2.

We remind the reader that the solution of Equation (3.1) is given by

$$\mathbf{x}(t) = A^t \mathbf{x}_0. \qquad (3.2)$$

There are several methods to compute A^t and we refer the reader to the paper by Elaydi and Harris [121] Here we will present another method to find the solution $\mathbf{x}(t)$. If λ is an eigenvalue of the matrix A and $\mathbf{v}$ is the corresponding eigenvector, then $A\mathbf{v} = \lambda \mathbf{v}$. We claim that $\mathbf{x}(t) = \lambda^t \mathbf{v}$ is a solution of Equation (3.2). To show this, note that

$$\mathbf{x}(t+1) = \lambda^{t+1}\mathbf{v} = \lambda^t(\lambda \mathbf{v}) = \lambda^t A \mathbf{v} = A(\lambda^t \mathbf{x}(t)) = A\mathbf{x}(t).$$

Recall from Section 2.3 that if $\lambda_1, \lambda_2, \ldots, \lambda_n$ are the eigenvalues of A and there are corresponding n linearly independent eigenvectors $\mathbf{v}_1, \mathbf{v}_2, \ldots, \mathbf{v}_n$, then a general solution of Equation (3.1) is given by

$$\mathbf{x}(t) = c_1 \lambda_1^t \mathbf{v}_1 + c_2 \lambda_2^t \mathbf{v}_2 + \ldots + c_n \lambda_n^t \mathbf{v}_n. \tag{3.3}$$

Note that, if all eigenvalues are distinct, then the corresponding eigenvectors are linearly independent. However, if some eigenvalues are repeated, then we may not have n linearly independent eigenvectors. For example, the matrix

$$\begin{pmatrix} 2 & 1 \\ 0 & 2 \end{pmatrix}$$

has a repeated eigenvalue $\lambda_1 = \lambda_2 = 2$ with only one associated eigenvector $\mathbf{v} = (1,0)^T$. By contrast, the matrix

$$\begin{pmatrix} 2 & 0 \\ 0 & 2 \end{pmatrix}$$

has a repeated eigenvalue $\lambda_1 = \lambda_2 = 2$ with two corresponding eigenvectors $\mathbf{v}_1 = (1,0)^T$, $\mathbf{v}_2 = (0,1)^T$, respectively. In this latter case, the solution follows from formula (3.3),

$$\mathbf{x}(t) = c_1 \lambda^t \mathbf{v}_1 + c_2 \lambda^t \mathbf{v}_2.$$

However, in the first case, the solution is given by

$$\mathbf{x}(t) = c_1 \lambda^t \mathbf{v} + c_2 (t\lambda^{t-1}\mathbf{v}_1 + \lambda^t \mathbf{w}),$$

where $\mathbf{w}$ is given by the formula

$$(A - \lambda I)\mathbf{w} = \mathbf{v}. \tag{3.4}$$

$\mathbf{w}$ is called a generalized eigenvector. To compute $\mathbf{w} = (x_1, x_2)^T$, we solve Equation (3.4) which gives

$$\begin{pmatrix} 0 & 1 \\ 0 & 0 \end{pmatrix} \begin{pmatrix} x_1 \\ x_2 \end{pmatrix} = \begin{pmatrix} 1 \\ 0 \end{pmatrix}.$$

This yields $x_2 = 1$, and x_1 free. Hence, $\mathbf{w} = (0,1)^T$. Therefore,

$$\mathbf{x}(t) = c_1 2^t \begin{pmatrix} 1 \\ 0 \end{pmatrix} + c_2 \left(t 2^{t-1} \begin{pmatrix} 1 \\ 0 \end{pmatrix} + 2^t \begin{pmatrix} 0 \\ 1 \end{pmatrix} \right)$$

$$\mathbf{x}(t) = A^t \mathbf{x}_0. \tag{3.5}$$

For an n-dimensional linear system, if $\lambda_1 = \lambda_2 = \cdots = \lambda_k = \lambda$ and there is only one corresponding eigenvector $\mathbf{v}_1 = \mathbf{v}_2 = \cdots = v_k = v$, the solution of Equation (3.1) is given by

$$\mathbf{x}(t) = c_1 \lambda^t \mathbf{v} + \hat{c}_1 \left(\binom{t}{k-1} \lambda^{t-k+1} \mathbf{v} + \binom{t}{k-2} \lambda^{t-k+2} \mathbf{w}_1 + \cdots + \lambda^t \mathbf{w}_{k-1} \right) + \sum_{i=k+1}^{n} c_i \lambda_i^t \mathbf{v}_i,$$

3.2. LINEAR SYSTEMS

where $A\mathbf{w}_1 = \lambda \mathbf{v}, A\mathbf{w}_2 = \lambda \mathbf{w}_1, \ldots, A\mathbf{w}_{k-1} = \lambda \mathbf{w}_{k-2}$.[1]

Next, we will discuss the case when we have complex eigenvalues. Let $\lambda = \alpha + i\beta$ be an eigenvalue of the matrix A. Then the corresponding eigenvector $\mathbf{v}$ is also a complex vector which may be written as $\mathbf{v} = \mathbf{u} + i\mathbf{w}$. Then a solution of system (3.1) can be written as

$$\mathbf{x}(t) = \lambda^t \mathbf{v} = (\alpha + i\beta)^t (\mathbf{u} + i\mathbf{w}).$$

Note that one may write $\alpha + i\beta = re^{i\theta}$, where $|\lambda| = \sqrt{\alpha^2 + \beta^2}$, and $\theta = \tan^{-1}(\beta/\alpha)$. Moreover, $\lambda^t = r^t e^{it\theta} = r^t(\cos(t\theta) + i\sin(t\theta))$. Thus,

$$\mathbf{x}(t) = r^t (\cos(t\theta) + i\sin(t\theta)) (\mathbf{u} + i\mathbf{w})$$
$$= r^t [\cos(t\theta)\mathbf{u} - \sin(t\theta)\mathbf{w}] + ir^t [\sin(t\theta)\mathbf{u} + \cos(t\theta)\mathbf{w}].$$

One may show that both the real part and the imaginary part of $\mathbf{x}(t)$ above are solutions of Equation (3.1). Hence, the general solution of Equation (3.1) if A is a 2×2 matrix is given by

$$\mathbf{x} = c_1 r^t (\cos(t\theta)\mathbf{u} - \sin(t\theta)\mathbf{w}) + c_2 r^t (\sin(t\theta)\mathbf{u} + \cos(t\theta)\mathbf{w}). \tag{3.6}$$

Example 47 *Consider the system*

$$\mathbf{x}(t+1) = A\mathbf{x}(t), \tag{3.7}$$

where

$$A = \begin{pmatrix} 1 & 1 \\ -1 & 1 \end{pmatrix}.$$

The eigenvalues are the complex conjugates $\lambda = 1 \pm i$ and the corresponding eigenvector is $(1, i)^T = \mathbf{u} + i\mathbf{w} = (1, 0)^T + i(0, 1)^T$ with $\mathbf{u} = (1, 0)^T$, $\mathbf{w} = (0, 1)^T$. Thus, the solution of the system (3.7) is given by

$$\mathbf{x} = c_1 r^t (\cos(t\theta)\mathbf{u} - \sin(t\theta)\mathbf{w}) + c_1 r^t (\sin(t\theta)\mathbf{u} + \cos(t\theta)\mathbf{w}).$$

Note that $r = \sqrt{\mathbf{w}}$, and $\theta = \tan^{-1}(1) = \pi/r$. Then

$$\mathbf{x}(t) = c_1 2^{\frac{t}{2}} \left(\cos\left(\frac{\pi t}{4}\right)\mathbf{u} - \sin\left(\frac{\pi t}{4}\right)\mathbf{w} \right) + c_2 2^{\frac{t}{2}} \left(\sin\left(\frac{\pi t}{4}\right)\mathbf{u} + \cos\left(\frac{\pi t}{4}\right)\mathbf{w} \right)$$
$$= 2^{\frac{t}{2}} \left[\left(c_1 \cos\left(\frac{\pi t}{4}\right) + c_2 \sin\left(\frac{\pi t}{4}\right) \right)\mathbf{u} + \left(-c_1 \sin\left(\frac{t\pi t}{4}\right) + c_2 \cos\left(\frac{\pi t}{4}\right) \right)\mathbf{w} \right]$$
$$= 2^{\frac{t}{2}} \begin{pmatrix} c_1 \cos\left(\frac{\pi t}{4}\right) + c_2 \sin\left(\frac{\pi t}{4}\right) \\ -c_1 \sin\left(\frac{\pi t}{4}\right) + c_2 \cos\left(\frac{\pi t}{4}\right) \end{pmatrix}.$$

The spectral radius of $\rho(A)$ of the matrix A is defined as

$$\rho(A) = \max\{|\lambda_i| : i = 1, 2, \ldots, n\}.$$

---

[1] $\binom{t}{r} = \frac{t(t-1)\cdots(t-r+1)}{r!}$

Without loss of generality, let us assume that $\rho(A) = |\lambda_1|$. Then

$$\mathbf{x}(t) = \lambda_1^t \left[c_1 \mathbf{v}_1 + c_2 \left(\frac{\lambda_2}{\lambda_1}\right)^t \mathbf{v}_2 + \cdots + c_n \left(\frac{\lambda_n}{\lambda_1}\right)^t \mathbf{v}_n \right]$$

$$\lim_{t \to \infty} \mathbf{x}(t) = \lim_{t \to \infty} \lambda_1^t c_1 \mathbf{v}_1.$$

Hence, if $\rho(A) = |\lambda_1| < 1$, $\lim_{t \to \infty} \mathbf{x}(t) = 0$ and if $\rho(A) = |\lambda_1| > 1$, $\lim_{t \to \infty} \mathbf{x}(t) = \infty$.

It is important to note that, unlike nonlinear systems, for linear systems, local asymptotic stability implies global asymptotic stability.

To summarize, we have the following result.

Theorem 19 *The following statements hold true.*

(i) If $\rho(A) < 1$, then the zero solution of Equation (3.1) is globally asymptotically stable,

(ii) If $\rho(A) > 1$, then the zero solution of Equation (3.1) is unstable,

(iii) If $\rho(A) = 1$, then the zero solution of Equation (3.1) may be stable or unstable.

3.2.2 Variation of Constants Formula

Consider the nonautonomous n-dimensional linear system

$$\mathbf{x}(t+1) = A(t)\mathbf{x}(t), \tag{3.8}$$

where $A(t)_{n \times n} = (a_{ij})_{n \times n}$ is an $n \times n$ matrix and $\mathbf{x} = (x_1, x_2, \ldots, x_n)^T \in \mathbb{R}^n$. Note that we have encountered a special class of these systems, the structured population models, in Chapter 2.

If $\mathbf{x}(0) = \mathbf{x}_0$, then, by iteration, one may show that the solution of Equation (3.8) is given by

$$\mathbf{x}(t) = \prod_{j=t+1}^{t} A(j))\mathbf{x}_0. \tag{3.9}$$

Next, consider the n-dimensional linear system with migration/immigration

$$\mathbf{x}(t+1) = A(t)\mathbf{x}(t) + \mathbf{m}(t). \tag{3.10}$$

Mimicking what we did in Chapter 1, Formula (1.17), we obtain the following variation of constants formula for system (3.10)

$$\mathbf{x}(t) = \mathbf{x}_0 \prod_{i=0}^{t-1} A(i) + \sum_{i=0}^{t-1} (\prod_{j=i+1}^{t-1} A(j))\mathbf{m}(i), \tag{3.11}$$

where

$$\prod_{j=t+1}^{t} A(j) = I, \tag{3.12}$$

where I is the identity matrix.

3.2.3 Phase Space Analysis

One of the main tools in the study of the qualitative behavior of solutions of linear and nonlinear systems of difference equations is the phase space diagram. In this diagram, we draw the orbits (solutions) of as many points in the domain as possible. In the last section, we have investigated the behavior of solutions when $\rho(A) < 1$ and $\rho(A) > 1$. In this section, we will get a little deeper in our analysis of the behavior of all solutions. Assuming $\rho(A) > 1$, we arrange the eigenvalues of A into two sets $\mathcal{S} = \{\lambda_1, \lambda_2, \ldots, \lambda_r\}$, $G = \{\lambda_{r+1}, \lambda_{r+2}, \ldots, \lambda_n\}$ with $|\lambda_i| < 1$ if $\lambda_i \in \mathcal{S}$ and $|\lambda_j| > 1$ if $\lambda_j \in G$. Let E^s be the eigenspace spanned by the eigenvectors and the generalized eigenvectors of the eigenvalues in $\mathcal{S}$, and let E^u be the eigenspace spanned by the eigenvectors and the generalized eigenvectors of the eigenvalue in G. The $\mathbb{R}^n = E^{\mathcal{S}} \oplus E^u$. The sets E^s and E^s are called the stable and the unstable subspaces of the zero solution of Equation (3.1). Note that both E^s and E^u are invariant subspaces, that is, if $\mathbf{x}_0 \in \mathcal{S}$, then $\mathbf{x}(t) \in \mathcal{S}$, for all t, and similarly for E^u.

For if $\mathbf{x}_0 \in E^s$, then $\mathbf{x}_0 = c_1 \mathbf{v}_1 + c_2 \mathbf{v}_2 + \ldots + c_r \mathbf{v}_r$. Then

$$\begin{aligned}\mathbf{x}(t) = A^t \mathbf{x}_0 &= A^t(c_1 \mathbf{v}_1 + c_2 \mathbf{v}_2 + \ldots + c_r \mathbf{v}_r) \\ &= c_1 A^t \mathbf{v}_1 + c_2 A^t \mathbf{v}_2 + \ldots + c_r A^t \mathbf{v}_r \\ &= c_1 \lambda_1^t \mathbf{v}_1 + c_2 \lambda_2^t \mathbf{v}_2 + \ldots + c_r \lambda_r^t \mathbf{v}_r \in E^s.\end{aligned}$$

Similarly, one may show that E^u is invariant. Moreover, if $\mathbf{x}_0 \in E^s$, then $\lim_{t \to \infty} \mathbf{x}(t) = 0$, and all orbits (solutions) that start in E^s will converge to the zero solution. And if $\mathbf{x}_0 \in E^u$, then $\lim_{t \to \infty} \mathbf{x}(t) = \infty$. The fixed point is called a saddle if $E^u \neq \emptyset$ and $E^s \neq \emptyset$.

Next, we will explore the phase diagrams of two-dimensional linear system. There are three generic cases to consider.

(I) Distinct real eigenvalues, where

$$A = \begin{pmatrix} \lambda_1 & 0 \\ 0 & \lambda_2 \end{pmatrix}.$$

Here, we have several cases: (a) $|\lambda_1| > |\lambda_2| > 1$: The zero solution is an unstable node, (b) $|\lambda_1| < |\lambda_2| < 1$: The zero solution is a stable node (asymptotically stable), (c) $|\lambda_1| < 1$ and $|\lambda_2| > 1$: The zero solution is a saddle. Moreover, the stable subspace is the x-axis and the unstable subspace is the y-axis (Figure 3.1).

(II) Repeated real eigenvalues

There are two cases to consider.

(a) If $\lambda_1 = \lambda_2 = \lambda \neq 1$ and the eigenspace has a basis of two linearly independent eigenvectors $\mathbf{v}_1$ and $\mathbf{v}_2$. Then the solution is given $\mathbf{x}(t) = c_1 \lambda^t \mathbf{v}_1 + c_2 \lambda^t \mathbf{v}_2$. The phase space diagram is similar to type (I).

(b) If $\lambda_1 = \lambda_2 = \lambda \neq 1$, but the eigenspace is spanned by one eigenvector $\mathbf{v}_1$. In this case, we have only a stable or unstable subspace. If $|\lambda| < 1$, then $\lim_{t \to \infty} \mathbf{x}(t) = \lim c_1 \lambda^t \vec{\mathbf{v}_1} + c_2(t \lambda^{t-1} \vec{\mathbf{v}_1} + \lambda^t \vec{\mathbf{v}_2})$. In this case, the zero solution is called a degenerate sink. If $|\lambda| > 1$, then $\lim_{t \to \infty} |\mathbf{x}(t)| = \infty$. In this case, the zero solution is called a degenerate source (Figure 3.2).

(c) If $\lambda_1 = \lambda_2 = 1$, then all points are equilibrium points.

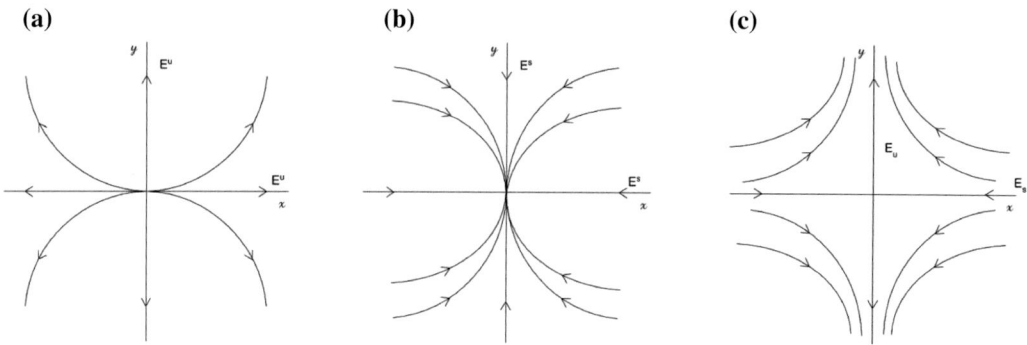

Figure 3.1: (a) unstable node: $\lambda_1 > \lambda_2 > 1$. (b) stable node: $-1 < \lambda_1 < \lambda_2 < 1$. (c) saddle: $-1 < \lambda_1 < 1, \lambda_2 < 1$.

Figure 3.2: (a) a degenerate sink: $\lambda_1 = \lambda_2 = \lambda, \lambda < 1$. (b) a degenerate source: $\lambda_1 = \lambda_2 = \lambda, \lambda > 1$.

(d) If $\lambda_1 = 1$ and $\lambda_2 < 1$, then all points on the x-axis are equilibrium points. $\lim_{t \to \infty}(x(t), y(t))^T = (x_0, 0)^T$ for all $(x_0, y_0)^T \in \mathbb{R}^2$ and if $\lambda_2 > 1$, then $\lim_{t \to \infty}(x(t), y(t))^T = (x_0, \infty)^T$.

(e) If $\lambda_1 = \lambda_2 = -1$, then $(x_0, y_0)^T \to (-x_0, -y_0)^T \to (x_0, y_0)^T$ and every point except the origin is periodic of period 2.

(III) Complex Eigenvalues

The canonical form of the matrix with complex conjugate eigenvalues $\lambda_1 = \alpha_1 - i\beta$ and $\lambda_1 = \alpha_1 + i\beta$ is given by

$$A = \begin{pmatrix} \alpha & \beta \\ -\beta & \alpha \end{pmatrix}.$$

The corresponding eigenvectors are $\mathbf{v}_1 = (1, -i)^T$, $\mathbf{v}_2 = (1, i)^T$. Let $|\lambda| = |\lambda_1| = |\lambda_2| = \sqrt{\alpha^2 + \beta^2}$ and $\theta = \tan^{-1}(\beta/\alpha)$, then $\lambda_1 = |\lambda|e^{-i\theta}$, $\lambda_2 = |\lambda|e^{i\theta}$. Hence,

3.2. LINEAR SYSTEMS

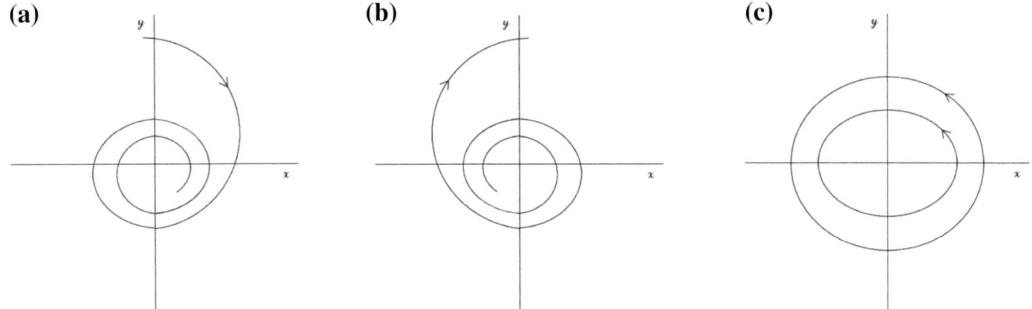

Figure 3.3: (a) stable focus: $|\lambda| < 1$. (b) source: $|\lambda| > 1$. (c) center: $|\lambda| = 1$.

$$\lambda_1^t = |\lambda|^t e^{-it\theta} = |\lambda|^t [\cos(t\theta) - i\sin(t\theta)]$$
$$\lambda_2^t = |\lambda|^t e^{it\theta} = |\lambda|^t [\cos(t\theta) + i\sin(t\theta)].$$

and
$$\mathbf{x}(t) = c_1 |\lambda|^t (\cos(t\theta) - i\sin(t\theta))\mathbf{v}_1 + c_2 |\lambda|^t (\cos(t\theta) + i\sin(t\theta)))]\mathbf{v}_2.$$

Writing $k_1 = c_1 + c_2$ and $k_2 = i(c_2 - c_1)$, then

$$\mathbf{x}(t) = |\lambda|^t \begin{pmatrix} k_1 \cos(t\theta) + k_2 \sin(t\theta) \\ k_2 \cos(t\theta) - k_1 \sin(t\theta) \end{pmatrix}. \tag{3.13}$$

There are three cases here (Figure 3.3)

(a) If $|\lambda| < 1$, then the zero solution is a stable focus.

(b) If $|\lambda| > 1$, then the zero solution is a source.

(c) If $|\lambda| = 1$, then the zero solution is a center.

One of the most effective methods to check that $\rho(A) < 1$, that is all of the eigenvalues of A are inside the unit-disk in the complex plane, is the Jury test [179]. In the sequel, we are going to describe this test. We begin with recalling that the eigenvalues of A are the zeros of the characteristic polynomial of A given by

$$p(\lambda) = \det(A - \lambda I) = a_0 \lambda^n + a_1 \lambda^{n-1} + a_2 \lambda^{n-2} + \ldots + a_{n-1}\lambda + a_n, a_0 > 0. \tag{3.14}$$

The Jury table is constructed as follows.

The first row consists of the coefficients of $p(\lambda)$ arranged in ascending order of powers of λ. The second row is the reverse order of the first row. All even-numbered rows are the reverse of the preceding odd-numbered rows. The elements in the third row are defined as follows:

$$b_i = \det \begin{pmatrix} a_n & a_{n-i-1} \\ a_0 & a_{i+1} \end{pmatrix}, i = 0, 1, \ldots, n-1.$$

Row	λ^0	λ^1	λ^2	$\cdots$	λ^k	$\cdots$	λ^{n-2}	λ^{n-1}	λ^n
1	a_n	a_{n-1}	a_{n-2}	$\cdots$	a_{n-k}	$\cdots$	a_2	a_1	a_0
2	a_0	a_1	a_2	$\cdots$	a_k	$\cdots$	a_{n-2}	a_{n-1}	a_n
3	b_{n-1}	b_{n-2}	b_{n-3}	$\cdots$	b_{n-k-1}	$\cdots$	b_1	b_0	
4	b_0	b_1	b_2	$\cdots$	b_k	$\cdots$	b_{n-2}	b_{n-1}	
5	c_{n-2}	c_{n-3}	c_{n-4}	$\cdots$	c_{n-k-2}	$\cdots$	c_0		
6	c_0	c_1	c_2	$\cdots$	c_k	$\cdots$	c_{n-2}		
$\vdots$									
$2n-5$	p_3	p_2	p_1	p_0					
$2n-4$	p_0	p_1	p_2	p_3					
$2n-3$	q_2	q_1	q_0						

The fifth row is defined as follows.
$$c_{ii} = \det\begin{pmatrix} b_{n-1} & b_{n-i-1} \\ b_0 & b_{i+1} \end{pmatrix}, i = 0, 1, \ldots, n-2.$$

The pattern continues until finally, we get only three terms in the last row, row $2k-3$
$$q_i = \det\begin{pmatrix} p_3 & p_{2-i} \\ p_0 & p_{i+1} \end{pmatrix}, i = 0, 1, \ldots, n-1.$$

Theorem 20 (Jury [179]) *The roots of the characteristic polynomial (3.14) all lie inside the unit disk if and only if the following conditions hold:*

(i) $p(1) > 0$ and $(-1)^n p(1) > 0$

(ii) $|a_n| < a_0$

(iii) $\begin{aligned} |b_{n-1}| &> |b_0| \\ |c_{n-2}| &> |c_0| \\ &\vdots \\ |q_2| &> |q_0| \end{aligned}$

Example 48 *Apply Jury's test to find conditions under which the spectral radius of a 2×2 matrix $\rho(A) < 1$.*

Let us start with a 2×2 matrix
$$A = \begin{pmatrix} a_{11} & a_{12} \\ a_{21} & a_{22} \end{pmatrix}.$$

The characteristic equation of A is given by
$$\begin{aligned} p(\lambda) &= \det(A - \lambda I) = \lambda^2 - (a_{11} + a_{22})\lambda + (a_{11}a_{22} - a_{12}a_{21}) \\ p(\lambda) &= \lambda^2 - \text{tr}\,(A)\lambda + \det A, \end{aligned} \qquad (3.15)$$

where $\text{tr}\,(A)$ = trace of $A = a_{11} + a_{22}$. Here $a_0 = 1$, $a_1 = -\text{tr}\,(A)$, $a_2 = \det A$. The Jury table is only one row.
$$a_2 \quad a_1 \quad a_0.$$

Hence, the necessary and sufficient conditions for $\rho(A) < 1$ is given by

3.2. LINEAR SYSTEMS

(i) $p(1) = 1 - \text{tr}(A) + \det(A) > 0$ *or* $\det(A) > \text{tr}(A) - 1$

$(-1)^2 p(-1) = 1 + \text{tr}\, A + \det A > 0$ *or* $\det A > -\text{tr}(A) - 1$

(ii) $\det A < 1$

Combining the three conditions we have

$$|\text{tr}\, A| < 1 + \det A < 2. \qquad (3.16)$$

Figures 3.4(a), 3.4(b), 3.4(c) shows the stability regions of the zero solution, where $\rho(A) < 1$ for a 2×2 matrix A.

Remark 3 *Note that, if*

- *if* $\text{tr}\, A = 1 + \det A$, *then the eigenvalues of A are $\lambda_1 = 1$ and $\lambda_2 = \det A$.*
- *if* $\text{tr}\, A = -1 - \det A$, *then the eigenvalues of A are $\lambda_1 = \det A$ and $\lambda_2 = 1$.*

This may be seen in Figures 3.4(a), 3.4(b).

Example 49 *Apply Jury's test to find conditions under which the spectral radius of a 3×3 matrix $\rho(A) < 1$.*

The characteristic polynomial of the matrix

$$A = \begin{pmatrix} a_{11} & a_{12} & a_{13} \\ a_{21} & a_{22} & a_{23} \\ a_{31} & a_{32} & a_{33} \end{pmatrix}$$

is given by

$$p(\lambda) = \det(A - \lambda I) = \lambda^3 - \text{tr}(A)\lambda^2 + \left(\sum_{i=1}^{3} M_{ii}\right)\lambda - \det(A), \qquad (3.17)$$

where $\text{tr}(A) = a_{11} + a_{22} + a_{33}$, $\det(A) =$ determinant of A, and M_{ii}, $i = 1, 2, 3$, are the principle minors of A defined as the determinants of the following submatrices

$$M_{11} = \det\begin{pmatrix} a_{22} & a_{23} \\ a_{32} & a_{33} \end{pmatrix}, \quad M_{22} = \det\begin{pmatrix} a_{11} & a_{13} \\ a_{31} & a_{33} \end{pmatrix}, \quad M_{33} = \det\begin{pmatrix} a_{11} & a_{12} \\ a_{21} & a_{22} \end{pmatrix}.$$

Here

$$a_0 = 1, \quad a_1 = -\text{tr}(A), \quad a_2 = \sum_{i=1}^{3} M_i, \quad a_3 = -\det(A).$$

The Jury table is given below.

$$\begin{array}{cccc} a_3 & a_2 & a_1 & a_0 \\ a_0 & a_1 & a_2 & a_3, \\ b_2 & b_1 & b_0 & \end{array}$$

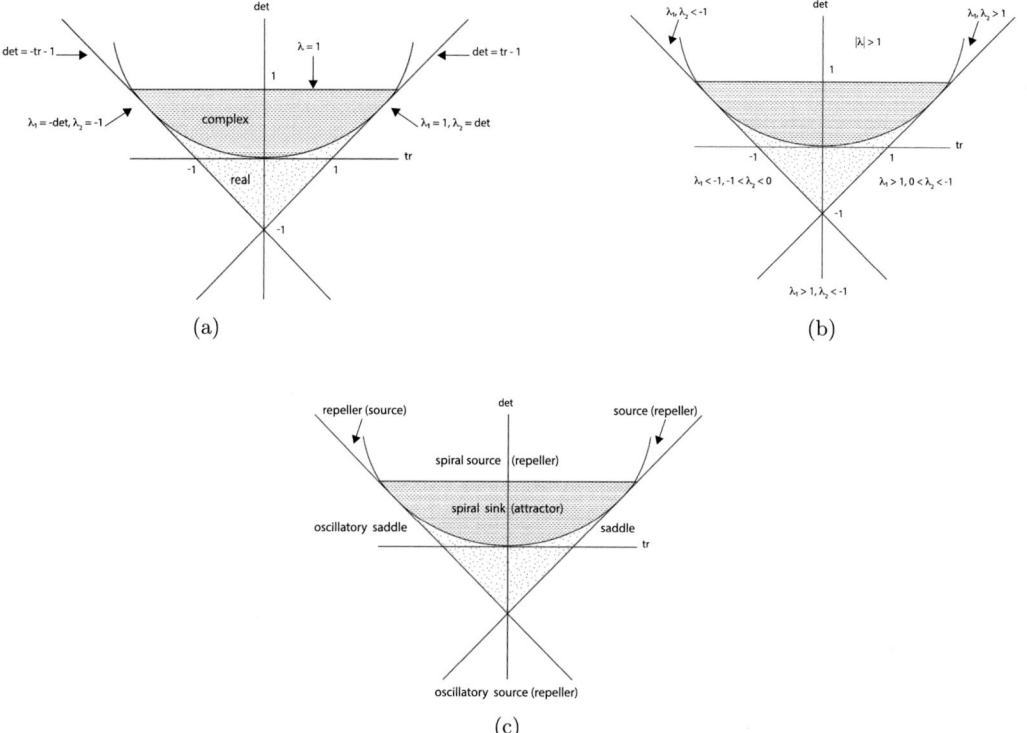

Figure 3.4: The shaded area is the stability region. Complex eigenvalues are inside the parabola $\det = \frac{1}{4}\operatorname{tr}^2$ and real eigenvalues are outside the parabola $\det = \frac{1}{4}\operatorname{tr}^2$. Note that, if $\operatorname{tr} A = 1 + \det A$, then the eigenvalues of A are $\lambda_1 = 1$ and $\lambda_2 = \det A$. On the other hand, if $\operatorname{tr} A = -1 - \det A$, then the eigenvalues of A are $\lambda_1 = -\det A$ and $\lambda_2 = -1$.

where

$$b_2 = \det \begin{pmatrix} a_3 & a_0 \\ a_0 & a_3 \end{pmatrix} = a_3^2 - a_0^2 = (\det A)^2 - 1$$

$$b_1 = \det \begin{pmatrix} a_3 & a_1 \\ a_0 & a_2 \end{pmatrix} = a_3 a_2 - a_0 a_1 = -\det(A) \sum_{i=1}^{3} M_{ii} + \text{tr}(A)$$

$$b_0 = \det \begin{pmatrix} a_3 & a_2 \\ a_0 & a_1 \end{pmatrix} = a_3 a_1 - a_0 a_2 = \det(A) \text{tr}(A) - \sum_{i=1}^{3} M_{ii}.$$

Hence, the necessary and sufficient conditions for $\rho(A) < 1$ are

(i_a) $p(1) = 1 - \text{tr}(A) + \sum_{i=1}^{3} M_{ii} - \det A > 0$

(i_b)

$$(-1)^3 p(-1) = -[-1 - \text{tr}(A) - \sum_{i=1}^{3} M_{ii} - \det(A)]$$

$$= 1 + \text{tr}(A) + \sum_{i=1}^{3} M_{ii} + \det(A) > 0$$

(ii) $|\det(A)| < 1$

(iii) $|(\det A)^2 - 1| > \left| \text{tr}(A) \det(A) - \sum_{i=1}^{3} M_{ii} \right|$

Summarizing, the necessary and sufficient conditions for $\rho(A) < 1$ are

Theorem 21 (i) $|\det(A)| < 1$

(ii) $-\left(\sum_{i=1}^{3} M_{ii} + 1 \right) < \text{tr}(A) + \det(A) < \left(\sum_{i=1}^{3} M_{ii} + 1 \right)$

(iii) $\sum_{i=1}^{3} M_{ii} - \text{tr}(A) \det(A) + [\det(A)]^2 < 1$

Exercises 3.2

1. Consider the system of difference equations

$$x(t+1) = x(t) + y(t)$$
$$y(t+1) = \frac{1}{4} x(t) + y(t)$$

(a) Find the eigenvalues and the eigenvectors associated with the system and find the equation of the stable and unstable manifolds.

(b) Write down the solution of the system if $x(0) = x_0$ and $y(0) = y_0$.

(c) Draw the phase space diagram, showing the stable and unstable manifolds.

2. A model of red blood cell production in the circulatory system, the red blood cells (RBCs) are constantly being destroyed and replaced. Since these cells carry oxygen throughout the body, their number must be maintained at some fixed level. Assume that the spleen filters out and destroys a certain fraction of the cells daily and that the bone marrow produces a number proportional to the number lost on the previous day.

   Let

   $$R(t) = \text{number of RBCs in circulation on day } t$$
   $$M(t) = \text{number of RBCs produced by marrow on day } t$$
   $$\alpha = \text{fraction of RBCs removed by the spleen}$$
   $$\beta = \text{production constant (number produced per number lost)}.$$

   Then the equation for $R(t)$ and $M(t)$ are

   $$R(t+1) = (1-\alpha)R(t) + M(t)$$
   $$M(t+1) = \alpha\beta R(t).$$

   (a) Explain how we arrived at these equations.

   (b) Find the eigenvalues of the system.

   (c) For homeostasis in the red cell count, the total number of red blood cells, $R(t)$, should remain roughly constant. Show that one way of achieving this is by letting $\lambda_1 = 1$. What does this imply about β?

   (d) Show that the second eigenvalue is then given by $\lambda_2 = -\alpha$. What then is the behavior of the solution $R(t)$?

3. [125] Let $p(t)$ be the size of a certain annual plant at time t. Then a model for the growth of the annual plant may be given by the second difference equation

   $$p(t+2) = \alpha\gamma\sigma p(t+1) + \beta\gamma\sigma^2(1-\alpha)p(t),$$

   where

   $$\gamma = \text{number of seeds produced per plant in August}$$
   $$\alpha = \text{fraction of 1-year-old seeds that germinate in May } (\alpha < 1)$$
   $$\beta = \text{fraction of 2-year-old seeds that germinate in May}$$
   $$\sigma = \text{fraction of seeds that survive a given winter}$$

   By letting $p(t) = u(t)$, $p(t+1) = v(t)$, we obtain the two-dimensional system

   $$u(t+1) = v(t)$$
   $$v(t+1) = \alpha\gamma\sigma v(t) + \beta\gamma\sigma^2(1-\alpha)u(t)$$

(a) Find the eigenvalues and the eigenvectors of the system. Show that $\lambda_1 > 0$ and $\lambda_2 < 0$.

(b) Find conditions under which the plants propagate, i.e., $\lambda_1 > 1$. Give a biological interpretation of your answer.

4. The graph depicts a life cycle for an age-classified life cycle. This is represented by the system $\mathbf{n}(t+1) = A\mathbf{n}(t)$,

where $A = \begin{pmatrix} 0 & F_2 & F_3 \\ P_1 & 0 & 0 \\ 0 & P_2 & 0 \end{pmatrix}$.

Find conditions such that the population will survive, i.e., the extinction equilibrium is unstable (a repeller).

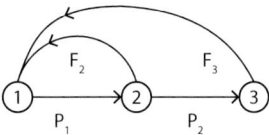

Figure 3.5: The chart depicts three age classes n_1, n_2 and n_3, where P_i is the probability an individual in age class i, $1 \leq i \leq 2$, is alive at time $t+1$, and F_i is the fertility rate of n_i.

5. Repeat the problem for the life cycle represented by the system $\mathbf{n}(t+1) = A\mathbf{n}(t)$,

where $A = \begin{pmatrix} 0 & F_2 & F_3 & F_4 \\ P_1 & 0 & 0 & 0 \\ 0 & P_2 & 0 & 0 \\ 0 & 0 & P_3 & 0 \end{pmatrix}$.

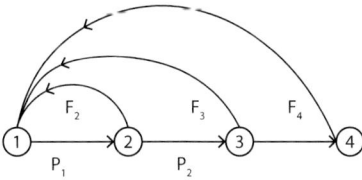

Figure 3.6: The chart depicts four age classes n_1, n_2, n_3 and n_4, where P_i is the probability an individual in age class i, $1 \leq i \leq 3$, is alive at time $t+1$, and F_i is the fertility rate of n_i.

6. Consider the life cycle of a plant which may be represented by $\mathbf{n}(n+1) = A\mathbf{n}(t)$ where

$A = \begin{bmatrix} 0.088 & 0 & 46 \\ 0.789 & 0.095 & 0 \\ 0 & 0.856 & 0.806 \end{bmatrix}$.

Determine whether the population survives or not.

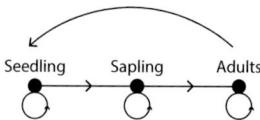

Figure 3.7: The chart depicts the three age classes of a plant, seedling, sapling, and adult.

In problems 7–9, draw the phase space diagram of the following systems.

7.
$$x(t+1) = x(t) - 5y(t)$$
$$y(t+1) = y(t) - x(t)$$

8.
$$x(t+1) = x(t) + y(t)$$
$$y(t+1) = 0.25x(t) + y(t)$$

9.
$$x(t+1) = x(t) + 3y(t)$$
$$y(t+1) = -x(t) + y(t)$$

10. [41] Let A be a 4×4 matrix. Show that a necessary and sufficient condition that $\rho(A) < 1$ are

    (i) $|\det(A)| < 1$

    (ii) $\left| 1 - (\det(A))^2 \left(\sum S_{ij}(A) \right) + \left(\sum_{i=1}^{4} M_{ii}(A) \right)^2 + \det(A)(\operatorname{tr}(A))^2 \right| <$
    $1 + \left[\operatorname{tr}(A) \sum_{i=1}^{4} M_{ii} - \det(A) \right] (1 + \det(A)) + (\det(A))^3,$

    (iii) $\left| \operatorname{tr}(A) + \sum_{i=1}^{4} M_{ii} \right| < 1 + \sum S_{ij}(A) + \det(A)$,

    (iv) $\frac{1}{2} \sum S_{ij}(A) - \frac{3}{16} \operatorname{tr}(A) \sum_{i=1}^{4} M_{ii}(A) + \frac{1}{16} \left(\sum M_{ii}(A) \right)^2 < 1$

    (v) $\left| 3 \operatorname{tr}(A) + \sum_{i=1}^{4} M_{ii}(A) \right| < 4 + 2 \sum S_{ij}(A)$

    where $\sum S_{ij}(A)$ is the sum of the six determinants of the principle 2×2 submatrices of A. Here, one must add the six determinants of the six 2×2 matrices that result from two rows and two columns of A. Note that $\sum_{i=1}^{4} M_{ii}(A)$ is the sum of the four principle minors of A obtained by deleting the ith row and the ith column.

11. Consider a flour-beetle model

$$L(t+1) = bA(t)$$
$$P(t+1) = (1-\mu_L)L(t)$$
$$A(t+1) = (1-\mu_P)P(t) + (1-\mu_A)A(t)$$

where $L(t)$, $P(t)$, $A(t)$ are the larvae, pupae, and adult populations at time, respectively. The parameter b is the larval recruitment rate per adult in unit time and μ_L, μ_P, μ_A are the death rates in the respective stages. Find conditions under which the population would survive (Figure 3.5).

3.3 Nonlinear Competition Models

There are two types of competition, one occurs among individuals of the same species, and is called intraspecific competition, and the other occurs among two (or more) species and is called interspecific competition. The intraspecific competition was accounted for in one-dimensional models such as the discrete logistic (Beverton–Holt) and exponential (Ricker) models. In this chapter, we focus on interspecific competition among multiple species. Note that interspecific competition is generally asymmetric—the consequences are not the same for all species. For example, interspecific competition is asymmetric in the competition between *Paramecium Aurelia* and *Paramecium Caudatum*. However, it is symmetric for *Paramecium Aurelia* and *Paramecium Bursaria* since both species are equally affected by the competition (Figure 3.6).

Two historical laboratory experiments were conducted to understand interspecific competition. The first was conducted by G. F. Gause [145–147] in 1935 on three different species of *Paramecium*, *P. Aurelia*, *P. Caudatum* and *P. Bursaria*. The other laboratory experiment was conducted by T. Park [248, 249] in 1954 on two species of flour beetles in the genus *Tribolium*, namely, *T. Castaneum* and *T. Confusum*. These experiments were influential in the formulation of what became known as the *competitive exclusion principle* or *Gause's Law*. This principle states that two species cannot occupy the same niche, that is to say, in order to coexist they must sufficiently differ in their utilization of limited resources (food, habitat, etc.). In other words, when two species are competing for the same resources, the one that is best adapted or that has an advantage over the other will survive and the other may become extinct. Hence, in order for two species to coexist, they must find a way to decrease their competition for resources (Figure 3.7).

Gause conducted an experiment growing two species of *Paramecium*, *P. Aurelia* and *P. Caudatum*. He placed *P. Caudatum* and *P. Aurelia* together in a test tube with the same food supply. *P. Aurelia* grew faster than *P. Caudatum* and when grown together, *P. Aurelia* out-multiplied and eliminated *P. Caudatum*. As shown in Figure 3.8, *P. Caudatum* is 2.5 larger than *P. Aurelia*. However, its carrying capacity is less than one-half of *P. Aurelia* (Figure 3.9).

Growing *P. Caudatum* together with *P. Aurelia* led to the extinction of the former (Figure 3.10(a)).

In another experiment, Gause grew P. Aurelia with the equally small *P. Bursaria*. This time both species coexisted (Figure 3.10(b)).

Density-dependent competition models of n-species are based on the same principles we followed for the one-species models of Chapter 1. The main difference between multi-species models and single-species models is the introduction of new parameters called interspecific parameters, that represent competition between different species. For instance, in modeling the competition of

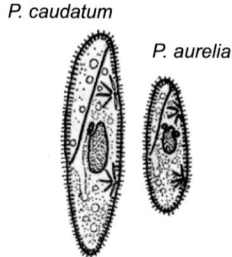

Figure 3.8: *Paramecium*

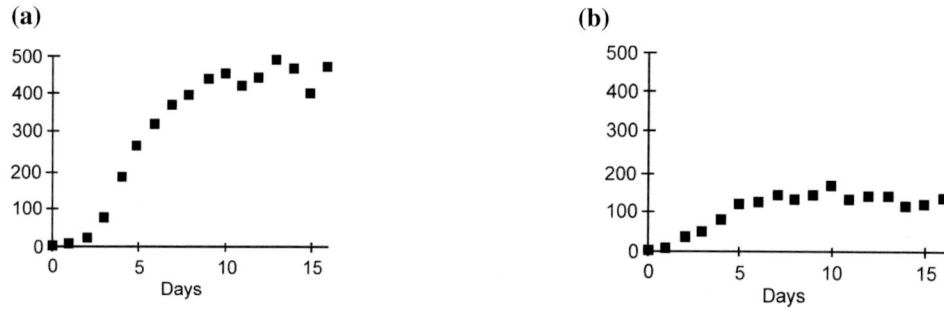

Figure 3.9: (a) Time series for the growth of *P. Aurelia* alone. (b) Time series for the growth of *P. Caudatum* alone.

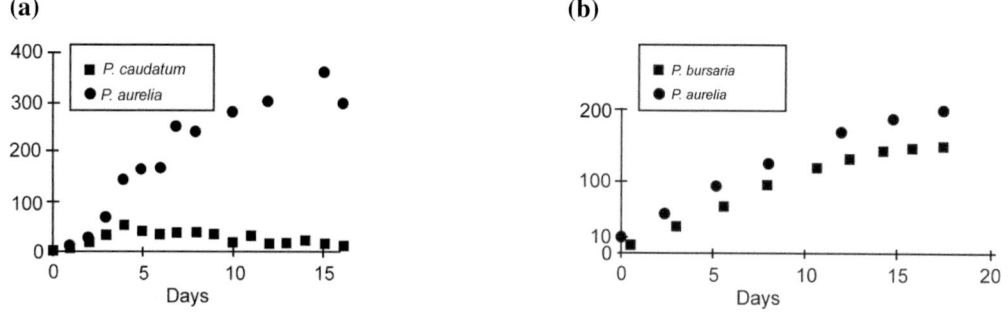

Figure 3.10: (a) *P. Aurelia* out competes *P. Caudatum*. (b) *P. Aurelia* coexists with *P. Bursaria*.

3.3. NONLINEAR COMPETITION MODELS

two species x_1 and x_2, one would introduce two types of parameters, intraspecific parameters representing the negative impact of competition among individuals of the same species, and interspecific parameters representing the negative impact competition among species x_1 and x_2. Now if we have n-species, $x_1, x_2, \ldots x_n$, with intraspecific parameters $a_1, a_2, \ldots, a_n$ and interspecific parameters b_{ij}, $i = 1, 2, \ldots, n$; $j = 1, 2, \ldots, n$, then the competition model is given by the following system of difference equations

$$\begin{aligned} x_1(t+1) &= f_1(a_1, b_{12}, b_{12}, \ldots, b_{1n}, x_1(t), x_2(t), \ldots, x_n(t)) \\ x_2(t+1) &= f_2(a_2, b_{21}, b_{22}, \ldots, b_{2n}, x_1(t), x_2(t), \ldots, x_n(t)) \\ &\vdots \\ x_n(t+1) &= f_n(a_n, b_{n1}, b_{n2}, \ldots, b_n, x_1(t), x_2(t), \ldots, x_n(t)) \end{aligned} \tag{3.18}$$

or in a vector form

$$\mathbf{x}(t+1) = F(\mathbf{x}(t)), \tag{3.19}$$

where $\mathbf{x} = (x_1, x_2, \ldots, x_n)^T$ and

$$\mathbb{R}_+^n = \{\mathbf{x} : x_i \geq 0, i = 1, 2, \ldots, n\}$$

and where $F = (f_1, f_2, \ldots, f_n)^T$, $F : D \to D$ where the domain D is a subset of $\mathcal{R}^n+$.

Competition models usually have the form

$$f_i(x_1, x_2, \ldots, x_n) = x_i r_i(x_1, x_2, \ldots, x_n),$$

where r_i is the population growth rate of species i. Such model equations are said to be of Kolmogorov type. The equilibrium points are obtained as the fixed points $\mathbf{x}^*$ of the map F, i.e., $F(\mathbf{x}^*) = \mathbf{x}^*$ or

$$\begin{aligned} f_1(x_1, x_2, \ldots, x_n) &= x_1 \\ f_2(x_1, x_2, \ldots, x_n) &= x_2 \\ &\vdots \\ f_n(x_1, x_2, \ldots, x_n) &= x_n. \end{aligned}$$

The solution set of (3.20) in $\mathbb{R}_+^n$ gives the set of nonnegative equilibria of the system. For Kolmogorov-type equations, the positive solutions $x_i > 0$ of the equations

$$\begin{aligned} r_1(x_1, x_2, \ldots, x_n) &= 1 \\ r_2(x_1, x_2, \ldots, x_n) &= 1 \\ &\vdots \\ r_n(x_1, x_2, \ldots, x_n) &= 1 \end{aligned} \tag{3.20}$$

gives the set of positive equilibria of the system. In addition to positive equilibria, we have the extinction equilibrium $(0, 0, \cdots 0)^T$ and possibly other equilibrium points in which one or more of the components is zero. Examples of non-Kolmogorov models include predator–prey models, models with migration/immigration and structured population models.

In the sequel, we give examples of competition models that are widely used in population biology.

3.3.1 Examples of Competition Models

Example 50 ***Leslie–Gower Model.*** *A general Leslie-Gower model, for multi-species, is given by*

$$x_i(t+1) = \frac{b_i x_i(t)}{1 + \sum_{j=1}^{n} c_i j x_j(t)}, i = 1, 2, \ldots, n, \tag{3.21}$$

where $x_i(t)$ represents the size or density of species x_i at time t, c_{ii} the intraspecific parameters for species x_i, and c_{ij} the interspecific parameter between species x_i and x_j. For $n = 2$ species, we have

$$\begin{aligned} x_1(t+1) &= \frac{b_1 x_1(t)}{1 + c_{11} x_1(t) + c_{12} x_2(t)} \\ x_2(t+1) &= \frac{b_2 x_2(t)}{1 + c_{21} x_1(t) + c_{22} x_2(t)}. \end{aligned} \tag{3.22}$$

Remark 4 *It should be noted that the Leslie–Gower model is the discrete analog of the famous continuous competition model ([227]), the Lotka–Volterra the continuous model is given by*

$$\begin{aligned} x_1'(t) &= a x_1(t) - b x_1(t) x_2(t) \\ x_2'(t) &= -c x_2(t) + d x_1(t) x_2(t). \end{aligned} \tag{3.23}$$

Example 51 ***The Ricker Competition Model.*** *A general Ricker competition model for n species based on Ricker dynamics is given by the equations*

$$x_i(t+1) = x_i(t) \exp\left(\alpha_i - \Sigma_{j=1}^{n} c_{ij} x_j(t)\right), \ i = 1, 2, \ldots, n. \tag{3.24}$$

And for the two-species competition model

$$\begin{aligned} x_1(t+1) &= x_1(t) \exp\left(\alpha_1 - c_{11} x_1(t) - c_{12} x_2(t)\right) \\ x_2(t+1) &= x_2(t) \exp\left(\alpha_2 - c_{12} x_1(t) - c_{22} x_2(t)\right) \end{aligned} \tag{3.25}$$

Example 52 ***The Quadratic Competition Model.*** *A general model of this type is given by*

$$x_i(t+1) = \frac{b_i x_i(t)[1 - x_i(t)]}{1 + \sum_{\substack{j=1 \\ j \neq i}}^{n} c_i j x_j(t)}, i = 1, 2, \ldots, n. \tag{3.26}$$

For the case of $n = 2$ species the model becomes

$$\begin{aligned} x_1(t+1) &= \frac{b_1 x_1(t)(1 - x_1(t))}{1 + c_{12} x_2(t)} \\ x_2(t+1) &= \frac{b_2 x_2(t)(1 - x_2(t))}{1 + c_{21} x_1(t)}. \end{aligned} \tag{3.27}$$

In these models we take the domain D to be the unit square $[0, 1] \times [0, 1]$. We leave it to the reader to verify that the right sides of these model equations map D into D if $b_i < 4$.

We now explore the role of the isoclines in determining the behavior of the orbits. In Equation (3.24), the isoclines are given by

$$1 + c_{11} x_1 + c_{12} x_2 = b_1 \text{ or } x_2 = -\frac{c_{11}}{c_{12}} x_1 + \frac{b_1 - 1}{c_{12}} \tag{3.28}$$

3.3. NONLINEAR COMPETITION MODELS

and
$$1 + c_{21}x_1 + c_{22}x_2 = b_2 \text{ or } x_2 = -\frac{c_{21}}{c_{22}}x_1 + \frac{b_2 - 1}{c_{22}}. \quad (3.29)$$

It is clear that $E_1^* = (0,0)$ is the extinction equilibrium point. Solving equations (3.28) and (3.29), give the remaining equilibrium points: The exclusion equilibria
$$E_2^* = \left(\frac{b_1 - 1}{c_{11}}, 0\right)^T, \quad E_3^* = \left(0, \frac{b_2 - 1}{c_{22}}\right)^T$$

and the co-existence equilibrium point
$$E_4^* = \left(\frac{(b_1 - 1)c_{22} - (b_2 - 1)c_{12}}{c_{11}c_{22} - c_{12}c_{21}}, \frac{(b_2 - 1)c_{11} - (b_1 - 1)c_{21}}{c_{11}c_{22} - c_{12}c_{21}}\right)^T$$

Now E_2^* and E_3^* are positive equilibria if $b_1 > 1$ and $b_2 > 1$, respectively, and E_4^* is positive if either
$$\frac{c_{21}}{c_{11}} < \frac{b_2 - 1}{b_1 - 1} < \frac{c_{22}}{c_{12}} \quad (3.30)$$
or
$$\frac{c_{22}}{c_{12}} < \frac{b_2 - 1}{b_1 - 1} < \frac{c_{21}}{c_{11}}. \quad (3.31)$$

From Equation (3.24) we see that
$$x_1(t+1) > x_1(t) \text{ if } x_2 < -\frac{c_{11}}{c_{12}}x_1 + \frac{b_1 - 1}{c_{12}}$$
and
$$x_1(t+1) < x_1(t) \text{ if } x_2 > -\frac{c_{11}}{c_{12}}x_1 + \frac{b_1 - 1}{c_{12}}.$$

Moreover,
$$x_2(t+1) > x_2(t) \text{ if } x_2 < -\frac{c_{21}}{c_{22}}x_1 + \frac{b_2 - 1}{c_{22}}$$
and
$$x_2(t+1) < x_2(t) \text{ if } x_2 > -\frac{c_{21}}{c_{22}}x_1 + \frac{b_2 - 1}{c_{22}}.$$

This information is summarized in Figures 3.11(a) and 3.11(b).

When x_1 is increasing, i.e., $x_1(t+1) > x_1(t)$, we represent this by an arrow to the right $\to$ and when it is decreasing we denote it by an arrow to the left $\leftarrow$. Similarly, if x_2 is increasing, which is represented by a vertical arrow pointing upward $\uparrow$ and if it is decreasing the arrow is reversed $\downarrow$. Note that, in Figure 3.11(a), the arrows in regions $\mathbb{R}_2$ and $\mathbb{R}_4$ are pointing toward the positive co-existence equilibrium point E_4^* and so do the arrows in regions $\mathbb{R}_1$ and $\mathbb{R}_3$. Hence, there is a possibility that all orbits would converge to E_4^*. However, in Figure 3.11(b), the arrows in regions $\mathbb{R}_2$ and $\mathbb{R}_4$ are pointing away from E_4^*, and E_4^* is definitely not attracting the orbits around it.

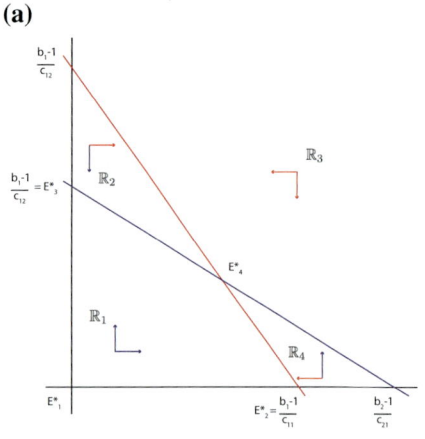 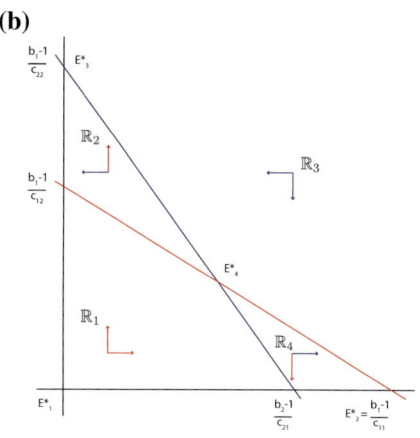

Figure 3.11: (a) In the first scenario, $\frac{c_{12}}{c_{22}} < \frac{b_1-1}{b_2-1} < \frac{c_{11}}{c_{21}}$, the orbits stay in the neighborhood of the equilibrium point E_4^*. (b) In the second scenario, $\frac{c_{11}}{c_{21}} < \frac{b_1-1}{b_2-1} < \frac{c_{12}}{c_{22}}$, the orbits do not stay in the neighborhood of the equilibrium point E_4^*, and thus E_4^* is unstable.

3.3.2 Stability Analysis

In this section, we will introduce the various notions of stability as an extension of the analogous notions of single-species models of Chapter 1.

Consider the difference systems
$$\mathbf{x}(t+1) = F(\mathbf{x}(t)), \tag{3.32}$$
where $F : \Gamma \to \Gamma$ is a C^1 and Γ is an open subset of $\mathbb{R}^n$.

Definition 14 *For each $\mathbf{x} = (x_1, x_2, \ldots, x_n)^T \in \mathbb{R}^n$, one may define a norm $\|\mathbf{x}\|$, as an extension of the absolute value $\mathbb{R}$, with the following properties.*

(i) $\|\mathbf{x}\| \geq 0$ and $\|\mathbf{x}\| = 0$ if and only if $\mathbf{x} = 0$ (the zero vector)

(ii) $\|k\mathbf{x}\| = |k|\|\mathbf{x}\|$, for any scalar $k \in \mathbb{R}$, $\mathbf{x} \in \mathbb{R}^n$

(iii) $\|\mathbf{x} + \mathbf{y}\| \leq \|\mathbf{x}\| + \|\mathbf{y}\|$ (The triangle inequality vector)

Now we will use three norms on vectors in $\mathbb{R}^n$.

(a) $\|\mathbf{x}\|_1 = \sum_{i=1}^{n} |x_i|$, the ℓ_1-norm

(b) $\|\mathbf{x}\|_2 = \left(\sum_{i=1}^{n} x_i \right)^{\frac{1}{2}}$, the ℓ_2-norm

(c) $\|\mathbf{x}\|_\infty = \max\{|x_i| : 1 \leq i \leq n\}$, the ℓ_∞-norm

3.3. NONLINEAR COMPETITION MODELS

These three vector norms are equivalent. Hence, one may use any one of them in studying the dynamics of (3.32) and obtain the same results. Note that we used the ℓ_1-norm in Chapter 2.

Next, we introduce the various notions of stability analogous to those introduced in Chapter 1.

Definition 15 *Let $\mathbf{x}^*$ be an equilibrium point of Equation (3.32). Then*

(i) The equilibrium point $\mathbf{x}^*$ is stable if for any $\varepsilon > 0$, there exists $\delta > 0$ such that $\|F^t(\mathbf{x}_0) - \mathbf{x}^*\| < \varepsilon$, whenever $\|\mathbf{x}_0 - \mathbf{x}^*\| < \delta$. In other words, $\|\mathbf{x}(t) - \mathbf{x}^*\| < \varepsilon$, whenever $\|\mathbf{x}_0 - \mathbf{x}^*\| < \delta$, where $\mathbf{x}(0) = \mathbf{x}_0$.

(ii) The equilibrium point $\mathbf{x}^*$ is attracting if there exists $\nu > 0$ such that $\lim_{t \to \infty} F^t(\mathbf{x}_0) = \mathbf{x}^*$, i.e., $\lim_{t \to \infty} \mathbf{x}(t) = \mathbf{x}^*$, whenever $\|\mathbf{x}_0 - \mathbf{x}^*\| < \nu$.

(iii) The equilibrium point $\mathbf{x}^*$ is locally asymptotically stable (LAS) if it is stable and attracting.

(iv) Let D be a set in the domain of F. The equilibrium point $\mathbf{x}^*$ is globally asymptotically stable on D if it is locally asymptotically stable and $\lim_{t \to \infty} F^t(\mathbf{x}_0) = \mathbf{x}^*$, i.e., $\lim_{t \to \infty} \mathbf{x}(t) = \mathbf{x}^*$, for all $\mathbf{x}_0 \in D$.

One of the main tools in the study of the stability of nonlinear systems of multi-species interacting populations is a linearization at equilibria, just as in the case of single-species models. Thus, the behavior of solutions or orbits near an equilibrium point is determined by the behavior of solutions or orbits of the linearized system at the equilibrium.

Let us now consider the linearization of Equation (3.32) around the equilibrium point $\mathbf{x}^*$. We make the change of variables $\mathbf{y} = \mathbf{x} - \mathbf{x}^*$, obtaining the system

$$\mathbf{y}(t+1) = F(\mathbf{y}(t) + \mathbf{x}^*) - \mathbf{x}^*.$$

Using Taylor's Theorem for functions of several variables, we write

$$F(\mathbf{y} + \mathbf{x}^*) = F(\mathbf{x}^*) + JF(\mathbf{x}^*)\mathbf{y} + G(\mathbf{y}),$$

where $JF(\mathbf{x}^*)$ is the derivative of F evaluated at $\mathbf{x}^*$, and $G(\mathbf{x})$ is small in the sense that $\lim_{\|\mathbf{y}\| \to 0} \frac{\|G(\mathbf{y})\|}{\|\mathbf{y}\|} = 0$. Recall that the Jacobian matrix JF evaluated at $\mathbf{x}^*$ is given by

$$JF(\mathbf{x}^*) = \begin{pmatrix} \frac{\partial f_1}{\partial x_1}(\mathbf{x}^*) & \frac{\partial f_1}{\partial x_2}(\mathbf{x}^*) & \cdots & \frac{\partial f_1}{\partial x_n}(\mathbf{x}^*) \\ \frac{\partial f_2}{\partial x_1}(\mathbf{x}^*) & \frac{\partial f_2}{\partial x_2}(\mathbf{x}^*) & \cdots & \frac{\partial f_2}{\partial x_n}(\mathbf{x}^*) \\ & & \vdots & \\ \frac{\partial f_n}{\partial x_1}(\mathbf{x}^*) & \frac{\partial f_n}{\partial x_2}(\mathbf{x}^*) & \cdots & \frac{\partial f_n}{\partial x_n}(\mathbf{x}^*) \end{pmatrix}.$$

Hence, we have

$$\mathbf{y}(t+1) = JF(\mathbf{x}^*)\mathbf{y}(t) + G(\mathbf{y}(\mathbf{t}))$$

and the linear part is given by

$$\mathbf{y}(t+1) = JF(\mathbf{x}^*)\mathbf{y}(t). \tag{3.33}$$

As we have seen in Chapter 1, the main tool of investigating the local asymptotic stability of the equilibria of the nonlinear system (3.32) is to examine the asymptotic stability of the linear part of the system (3.33). This scheme is called The linearization principle and is given in the next result. But before stating this main result, let us recall some matrix terminology. Let $\{\lambda_1, \lambda_2, \lambda_3, \ldots \lambda_n\}$ be the eigenvalues of an $n \times n$ matrix A. Then the spectral radius ρ of A is given by $\rho(A) = \max\{|\lambda_i| : i = 1, 2, \ldots, n\}$.

Theorem 22 *The following statements hold true.*

(i) *If $\rho(JF(\mathbf{x}^*)) < 1$, then the equilibrium point $\mathbf{x}^*$ of system (3.32) is locally asymptotically stable,*

(ii) *If $\rho(JF(\mathbf{x}^*)) > 1$, then the equilibrium point $\mathbf{x}^*$ of system (3.32) is unstable,*

(iii) *If $\rho(JF(\mathbf{x}^*)) = 1$, then the equilibrium point $\mathbf{x}^*$ of system (3.32) may be stable or unstable.*

Remark 5 *Now if $\rho(JF(\mathbf{x}^*)) \neq 1$, then $\mathbf{x}^*$ is called a hyperbolic fixed point of the map F and if $\rho(JF(\mathbf{x}^*)) = 1$, then $\mathbf{x}^*$ is called a nonhyperbolic fixed point of the map F. The above theorem does not address the stability of nonhyperbolic fixed points and further analysis is needed in this case. There are three scenarios to consider when dealing with nonhyperbolic fixed points, which lead to three types of bifurcation. Let λ be the eigenvalue of $JF(\mathbf{x}^*)$ with the largest modulus. If (i) $\lambda = 1$, we get a saddle-node bifurcation, (ii) if $\lambda = -1$, we get a period-doubling bifurcation and (iii) if $\lambda = \alpha + i\beta$, then we get a Neimark–Sacker bifurcation. These bifurcation types will be discussed later in Section 3.10.*

In order to put this theorem into use, we need to explore the stability and the dynamics of linear systems.

3.4 Local Stability of Competition Systems

In this section we will investigate the stability analysis of multi-species models.

Example 53 *Consider again the Leslie–Gower 2-species competition model*

$$\begin{aligned} x(t+1) &= \frac{b_1 x(t)}{1 + c_{11} x(t) + c_{12} y(t)} = f(x,y) \\ y(t+1) &= \frac{b_2 y(t)}{1 + c_{21} x(t) + c_{22} y(t)} = g(x,y) \end{aligned}, \quad (3.34)$$

where all the parameters $b_1, b_2, c_{ij} > 0$, $i = 1,2;\ j = 1,2$. The equilibria are

$$E_1^* = \mathbf{0} = (0,0)^T, \quad E_2^* = \left(\frac{b_1 - 1}{c_{11}}, 0\right)^T, \quad E_3^* = \left(0, \frac{b_2 - 1}{c_{22}}\right)^T, \quad E_4^* = (x^*, y^*)^T,$$

where

$$\begin{aligned} x^* &= \frac{(b_1 - 1)c_{22} - (b_2 - 1)c_{12}}{c_{11} c_{22} - c_{12} c_{21}} \\ y^* &= \frac{(b_2 - 1)c_{11} - (b_1 - 1)c_{21}}{c_{11} c_{22} - c_{12} c_{21}}. \end{aligned}$$

Note that the extinction equilibrium E_1^* exists for all parameter values; E_2^* is nonnegative if $b_1 > 0$ and E_3^* is nonnegative if $b_2 > 1$. However, in order for the coexistence equilibrium point E_4^* to be positive, we must assume that $b_1 > 1$ and $b_2 > 1$ and either (3.30) or (3.31). We first consider (3.3), i.e. we assume

$$\frac{c_{21}}{c_{11}} < \frac{b_2 - 1}{b_1 - 1} < \frac{c_{22}}{c_{12}}. \quad (3.35)$$

3.4. LOCAL STABILITY OF COMPETITION SYSTEMS

To start our stability analysis we linearize our system and find the Jacobian matrix at the equilibrium points.

Now

$$JF(\mathbf{x}) = \begin{pmatrix} \frac{\partial f}{\partial x} & \frac{\partial f}{\partial y} \\ \frac{\partial g}{\partial x} & \frac{\partial g}{\partial y} \end{pmatrix}$$

$$= \begin{pmatrix} \frac{(1+c_{11}x+c_{12}y)b_1 - b_1 c_{11} x}{(1+c_{11}x+c_{12}y)^2} & \frac{-b_1 c_{12} x}{(1+c_{11}x+c_{12}y)^2} \\ \frac{-b_2 c_{21} y}{(1+c_{21}x+c_{22}y)^2} & \frac{(1+c_{21}x+c_{22}y)b_2 - b_2 c_{22} y}{(1+c_{21}x+c_{22}y)^2} \end{pmatrix}$$

and

$$JF(\mathbf{0}) = \begin{pmatrix} b_1 & 0 \\ 0 & b_2 \end{pmatrix}.$$

Hence, the extinction equilibrium point $E_1^*(0,0)^T$ is unstable (repeller) if $\lambda_1 = b_1 > 1$ and $\lambda_2 = b_2 > 1$. This means that both species will survive in this case. However, if both $b_1, b_2 < 1$, then the extinction equilibrium is asymptotically stable and for small initial values of both species go extinct asymptotically.

Next, we examine the stability of E_2^* by considering the Jacobian

$$JF(E_2^*) = \begin{pmatrix} \frac{1}{b_1} & -\frac{c_{12}}{c_{11}}\left(\frac{b_1-1}{b_1}\right) \\ 0 & \frac{b_2}{1+c_{21}\left(\frac{b_1-1}{c_{11}}\right)} \end{pmatrix}.$$

Since $b_1 > 1$, it follows that $\frac{1}{b_1} < 1$, and the eigenvalue $\lambda_1 < 1$. Under conditions (3.35),

$$\lambda_2 = \frac{b_2}{1+c_{21}\left(\frac{b_1-1}{c_{11}}\right)} > 1.$$

Hence, E_2^* is a saddle.

For the equilibrium E_3^* we examine the Jacobian matrix

$$JF(E_3^*) = \begin{pmatrix} \frac{b_1}{1+c_{12}\left(\frac{b_2-1}{c_{22}}\right)} & 0 \\ \frac{c_{21}}{c_{22}}\left(\frac{b_2-1}{b_2}\right) & \frac{1}{b_2} \end{pmatrix}.$$

Since $b_2 > 1$, it follows that $\lambda_2 = \frac{1}{b_2} < 1$. Under conditions (3.35) we find that

$$\lambda_1 = \frac{b_1}{1+c_{12}\left(\frac{b_2-1}{c_{22}}\right)} > 1.$$

Hence, E_3^* is a saddle.

Finally, for $E_4^* = (x^*, y^*)^T$, noting that $1 + c_{11}x^* + c_{12}y^* = b_1$ and $1 + c_{21}x^* + c_{22}y^* = b_2$, we have

$$JF(E_4^*) = \begin{pmatrix} 1 - \frac{c_{11}x^*}{b_1} & \frac{-c_{12}x^*}{b_1} \\ \frac{-c_{21}y^*}{b_2} & 1 - \frac{c_{22}y^*}{b_2} \end{pmatrix}.$$

By the Jury conditions E_4^* is asymptotically stable if

$(i)\ \det JF < 1, \quad (ii)\ \det(JF) > \operatorname{tr}(JF) - 1, \quad (iii)\ \det(JF) > -\operatorname{tr}(JF) - 1$

(formula (3.16)) where

$$\det = \det(JF(E_4^*)) = 1 - \frac{c_{11}x^*}{b_1} - \frac{c_{22}y^*}{b_2} - \frac{c_{12}c_{21}x^*y^*}{b_1b_2} + \frac{c_{11}c_{22}x^*y^*}{b_1b_2}$$

$$\operatorname{tr} = \operatorname{tr}(JF(E_4^*)) = 2 - \frac{c_{11}x^*}{b_1} - \frac{c_{22}y^*}{b_2}.$$

To show condition (i) holds we calculate

$$\det = 1 - \frac{c_{11}b_2 x^* + c_{22}b_1 y^* + c_{12}c_{21}x^*y^* - c_{11}c_{22}x^*y^*}{b_1b_2}$$

$$= 1 - \frac{c_{11}x^*(1 + c_{21}x^* + c_{22}y^*) + c_{22}y^*(1 + c_{11}x^* + c_{12}y^*) + c_{12}c_{21}x^*y^* - c_{11}c_{22}x^*y^*}{b_1b_2}$$

$$= 1 - \frac{c_{11}x^*(1 + c_{21}x^*) + c_{22}y^*(1 + c_{11}x^* + c_{12}y^*) + c_{12}c_{21}x^*y^*}{b_1b_2} < 1.$$

Condition (ii) holds since

$$\frac{c_{11}c_{22}x^*y^*}{b_1b_2} - \frac{c_{12}c_{21}x^*y^*}{b_1b_2} = \frac{x^*y^*(c_{11}c_{22} - c_{12}c_{21})}{b_1b_2} > 0.$$

Condition (iii) holds as can be verified by again using $1 - c_{11}x^* + c_{12}y^* = b_1$ *and* $1 + c_{21}x^* + c_{22}y^* = b_2$.

We have shown that under conditions (3.30) (i.e. (3.35)) the positive equilibrium E_4^* is stable. See Figure 3.12(a).

We leave it for the reader to show that under the alternative condition (3.31) that:

(i) E_4^* is a saddle point.

(ii) E_2^* and E_3^* are locally asymptotically stable.

See Figure 3.12(b).

Finally, we consider the cases when E_4^* is not positive. First, if

$$c_{11}c_{22} - c_{12}c_{21} > 0, \quad \frac{b_2 - 1}{b_1 - 1} < \frac{c_{21}}{c_{11}}, \quad \frac{c_{22}}{c_{12}} > \frac{b_2 - 1}{b_1 - 1},$$

then E_4^* does not exist (as a point in the interior of $\mathbb{R}_+^2$). Moreover, E_3^* is a saddle and E_2^* is locally asymptotically stable (Figure 3.13).

On the other hand, if

$$c_{11}c_{22} - c_{12}c_{21} > 0, \quad \frac{b_2 - 1}{b_1 - 1} > \frac{c_{22}}{c_{12}}, \quad \frac{b_2 - 1}{b_1 - 1} > \frac{c_{21}}{c_{11}},$$

then E_4^* does not exist (as a point in the interior of $\mathbb{R}_+^2$), E_2^* is a saddle, and E_3^* is locally asymptotically stable, Figure 3.13.

3.4. LOCAL STABILITY OF COMPETITION SYSTEMS

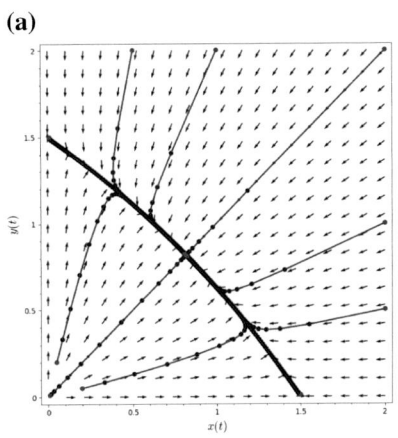 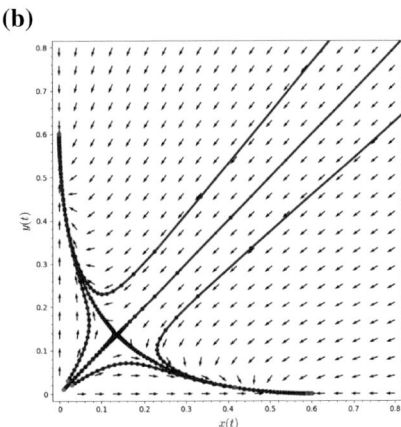

Figure 3.12: (a) The phase space diagram of the Leslie–Gower model: E_1^* is a repeller (unstable), E_2^* and E_3^* are saddle points, while E_4^* is asymptotically stable. (b) The phase space diagram of the Leslie–Gower model: E_4^* is a saddle, E_2^* and E_3^* are locally asymptotically stable.

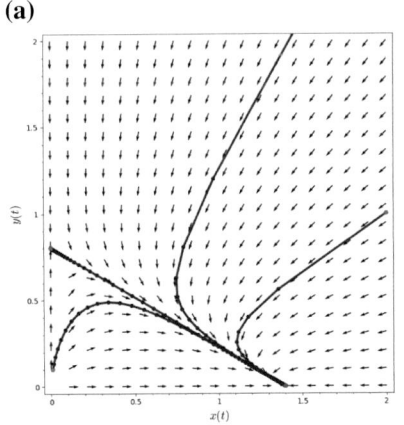 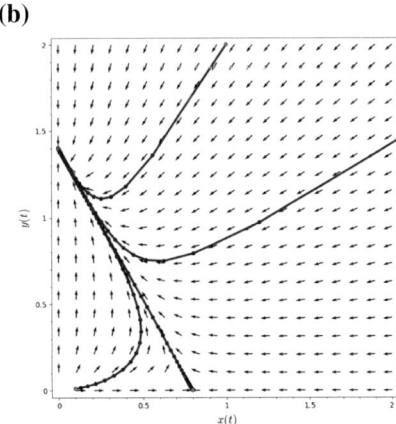

Figure 3.13: (a) The phase space diagram of the Leslie–Gower model: The equilibrium points E_2^* is locally asymptotically stable and E_3^* is a saddle. (b) The phase space diagram of the Leslie–Gower model: The equilibrium points E_3^* is locally asymptotically stable and E_2^* is a saddle.

Example 54 *Next, we investigate the more complex model, the Ricker competition model given by*

$$x(t+1) = x(t)e^{\alpha - c_{11}x - c_{12}y}$$
$$y(t+1) = y(t)e^{\beta - c_{21}x - c_{22}y} \tag{3.36}$$

or by the map

$$F(x,y) = (xe^{\alpha - c_{11}x - c_{12}y}, ye^{\beta - c_{21}x - c_{22}y})^T.$$

There are four equilibrium points

$$E_1^* = (0,0)^T, \quad E_2^* = \left(\frac{\alpha}{c_{11}}, 0\right)^T, \quad E_3^* = \left(0, \frac{\beta}{c_{22}}\right)^T$$

$$E_4^* = (x^*, y^*)^T = \left(\frac{\alpha c_{22} - \beta c_{12}}{c_{11}c_{22} - c_{12}c_{21}}, \frac{\beta c_{11} - \alpha c_{21}}{c_{11}c_{22} - c_{12}c_{21}}\right)^T.$$

E_4^* *is in the interior of* $\mathbb{R}_+^2$*, i.e., positive, if either one of the following assumptions is satisfied.*

$(H_1): \frac{c_{12}}{c_{22}} < \frac{\alpha}{\beta} < \frac{c_{11}}{c_{21}}$

or

$(H_2): \frac{c_{11}}{c_{21}} < \frac{\alpha}{\beta} < \frac{c_{12}}{c_{22}}$

The eigenvalues of $JF(E_2^*)$ are

$$\lambda_1 = e^{\alpha - \frac{r_2 c_{12}}{c_{22}}}, \quad \lambda_2 = 1 - \beta.$$

We have the following cases.

(a) *Note that* $0 < \lambda_1 < 1$ *if* $\frac{\alpha}{\beta} < \frac{c_{12}}{c_{22}}$*, which is condition* (H_2)*. Moreover,* $|\lambda_2| < 1$ *if and only if* $|1 - \beta| < 1$ *or* $0 < \beta < 2$*. In this case,* E_2^* *is asymptotically stable.*

(b) *Note that* $\lambda_1 > 1$ *if* $\frac{\alpha}{\beta} > \frac{c_{12}}{c_{22}}$*, which is condition* (H_1)*, and* $|\lambda_2| < 1$ *if* $0 < \beta < 2$*. In this case,* E_2^* *is a saddle point.*

(c) *Note that* $\lambda_1 > 1$ *and* $|\lambda_2| > 1$ *if condition* (H_1) *holds and* $\beta < 0$ *or* $\beta > 2$*. In this case,* E_2^* *is unstable.*

(d) *Note that* $0 < \lambda_1 < 1$ *and* $|\lambda_2| > 1$ *if condition* (H_2) *holds and* $\beta < 0$ *or* $\beta > 2$*. In this case,* E_2^* *is a saddle.*

The Jacobian matrix at $E_3^* = \left(\frac{\alpha}{c_{11}}, 0\right)^T$ *is given by*

$$JF(E_3^*) = \begin{pmatrix} 1 - c_{11}\left(\frac{\alpha}{c_{11}}\right) e^{\alpha - c_{11}\left(\frac{\alpha}{c_{11}}\right)} & -c_{12}\left(\frac{\alpha}{c_{11}}\right) e^{\alpha - c_{11}\left(\frac{\alpha}{c_{11}}\right)} \\ 0 & e^{\beta - c_{12}\left(\frac{\alpha}{c_{11}}\right)} \end{pmatrix}$$

$$= \begin{pmatrix} 1 - \alpha & -\frac{r_1 c_{12}}{c_{11}} \\ 0 & e^{\beta - \frac{r_1 c_{12}}{c_{11}}} \end{pmatrix}.$$

The analysis of E_3^* *is similar to the of* E_2^**.*

3.4. LOCAL STABILITY OF COMPETITION SYSTEMS

(a) $|\lambda_1| < 1$ if $0 < \alpha < 2$, and $0 < \lambda_2 < 1$ if $\frac{\alpha}{\beta} > \frac{c_{11}}{c_{21}}$ (Condition H_1). In this case, E_3^* is asymptotically stable (attractor).

(b) $|\lambda_1| < 1$ if $0 < \alpha < 2$, and $\lambda_2 > 1$ if $\frac{c_{11}}{c_{12}} < \frac{\alpha}{\beta}$ (Condition H_2). In this case, E_3^* is a saddle.

(c) $|\lambda_1| > 1$ if $\alpha < 0$ or $\alpha < 2$, and $\lambda_2 > 1$ if $\frac{c_{11}}{c_{12}} > \frac{\alpha}{\beta}$ (Condition H_2). In this case, E_3^* is unstable (a repeller).

(d) $|\lambda_1| > 1$ if $\alpha < 0$ or $\alpha > 2$ and $0 < \lambda_2 < 1$ if (Condition H_1) holds.

Finally, we investigate the stability of the positive equilibrium E_4^* by examining the Jacobian.

$$JF(x^*, y^*) = \begin{pmatrix} 1 - c_{11}x^* & -c_{12}x^* \\ -c_{21}y^* & 1 - c_{22}y^* \end{pmatrix}.$$

The simplification and the disappearance of the exponential term is due to the fact that $(x^*, y^*)^T$ satisfies the isocline equations $e^{\alpha - c_{11}x^* - c_{12}y^*} = 1$ and $e^{\beta - c_{21}x^* - c_{22}y^*} = 1$. A calculation of the eigenvalues of this Jacobian and determining whether their moduli less than 1 or greater than is a very tedious. Instead, we will use formula (3.16) which states that $|\lambda_1| < 1$ and $|\lambda_2| < 1$ if $|\operatorname{tr} J| < 1 + \det J < 2$. Now

$$\operatorname{tr} J = 2 - c_{11}x^* - c_{22}y^*$$
$$\det J = 1 + c_{11}c_{22}x^*y^* - c_{11}x^* - c_{22}y^* - c_{12}c_{21}x^*y^*.$$

Hence, the condition for the local asymptotic stability of E_4^* is

$$|2 - c_{11}x^* - c_{22}y^*| < 2 + c_{11}c_{22}x^*y^* - c_{11}x^* - c_{22}y^* - c_{12}c_{21}x^*y^* < 2$$

or

$$\left| 2 - c_{11}\left(\frac{\alpha c_{22} - \beta c_{12}}{c_{11}c_{22} - c_{12}c_{21}}\right) - c_{22}\left(\frac{\beta c_{11} - \alpha c_{21}}{c_{11}c_{22} - c_{12}c_{21}}\right) \right| <$$
$$2 + \frac{(\alpha c_{22} - \beta c_{21})(\beta c_{11} - \alpha c_{21})}{c_{11}c_{22} - c_{12}c_{21}} - c_{11}\left(\frac{\alpha c_{22} - \beta c_{12}}{c_{11}c_{22} - c_{21}c_{12}}\right) - c_{22}\left(\frac{\beta c_{11} - \alpha c_{21}}{c_{11}c_{22} - c_{12}c_{21}}\right) < 2.$$

An example of when this holds is seen in (Figure 3.14).

Case 1: In this case, species y goes to extinction while species x thrives and in fact the equilibrium point E_2^* is locally asymptotically stable, as seen in (Figure 3.15). In this case, it is assumed that

$$\frac{\alpha}{\beta} > \frac{c_{11}}{c_{21}}, \quad \frac{\alpha}{\beta} > \frac{c_{12}}{c_{22}}, \quad 0 < \alpha < 2.$$

Case 2: In this case, we assume that $\frac{\alpha}{\beta} < \frac{c_{12}}{c_{22}}$, $\frac{\alpha}{\beta} < \frac{c_{11}}{c_{21}}$, $0 < \beta < 2$.

Here species x goes to extinction, while species y thrives and, in fact, the equilibrium point E_3^* is locally asymptotically stable, as seen in (Figure 3.16).

Note that, in both cases, Cases 1 and 2 above, the coexistence equilibrium E_4^* does not exist.

(a) 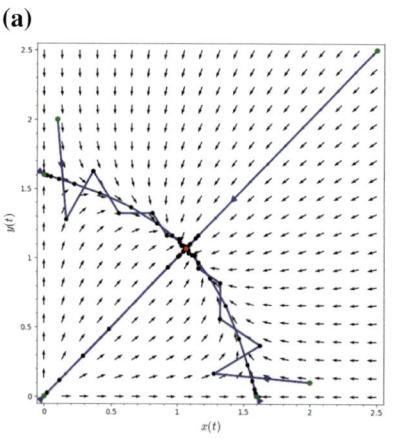 **(b)**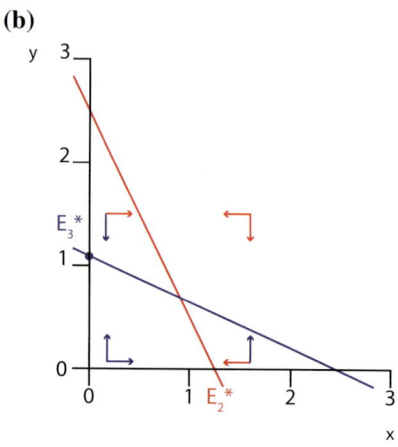

Figure 3.14: (a) The equilibrium points E_4^* is locally asymptotically stable and both E_2^* and E_3^* are saddle points. (b) The isoclines intersect at the coexistence equilibrium point E_4^*. The orientation of the orbits indicate that E_4^* is most likely asymptotically stable.

(a) 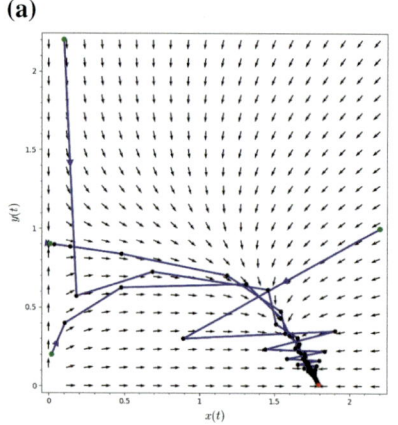 **(b)**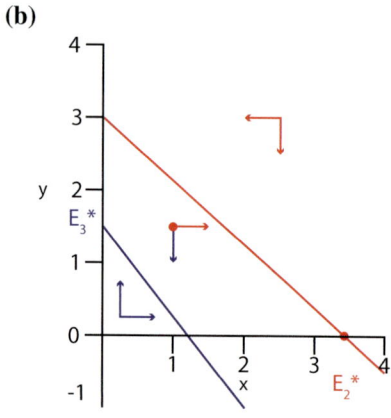

Figure 3.15: (a) The equilibrium points E_2^* is locally asymptotically stable and E_3^* is a saddle. (b) The isoclines do not intersect in the interior of R_+^2. The orientation of the orbits indicate that E_2^* is most likely asymptotically stable.

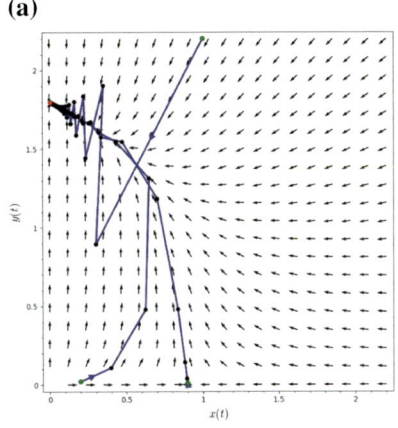 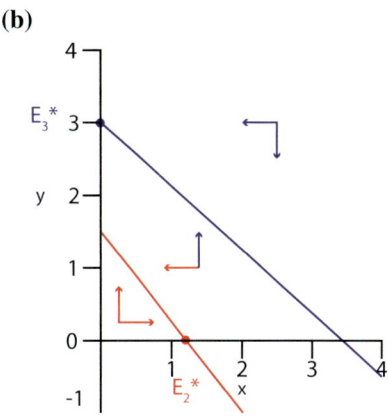

Figure 3.16: (a) The phase space diagram of the Ricker model: The equilibrium points E_3^* is locally asymptotically stable and E_2^* is a saddle. (b) The isoclines do not intersect in the interior of R_+^2. The orientation of the orbits indicate that E_3^* is most likely asymptotically stable.

3.5 Predator–Prey Models

3.5.1 Nicholson–Bailey Model

We begin with the Host–Parasitoid model proposed by Nicholson and Bailey in 1935 [245]. Let $H(t)$ and $P(t)$ be the density (or number) of hosts and parasitoids in generation t (or time period t), $f(H(t), P(t))$ be the fraction of hosts that are not parasitized, λ be the net reproductive rate of the hosts and c is the "clutch" size of the parasitoids. Note that a parasitoid is a parasite which is free living as an adult but lays eggs in the host. Hosts that are not parasitized give rise to their own progeny, while those who are parasitized die. However, the eggs laid by the parasitoid may survive to the next generation.

The general model is given by

$$H(t+1) = H(t)f(H(t), P(t))$$
$$P(t+1) = cH(t)[1 - f(H(t), P(t))]. \qquad (3.37)$$

In the Nicholson–Bailey model, it is assumed that the number of encounters E between potential hosts and parasitoids is proportional to the product of their densities,

$$E = aH(t)P(t). \qquad (3.38)$$

The second assumption in the model is that the encounters are distributed randomly, according a Poisson distribution among the available hosts.

If μ is the average encounter rate, then the probability of n encounters, according to the Poisson distribution, is given by

$$p_n = \frac{e^{-\mu}\mu^n}{n!}. \qquad (3.39)$$

But by (3.38), $\mu = \frac{E}{H(t)} = aP(t)$. Hence,

$$p_n = \frac{e^{-aP(t)} \cdot (a(P(t))^n}{n!}. \tag{3.40}$$

Thus, probability p_0 that a host will escape parasitism is given by

$$f(H(t), P(t)) = p_0 = e^{-aP(t)}. \tag{3.41}$$

One more assumption is needed for the Nicholson–Bailey model. It is assumed that only the first encounter, between a host and a parasitoid, is important. In particular, we will assume that a host that has been parasitized will bear exactly c parasitoid progeny regardless of the number of encounters.

The above discussion leads to the historic Nicholson–Bailey model

$$\begin{aligned} H(t+1) &= bH(t)e^{-aP(t)} \\ P(t+1) &= cH(t)(1 - e^{-aP(t)}). \end{aligned} \tag{3.42}$$

This may be written as the difference equation

$$\mathbf{x}(t+1) = F(\mathbf{x}(t)),$$

where $\mathbf{x}(t) = (P(t), H(t))^T$ and

$$F(\mathbf{x}) = \begin{pmatrix} bHe^{-aP} \\ cH(1 - e^{-aP}) \end{pmatrix}.$$

There are two equilibrium points

$$E_1^* = (0,0)^T \text{ and } E_2^* = (H^*, P^*)^T.$$

To find E_2^*, we let $H(t+1) = H(t)$ and $P(t+1) = P(t)$, that is

$$bHe^{-aP} = H$$

which gives

$$P^* = \frac{1}{a} \ln b$$

and

$$cH^*(1 - e^{-aP}) = P^*$$

which implies

$$H^* = \frac{b \ln b}{ac(b-1)}.$$

To determine the stability of E_1^* and E_2^* by means of the linearization principle, we evaluate the Jacobian of $F(\mathbf{x})$

$$JFF(\mathbf{x}) = \begin{pmatrix} \lambda e^{-aP} & -baHe^{-aP} \\ c(1 - e^{-aP}) & caHe^{-aP} \end{pmatrix}$$

3.5. PREDATOR–PREY MODELS

at each of these equilibrium. First, we have

$$JF(E_1^*) = \begin{pmatrix} b & 0 \\ 0 & 0 \end{pmatrix}$$

whose eigenvalues are $\lambda_1 = b$ and $\lambda_2 = 0$. Thus, E_1^* is asymptotically stable if $b < 1$ and a saddle if $b > 1$ (Figure 3.19).

Note that E_2^* is biologically feasible (i.e., is positive) if $b > 1$, which we will assume. To examine its stability, we consider the Jacobian

$$JF(E_2^*) = \begin{pmatrix} 1 & \frac{-b \ln b}{c(b-1)} \\ c\left(\frac{b-1}{b}\right) & \frac{\ln b}{b-1} \end{pmatrix}.$$

Note that the determinant of this matrix is

$$\det JF(E_2^*) = \ln b \left(\frac{b}{b-1}\right).$$

Now $b > 1$ implies $\left(\frac{b}{b-1}\right) \ln b > 1$. To see this, let $h(b) = b \ln b - (b-1)$. Then $h(1) = 0$ and $h'(b) = \ln b > 0$. Hence, $h(\lambda) > 0$ and, consequently, $\det JF(E_2^*) > 1$, which implies E_2^* is unstable.

We conclude that this historical simple model is not a practical representation of real host–parasitoid interactions and there is a real need to modify the model. And in 1975, Beddington et al. [26] did provide a revised model.

3.5.2 Beddington et al. Model

In this revised model, it is assumed that the growth rate is density-dependent, i.e., $b = b(H(t))$.

In the revised model, the reproductive rate follows the Ricker model and is thus given by

$$b(H) = e^{r\left(1 - \frac{H(t)}{K}\right)}.$$

The revised model is given by

$$\begin{aligned} H(t+1) &= H(t) \exp\left(r\left(1 - \frac{H(t)}{K}\right) - aP(t)\right) \\ P(t+1) &= cH(t)\left(1 - e^{-aP(t)}\right). \end{aligned} \quad (3.43)$$

There are three equilibrium points $E_1^* = (0,0)$, $E_2^* = (K,0)$, and $E_3^* = (H^*, P^*)$. To determine the stability of these equilibria via linearization, we compute the Jacobian matrix

$$JF(H,P) = \begin{pmatrix} \left[1 + H\left(-\frac{r}{K}\right)\right] e^{r\left(1 - \frac{H}{K}\right) - aP} & \vdots & -aHe^{r\left(1 - \frac{H}{K}\right) - aP} \\ c\left(1 - e^{-aP}\right) & \vdots & +caHe^{-aP} \end{pmatrix}.$$

1. $E_1^* = (0,0)$: The eigenvalues of

$$JF(0,0) = \begin{pmatrix} e^r & 0 \\ 0 & 0 \end{pmatrix}$$

are $\lambda_1 = e^r$ and $\lambda_2 = 0$. Thus, E_1^* is asymptotically stable if $r < 0$ and a saddle if $r > 0$.

2. $E_2^* = (K, 0)$: The eigenvalues of

$$JF(E_2^*) = \begin{pmatrix} 1-r & -aK \\ 0 & caK \end{pmatrix}$$

are $\lambda_1 = 1-r$ and $\lambda_2 = caK$. Hence, E_2^* is asymptotically stable if $0 < r < 2$ and $caK < 1$ it is a saddle if $0 < r < 2$ and $caK > 1$.

3. $E_3^* = (H^*, P^*)$: From the first Equation of (3.43) we get $H = K - \frac{a}{r}P$. From the second Equation of (3.43) we get

$$H = \frac{Pe^{ap}}{ca(e^{ap}-1)}.$$

The following lemma addresses the positivity of the coexistence equilibrium (H^*, P^*). The following Lemma answers this question.

Lemma 4 *If $a > \frac{1}{cK}$, then Equation (3.43) has a unique positive equilibrium (H^*, K^*).*

Proof. The two isoclines of Equation (3.43) are given by

$$f_1(P) = K - \frac{aP}{r}, \quad f_2(P) = \frac{Pe^{aP}}{c(e^{aP}-1)}.$$

Now $f_1(0) = K$ and

$$f_2(0) = \lim_{P \to 0} \frac{Pe^{aP}}{c(e^{aP}-1)} = \frac{1}{ca}$$

since $a > \frac{1}{cK}$, $\frac{1}{ca} < K$. Now the graph of f_1 is a straight line with negative slope and intersects the P-axis at $P + \frac{rK}{a}$. Note that the function f_2 is positive and increasing to $+\infty$ as P increases for

$$f_2'(P) = \frac{ce^{aP}(e^{aP} - a - ap)}{c^2 a^2 (e^{aP} - 1)^2} > 0$$

for $P > 0$. Hence, f_1 and f_2 must intersect in the first quadrant (see Figure 3.17).

Now the Jacobian matrix at $E_3^* = (H^*, P^*)^T$ is given by

$$JF(E_3^*) = \begin{pmatrix} 1 - r\frac{H^*}{K} & -arH^* \\ c\left(1 - e^{-r\left(1-\frac{H^*}{K}\right)}\right) & caH^* e^{-r\left(1-\frac{H^*}{K}\right)} \end{pmatrix}.$$

By the det-tr criteria for asymptotic stability, E_3^* is asymptotically stable if and only if $|\text{tr } JF| < 1 + \det JF < 2$ or

$$\left| 1 - \frac{rH^*}{K} + caH^* e^{-r\left(1-\frac{H^*}{K}\right)} \right| < 1 + \left(1 - \frac{H^*}{K}\right) caH^* e^{-r\left(1-\frac{H^*}{K}\right)} + car\frac{H^*}{K}\left(1 - e^{-r\left(1-\frac{H^*}{K}\right)}\right)$$

$$< 2.$$

Figure 3.18 shows the parameter space (c, r)-bifurcation diagram in which the shaded area is the stability region of the positive equilibrium point E_3^*. ∎

The dynamics of the Beddington et al. model are illustrated by the phase space diagrams (Figures 3.19(a), 3.19(b), 3.19(c)).

More detailed analysis of this model may be found in Kapcak, Ufuktepe, Elaydi [185] [186].

3.5. PREDATOR–PREY MODELS

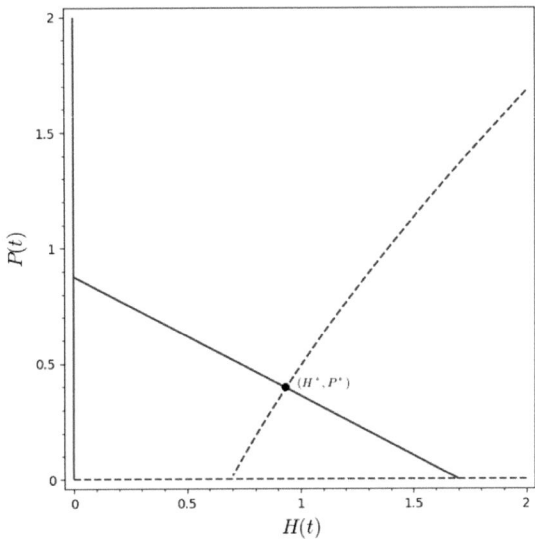

Figure 3.17: Numerical computation of the coexistence equilibrium point (H^*, P^*).

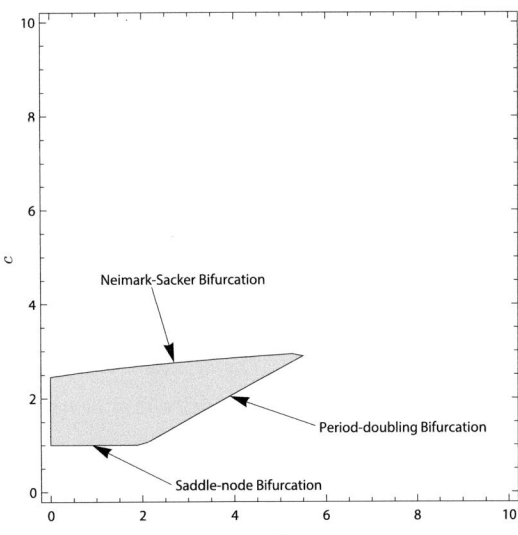

Figure 3.18: The figure depicts the parameter space $r - c$ bifurcation diagram. The equilibrium $E_3^* = (H^*, P^*)^T$ is asymptotically stable when the parameters r and c are in the shaded region.

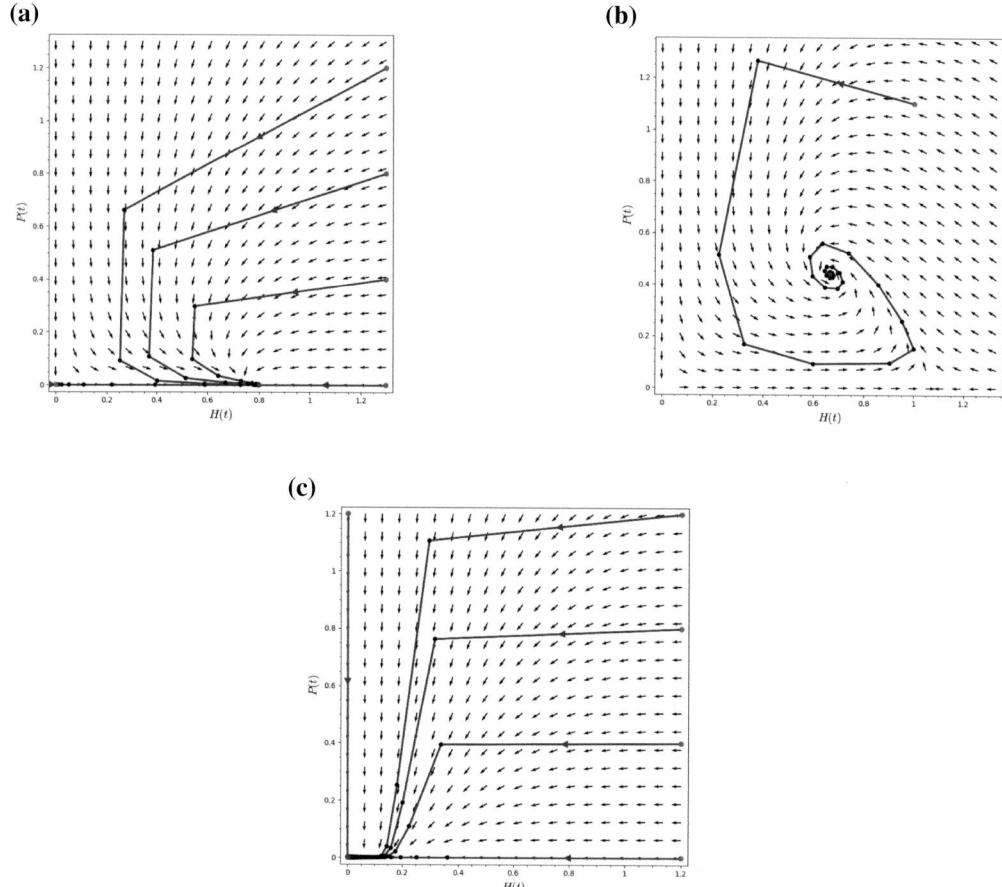

Figure 3.19: (a) $0 < r < 2, caK < 1, E_2^*$ is asymptotically stable. (b) $0 < r < 2, caK > 1, E_2^*$ is a saddle and E_3^* is asymptotically stable. (c) $E_1^* = (0,0)$ is asymptotically stable.

3.5.3 May Model

Another revision of the Nicholson–Bailey model was proposed by May [231]. In this revision, the random distribution of parasitoid attacks is replaced with an over-dispersed distribution, described by a negative binomial with the clumping parameter k. The other biological assumptions such as the mean attack rate remain as the Nicholson–Bailey Model and is equal to aP. Hence, the probability that the host escapes parasitism is the zero term in the negative binomial which is equal to

$$\left(1 + a\frac{P}{k}\right)^{-k}.$$

Note that as $k \to \infty$, we obtain the Poisson distribution e^{-aP}, and we get the Nicholson–Bailey model. May's model is now given as follows:

$$\begin{aligned} H(t+1) &= bH(t)\left(1 + a\frac{P(t)}{k}\right)^{-k} \\ P(t+1) &= cH(t)\left[1 - \left(1 + a\frac{P(t)}{k}\right)^{-k}\right]. \end{aligned} \qquad (3.44)$$

The stability analysis of this model is left as Problem 6, Section 3.5.

Exercises 3.5

1. Assume that the population P and its host population H are modeled by the system

$$\begin{aligned} P(t+1) &= \alpha H(t)(1 - e^{-P(t)}) \\ H(t+1) &= \alpha H(t)(e^{-aP(t)}). \end{aligned}$$

   (a) Find the equilibrium population sizes.
   (b) Show that there exists a positive equilibrium point if $\alpha > 1$.
   (c) Show that this positive equilibrium point is unstable.
   (d) Draw the phase space portrait for $\alpha = 2$, $a = 1.2$ to confirm your conclusion in part (c).
   (e) Analyze the isoclines.

2. Consider the single species, age-structured model

$$\begin{aligned} x(t+1) &= y(t)\exp(r - ax(t) - y(t)) \\ y(t+1) &= x(t), \end{aligned}$$

   where a, r are positive.

   (a) Find the equilibrium points and determine their stability.
   (b) Draw the phase space portrait for $r = 1.5$, $a = 1$ to confirm your conclusion in part (a).

3. Consider the modified Lagged Beverton–Holt equation
$$x(t+1) = \frac{rK[\alpha x(t) + \beta x(t-1)]}{K + (r-1)x(t-1)}$$
with $\alpha + \beta = 1$ and $\alpha, \beta > 0$.

   (a) Show that the zero solution (Equilibrium point $(0,0)$) is stable if $0 < r < 1$.

   (b) Draw the phase space portrait for $r = 0.8$, $K = 2$, $\alpha = 0.4$, $\beta = 0.6$ to confirm your conclusion in part (a).

4. The following equation describes the growth of a mosquito population
$$P(t+1) = (aP(t) + bP(t-1))e^{-P(t)},$$
where $0 \leq a < 1$, $b > 0$.

   (a) Show that the zero solution (Equilibrium point $(0,0)$) is stable if $a + b \leq 1$ and $b < 1$.

   (b) Draw the phase space portrait for $a = 0.5$, $b = 0.3$, to confirm your conclusion in part (a).

5.* Consider the Ricker competition model (3.36).

   (a) Find all equilibrium points.

   (b) Determine the stability of the equilibrium points.

   (c) Draw the phase space portrait to confirm your conclusion in part (b).

6.* Investigate the stability of the equilibrium points of May's model (3.44).

7. Consider the Leslie–Gower model (3.34) with the symmetry assumption $b_1 = b_2$. Show that

   (i) if $c_{12} - c_{22} < 0$ and $c_{21} - c_{11} > 0$, then the equilibrium point
$$E_2^* = \begin{pmatrix} \frac{b_1 - 1}{c_{11}} \\ 0 \end{pmatrix}$$
   is asymptotically stable, species x survives while species y goes to extinction.

   (ii) if $c_{12} - c_{22} > 0$ and $c_{21} - c_{11} > 0$, then the interior equilibrium
$$E_4^* = \begin{pmatrix} x^* \\ y^* \end{pmatrix}$$
   is a saddle.

   (iii) if $c_{12} - c_{22} < 0$ and $c_{21} - c_{11} < 0$, then species x and y coexist.

   (iv) if $c_{12} - c_{22} > 0$ and $c_{21} - c_{11} < 0$, then species y survives and species x goes to extinction.

8. Consider the following herbivore-plant model
$$P(t+1) = P(t)e^{r(1-P(t))-aH(t)}$$
$$H(t+1) = P(t)e^{r(1-P(t))}[1 - e^{-aH(t)}].$$

   (a) Determine the stability of equilibrium points.

   (b) Draw the phase space diagrams to confirm your conclusion in part (a).

9. (Research Project) Investigate the dynamics of May's model (3.44).

10. (Research Project) Consider the flour-beetle model
$$L(t+1) = bA(t)\exp(-c_{EA}A(t) - c_{EL}L(t))$$
$$P(t+1) = (1 - \mu_L)L(t)$$
$$A(t+1) = P(t)\exp(-c_{PA}A(t)) + (1 - \mu_A)A(t),$$

    where $L(t)$, $P(t)$, $A(t)$ are the size of the larvae, pupae and adult populations, respectively. The constant μ_L, μ_A are the larval and adult probability of dying from causes other than cannibalism, respectively. Thus, $0 \le \mu_L \le 1$, $0 \le \mu_A \le 1$. The terms $\exp(-c_{EA}A(t))$, $\exp(-c_{EL}L(t))$ represent the probability that an egg is not eaten in the presence of $A(t)$ adults, and the probability that an egg is not eaten in the presence of $L(t)$ larvae.

    (a) Assume $c_{EL} = 0$. Find the equilibrium points and determine their stability. This is the case when there is no cannibalism of eggs by larvae.

    (b) Assume $c_{EL} \ne 0$. Find the equilibrium points and determine their stability.

    (c) Define the reproduction number $\mathcal{R}_0 = \frac{b(1-\mu_L)}{\mu_A}$. Show that the trivial equilibrium is asymptotically stable if $\mathcal{R}_0 < 1$. What happens if $\mathcal{R}_0 > 1$?

3.6 Population Cycles: Periodic Orbits

We have seen in Chapter 1, Section 1.4.3, that a single-species population may exhibit complex patterns from equilibrium points to cyclic oscillations and even chaotic dynamics. The dynamics of multi-species models or structured models is much more complicated and challenging. In this section, we will study the stability of periodic orbits or cycles using an extension of the linearization principle in Theorem 22.

Definition 16 *Consider the n-dimensional difference system*
$$\mathbf{x}(t+1) = F(\mathbf{x}(t)). \tag{3.45}$$

Then
$$c_k = \{\overline{\mathbf{x}}_1, \overline{\mathbf{x}}_2, \cdots, \overline{\mathbf{x}}_k\}$$

is called a k-cycle of F or a k-periodic orbit of Equation (3.45) *if*
$$F(\overline{\mathbf{x}}_i) = \overline{\mathbf{x}}_{i+1} \quad \text{for each } i \in \{1, 2, \cdots, (k-1)\} \quad \text{and} \quad F(\overline{\mathbf{x}}_k) = \overline{\mathbf{x}}_1.$$

Note that a k-cycle is the orbit of a periodic point of period k. Moreover, $F^k(\overline{\mathbf{x}}_i) = \overline{\mathbf{x}}_i$, for all $1 \leq i \leq k$.

The stability of a cycle may be determined by extending the linearization principle that was used to determine the stability of equilibrium points. Viewing every point $\mathbf{x}_i$ in the cycle c_k, as a fixed point of the composition map $G = F^k$, we may define local asymptotic stability of the cycle c_k as follows.

Definition 17 Let $c_k = \{\overline{\mathbf{x}}_1, \overline{\mathbf{x}}_2, \cdots, \overline{\mathbf{x}}_k\}$. Then

- The cycle c_k is stable if for each $\epsilon > 0$, there exists $\delta > 0$ such that $\|\mathbf{x}_0 - \overline{\mathbf{x}}_i\| < \delta$ implies $\|F^{tk+i-1}(\mathbf{x}_0) - \overline{\mathbf{x}}_i\| < \epsilon$ or $\|\mathbf{x}(tk+i-1) - \overline{\mathbf{x}}_i\| < \epsilon$, for $1 \leq i \leq k$ and $t > 0$, where $\mathbf{x}(0) = \mathbf{x}_0$

- The cycle c_k is attracting if there exists $\mu > 0$ such that, if $\|\mathbf{x}_0 - \overline{\mathbf{x}}_i\| < \mu$, then
$$\lim_{t \to \infty} F^{tk+i-1}(\mathbf{x}_0) = \overline{\mathbf{x}}_i \text{ or } \lim_{t \to \infty} \mathbf{x}(tk+i-1) = \overline{\mathbf{x}}_i, \text{ for } 1 \leq i \leq k, \text{ where } \mathbf{x}(0) = \mathbf{x}_0.$$

- The cycle c_k is locally asymptotic stable (LAS) if it is stable and attracting.

- Let D be a subset of the domain of F. The cycle c_k is globally asymptotically stable on D if $\lim_{t \to \infty} F^{tk+i-1}(\mathbf{x}_0) = \overline{\mathbf{x}}_i$ for all $\mathbf{x}_0 = \mathbf{x}(0) \in D$ and $1 \leq i \leq k$.

One may ask whether a k-cycle $k > 1$ can be globally asymptotically stable. This question was answered negatively in the following theorem. (See Theorem 4.11 in [119].)

Theorem 23 Let $F : \mathbb{R}^n \to \mathbb{R}^n$ be a continuous map and c_k be a k-cycle, $k \geq 2$, of Equation (3.45). Then c_k cannot be globally asymptotically stable.

The basin of attraction of a cycle c_k is the union of the basin of attractions of the points $\overline{\mathbf{x}}_i$ in c_k as fixed points of the composite maps $G = F^k$. Explicitly, the basin of attraction of a cycle is given by

$$\mathcal{B}(c_k) = \left\{ \mathbf{x}(0) \in \mathbb{R}_+^n : \lim_{t \to \infty} F^{tk+i}(\mathbf{x}(0)) = \overline{\mathbf{x}}_i \text{ for some } \overline{\mathbf{x}}_i \text{ in } c_k \right\}$$
$$= \cup_{i=1}^k \mathcal{B}(\overline{\mathbf{x}}_i),$$

where $\mathcal{B}(\overline{\mathbf{x}}_i)$ is the basin of attraction of $\overline{\mathbf{x}}_i$ as a fixed point of the composite map G.

Next, we will develop the linearization principle for cycles. The differentials of the maps F and G are the Jacobian matrices JF and JG, respectively. By the chain rule,

$$JG(\overline{\mathbf{x}}_1) = JF^k(\overline{\mathbf{x}}_1) = JF(\overline{\mathbf{x}}_k) JF(\overline{\mathbf{x}}_{k-1}) \cdots JF(\overline{\mathbf{x}}_1),$$

and in general,

$$JG(\overline{\mathbf{x}}_i) = JF^k(\overline{\mathbf{x}}_i) = JF(\overline{\mathbf{x}}_{i-1}) JF(\overline{\mathbf{x}}_{i-2}) \cdots JF(\overline{\mathbf{x}}_1) JF(\overline{\mathbf{x}}_k) JF(\overline{\mathbf{x}}_{k-1}) \cdots JF(\overline{\mathbf{x}}_i).$$

Let ρ_i be the spectral radius of $JG(\overline{\mathbf{x}}_i)$. Notice that $\rho_1 = \rho_2 = \cdots = \rho_k$ and we will denote all of them by ρ.

Theorem 24 (The Linearization Principle of Cycles) Let $F : \Omega \to \Omega$ be a continuously differentiable map, where Ω is an open set in $\mathcal{R}^n$ containing a k-cycle c_k, $k \geq 2$ of Equation (3.45) The following statements hold true.

3.6. POPULATION CYCLES: PERIODIC ORBITS

(i) If $\rho < 1$, then the cycle c_k is asymptotically stable.

(ii) If $\rho > 1$, then the cycle c_k is unstable.

(iii) If $\rho = 1$, then the cycle c_k may be stable or unstable.

Corollary 4 *A k-cycle c_k is locally asymptotically stable (or unstable) if and only if every $\bar{x}_i$ is locally asymptotically stable (or unstable) as a fixed point of the composite map $G = F^k$.*

Example 55 *Consider again the Ricker competition model.*
$$x_1(t+1) = x_1(t)\exp\left(\alpha - c_{11}x_1(t) - c_{12}x_2(t)\right)$$
$$x_2(t+1) = x_2(t)\exp\left(\beta - c_{21}x_1(t) - c_{22}x_2(t)\right).$$

To simplify the mathematical analysis, we make changes of variable $c_{11}x_1(t) = x(t)$ and $c_{22}x_2(t) = y(t)$ and we let $\frac{c_{12}}{c_{22}} = a$, $\frac{c_{21}}{c_{11}} = b$. This yields the system

$$\begin{aligned} x(t+1) &= x(t)e^{\alpha - x(t) - ay(t)} \\ y(t+1) &= y(t)e^{\beta - y(t) - bx(t)}. \end{aligned} \tag{3.46}$$

This may be written in a vector form as $\mathbf{x}(t+1) = F(\mathbf{x}(t))$. Recall that there are four equilibrium points

$$E_1^* = \begin{pmatrix} 0 \\ 0 \end{pmatrix}, \quad E_2^* = \begin{pmatrix} \alpha \\ 0 \end{pmatrix}, \quad E_3^* = \begin{pmatrix} 0 \\ \beta \end{pmatrix}, \quad E_4^* = \begin{pmatrix} x^* \\ y^* \end{pmatrix} = \begin{pmatrix} \frac{\alpha - a\beta}{1 - ab} \\ \frac{\beta - b\alpha}{1 - ab} \end{pmatrix}.$$

The coexistence equilibrium exists if $b\alpha < \beta < \frac{\alpha}{a}$ and $ab < 1$. We will consider here the symmetric case, where $a = b$.

Simulation shows that when $a = b = 0.5$, a 2-cycle appears when $\alpha > 2$ and $\beta > 2$, and the equilibrium point $E_4^* = (x^*, y^*)^T$ loses its stability. Simulation shows that, in the symmetric case, $a = b$, chaos occurs at $(\alpha, \beta) \approx (2.692, 2.692)$, which the same numbers we got for the one-dimensional Ricker map. To illustrate the occurrence of a 2-cycle, we take $a = 2.4$ and, $\beta = 2.2$. Solving the equation $F^2(x,y) = (x,y)^T$ numerically, we get the 2-cycle $c_2 = \{\bar{\mathbf{x}}_1, \bar{\mathbf{x}}_2\}$ with

$$\bar{\mathbf{x}}_1 = \begin{pmatrix} 2.81622 \\ 2.09854 \end{pmatrix}, \quad \bar{\mathbf{x}}_2 = \begin{pmatrix} 0.65045 \\ 0.56813 \end{pmatrix}$$

(see Figure 3.20). The Jacobians at $\bar{\mathbf{x}}_1, \bar{\mathbf{x}}_2$ are

$$JF(\bar{\mathbf{x}}_1) = \begin{pmatrix} -0.419482 & -0.325228 \\ -0.284045 & 0.297402 \end{pmatrix}, \quad JF(\bar{\mathbf{x}}_2) = \begin{pmatrix} 1.513426 & -1.408108 \\ -1.049260 & 1.595224 \end{pmatrix}$$

which yield

$$JF(c_r) = JF(\bar{\mathbf{x}}_2)JF(\bar{\mathbf{x}}_1) = \begin{pmatrix} -0.23489 & 0.91098 \\ -0.12969 & 0.81567 \end{pmatrix}$$

whose eigenvalues are $\lambda_1 = -0.246018 < 1$ and $\lambda_2 = 0.826798 < 1$. Hence, by Theorem 24 the 2-cycle c_2 is locally asymptotically stable. See Figures 3.20, 3.21.

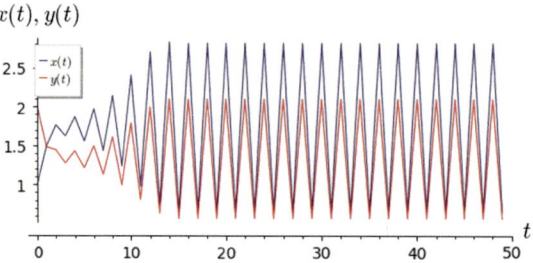

Figure 3.20: Time series showing a 2-cycle.

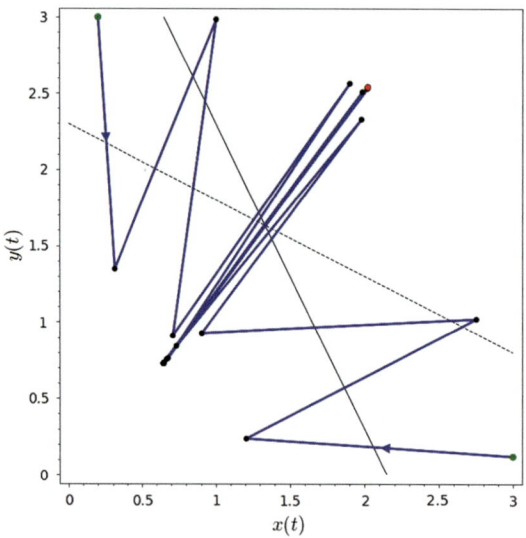

Figure 3.21: A phase space diagram showing a locally asymptotically 2-cycle.

3.6. POPULATION CYCLES: PERIODIC ORBITS

Example 56 *We consider a special case of a competition model in which species x is modeled by a Leslie–Gower type equation, while species y is modeled by a Ricker type equation [140]. Hence, we have*

$$x(t+1) = \frac{bx(t)}{a + x(t) + y(t)} \qquad (3.47)$$
$$y(t+1) = y(t)\exp(\beta - c(x(t) + y(t))).$$

The equilibrium points are

$$E_1^* = \begin{pmatrix} 0 \\ 0 \end{pmatrix}, \quad E_2^* = \begin{pmatrix} b-a \\ 0 \end{pmatrix}, \quad E_3^* = \begin{pmatrix} 0 \\ \frac{\beta}{c} \end{pmatrix}.$$

Let us represent this system by the map $F : \mathbb{R}_+^2 \to \mathbb{R}_+^2$. So to find the 2-cycles, we solve the equation $F^2(x,y) = (x,y)^T$, when $x \neq 0$ and $y \neq 0$. This yields the 2-cycle

$$c_2 = \left\{ \begin{pmatrix} \overline{x}_1 \\ \overline{y}_1 \end{pmatrix}, \begin{pmatrix} \overline{x}_2 \\ \overline{y}_2 \end{pmatrix} \right\},$$

where

$$\begin{pmatrix} \overline{x}_1 \\ \overline{y}_1 \end{pmatrix} = \begin{pmatrix} u + \frac{bu - \left(\frac{2\beta}{c} - u\right)(u+a)}{-b+(u+a)e^{\beta-cu}} \\ \left(\frac{2\beta}{c} - u\right)(u+a) - b(u+1) + (u+a)e^{\beta-cu} \end{pmatrix}$$

$$u = \frac{\beta}{c} + \sqrt{\left(\frac{\beta}{c}\right)^2 - b^2}$$

and

$$\begin{pmatrix} \overline{x}_2 \\ \overline{y}_2 \end{pmatrix} = \begin{pmatrix} u + \frac{bu - \left(\frac{2\beta}{c} - u\right)(u+a)}{-b+(u+a)e^{\beta-cu}} \\ \left(\frac{2\beta}{c} - u\right)(u+a) - b(u+1) + (u+a)e^{\beta-cu} \end{pmatrix}$$

$$u = \frac{\beta}{c} - \sqrt{\left(\frac{\beta}{c}\right)^2 - b^2}.$$

As an example, let $a = 1$, $b = 21$, $\beta = 2.5$, $c = 0.1$. In this case,

$$E_1^* = \begin{pmatrix} 0 \\ 0 \end{pmatrix}, \quad E_2^* = \begin{pmatrix} 20 \\ 0 \end{pmatrix}, \quad E_3^* = \begin{pmatrix} 0 \\ 25 \end{pmatrix}$$

are the only nonnegative equilibrium points. We have the periodic cycle c_2 given by the two vectors

$$\begin{pmatrix} \overline{x}_1 \\ \overline{y}_1 \end{pmatrix} = \begin{pmatrix} 3.297 \\ 37.03 \end{pmatrix}, \quad \begin{pmatrix} \overline{x}_2 \\ \overline{y}_2 \end{pmatrix} = \begin{pmatrix} 1.675 \\ 7.995 \end{pmatrix}.$$

The eigenvalues of $JF^2(\overline{x}_1, \overline{y}_1) = JF(\overline{x}_2, \overline{y}_2)JF(\overline{x}_1, \overline{y}_1)$ are $\lambda_1 = 0.785$ and $\lambda_2 = -0.154$. Thus, this 2-cycle is locally asymptotically stable, and the two species coexist even though there are no interior equilibrium points.

Remark 6 *(i) The model in Example 56 differs from all the competition models considered in previous sections because the two species coexist even though there exist no positive (coexistence) equilibrium points. This is unlike the Leslie–Gower and the Ricker competition models where species will not coexist if there are no interior equilibrium points. In fact, in this case, we have one species survive and the other goes to extinction (the competitive exclusion principle) or both species go extinct.*

(ii) Note that the phenomena in the above example do not occur in systems of differential equations.

We have seen in Chapter 1 (Theorem 10), a continuous map on the real line cannot have a globally asymptotically stable k-cycle, $k > 1$. We now extend this theorem to continuous maps on $\mathbb{R}^n$.

Theorem 25 (Elaydi and Yakubu [134]) *Let $F : \mathbb{R}^n \to \mathbb{R}^n$ be a continuous map. Then no k-cycle, $k > 1$, can be globally asymptotically stable.*

Remark 7 *You may be wondering whether or not there are systems of difference equations, where k-cycle is globally asymptotically stable. The answer is no and yes. No, for all autonomous systems (maps) that we have dealt with in the book. Yes, for nonautonomous difference equations. In [126], the authors showed that the periodic discrete logistic (Beverton–Holt) equation, where the carrying capacity $K(t)$ is periodic with period p, that is $K_{t+p} = K_t$ has a globally asymptotically stable p-cycle. Moreover, the authors proved that the 2-periodic discrete logistic (Beverton–Holt) equation,*

$$x(t+1) = \frac{rK(t)x(t)}{K(t) + (r-1)x(t)},$$

with $K(t+2) = K(t)$, for all $g \in \mathbb{Z}^+$, has a globally asymptotically stable 2-cycle. Nonautonomous difference equations is the subject of Chapter 8.

Exercises 3.6

Problem 1 - refer to Figure 3.32 and equation (3.46).

1. Show that, for the Ricker competition model, if $0 < \alpha < 2$ and $\beta < b\alpha$, then the equilibrium point $(\alpha, 0)^T$ is locally asymptotically stable.

2. Show that, for the Ricker competition model, if $0 < \beta < 2$ and $\alpha < a\beta$, then the equilibrium point $(0, \beta)^T$ is locally asymptotically stable.

3. If $a = b = 0.5$, $\alpha = 2.4$, $\beta = 1.4$, find the 2-cycle, and show it is locally asymptotically stable.

4. Consider the hierarchal model

$$x(t+1) = x(t)e^{\alpha - x(t)}$$
$$y(t+1) = y(t)e^{\beta - y(t) - bx(t)},$$

Where species x is the dominant species and the presence of species y does not affect its dynamics, while species is the weaker species in which the interspecific parameter $b > 0$. Use numerical computation to find a 2-cycle if $\alpha, \beta > 0$. Let $\alpha = 2.2$, $\beta = 2.2$, $b = 0.5$. Find the 2-cycle and then show it is locally asymptotically stable.

5. (Burton & Henson). The ovulation synchrony in gulls may be modeled as
$$x(t+1) = be^{-cx(t)} + py(t)$$
$$y(t+1) = x(t),$$
where $b > 0$ is the number of birds per day to commence ovulation, $x(t)$ is the number of hens is the first 24h of an ovulation cycle and $y(t)$ is the number of hens in the second 24h of an ovulation cycle. A hen in the y-class that is ready to enter the x-class has a probability $1 - e^{-(x(t))}$. The time step is one day.

   (a) Show that, if $0 < c < \frac{e(1-p)}{b}$, then the positive equilibrium is locally asymptotically and unstable if $c > \frac{e(1-p)}{b}$.

   (b) Show that, if $c > \frac{e(1-p)}{b}$, then there is a 2-cycle that is locally asymptotically stable.

   (c) Study the dynamics of the model when c is sufficiently large. Show that
$$\left\{ \left(0, \frac{b}{1-p}\right)^T, \left(\frac{b}{1-p}, 0\right)^T \right\}$$
is a 2-cycle. Is this cycle asymptotically stable?

6. (Elaydi–Yakubu) Consider the Ricker competition model in which species x is stocked at the constant per-capita stocking rate p per generation
$$x(t+1) = x(t)e^{\alpha_1 - q_1(x(t)+y(t))}$$
$$y(t+1) = y(t)e^{\alpha_2 - q_2(x(t)+y(t))+px(t)}.$$

   (a) Show that the system has a positive 2-cycle if $\frac{(\alpha_1 - \ln(1-\alpha))}{\alpha_2} \leq \alpha_2 \leq 3.411822071$.

   (b) If $\alpha_1 = 1.5$, $\alpha_2 = 2.2$, $q_1 = q_2 = 1$, $p = 0.5$, show that the 2-cycle $\{(0.4063, 1.9592), (0.3741, 1.6604)\}$ is locally asymptotically stable.

7. In [183], the authors studied the following simplified version of the model in Example 56,
$$x(t+1) = \frac{bx(t)}{x(t) + y(t)} \tag{3.48}$$
$$y(t+1) = y(t)e^{\beta - (x(t)+y(t))}. \tag{3.49}$$

Note that Equation (3.48) is the nonoverlapping lottery model with a singularity at the origin. The lottery model emphasizes the role of chance. It assumes that resources are captured at random by recruits from a larger pool of potential colonists.

   (a) Show that, if $b > \beta$

8. Consider the Ricker competition model of Example 55, with $a = b = 0.5$, $\alpha_1 = 2.5265$, $\beta = 2.5266$. Numerical computation shows that
$$c_4 = \left\{ \begin{pmatrix} 2.91238 \\ 2.91277 \end{pmatrix}, \begin{pmatrix} 0.461491 \\ 0.461509 \end{pmatrix}, \begin{pmatrix} 2.88915 \\ 2.88952 \end{pmatrix}, \begin{pmatrix} 0.474047 \\ 0.474067 \end{pmatrix} \right\}$$
$$= \left\{ \begin{pmatrix} \bar{x}_1 \\ \bar{y}_1 \end{pmatrix}, \begin{pmatrix} \bar{x}_2 \\ \bar{y}_2 \end{pmatrix}, \begin{pmatrix} \bar{x}_3 \\ \bar{y}_3 \end{pmatrix}, \begin{pmatrix} \bar{x}_4 \\ \bar{y}_4 \end{pmatrix} \right\}$$
is a 4-cycle.

(a) Show that the 4-cycle is locally asymptotically stable.

(b) Show that the 2-cycle in Example 55 becomes unstable with the above parameter values.

9. (Guzowska, Luis, Elaydi) Consider the quadratic competition model with two species x and y for coefficients $0 < a, b < 4$ (so that the equations map the unit square $[0, 1] \times [0, 1]$ into itself).

$$x(t+1) = a \frac{1-x(t)}{1+cy(t)} x(t)$$

$$y(t+1) = b \frac{1-y(t)}{1+cx(t)} y(t)$$

(a) Find the values of a and b at which a 2-cycle appears. Find conditions under which the 2-cycle is asymptotically stable.

(b) Estimate the values of a and b at which a 4-cycle appears. Find conditions under which the 4-cycle is asymptotically stable.

3.7 The Stable Manifold Theorem

Consider the nonlinear difference equation

$$\mathbf{x}(t+1) = F(\mathbf{x}(t)) \tag{3.50}$$

such that F has an equilibrium point $\mathbf{x}^* \in \mathbb{R}^n$, and F is a C^2-map in an open neighborhood of $\mathbf{x}^*$. Let $A = JF(\mathbf{x}^*)$ be the Jacobian matrix associated with F. Then the linear part of (3.50) is given by

$$\mathbf{x}(t+1) = A\mathbf{x}(t). \tag{3.51}$$

Assume that the map F is hyperbolic, that is all the eigenvalues of A are off the unit circle in the complex plane. We arrange the eigenvalues of A into two sets, $S = \{\lambda_1, \lambda_2, \ldots, \lambda_r\}$ and $T = \{\lambda_{r+1}, \lambda_{r+2}, \ldots, \lambda_n\}$, where $|\lambda_i| < 1$ if $\lambda_i \in S$, and $|\lambda_j| > 1$ if $\lambda_i \in T$. Let E^s be the eigenspace spanned by the generalized eigenvectors corresponding to S and let E^u be the eigenspace spanned by the generalized eigenvectors corresponding to T. Then $\mathbb{R}^n = E^s \oplus E^0$. The local stable manifold of $\mathbf{x}^*$ is defined as

$$W_\ell^S(\mathbf{x}^*) = \left\{ \mathbf{x}_0 \in G : O(\mathbf{x}_0) \subset G \text{ and } \lim_{t \to \infty} F^t(\mathbf{x}_0) = \mathbf{x}^* \right\},$$

where G is an open neighborhood $B(\mathbf{x}^*, \varepsilon)$ of $\mathbf{x}^*$ and $O(x_o)$ is the orbit with initial condition x_o. To define the unstable manifold, we need to look at the negative orbits. Since F is not assumed to be invertible, we need to define a principle negative orbit

$$\{O^-(\mathbf{x}_0) = \{\ldots, \mathbf{x}(-t), x(-t+1), \ldots, \mathbf{x}(-1), x_0\},$$

where $F(\mathbf{x}(-t)) = \mathbf{x}_{-t+1}, t \in \mathbb{Z}^+$. The local unstable manifold of $\mathbf{x}^*$ is given by

$$W_\ell^u(\mathbf{x}^*) = \left\{ \mathbf{x}_0 \in G : \text{there exists a principle negative orbit } O^-(\mathbf{x}_0) \text{ such that } \lim_{t \to \infty} \mathbf{x}(-t) = \mathbf{x}^* \right\}.$$

The following result connects the eigenspaces E^s and E^u with the local stable and unstable manifolds W_ℓ^s and W_ℓ^u, respectively.

3.7. THE STABLE MANIFOLD THEOREM   145

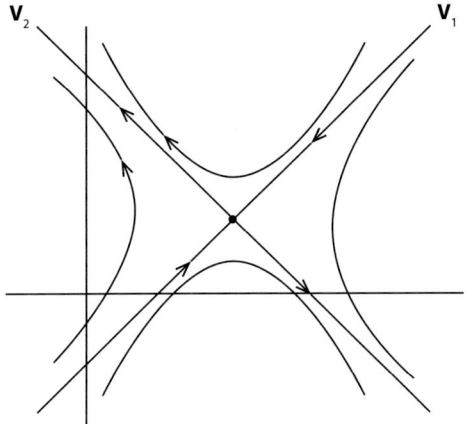

(a) The phase space diagram of the linearized system $\mathbf{x}(t+1) = A\mathbf{x}(t)$. Hence the origin is a saddle.

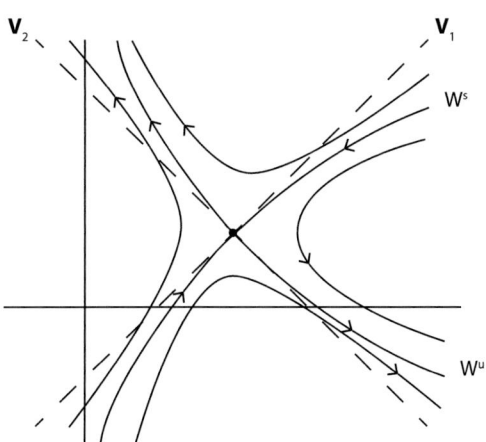

(b) The phase space diagram of the nonlinear system $\mathbf{x}(t+1) = F(\mathbf{x}(t))$.

Figure 3.22: The phase space diagram of the linearized system $\mathbf{x}(t+1) = A\mathbf{x}(t)$. Hence, the origin is a saddle.

Theorem 26 (The Stable Manifold Theorem) *Let $F : \mathbb{R}^2 \to \mathbb{R}^2$ be a hyperbolic C^2-map such that $\mathbf{x}^*$ is a fixed point of F. Then, in an open neighborhood G of $\mathbf{x}^*$, there exist two manifolds, the stable manifold $W_\ell^s(\mathbf{x}^*)$ of the same dimension as E^s and the unstable manifold $W_\ell^u(\mathbf{x}^*)$ of the same dimension as E^u, both passing through $\mathbf{x}^*$ such that*

(i) $W_\ell^s(\mathbf{x}^*)$ is positively invariant, E^s is tangent to $W_\ell^s(\mathbf{x}^*)$ at $\mathbf{x}^*$, and for any $\mathbf{x}_0 \in W_\ell^s(\mathbf{x}^*)$, $\lim_{t \to \infty} \mathbf{x}(t) = \mathbf{x}^*$. In other words, $\lim_{t \to \infty} F^t(\mathbf{x}_0) = \mathbf{x}^*$.

(ii) $W_\ell^u(\mathbf{x}^*)$ is positively invariant, E^u is tangent to $W_\ell^u(\mathbf{x}^*)$ at $\mathbf{x}^*$, and if $\mathbf{x}_0 \in W_\ell^u(\mathbf{x}^*)$ there exists a principle negative orbit $O^-(\mathbf{x}_0)$, such that $\lim_{t \to \infty} \mathbf{x}(-t) = \mathbf{x}^*$.

In two dimensions, the stable manifold $W_\ell^s(\mathbf{x}^*)$ is a curve in the plane, to which the eigenvector $\mathbf{v}_1$ corresponding to the eigenvalue $|\lambda_1| < 1$ (or $|\lambda_2| < 1$) is tangent to at $\mathbf{x}^*$, whereas the unstable manifold $W_\ell^u(\mathbf{x}^*)$ is a curve in the plane, where the eigenvector $\mathbf{v}_2$ corresponding to the eigenvalue $|\lambda_2| > 1$ (or $|\lambda_1| > 1$) is tangent to at $\mathbf{x}^*$ (Figure 3.22).

The stable manifold $W_\ell^s(\mathbf{x}^*)$ is tangent to the eigenvector $\mathbf{v}_1$ at the equilibrium point $\mathbf{x}^*$ and the unstable manifold $W_\ell^u(\mathbf{x}^*)$ is tangent to the eigenvector $\mathbf{v}_2$. Both $W_\ell^s(\mathbf{x}^*)$ and $W_\ell^u(\mathbf{x}^*)$ are positively invariant.

The map h is called the conjugacy map between the nonlinear system and its linear part. Roughly speaking, the theorem says that, in an open neighborhood of $\mathbf{x}^*$, the solutions of the nonlinear system behave similarly to the solutions of the linear system in an open neighborhood of the origin.

The following example illustrates the above theorem.

Example 57 Consider the difference equation
$$x(t+1) = 2x(t)$$
$$y(t+1) = \frac{1}{2}y(t) + y^2(t).$$

There are two fixed points $E_1^* = (0,0)$, and $E_2^* = (0, \frac{1}{2})$. The linearized system around E_1^* is given by
$$\begin{pmatrix} x(t+1) \\ y(t+1) \end{pmatrix} = A \begin{pmatrix} x(t) \\ y(t) \end{pmatrix}, \text{ where } A = \begin{pmatrix} 2 & 0 \\ 0 & \frac{1}{2} \end{pmatrix}.$$

Let
$$h(x,y) = \begin{pmatrix} x \\ ay + by^2 + cy^3 + O(y^4) \end{pmatrix}.$$

Now
$$F(h(x)) = \begin{pmatrix} 2x \\ \frac{1}{2}(ay + by^2 + cy^3 + O(y^4)) + (ay + by^2 + cy^3 + O(y^4)) \end{pmatrix}$$
$$= h\left(A\begin{pmatrix} x \\ y \end{pmatrix}\right) = h\begin{pmatrix} 2x \\ \frac{1}{2}y \end{pmatrix}$$
$$= \begin{pmatrix} 2x \\ \frac{a}{2}y + \frac{b}{4}y^2 + \frac{c}{8}y^3 + O(y^4) \end{pmatrix}.$$

Comparing the coefficients of y, y^2, y^3, we get
$$\frac{b}{4} = \frac{1}{2}b + a^2$$
$$\frac{c}{8} = \frac{1}{2}c + 2ab.$$

If we let $a = 1$, then $b = -4$, $c = \frac{64}{3}$ and
$$h(x,y) = \begin{pmatrix} x \\ y^x - 4y^2 + \frac{64}{3}y^3 + O(y^4) \end{pmatrix}.$$

An important question that may be raised here: what is the behavior of solutions $\mathbf{x}(t)$ whose initial points $\mathbf{x}(0) = \mathbf{x}_0$ do not lie on either the stable manifold $W_\ell^u(\mathbf{x}^*)$? The answer to this question may be addressed by the classical Hartman–Grobman Theorem.

Theorem 27 (Hartman–Grobman Theorem) [120] Let $F : \mathbb{R}^n \to \mathbb{R}^n$ be a C^1-diffeomorphism which is hyperbolic at the fixed point $\mathbf{x}^*$. Then there exist an open neighborhood U of $\mathbf{x}^*$, and a homeomorphism $h : U \to U$ such that $F(h(\mathbf{x})) = h(A\mathbf{x})$, where $A = JF(\mathbf{x}^*)$ is the Jacobian matrix of the map F (see Figure 3.23).

Remark 8 Note that, Hartman–Grobman Theorem is also valid for locally diffeomorphism maps. A map $F : \mathbb{R}^n \to \mathbb{R}^n$ is a local diffeomorphism if for every $\mathbf{x} \in \mathbb{R}^n$, there exists an open neighborhood G of $\mathbf{x}$ such that the restriction of F to G, $F|_U$, is a diffeomorphism from U to $F(G)$.

3.7. THE STABLE MANIFOLD THEOREM

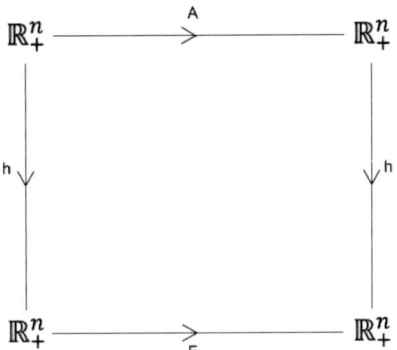

Figure 3.23: The map F is conjugate to $A = JF(\mathbf{x}^*)$, that is, $F(h(\mathbf{x})) = h(A\mathbf{x})$.

Example 58 *Consider the 2-species Ricker Competition model.*

$$x(t+1) = x(t)e^{r-c_{11}x-c_{12}y}$$
$$y(t+1) = y(t)e^{s-c_{21}x-c_{22}y}.$$

One may simplify this model by letting

$$u = c_{11}x, \quad v = c_{22}y, \quad a = \frac{c_{12}}{c_{22}}, \quad b = \frac{c_{21}}{c_{11}}.$$

Then we have

$$u(t+1) = u(t)e^{r-u-av}$$
$$v(t+1) = u(t)e^{s-v-bu}.$$

This may be represented by the map

$$F(\mathbf{u}) = \begin{pmatrix} ue^{r-u-av} \\ ve^{s-v-bu} \end{pmatrix}.$$

The determinant of the Jacobian, $JF(x)$, is either positive or negative for all $\mathbf{u} \in \mathbb{R}_+^2$ except at the critical (singular) points of JF, where $\det(JF(x)) = 0$. By the inverse function theorem, the map F is a local diffeomorphism except at the set of critical (singular) points in $\mathbb{R}_+^2$. The interior equilibrium is given by

$$(u^*, v^*) = \left(\frac{r-as}{1-ab}, \frac{s-rb}{ab-1} \right)^T.$$

At this equilibrium point, $\det(JF(u^*, v^*)^T) \neq 0$ except when $s = r = 1$. Hence, Hartman–Grobman applies for all values of r and s except when $r = s = 1$.

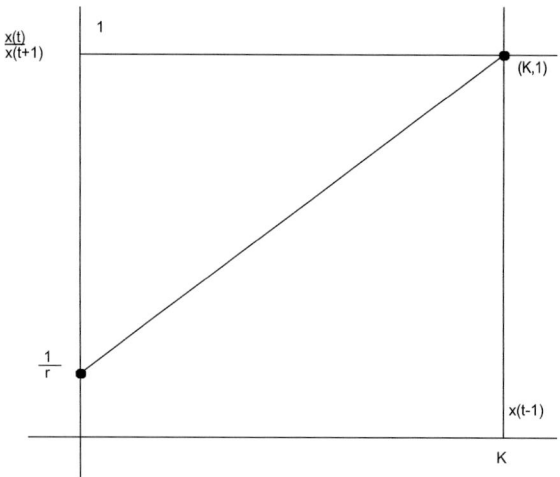

Figure 3.24: The graph depicts the reciprocal of the growth rate of the population r versus $x(t-1)$ on the x-axis.

Example 59 *The Lagged Beverton–Holt model. Suppose that the density of generation $(t+1)$ depends not only on the density of generation t but also on the density of generation $(t-1)$.*

Then from Figure 3.24 we have

$$\frac{x(t)}{x(t+1)} - \frac{1}{r} = \frac{\left(1 - \frac{1}{r}\right)}{K} x(t-1).$$

Hence,

$$x(t+1) = \frac{rKx(t)}{K + (r-1)x(t-1)}. \tag{3.52}$$

Equation (3.52) will be referred to as the Lagged Beverton–Holt model.

We may convert Equation (3.52) into a system of two equations by letting $x(t) = x(t-1)$, $y(t) = x(t)$. Then $x(t+1) = x(t) = y(t)$ and

$$y(t+1) = x(t+1) = \frac{rKx(t)}{K + (r-1)x(t-1)} = \frac{rKy(t)}{K + (r-1)x(t)}.$$

This gives us the system that is equivalent to Equation (3.52).

$$\begin{aligned} x(t+1) &= y(t) \\ y(t+1) &= \frac{rKy(t)}{K + (r-1)x(t)}. \end{aligned} \tag{3.53}$$

From the first equation, we have $x = y$. From the second equation, we have

$$\frac{rKy}{K + (r-1)x} = y.$$

Thus, either $y = 0$ or

$$\frac{rK}{K + (r-1)x} = 1$$

or $x = K$. Hence, we have two equilibrium points

$$E_1^* : \mathbf{x}_1^* = \begin{pmatrix} 0 \\ 0 \end{pmatrix}, \quad E_2^* : \mathbf{x}_2^* = \begin{pmatrix} K \\ K \end{pmatrix}$$

as expected. The Jacobian matrix is given by

$$J(\mathbf{x}) = \begin{pmatrix} 0 & 1 \\ \frac{-(r-1)rKy}{[K+(r-1)x]} & \frac{rK}{(K+(r-1)x)} \end{pmatrix}.$$

From the Jacobian

$$J(E_1^*) = \begin{pmatrix} 0 & 1 \\ 0 & r \end{pmatrix}$$

we find that E_1^* is stable if $r < 1$ and unstable if $r > 1$. The Jacobian evaluated at E_2^* is

$$J(E_2^*) = \begin{pmatrix} 0 & 1 \\ \frac{1-r}{r} & 1 \end{pmatrix}.$$

An application of the Jury conditions (Section 3.2.3) shows the equilibrium point E_2^* is unstable for $r < 1$ and asymptotically stable if $r < 1$.

3.8 Global Stability

Up to this point, we have discussed the local stability of equilibrium points and cycles. This is due to the fact we have used the linearization principle which depends on the behavior of the linear part of the nonlinear model under discussion. And by the Hartman–Grobman Theorem, for hyperbolic equilibrium points, the nonlinear system and its linear part agree only on a neighborhood of the given equilibrium point or cycle. In other words, what we can tell from the linearization principle is what happens in close proximity to an equilibrium or a cycle. Noting that, in discrete models, we may have large jumps of the population from one step to another, it is rather difficult to determine the global stability of systems of difference equation of dimension $n \geq 2$. From a biological point of view, it is can be of interest to know if, within the state space, all solutions approach the same locally asymptotically stable equilibrium point. That is to say if a locally asymptotically stable equilibrium is globally asymptotically stable. We have seen in Chapter 1, model (1.53), we may have a locally but not globally asymptotically stable. This may happen if we have multi-stable equilibrium points as in model (1.53) or due to the presence of a 2-cycle.

In contrast to a single species where we have powerful tools to determine the global asymptotic stability of equilibrium, in higher dimensional systems this is rather difficult. Indeed, it is an area of many important unsolved problems, particularly in discrete-time models. This would be a surprise to many researchers that deal only with continuous-time models (differential equations).

Next, we will explore several methods that have been used, with partial success, to study the global stability of equilibrium points. We will embark on this endeavor by starting with the most classical method in the literature, namely the use of Liapunov functions.

3.8.1 Liapunov Functions

One of the tools used to investigate the qualitative behavior of solutions of both difference and differential equations is the use of Liapunov functions. The method was introduced in 1892 by the Russian mathematician A.M. Liapunov in his famous memoir "Problemè général de la stabilité du movement" [226]. Later on, LaSalle extended Liapunov functions to difference equations [207, 208]. Moreover, he extended his theory to the famed LaSalle Invariance Principle. The method of Liapunov and its extensions is used either to find the basin of attraction of a fixed point or a cycle. However, in mathematical biology, it is mainly used to prove the global asymptotic stability of equilibrium points. The drawback, however, for the method of Liapunov is to find the appropriate Liapunov function, particularly, in competition and predator–prey models. However, we will show in Chapter 4 that the Liapunov method can be effective in an infectious disease model.

Now consider the difference system

$$\mathbf{x}(t+1) = F(\mathbf{x}(t)), \tag{3.54}$$

where $F : \mathbb{R}_+^n \to \mathbb{R}_+^n$. Let $V : \mathbb{R}_+^n \to \mathbb{R}$ be a real-valued function. Then the variation of V along the solutions of Equation (3.54) is given by $\Delta V(x(t)) = V(x(t+1)) - V(x(t))$. Note that, if $\Delta V(x(t)) \leq 0$, then the function V is non-increasing along the solution $x(t)$ of Equation (3.54).

Definition 18 *Let $\mathbf{x}^*$ be an equilibrium point of Equation (3.54). Then the real-valued function $V : \mathbb{R}_+^n \to \mathbb{R}$ is called a Liapunov function if for some neighborhood G of x^*, the following conditions hold:*

(i) V is continuous on G.

(ii) If $\mathbf{x} \in G$, then $F(\mathbf{x}) \in G$, and $\Delta V(\mathbf{x}(t)) \leq 0$.

We say that a real-valued function V is positive definite on a neighborhood G of an equilibrium point $\mathbf{x}^*$ if $V(\mathbf{x}) = 0$ and $V(\mathbf{x}) > 0$, for all $\mathbf{x} \neq \mathbf{x}^*$ in G. The following main result in the setting of difference equations is due to LaSalle [207, 208].

Theorem 28 (Liapunov Stability Theorem) *Let V be a positive-definite Liapunov function of Equation (3.54) on G. If $\Delta V(\mathbf{x}) < 0$, for all $\mathbf{x} \in G$, then $\mathbf{x}^*$ is locally asymptotically stable. Moreover, if $G = \mathbb{R}_+^n$ and $\lim_{||\mathbf{x}|| \to \infty} V(\mathbf{x}) = \infty$, then $\mathbf{x}^*$ is globally asymptotically stable on G.*

A proof of this theorem may be found in LaSalle [207, 208] and Elaydi [120].

To illustrate the utilization of the Liapunov Stability Theorem, we begin with a simple model, namely, the single-species Ricker model.

Example 60 *Consider the single-species Ricker model (1.30)*

$$x(t+1) = x(t)\exp(\alpha - cx(t)) \tag{3.55}$$

We have two cases to consider

- *If $\alpha < 0$, then $x^* = 0$ is the only nonnegative equilibrium point. We let $V(x) = x$. Then $\Delta V(x) = x(exp(\alpha - cx) - 1) < 0$ if $x \neq 0$. Then V is a Liapunov function on $G = [0, \infty)$, with $\lim_{||x|| \to \infty} V(\mathbf{x}) = \infty$. Hence, by the Liapunov stability theorem 28, the extinction equilibrium point is globally asymptotically stable.*

- If $0 < \alpha \leq 2$, then we have two equilibrium points $x_1^* = 0$ and $x_2^* = \alpha/c$. Let $V(x) = (x - x_2^*)^2$. Then

$$\Delta V(x) = (x \exp(\alpha - cx) - x_2^*)^2 - (x - x_2^*)^2$$
$$= x(e^{\alpha-cx} - 1)(xe^{\alpha-x} + x - 2x_2^*)$$

Letting $\phi(x) = xe^{\alpha-x} + x - 2x_2^*$ we have

$$\phi'(x) = e^{\alpha-cx}(e^{\alpha(x-x_2^*)} + 1 - cx)$$
$$\geq e^{\alpha-cx}(1 + c(x - x_2^*) + 1 - cx)$$
$$= e^{\alpha-cx}(2 - \alpha) \geq 0$$

if $\alpha \leq 2$. Hence, when $\alpha < 2$, ϕ is a strictly increasing function that vanishes at $x = x_2^*$. When $\alpha = 2$, $\phi'(x) > 0$ if $x \neq x_2^*$. Hence, for $\alpha \leq 2$, ϕ has a single zero at $x = x_2^*$. Since $e^{\alpha-cx}$ is decreasing we see that $\Delta V(x) \leq 0$ with equality if and only if $x = 0$ or $x = x_2^*$. Thus, $V(x)$ is a Liapunov function on $G = (0, \infty)$ with $\lim_{\|x\| \to \infty} V(\mathbf{x}) = \infty$. Hence, by the Liapunov stability theorem 28, the x_2^* is globally asymptotically stable.

Next, we will introduce one of the significant results on global stability, namely, LaSalle Invariance Principle [208]. But before stating the theorem, we need the following definition of invariance.

Definition 19 *A set M is positively invariant if $F(M) \subseteq M$, and invariant if $F(M) = M$. In case that F is injective (one-to-one), then M is negatively invariant if $F^{-1}(M) \subseteq M$.*

Theorem 29 (LaSalle Invariance Principle) *Consider the difference equation*

$$\mathbf{x}(t+1) = F(\mathbf{x}(t)), \tag{3.56}$$

where $F : \mathbb{R}_+^n \to \mathbb{R}_+^n$ is continuous on a subset G of $\mathbb{R}_+^n$. Suppose there is a Liapunov function $V : G \to \mathbb{R}$ such that V is continuous on the closure $\overline{G}$ of G. Let $E = \{\mathbf{x} : \Delta V(\mathbf{x}(t)) = 0\}$ and M is the largest positively invariant subset of E. Assume that, for every point $\mathbf{x} \in G$, its orbit $O(\mathbf{x})$ is bounded and is a subset of G. Then there exists $c \in R$ such that, for every $x \in G$, the omega limit set of orbit $O(x)$ satisfies $\omega(x) \subset M \cap V^{-1}(c)$.

For proof, we refer the reader to the Appendix 3.11.

Remark 9 *Note that, if M is an equilibrium point $\mathbf{x}^*$ in Theorem 29, then it is globally asymptotically stable on G.*

The following example illustrates the effectiveness of the LaSalle Invariance Principle.

Example 61 *([18]) Consider again the Ricker competition model*

$$x(t+1) = x(t) \exp\left(\alpha - c_{11}x(t) - c_{12}y(t)\right)$$
$$y(t+1) = y(t) \exp\left(\beta - c_{22}y(t) - c_{21}x(t)\right).$$

Define a Liapunov function $V: \mathbb{R}_+^2 \to \mathbb{R}_+$ as

$$V(x,y) = \frac{c_{21}}{c_{11}}x^2 + \frac{c_{12}}{c_{22}}y^2 + 2\frac{c_{12}c_{21}}{c_{11}c_{22}} - 2\alpha\frac{c_{21}}{c_{11}}x - 2\beta\frac{c_{12}}{c_{22}}y.$$

Using this quadratic Liapunov function, the authors of ([18]) obtained the following remarkable result.

Theorem 30 *For the Ricker map with $\alpha, \beta \in (0,2]$. Then the following statements hold true,*

(a) *If $\frac{c_{12}}{c_{22}} < \frac{\alpha}{\beta} < \frac{c_{11}}{c_{22}}$, then the unique interior equilibrium point (x^*, y^*) is globally asymptotically stable in the interior of $\mathbb{R}_+^2$ and each of the axial equilibrium points $(\alpha, 0)^T$ and $(0, \beta)^T$ is a saddle point with the positive half-axis as its stable manifold and the (so-called heteroclinic) orbit from this fixed point to (x^*, y^*) as its unstable manifold.*

(b) *If $\frac{c_{12}}{c_{22}} > \frac{\alpha}{\beta} > \frac{c_{11}}{c_{22}}$, then the unique interior fixed point $(x^*, y^*)^T$ is a saddle point with orbits from $(0,0)^T$ to $(x^*, y^*)^T$ as part of the stable manifold $W^s(x^*, y^*)$, which divides $\mathbb{R}_+^2 \setminus \{(0,0)\}$ into two open disjoint regions R_1, R_2 with $\mathbb{R}_+^2 \setminus \{(0,0)^T\} = R_1 \cup W^s(x^*, y^*) \cup R_2$, where $(0, \beta)^T \in R_1$ and $(\alpha, 0)^T \in R_2$. Each of the axial fixed points is asymptotically stable with R_1 or R_2 as its basin of attraction.*

(c) *If $\frac{c_{21}}{c_{11}} \leq \frac{\alpha}{\beta} \leq \frac{c_{12}}{c_{22}}$ but not $\frac{c_{21}}{c_{11}} = \frac{\alpha}{\beta} = \frac{c_{12}}{c_{22}}$, then there is no interior fixed point, $(\alpha, 0)^T$ is globally asymptotically stable in $\mathbb{R}_+^2$ excluding y-axis, and $(0, \beta)^T$ is a saddle point with the positive half y-axis as its stable manifold $W^s(0, \beta)$ and the orbits from $(0, \beta)^T$ to $(\alpha, 0)^T$ as its unstable manifold.*

(d) *If $\frac{c_{12}}{c_{22}} \leq \frac{\alpha}{\beta} \leq \frac{c_{11}}{c_{21}}$ but not $\frac{c_{12}}{c_{22}} = \frac{\alpha}{\beta} = \frac{c_{11}}{c_{21}}$, then there is no interior fixed point, the exclusion equilibrium $(0, \beta)^T$ is globally asymptotically stable in $\mathbb{R}_+^2$ excluding x-axis, and the other exclusion equilibrium $(\alpha, 0)^T$ is a saddle point with the positive half x-axis as its stable manifold $W^s(\alpha, 0)$ and the orbits from $(\alpha, 0)^T$ to $(0, \beta)^T$ as its unstable manifold.*

(e) *If $\frac{c_{12}}{c_{22}} = \frac{\alpha}{\beta} = \frac{c_{11}}{c_{21}}$, then, for every point $p \in_+ \setminus \{(0,0)\}$, $\{F^t(p)\}$ converges to a fixed point on $L_0 \cap \mathbb{R}_+^2$, where L_0 denotes the line $\alpha = x + \frac{c_{12}}{c_{22}}y$.*

Proof. The proof can be found in ([18]). ∎

Remark 10 *The Liapunov function method here does not prove that local stability in the Ricker model implies global stability as our Liapunov function is only for $0 < \alpha, \beta \leq 2$. For example, when $c_11 = 0.5$, $c_12 = 0.2$, $c_22 = 0.5$, $c_12 = 0.2$, $\alpha = 2.2$, $\beta = 1.6$, interior orbits globally converge to $(1.696, 1.261)^T$.*

It has been widely conjectured (e.g., [20]) that, for the Ricker model, local asymptotic stability of the interior equilibrium point implies its global asymptotic stability, but this remains an open problem.

Next, we give another example to demonstrate the effectiveness of the Theorem 29 (Figure 3.25).

Example 62 (A model of gonorrhea infection) *[207] Consider two heterosexual populations P_1 and P_2, where infected members of one population can transmit the disease to susceptibles in the other population. Recovery is possible but there is no immunity, and the populations are assumed*

3.8. GLOBAL STABILITY

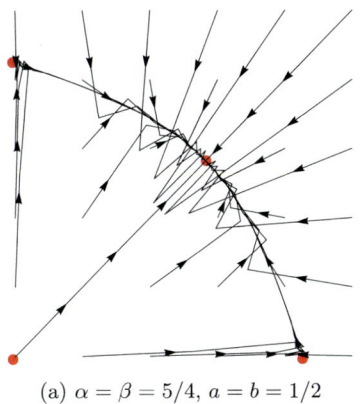
(a) $\alpha = \beta = 5/4$, $a = b = 1/2$

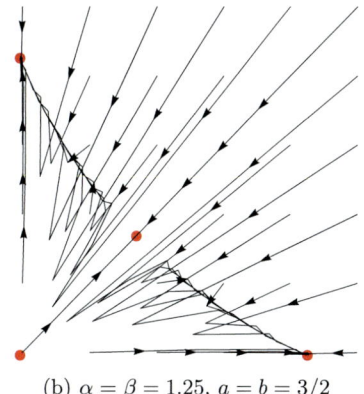
(b) $\alpha = \beta = 1.25$, $a = b = 3/2$

Figure 3.25: Examples of dynamics catalogued in Theorem 30.

to be constant. Let x_i be the fraction of the infected in population P_i, $i = 1, 2$. Then $(1 - x_i)$ is the fraction susceptible. This yields the following model.

$$\begin{aligned} x_1(t+1) &= a_1 x_2(t)(1 - x_1(t)) + (1 - b_1) x_1(t) \\ x_2(t+1) &= a_2 x_1(t)(1 - x_2(t)) + (1 - b_2) x_2(t) \\ &\text{or} \\ \mathbf{x}(t+1) &= F(\mathbf{x}(t)), \end{aligned} \tag{3.57}$$

where $0 < a_i < 1$, $0 < b_i < 1$. Note the F maps the unit square $G = [0, 1] \times [0, 1]$ into itself.
First, we will investigate the global asymptotic stability of the extinction equilibrium point

$$E_1^* = \begin{pmatrix} 0 \\ 0 \end{pmatrix}.$$

We will use the Liapunov function

$$V(x_1, x_2) = b_2 x_1 + b_1 x_2.$$

Then if $b_1 b_2 - a_1 a_2 \geq 0$, then

$$\Delta V(x_1, x_2) = -(b_1 b_2 - a_1 a_2) x_1 - (a_2 + b_2) a_1 x_1 x_2 \leq 0.$$

Then $\Delta V = 0$ if $x_1 = 0$ or in case $b_1 b_2 = a_1 a_2$, $x_2 = 0$. However, the only invariant set in the set

$$E = \left\{ \begin{pmatrix} x_1 \\ x_2 \end{pmatrix} : \text{ either } x_1 = 0 \text{ or } x_2 = 0 \right\}$$

is the origin E_1^*. Hence, by LaSalle invariance principle, E_1^* is globally asymptotically stable on G. Note that, if $b_1 b_2 - a_1 a_2 < 0$, then E_1^* is unstable (repeller).

Next, we investigate the coexistence equilibrium point

$$E_2^* = \begin{pmatrix} x_1^* \\ x_2^* \end{pmatrix}$$

$$x_1^* = \frac{1-\gamma_1\gamma_2}{1+\gamma_1}, \quad x_2^* = \frac{1-\gamma_1\gamma_2}{1+\gamma_2},$$

where
$$\gamma_i := \frac{b_i}{a_i}.$$

E_2^* is positive if $\gamma_1\gamma_s < 1$, i.e., if $b_1b_2 - a_1a_2 < 0$. It is convenient in this case to make the change of variables $u_i = x_i - x_i^*$, which yields new model equations

$$u_1(t+1) = (1-a_1\rho)u_1(t) + a_1(1-x_1(t))u_2(t)$$
$$u_2(t+1) = a_2(1-x_2(t))u_1(t) + (1-a_2\rho^{-1})u_2(t),$$

where $\rho = (1+\gamma_1)/(1+\gamma_2)$. Consider the Liapunov function

$$V(u_1, u_2) = a_2|u_1| + a_1\rho|u_2|.$$

If we assume that $1 - a_1\rho \geq 0$ and $1 - a_2\rho^{-1} \geq 0$, then

$$\Delta V(u_1, u_2) \leq -a_1a_2(x_1|u_2| + \rho x_2|u_1|).$$

Now $\Delta V = 0$ and E_1^* and E_2^* and the largest invariant set $M_0 = \{E_1^*, E_2^*\}$. Since E_1^* is unstable, all orbits in $G\backslash E_1^*$ converge to E_2^* and thus E_2^* is globally asymptotically stable on the interior of G.

The Competitive Exclusion Principle

In this section, we will elaborate on the Competitive Exclusion Principle that was described in Section 3.3. Consider the following exlusion principle: population x will displace population y if:

(i) Populations x and y are non-interbreeding and do the same thing, i.e., they occupy the same ecological "niche,"

(ii) Both populations occupy the same territory,

(iii) (added by Hardin [154]) Population x reproduces faster than population y.

The main objective here is to translate this competitive exclusion principle into concrete conditions on the parameters of the mathematical model under investigation. To accomplish this task, we start by stating a general result that will shed some light.

Consider the following competition model of two species x and y.

$$\begin{aligned} x(t+1) &= x(t)r_1(x(t), y(t)) \\ y(t+1) &= y(t)r_2(x(t), y(t)). \end{aligned} \tag{3.58}$$

Theorem 31 *If the per-capita growth rate of species x always exceeds that of species y, that is*

$$r_1(x,y) > r_2(x,y) > 0$$

for all point $(x,y)^T \in \mathbb{R}_+^2$, then species x drives species y to extinction.

3.8. GLOBAL STABILITY

Proof. Define a Liapunov functions as $V(x,y) = \frac{y}{x}$. Then

$$\Delta V(x,y) = \frac{y(t+1)}{x(t+1)} - \frac{y(t)}{x(t)}$$
$$= \frac{y(t)r_2(x(t),y(t))}{x(t)r_1(x(t),y(t))} - \frac{y(t)}{x(t)}$$
$$= \frac{y(t)}{x(t)}\left[\frac{r_2}{r_1} - 1\right] \leq 0.$$

Now $\Delta V(x,y) = 0$ if $y = 0$. Hence,

$$E = \{(x,y) : V(x,y) = 0\} = \{(x,0) : x \in (0,\infty)\}.$$

Let M be the largest positively invariant set in E. Then, by LaSalle Invariance Principle (Theorem 29), all orbits in the interior of $\mathbb{R}_+^2$ converge to a subset of M. Hence, species y goes to extinction. ∎

Corollary 5 *If the per-capita growth rate of species y exceeds that of species x, that is, $r_2(x,y) > r_1(x,y) > 0$ at each point $(x,y) \in \mathbb{R}_+^2$, then species y drives species x to extinction.*

The following example illustrates this theorem.

Example 63 *Consider the Ricker competition model*

$$x(t+1) = x(t)\exp(\alpha - c_{11}x(t) - c_{12}y(t))$$
$$y(t+1) = y(t)\exp(\beta - c_{21}x(t) - c_{22}y(t)).$$

Assume

$$\alpha > \beta, \quad 0 < c_{11} \leq c_{21}, \quad 0 < c_{12} \leq c_{22}.$$

Then

$$r_1(x,y) = \exp(\alpha c_{11}x - c_{12}y) > r_2(x,y) = \exp(\beta c_{21}x - c_{22}y)$$

and by Theorem 31, it follows that species y goes extinct.

3.8.2 Hierarchical Models

One type of higher dimensional model most amenable for determining global stability is a hierarchical model. By a hierarchical model, we mean a model with a network hierarchy of state variables [33, 35, 66, 125, 159]. In hierarchical models, it is assumed that one of the species is "silverback" species that gets first choice of resources and who is limited only by its own intraspecific competition, while other species are less dominant and their growth is limited by the presence of species that are above them in the hierarchy. A hierarchical model of n species $x_1, x_2, \ldots, x_n$ can be presented by a map $F : \mathbb{R}_+^n \to \mathbb{R}_+^n$, when

$$F(\mathbf{x}) = (f_1(x_1), f_2(x_1,x_2), \ldots, f_n(x_1,x_2,\ldots,x_n))^T,$$

where $\mathbf{x} = (x_1, x_2, \ldots, x_n)^T \in \mathbb{R}_+^n$. In the sequel, we will consider only Kolmogorov maps, where the scalar components of F have the form

$$f_1(x_1) = x_1 g_1(x_1)$$
$$f_2(x_1, x_2) = x_2 g_2(x_1, x_2)$$

$$f_n(x_1, x_2, \ldots, x_n) = x_n g_n(x_1, x_2, \ldots, x_n).$$

These types of maps are also called triangular maps [21] since their Jacobian is a triangular matrix.

A hyperspace H_k of dimension $k \leq n$ is defined as

$$H_k = \{(x_1, x_2, \ldots, x_k, 0, 0, \ldots, 0)^T : x_i > 0\}.$$

A fiber $\mathcal{F}_k$ in H_k is a one-dimensional subset of H_k. An example is the fiber

$$\mathcal{F}_k = \{(x_1^*, x_2^*, \ldots, x_{k-1}^*, x_k, 0, 0, \ldots, 0)^T : x_k > 0\},$$

where $(x_1^*, x_2^*, \ldots, x_{k-1}^*, x_k, 0, 0, \ldots, 0)^T$ is a fixed point in H_{k-1}. The dynamics on the fiber $\mathcal{F}_k$ is determined by the map f_k. Note that the map f_k on $\mathcal{F}$ may be regarded as a one-dimensional map since

$$F(x_1^*, x_2^*, \ldots, x_{k-1}^*, x_k, 0, 0, \ldots, 0) = (x_1^*, x_2^*, \ldots, x_{k-1}^*, f_k(x_1^*, x_2^*, \ldots, x_{k-1}^*, x_k, 0, 0, \ldots, 0))^T.$$

Hence, the dynamics on $\mathcal{F}_k$ is determined by the one-dimensional map

$$\tilde{f}_k(x_k) := f_k(x_1^*, x_2^*, \ldots, x_{k-1}^*, x_k)^T.$$

To state our main theorem, we will make the following assumptions.

(H_1) : All orbits of the system are bounded.

(H_2) : There are no 2-cycles on any fiber.

(H_3) : There are finitely many fixed points.

Theorem 32 [125] *Let $F : \mathbb{R}_+^n \to \mathbb{R}_+^n$ be a continuous triangular map such that assumption (H_1), (H_2), and (H_3) hold true. Then every orbit must converge to a fixed point of the map F in $\mathbb{R}_+^n$.*

Now suppose that we have only one interior equilibrium point that is locally asymptotically stable in a hierarchical model. Do we expect this equilibrium to be globally asymptotically stable? The answer is positive if the hierarchical model satisfies the three assumptions mentioned above. To illustrate this significant theorem, let us start with a 2-species hierarchical competition model of the Ricker type.

Example 64 *Consider the two species x and y, where species x is the dominant species. This may be given by the system*

$$x(t+1) = x(t)e^{\alpha - x(t)}$$
$$y(t+1) = y(t)e^{\beta - y(t) - bx(t)}. \tag{3.59}$$

Here we have three equilibrium points,

$$E_1^* = \begin{pmatrix} 0 \\ 0 \end{pmatrix}, \quad E_2^* = \begin{pmatrix} \alpha \\ 0 \end{pmatrix}, \quad E_3^* = \begin{pmatrix} 0 \\ \beta \end{pmatrix}, \quad E_4^* = \begin{pmatrix} x^* \\ y^* \end{pmatrix},$$

3.8. GLOBAL STABILITY

where $x^* = \alpha$, $y^* = \beta - b\alpha$. In order for E^* to be an interior (positive) equilibrium, we must assume $\beta > b\alpha$. Now we investigate the local stability of the three equilibria using the linearization principle. The Jacobian of one map $F(x,y) = \left(xe^{\alpha-x}, ye^{\beta-y-bx}\right)^T$ is given by

$$JF(x,y) = \begin{pmatrix} (1-x)e^{\alpha-x} & 0 \\ -bye^{\beta-y-bx} & (1-y)e^{\beta-y-bx} \end{pmatrix}.$$

Now

$$JF(E_1^*) = \begin{pmatrix} e^{\alpha} & 0 \\ 0 & e^{\beta} \end{pmatrix}.$$

Hence, the eigenvalue of $JF(0,0)$ are $\lambda_1 = e^{\alpha}$, $\lambda_2 = e^{\beta}$. Since $\alpha > 0$ and $\beta > 0$, the extinction equilibrium E_1^* is unstable.

The Jacobian

$$JF(E_2^*) = \begin{pmatrix} 1-\alpha & 0 \\ 0 & e^{\beta-b\alpha} \end{pmatrix}$$

associated with the equilibrium E_2^* has eigenvalues are $\lambda_1 = 1 - \alpha$ and $\lambda_2 = e^{\beta-b\alpha}$. If $0 < \alpha < 2$, then $|\lambda_1| < 1$. Since $\beta > b\alpha$, $\lambda_2 > 1$. Hence, the equilibrium point E_2^* is a saddle with the stable manifold lying on the x-axis and the unstable manifold lying in the interior of $\mathbb{R}_+^2$.

The Jacobian

$$JF(E_3^*) = JF\begin{pmatrix} 0 \\ \beta \end{pmatrix} = \begin{pmatrix} e^{\alpha} & 0 \\ -b\alpha & (1-\beta) \end{pmatrix}$$

associated with the equilibrium E_3^* has eigenvalues are $\lambda = e^{\alpha} > 1$ and $|\lambda_2| = |1-\beta| < 1$ if $- < \beta < 2$. Hence, E_3^* is a saddle where the stable manifold lies on the y-axis and the unstable manifold lies in the interior of $\mathbb{R}_+^2$.

Finally, the Jacobian

$$JF(E_4^*) = JF\begin{pmatrix} \alpha \\ \beta - b\alpha \end{pmatrix} = \begin{pmatrix} 1-\alpha & 0 \\ -by^* & 1-\beta+b\alpha \end{pmatrix},$$

associated with the equilibrium E_4^* has eigenvalues $\lambda_1 = 1 - \alpha$ and $\lambda_1 = 1 - (\beta - b\alpha)$. Thus, E_4^* is locally asymptotically stable if $0 < \alpha < 2$ and $\beta < 1 + b\alpha$.

Clearly, assumption (H_1) is satisfied as all solutions are bounded. Similarly, (H_3) is trivially satisfied as there are only four fixed points. Now the fiber $\mathcal{F}_1 = \{(\alpha, y) : y > 0\}$, the dynamics is controlled by the map $\tilde{f}_2(y) = ye^{(\beta-b\alpha)-y}$, which is a Ricker map, with a globally asymptotically stable equilibrium point (relative to the fiber $\mathcal{F}_1$) $(\alpha, \beta - b\alpha)$. Hence, there are no 2-cycles on the fiber $\mathcal{F}_1$ and Assumption (H_2) is satisfied. Hence, by Theorem 32, every orbit must converge to an equilibrium point. Since E_1^* is unstable, and the unstable manifolds of both E_2^* and E_3^* are in the interior of $\mathbb{R}_+^2$, it follows that equilibrium point E_4^* is globally asymptotically stable (see Figure 3.26).

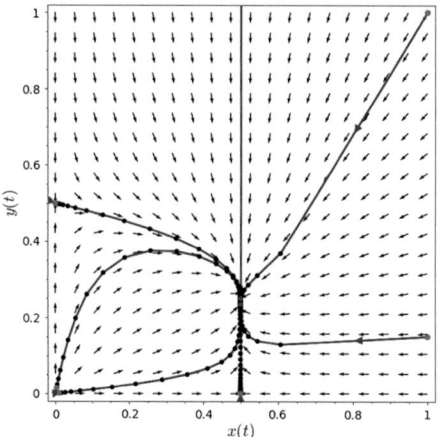

Figure 3.26: Phase space diagram of a hierarchical model, where the coexistence equilibrium E_3^* is globally asymptotically stable.

3.8.3 Competitive Systems

Competition models are competitive systems as given in the following formal definition. We will restrict our exposition to planar maps.

Definition 20 *Let $F : \mathbb{R}_+^2 \to \mathbb{R}_+^2$ be a continuous map, with $F(x, y) = (f(x, y), g(x, y))^T$. Then we say that the map F is competitive (strongly competitive), if $f(x, y)$ is non-decreasing (increasing) in x and non-increasing (decreasing) in y or non-increasing (decreasing) in x and non-decreasing (increasing) in y.*

A model represented by the difference equation

$$\mathbf{x}(t+1) = F(\mathbf{x}(t)), \tag{3.60}$$

$\mathbf{x} = (x, y)^T$ is called competitive (strongly competitive) if the map F is competitive (strongly competitive).

Note that, if the map F is differentiable, then it is strongly competitive if it's Jacobian is of the form $JF(\mathbf{x}) = \begin{pmatrix} + & - \\ - & + \end{pmatrix}$.

The following main global stability result is due to Smith [281].

Theorem 33 *[281] Assume that the map $F : \mathbb{R}_+^2 \to \mathbb{R}_+^2$ is C^1 and satisfies the following conditions:*

(i) $\det JF(\mathbf{x}) > 0$ for all $\mathbf{x} \in \mathbb{R}_+^2$,

(ii) F is competitive,

(iii) F is injective.

3.8. GLOBAL STABILITY

Then, if an orbit has a compact closure in $\mathbb{R}_+^2$, then it must converge to a fixed point of the map F.

Note that, for $\mathbf{x}_0 \in \mathbb{R}_+^2$, the orbit $O(\mathbf{x}_0)$ has a compact closure if it is contained in a closed and bounded subset of $\mathbb{R}_+^2$.

The proof of global injectivity of the map is computationally challenging. The following Theorem will simplify the proof.

Theorem 34 (Kestelman [188]) *Let $F: S \to \mathbb{R}_+^2$ be an open locally injective map such that $S \subset \mathbb{R}_+^2$ is compact, its boundary ∂S is connected and $F|_{\partial S}$ is injective. Then F is injective on S.*

Let us recall that a map F is open if it maps open sets to open sets. It is locally injective if for each $\mathbf{x} \in S$, there is an open neighborhood $B_{\mathbf{x}} = B(\mathbf{x}, r) \subset S$, $r > 0$, such that $F|_{B_{\mathbf{x}}}$ is injective, that is if $F: B_{\mathbf{x}} \to F(B_{\mathbf{x}})$ is injective. The closure of the orbit $O(\mathbf{x})$, denoted by $\overline{O(\mathbf{x})}$, is the set consisting of the orbit $O(\mathbf{x})$ and the set of its limit points. The closure of a set in $\mathbb{R}^n$ is always closed. Hence, if the orbit is bounded, then its closure is closed and bounded, and thus compact. The set of limit points of the orbit $O(\mathbf{x})$ is called the omega limit set and is denoted by $\omega(\mathbf{x})$. The omega limit set $\omega(\mathbf{x}) = \{\mathbf{y}: \lim_{t \to \infty} F^{i_t}(\mathbf{y}) = \mathbf{x}\}$ is closed and positively invariant.

So one may paraphrase the statement "If an orbit has a compact closure, then it converges to a fixed point" as "If an orbit has a compact closure, then its omega limit set is a fixed point."

Example 65 *The Leslie–Gower competition model is given by*

$$x(t+1) = \frac{b_1 x(t)}{1 + c_{11} x(t) + c_{12} y(t)}$$

$$y(t+1) = \frac{b_2 y(t)}{1 + c_{21} x(t) + c_{22} y(t)}$$

or $\mathbf{x}(t+1) = F(x(t), y(t))$. Recall from Example 53 that the interior equilibrium point

$$E_4^* = (x^*, y^*) = \left(\frac{(b_1-1)c_{22}\ \ (b_2-1)c_{12}}{c_{11}c_{11} - c_{12}c_{21}}, \frac{(b_2-1)c_{11} - (b_1-1)c_{12}}{c_{11}c_{11} - c_{12}c_{21}}\right)$$

is locally asymptotically stable if

$$b_1 > 1, \quad b_2 > 1, \quad c_{11}c_{22} - c_{12}c_{21} > 0, \quad \frac{c_{21}}{c_{11}} < \frac{b_2 - 1}{b_1 - 1} < \frac{c_{22}}{c_{12}}.$$

We will show that the map satisfies the conditions of Theorem 33. The Jacobian associated with this system is

$$JF(\mathbf{x}) = \begin{pmatrix} \frac{(1+c_{12}y)b_1}{(1+c_{11}x+c_{12}y)^2} & \frac{-b_1 c_{12} x}{(1+c_{11}x+c_{12}y)^2} \\ \frac{-b_2 c_{12} y}{(1+c_{21}x+c_{22}y)^2} & \frac{(1+c_{21}x)b_2}{(1+c_{11}x+c_{12}y)^2} \end{pmatrix}.$$

Hence, the map is strongly competitive (condition (ii)). Moreover,

$$\det JF(\mathbf{x}) = \frac{b_1 b_1 (1 + c_{12} y + c_{21} x)}{(1 + c_{11}x + c_{12}y)^2 (1 + c_{21}x + c_{22}y)^2} > 0$$

for all $(x, y)^T \in \mathbb{R}_+^2$, which verifies condition (i).

To prove injectivity, we are going to utilize Theorem 34. By simple computation, one may show that the inverse F^{-1} of the map is continuous. Hence, the map F is open. Moreover, since $\det JF(\mathbf{x}) > 0$, for all $\mathbf{x} \in \mathbb{R}_+^2$, the map is locally injective.

Next, we will apply Theorem 34 not on $\mathbb{R}_+^2$ but rather on a compact (closed and bounded) subset of $\mathbb{R}_+^2$ that contains $F(\mathbb{R}_+^2)$. Let the set

$$S = \left[\frac{b_1}{c_{11}}, 0\right] \times \left[0, \frac{b_2}{c_{22}}\right].$$

Then S is compact and contains $F(\mathbb{R}_+^2)$. Now from Chapter 1, we know that F is injective on the intervals $\left(\frac{b_1}{c_{11}}, 0\right)$ and $\left(0, \frac{b_2}{c_{22}}\right)$. So take two points on the line $x = \frac{b_1}{c_{11}}$, say $\left(\frac{b_1}{c_{11}}, y_1\right)^T$ and $\left(\frac{b_2}{c_{11}}, y_2\right)^T$ and assume that $F\left(\frac{b_1}{c_{11}}, y_1\right) = F\left(\frac{b_1}{c_{11}}, y_2\right)$. Then we get $y_1 = y_2$. Similarly, for points on the line $y = \frac{b_2}{c_{22}}$, if $F\left(x_1, \frac{b_2}{c_{22}}\right) = F\left(x_2, \frac{b_2}{c_{22}}\right)$, then $x_1 = x_2$. Hence, $F/\delta S$ is injective. By Theorem 34, F is injective on $S \supset F(\mathbb{R}_+^2)$, and assumption (iii) is proved. Hence, the equilibrium point is $E_4^* = (x^*, y^*)^T$ is globally asymptotically stable on the interior of $\mathbb{R}^+$.

To this end, we have considered only the less complicated models in which $\det JF \neq 0$. The most challenging cases are those when $\det JF(\mathbf{x}) = 0$ for some values of $\mathbf{x}$ in the domain of F. The set of points $\mathbf{x}$ for which $JF(\mathbf{x}) = 0$ is called the set of singular or critical points. In the population models that we have considered so far, the set of critical points, denoted by LC_{-1}, is a curve in $\mathbb{R}_+^2$, called the critical curve. For competitive or cooperative maps, one may impose conditions on the parameters of the model so that the critical curve is not in the range $F(\mathbb{R}_+^2)$. This point will be illustrated in the following widely studied in the literature, the Ricker competition model.

Example 66 *[258] Consider again the Ricker competition model*

$$x(t+1) = x(t)\exp\left(\alpha - x(t) - ay(t)\right)$$
$$y(t+1) = y(t)\exp\left(\beta - y(t) - bx(t)\right).$$

We assume that $0 < \alpha, \beta < 1$, $ab < 1$.

We will show that, under these conditions, the coexistence equilibrium

$$\mathbf{x} = \begin{pmatrix} x^* \\ y^* \end{pmatrix} = \begin{pmatrix} \frac{a\beta - \alpha}{ab - 1} \\ \frac{b\alpha - \beta}{ab - 1} \end{pmatrix}$$

is globally asymptotically stable. The Jacobian matrix of the Ricker map F is given by

$$JF(\mathbf{x}) = \begin{pmatrix} (1-x)e^{\alpha - x - ay} & -axe^{\alpha - x - ay} \\ -bye^{\beta - y - bx} & (1-y)e^{\beta - y - bx} \end{pmatrix}.$$

Hence, the determinant of JF is given by

$$\det JF(\mathbf{x}) = -(-1 + x + y - xy + abxy)\exp\left(\alpha + \beta - x - bx - y - ay\right).$$

Letting $\det JF(\mathbf{x}) = 0$ we obtain

$$y = \frac{1-x}{1-(1-ab)x}, \quad x \neq \frac{1}{1-ab}.$$

3.8. GLOBAL STABILITY

Hence, the critical curve is given by

$$LC_{-1} = \left\{(x,y) \in \mathbb{R}_+^2 \,|\, y = \frac{1-x}{1-(1-ab)x},\ x \neq \frac{1}{1-ab}\right\}$$

and is formed by two branches:

(i) LC_{-1}^1, a curve connecting the points $(0,1)^T$ and $(1,0)^T$ for $x < \frac{1}{1-ab}$,

(ii) LC_{-1}^2, an unbounded curve for $x > \frac{1}{1-ab}$.

It may shown in the region bounded by LC_{-1}^1, we have $\det JF(\mathbf{x}) > 0$ (Problem 7). Let $LC_0^1 = F(LC_{-1}^1)$. Now $F(1,0) = (e^{\alpha-1}, 0)^T$ and $F(0,1) = (0, e^{\beta-1})^T$ and since $0 < \alpha, \beta < 1$, we have $e^{\alpha-1} < 1$ and $e^{\beta-1} < 1$. Since the curve

$$y = \frac{1-x}{1-(1-a)x}$$

is concave down for, $x < \frac{1}{1-ab}$ and it follows that the curve LC_0^1 is below the critical curve LC_{-1}^1. Moreover, by Theorem 4.7 [20] it is shown that $F(\mathbb{R}_+^2)$ is contained in the region bounded by LC_0^1 (Figures 3.27, 3.28). Hence, assumptions (i) and (ii) of Theorem 33 hold.

It remains to prove the injectivity of the map on $F(\mathbb{R}_+^2) = \mathcal{D}$, where $\mathcal{D}$ is the region bounded by the curve LC_0^1 and the nonnegative x and y axes. To simplify the proof of injectivity, we use Theorem 35 below. To prepare for this topological result, we define a proper map as a map where the inverse image of a compact set is compact. And since $\mathcal{D}$ is compact, $F : \mathcal{D} \to \mathcal{D}$ is proper. This is due to the fact that, for a continuous map, the inverse image of a closed set is closed, and the inverse image of a compact (closed and bounded) set is also closed and bounded (compact).

Theorem 35 *[51] Suppose X and Y are metric spaces and $F : X \to Y$ is continuous and proper, and, for each $y \subset Y$, let $N(y)$ be the cardinal number of $F^{-1}(y)$. Then $N(y)$ is finite and constant on every connected component of $Y \setminus LC_0$.*

The proof of injectivity using the above Theorem is left to the reader as Problem 7b.

3.8.4 Comparison Methods

One of the simplest methods to prove global stability, especially for the extinction equilibrium is the comparison method. The idea is based on the squeeze theorem in analysis which states that given three sequences of real numbers $\{y(t)\}, \{x(t)\}, \{z(t)\}$ such that $y(t) \leq x(t) \leq z(t)$ and $\lim_{t\to\infty} y(t) = \lim_{t\to\infty} z(t) = L$, then $\lim_{t\to\infty} x(t) = L$.

We now illustrate the effectiveness of this method with the following examples.

Example 67 *Consider the Beddington Host–parasitoid model*

$$H(t+1) = H(t)\exp(\alpha - H(t) - aP(t))$$
$$P(t+1) = cH(t)(1 - e^{aP(t)}).$$

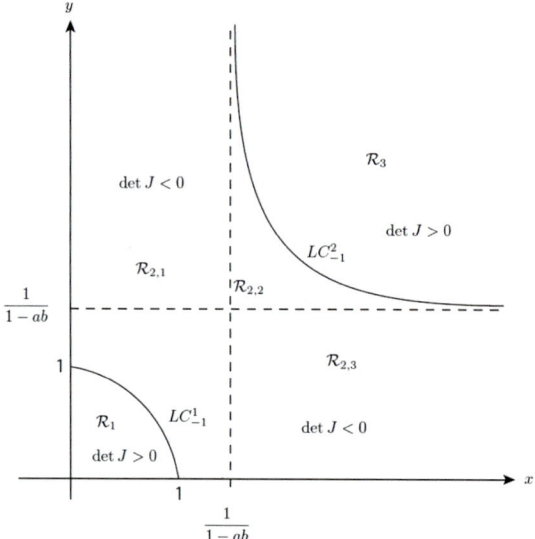

Figure 3.27: The set of singular points of the Ricker competition model consists of two branches, the lower curve LC^1_{-1} and the upper curve LC^2_{-1}.

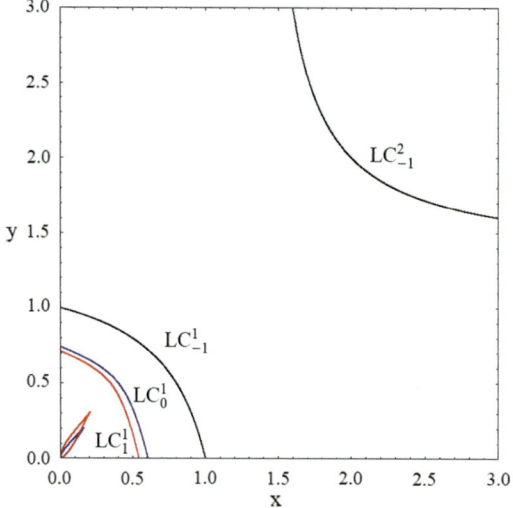

Figure 3.28: The images LC^1_0 and the upper curve LC^2_0 of LC^1_{-1} and the upper curve LC^2_{-1}, respectively, are shown.

3.8. GLOBAL STABILITY

Note that
$$0 \leq H(t+1) \leq H(t)e^{\alpha - H(t)}.$$

Let $y(t)$ be the solution of the initial value problem
$$y(t+1) = y(t)e^{\alpha - y(t)}, \; y(0) = H(0).$$

Then, by induction, one may show that
$$0 \leq H(t) \leq y(t),$$

for all $t \in \mathbb{Z}^+$. Now if $\alpha < 0$, then $y(t+1) < y(t)$ for all $t \in \mathbb{Z}^+$, and thus by the monotone convergence theorem and by Lemma 2 in Section 1.4.1, Chapter 1, $\lim\limits_{t \to \infty} y(t) = 0$. Hence, by the sandwiching theorem, $\lim\limits_{t \to \infty} H(t) = 0$ and, consequently, $\lim\limits_{t \to \infty} P(t) = 0$. Thus, the extinction equilibrium $\mathbf{x} = (0,0)^T$ is globally asymptotically stable on $\mathbb{R}^2_+$ if $\alpha < 0$.

In the next example, we use the comparison method for the extinction equilibrium for matrix models of structured populations of the form
$$\mathbf{x}(t+1) = P(\mathbf{x}(t))\mathbf{x}(t),$$
where $P(\mathbf{x})$ is nonnegative for all $\mathbf{x} \in \mathbb{R}^n_+$ and $P(0)$ is irreducible. Suppose
$$P(\mathbf{x}) \leq P(0) \tag{3.61}$$
for all $\mathbf{x} \in \mathbb{R}^n_+$. For any initial condition for all $\mathbf{x} \in \mathbb{R}^n_+$, we have
$$0 \leq \mathbf{x}(t+1) = P(\mathbf{x}(t))\mathbf{x}(t) \leq P(0)\mathbf{x}(t)$$
for all $t = 0, 1, 2, \cdots$. Let $\mathbf{y}(t)$ be the solution of the initial value problem
$$\mathbf{y}(t+1) = P(0)\mathbf{y}(t), \; \mathbf{y}(0) = \mathbf{x}(0).$$
Then an induction shows
$$0 \leq \mathbf{x}(t) \leq \mathbf{y}(t)$$
$t = 0, 1, 2, \cdots$. Then
$$\rho(P(0)) < 1 \implies \mathbf{y}(t) \to 0 \text{ as } t \to \infty \implies \mathbf{x}(t) \to 0 \text{ as } t \to \infty.$$
If $\mathcal{R}_0$ is defined using $P(0) = F + T$ in Chapter 2, then
$$\mathcal{R}_0 < 1 \implies \mathbf{x}(t) \to 0 \text{ as } t \to \infty.$$

The inequality (3.61) implies that the effect of population density on each entry in the projection matrix either decreases the entry or leaves it unaffected (i.e., cannot increase the entry).

Example 68 (LPA Model)
$$\begin{aligned} x_1(t+1) &= bx_3(t)\exp(-c_{el}x_2(t) - c_{ea}x_3(t)) \\ x_2(t+1) &= (1-\mu_l)x_1(t) \\ x_3(t+1) &= x_2(t)\exp(-c_{pa}x_3(t)) + (1-\mu_a)x_3(t) \end{aligned}$$

$$b, c_{el}, c_{ea}, c_{pa} \geq 0 \text{ and } 0 < \mu_l, \mu_a < 1$$

has coefficient matrix

$$P(\mathbf{x}) = \begin{pmatrix} 0 & 0 & b\exp(-c_{el}x_2 - c_{ea}x_3) \\ 1-\mu_l & 0 & 0 \\ 0 & \exp(-c_{pa}x_3) & 1-\mu_a \end{pmatrix}$$

$$\leq \begin{pmatrix} 0 & 0 & b \\ 1-\mu_l & 0 & 0 \\ 0 & 1 & 1-\mu_a \end{pmatrix} = P(0)$$

$P(0)$ is a Leslie matrix for which

$$\mathcal{R}_0 = b\frac{1-\mu_l}{\mu_a}$$

(see Chapter 2). Thus, if

$$\mathcal{R}_0 = b\frac{1-\mu_l}{\mu_a} < 1 \implies \mathbf{x}(t) \to 0 \text{ as } t \to \infty$$

for all nonnegative initial conditions.

3.8.5 Cooperative Systems

Not all species are competitive. As an example, aphids rely on symbiotic bacteria contained in specialized cells for essential amino acids lacking in their diet of sugary plant sap. In return, bacteria gain access to their host's offspring by entering aphid egg cells, being ingested by the offspring, or other mechanisms of transmission [114].

A population model of the two species x and y, represented by the difference equation

$$\mathbf{x}(t+1) = F(\mathbf{x}(t)), \quad \mathbf{x} = (x,y) \tag{3.62}$$

is said to be cooperative if $f(x,y)$ is non-decreasing in both the x and y and $g(x,y)$ is non-decreasing in both the x and y. This implies that the Jacobian matrix of F has the form

$$JF(\mathbf{x}) = \begin{pmatrix} + & + \\ + & + \end{pmatrix}$$

(allowing zero entries).

The following theorem is the main result concerning global stability of cooperative 2-species models [279].

Theorem 36 *Let $F : \mathbb{R}_+^2 \to \mathbb{R}_+^2$ be C^1-map such that the following statements hold true.*

(i) $\det JF(\mathbf{x}) < 0$, for all $\mathbf{x} \in \mathbb{R}_+^2$

(ii) F is cooperative

(iii) F is injective.

Then, if an orbit has a compact closure in $\mathbb{R}_+^2$, then its omega limit set is either a 2-cycle or a fixed point.

3.8. GLOBAL STABILITY

Example 69 *[157] Consider the model of juveniles x and adults y. Juveniles survive to become adults with density-dependent probability and adults leave juvenile offspring at a density-dependent rate*

$$x(t+1) = y(t)e^{r(1-y(t))}$$
$$y(t+1) = x(t)e^{-sx(t)}. \qquad (3.63)$$

This system may be represented by $\mathbf{x}(t+1) = F(\mathbf{x}(t))$, where $F : \mathbb{R}_+^2 \to \mathbb{R}_+^2$. We will work on the domain

$$F(\mathbb{R}_+^2) = \left[0, \frac{1}{r}e^{r-1}\right] \times \left[0, \frac{1}{s}e^{-1}\right].$$

We will show that F is cooperative on the larger domain $\left[0, \frac{1}{s}\right] \times \left[0, \frac{1}{r}\right]$ that contains $F(\mathbb{R}_+^2)$ assuming that $r < 2$, $re^{-1} < s < re^{1-r}$ (A_1).

There are two equilibrium points $E_1^* = (0,0)^T$ and $E_2^* = (x^*, y^*)^T$. To show that the existence of E_2^*, note the isoclines are given by the equations $r = sx + ry$ and $r - sx = rxe^{sx}$ with $0 < x < \frac{1}{s}$. Consider the function $h(x) = r - sx - rxe^{-sx}$. Now $h(0) = r > 0$ an

$$h\left(\frac{1}{s}\right) = r - 1 - \frac{r}{s}e^{-1}.$$

By assumption (A_1),

$$\frac{r}{s}e^{-1} > -1 > r - 1.$$

Thus, $h\left(\frac{1}{s}\right) < 0$. By the Intermediate Value Theorem, there is a positive root of this equation x^*, which gives, in turn, y^*. Hence, $E^* = (x^*, y^*)^T$ exists and is unique (Problem 8).

The Jacobian of the model system is

$$JF(\mathbf{x}) = \begin{pmatrix} 0 & (1-ry)e^{r(1-y)} \\ (1-sx)e^{-sx} & 0 \end{pmatrix}.$$

Hence, F is cooperative, with $\det JF(\mathbf{x}) = -(1-ry)(1-sx)e^{r(1-y)-sx} < 0$. Moreover, F is injective (Problem 6). Now for the extinction equilibrium $E_1^* = (0,0)^T$ we have the Jacobian

$$JF(E_1^*) = \begin{pmatrix} 0 & e^r \\ 1 & 0 \end{pmatrix}$$

with eigenvalues $\pm e^{\frac{r}{2}}$. Hence, E_1^* is unstable (a repeller).

Next, observe that $F(x^*, 0) = (0, y^*)^T$ and $F(0, y^*) = (x^*, 0)^T$. Hence, we have a 2-cycle $c_2 = \{(x^*, 0)^T, (0, y^*)^T\}$. Using Theorem 24 in Section 3.6, one may show that c_2 is a saddle fixed point of F^2, whose local stable manifold lies on the x-axis, and the unstable manifold is in the interior of $\mathbb{R}_+^2$.

By Theorem 33, the omega limit set of every point is a 2-cycle. Note that both the equilibrium point $E_1^* = (0,0)$ and the 2-cycle are repelling to points in the interior of $\mathbb{R}_+^2$, and thus cannot be the omega limit set of any interior point. Hence, we are left with one option, namely, the positive equilibrium E_2^* is the globally asymptotically stable on the interior of $\mathbb{R}_+^2$.

Exercises 3.8

1. Consider the discrete logistic (Beverton–Holt) model

$$x(t+1) = \frac{rKx(t)}{K + (r-1)x(t)}. \qquad (3.64)$$

   Use the Liapunov function $V(x) = |x - x^*| = |x - K|$ to show that $x^* = K$ is globally asymptotically stable for $x(0) > 0$.

2. In many biological models it is necessary to allow the growth rate of the variables to depend on the past history, rather than only the current values of the variables. These models require discrete delays. The delays or lags can represent gestation times, incubation periods, transport delays, or can simply lump complicated biological processes together, accounting only for the time required for these processes to occur. As an example, we introduce a population model of derived from the discrete logistic (Beverton–Holt) equation with one time unit time delay:

$$x(t+1) = \frac{bx(t-1)}{1 + cx(t)}.$$

   We convert this model to a two-dimensional system by letting $x_1(t) = x(t-1)$ and $x_2(t) = x(t)$. Then we have

$$x_1(t+1) = x_2(t)$$
$$x_2(t+1) = \frac{bx_1(t)}{1 + cx_2(t)}.$$

   (a) Find the equilibrium points.

   (b) Show that the interior (positive) equilibrium point exists and a saddle if $b > 1$.

   (c) Use the Liapunov function $V(x_1, x_2) = x_1^2 + x_2^2$ to show that the extinction equilibrium is globally asymptotically stable for $x(0) > 0$ if $0 < b < 1$ and unstable if $b > 1$.

3. [33] Consider the Hierarchical model of 2-species in which species x is the dominant (silverback) species and species y is the "wimpy" species.

$$x(t+1) = \frac{a_1 x(t)}{1 + b_1 x(t)}$$
$$y(t+1) = \frac{a_2 y(t)}{1 + b_2 (cx(t) + y(t))},$$

   where $a_1 > a_2 > 1$, $b_1 \leq b_2$ and $1 \leq c$.

   (a) Show that the per-capita growth rate of species x exceeds that of species y.

   (b) Show that that has a globally asymptotically equilibrium point for $x(0) > 0$.

4. [33] Consider a hierarchal model in which the dominant species has a scramble intraspecific competition, while the wimpy species has a contest intraspecific competition. The model is

given by
$$x(t+1) = x(t)e^{\alpha-x(t)}$$
$$y(t+1) = \frac{by(t)}{1+c(dx(t)+y(t))}$$

Show that, for $0 < \alpha < 2$, there is a globally asymptotically stable coexistence equilibrium point.

5. Investigate the global dynamics of the 3-species Hierarchical competition model of Ricker type.
$$x(t+1) = x(t)e^{\alpha-x(t)}$$
$$y(t+1) = y(t)e^{\beta-y(t)-b_1 x(t)}$$
$$z(t+1) = z(t) = e^{\alpha-z(t)-b_2 x(t) - cy(t)}$$

6. Investigate the global dynamics of the following hierarchical model of three species.
$$x(t+1) = \frac{ax(t)}{1+c_{11}x(t)}$$
$$y(t+1) = \frac{by(t)}{1+c_{21}x(t)+c_{22}y(t)}$$
$$z(t+1) = \frac{cz(t)}{1+c_{31}x(t)+c_{32}y(t)+c_{33}z(t)}$$

7. Consider the following model for competition between two species of nematodes feeding on a plant crop [178].
$$x(t+1) = a_x(n) + (1-a_1)\frac{b_1 x(t)}{c_1 + b_1 x(t) + y(t)}$$
$$y(n+1) = a_2 y(n) + (1-a_2)\frac{b_2 y(t)}{c_2 + x(t) + b_2 y(t)},$$
with $0 < a_1, a_2 < 1$, $b_1, b_2, c_1, c_2 > 0$. Show that the positive equilibrium point is globally asymptotically stable on the interior of $\mathbb{R}^2_+$.

8. Consider again the model of juveniles x and adults y model.
$$x(t+1) = y(t)e^{r(1-y(t))}$$
$$y(t+1) = x(t)e^{-sx(t)}.$$
Show that, if $0 < r < 2$, $re^{-1} < s < re^{1-4}$, then there exists a coexistence equilibrium point $E^* = (x^*, y^*)$.

9. (a) Consider the Ricker competition model in Example 66. Show that $\det JF(\mathbf{x}) > 0$ for all $\mathbf{x} \in \mathbb{R}_+^2$ in the region bounded by the lower branch of the critical curve LC_{-1}^1.

   (b) Prove that the map F is injective on the set $\mathcal{D} = F(\mathbb{R}_+^2)$.

10. [227] Consider the cooperative model of 2-species given by
$$x(t+1) = \frac{x(t)(r_1 + a_{12}y(t))}{1 + a_{11}x(t)}$$
$$y(t+1) = \frac{y(t)[r_2 + a_{21}x(t)]}{1 + a_{22}y(t)},$$
where all the parameters are positive and $a_{11}a_{22} > a_{12}a_{21}$, prove that the coexistence equilibrium point is globally asymptotically stable on the interior of $\mathbb{R}_+^2$.

11. [25] Consider the Ricker model of single species with delay $x(t+1) = x(t)\exp(\alpha - cx(t-1))$.

    (a) Write this equation as a system of two difference equations.

    (b) Prove that, if $0 < \alpha \leq 1$, then the positive equilibrium $(x^*, x^*)^T$ is globally asymptotically stable on the interior of $\mathbb{R}_+^2$.

12. (Research Project) Show that, in Problem 9, if $2 < b_1, b_2 < 3$, then the coexistence equilibrium point is globally asymptotically stable.

13.* [20] Show that, for the Ricker competition model, if $1 < \alpha, \beta < 2$, then the coexistence equilibrium is globally asymptotically stable on the interior of $\mathbb{R}_+^2$.

14. Consider the model
$$\begin{aligned} x_1(t+1) &= bx_2(t)\exp(-c_{11}x_1(t) - c_{12}x_2(t)) \\ x_2(t+1) &= sx_1(t)\exp(-c_{21}x_1(t) - c_{11}x_2(t)) \end{aligned} \quad (3.65)$$
which is a special case of Ebenman's juvenile–adult model
$$\begin{aligned} x_1(t+1) &= b\beta(x_1(t), x_2(t))x_2(t) \\ x_2(t+1) &= s\omega(x_1(t), x_2(t))x_2(t). \end{aligned} \quad (3.66)$$
Use the comparison method to show that

    (a) If $bs < 1$ in Equation (3.65), then the extinction equilibrium is globally asymptotically stable.

    (b) If $\beta(x_1, x_2) \leq 1$ and $\omega(x_1, x_2) \leq 1$ in Equation (3.66), then the extinction equilibrium is globally asymptotically stable on $\mathbb{R}_+^2$.

15. Consider the competition model

$$x(t+1) = x(t)r(c_{11}x(t) + c_{12}y(t))$$
$$y(t+1) = y(t)r(c_{21}x(t) + c_{22}y(t)).$$

Assume the $r(z)$ is a decreasing function and $c_{11} > c_{21}$ and $x_{12} > c_{22}$.

Biologically the inequalities imply species x feels more competition from its own individuals than from species y, and species y feels more competition from species x than from its own individuals.

Show that species x goes to extinction.

3.9 Persistence

Persistence is an important concept in ecology. This concept addresses the question of the long-term survival of species. Permanence, however, is concerned, in addition to persistence, with the control of population density. This is of particular importance in epidemic models, where the main concern is the prevention of pandemics.

Next, we give a general definition of persistence for maps. Let $F : \mathbb{R}_+^n \to \mathbb{R}_+^n$ be a map and let Ω be a subset of $\mathbb{R}_+^n$.

Definition 21 *[170] We say that F is uniformly persistent with respect to Ω if there exists $\eta > 0$ such that, for all $\mathbf{x} \in \mathbb{R}_+^n \setminus \Omega$,*

$$\liminf_{t \to \infty} d(F^t(\mathbf{x}, \Omega)) \geq \eta, \tag{3.67}$$

where

$$d(F^t(\mathbf{x}, \Omega)) = \sup_{z \in \mathbb{R}_+^n \setminus \Omega} \{||F^t\mathbf{x} - \mathbf{z}||\}.$$

Moreover, F is said to be uniformly permanent if, in addition to (3.67), there exits $\nu > 0$ such that $\limsup_{t \to \infty} d(F^t(\mathbf{x}, \Omega)) \leq \nu$.

In the sequel, we will focus on the case when the set Ω is the extinction equilibrium. In this case, uniform permanence of n species $x_1, x_2, \ldots, x_n$ with respect to Ω means that there exist two positive real numbers η and ν such that

$$\eta \leq \liminf_{t \to \infty} \min\{x_1(t), x_2(t), \ldots, x_n(t)\} \tag{3.68}$$

$$\nu \geq \limsup_{t \to \infty} \max\{x_1(t), x_2(t), \ldots, x_n(t)\} \tag{3.69}$$

with $x_i(0) > 0$, $1 \leq i \leq n$.

3.9.1 Single species

Consider the single-species population model

$$x(t+1) = x(t)f(x(t)). \tag{3.70}$$

Theorem 37 *Assume that $f : \mathbb{R}_+ \to \mathbb{R}_+$ is differentiable and satisfies the following conditions.*

(i) $f(x) > 0$ for all $x > 0$ and $f(0) > 1$

(ii) $\lim\limits_{x \to \infty} f(x) \leq \nu < 1$.

Then Equation (3.70) is uniformly permanent.

For a proof, the reader is referred to [181].
The following example illustrates this result.

Example 70 *Consider the exponential model*
$$x(t+1) = b_0 x(t) e^{-cx(t)} = x(t) f(x(t)),$$
where $f(x) = b_0 e^{-cx(t)}$. Then clearly (i) is satisfied if $f(0) = b_0 > 1$. Now (ii) is trivially satisfied since $\lim\limits_{x \to \infty} f(x) = 0 \leq \nu < 1$. Hence, the model is uniformly permanent if $b_0 > 1$.

3.9.2 Stage-structured Models

Consider the following stage-structured model
$$\mathbf{x}(t+1) = P(\mathbf{x}(t))\mathbf{x}(t), \tag{3.71}$$
where $P(\mathbf{x})$ is an $n \times n$ matrix whose elements are density dependent. Let us denote $P(\mathbf{0}) = (p_{ij})_{n \times n}$. The matrix $P(\mathbf{x})$ is assumed to be nonnegative for all $\mathbf{x} \in \mathbb{R}_+^n$. This assumption is always assumed in biological structured models. As a consequence, $\mathbf{x}(t) \in \mathbb{R}_+^n$, for all $\mathbf{x}(0) \in \mathbb{R}_+^n$. Recall from Chapter 2 that a matrix is irreducible if its life cycle graph contains a path (possibly with more than one step) from every node to every other node and we call such a graph strongly connected.

Now the notions of persistence and permanence are now adjusted, by biological consideration, to focus on total population size rather than the population size of each stage.

Definition 22 *Let $N(t) = \sum\limits_{i=1}^{n} x_i(t)$ be the total population density or size. Then model (3.71) is said to be uniformly persistent if*
$$\eta \leq \liminf_{t \to \infty} N(t) \tag{3.72}$$
and uniformly permanent if in addition to (3.72), we have
$$\limsup_{t \to \infty} N(t) \leq \nu \tag{3.73}$$
for some $\eta > 0$, $\nu > 0$ and for all $\mathbf{x}(0) \in \mathbb{R}_+^n$ with $N(0) > 0$.

To prepare for the main result, we need to introduce the concept of an absorbing set and dissipativity.

Definition 23 *A subset G of $\mathbb{R}_+^n$ is said to be an absorbing set of system (3.71) if for every $\mathbf{x}(0) \in \mathbb{R}_+^n$, there exists a positive real number $M = M(\mathbf{x}(0))$ such that $\mathbf{x}(0) \in G$, for all $t \geq M$.*

Definition 24 *System (3.71) is dissipative if it has a compact absorbing set.*

3.9. PERSISTENCE

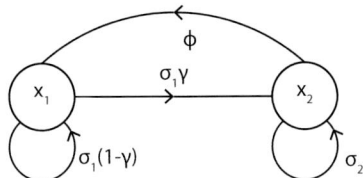

Figure 3.29: The life cycle graph of model (3.74).

Now we are ready to state the main result.

Theorem 38 *[198] Suppose that the map $P(\mathbf{x})\mathbf{x}$ is continuous and system (3.71) is dissipative, $P(\mathbf{0})$ is irreducible and $\mathbb{R}^n_+|\{0\}$ is positively invariant. Then system (3.71) is uniformly permanent if $P(\mathbf{0})$ has an eigenvalue λ with $|\lambda| > 1$.*

The following example, due to Neubert–Caswell, illustrates the preceding Theorem [198].

Example 71 *Let*

$$P(\mathbf{x}) = \begin{pmatrix} \sigma_1 f_1(\mathbf{x})(1 - \gamma f_3(\mathbf{x})) & \phi f_4(\mathbf{x}) \\ \sigma_1 f_1(\mathbf{x}) \gamma f_3(\mathbf{x}) & \sigma_2 f_2(\mathbf{x}) \end{pmatrix}, \qquad (3.74)$$

where $\sigma_1, \sigma_2, \gamma \in [0,1]$, $\phi > 0$, $f_i : \mathbb{R}^2_+ \to (0,1]$ are continuous functions with $f_1(0) = 1$, $1 \leq i \leq 4$.

The population has two stages, juveniles and adults. The functions f_i represent the density dependence part of the parameters $\gamma_1, \gamma_2, \sigma$ and ϕ, respectively. Now if $x = 0$, σ_1 and σ_2 are the density-free fractions of juveniles and adults that survive to the next generation, respectively. The parameter σ is the density-free fraction of surviving juveniles that mature to become adults and the parameter ϕ is the density-free number of juveniles produced by an adult. The life cycle graph of this model is given in Figure 3.29.

It is clear from Figure 3.29, matrix $P(\mathbf{0})$ is irreducible if and only if

$$\sigma_1 \gamma \phi > 0. \qquad (3.75)$$

It is straight forward to show that $\mathbb{R}^2_+|\{0\}$ is positively invariant if $\sigma_1 > 0$, or $\sigma_2 + \phi > 0$. Hence, it remains to show that the system is dissipative and the spectral radius $\rho(P(\mathbf{0})) > 1$.

Now if

$$\sigma_1(1 - \gamma) < 1 \text{ and } \sigma_2 < 1 \qquad (3.76)$$

and one of the terms $x_1 f_1(\mathbf{x})$, $x_1 f_3(\mathbf{x})$, or $x_2 f_4(\mathbf{x})$ is bounded above, the system (3.74) is dissipative. For proof, the reader is referred to [198].

Next, we consider the matrix

$$P(\mathbf{0}) = \begin{pmatrix} \sigma_1(1 - \gamma) & \phi \\ \sigma_1 \gamma & \sigma_2 \end{pmatrix}.$$

As stated earlier, $P(\mathbf{0})$ is irreducible if $\sigma_1 \gamma \phi > 0$, which we will assume in the sequel. Now $\rho(P(\mathbf{0})) > 1$ if and only if (i) $\det P(\mathbf{0}) < \operatorname{tr} P(\mathbf{0}) - 1$ or (ii) $\det P(\mathbf{0}) < -\operatorname{tr} P(\mathbf{0}) - 1$. Assumption (i) is equivalent to

$$\sigma_1 \gamma \phi > (1 - \sigma_2)[1 - \sigma_1(1 - \gamma)]. \qquad (3.77)$$

It is clear that Assumption (3.77) implies (3.75) and hence we have the following result: if (3.77) holds, then system (3.74) is uniformly permanent.

Next, we turn our attention to systems of Kolmogorov type in the following form

$$x(t+1) = x(t)f(x(t), y(t))$$
$$y(t+1) = y(t)g(x(t), y(t))$$
$$\text{or } \mathbf{x}(t+1) = F(\mathbf{x}(t)), \quad (3.78)$$

where $F(x,y) = (xf(x,y), yg(x,y))^T$. Note that Kolmogorov models can be written as structured models with

$$P(\mathbf{x}) = \begin{pmatrix} f(\mathbf{x}) & 0 \\ 0 & g(\mathbf{x}) \end{pmatrix},$$

where $\mathbf{x} = (x,y)^T$

We make the following assumptions:

A_1: $f(0,0) > 1$, $g(0,0) > 1$ and $f(x,y) > 0$, $g(x,y) > 0$ for all $x > 0$ and $y > 0$.

A_2: $\lim_{x \to \infty} f(x,0) = a_1 < 1$ and $\lim_{y \to \infty} g(0,y) = a_2 < 1$.

A_3: For the boundary equilibria $(\hat{x}, 0)^T$ and $(0, \hat{y})^T$ with $\hat{x} > 0$ and $\hat{y} > 0$, $f(0, \hat{y}) > 1$ and $g(\hat{x}, 0) > 1$.

A_4: $\frac{\partial f(x,0)}{\partial x} < 0$, $\frac{\partial f(x,y)}{\partial y} < 0$, $\frac{\partial g(x,y)}{\partial x} < 0$, $\frac{\partial g(0,y)}{\partial y} < 0$.

Condition A_1 ensures that the populations of species x and y move away from the origin and neither will go to extinction. Condition A_2 guarantees that species x and y have high enough intraspecific and interspecific competition which keeps both populations bounded. Condition A_3 ensures that both species have positive growth rates. Finally, condition A_4 implies that both species x and y suffer interspecific and intraspecific competition and also that the boundary equilibria $(x^*, 0)^T$ and $(0, y^*)^T$ are unique.

Next, we state the main result for Kolmogorov-type models.

Theorem 39 [181] *If system (3.78) satisfies conditions A_1-A_4 and both $(x, 0)^T$ and $(0, y)^T$ are globally asymptotically stable in their resident states, that is relative to interval $(0, \infty)$ on the x-axis and $(0, \infty)$ on the y-axis, respectively, then system (3.78) is uniformly permanent.*

For the proof, we refer the reader to [181].

We now illustrate this theorem by following the example of the competition between two-species.

Example 72 *Consider the Ricker competition model*

$$x(t+1) = x(t)e^{\alpha - x - ay}$$
$$y(t+1) = y(t)e^{\beta - y - bx}. \quad (3.79)$$

The conditions A_1, A_2 and A_4 are, clearly, satisfied. It remains to give conditions on the parameters to ensure that condition A_3 holds as well.

3.9. PERSISTENCE

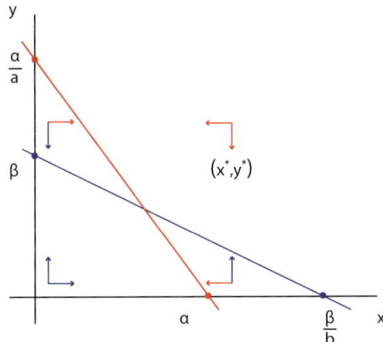

Figure 3.30: If $a\beta < \alpha < \frac{\beta}{b}$, the isoclines of system (3.79) intersect in the first quadrant guaranteeing the existence of a positive equilibrium $(x^*, y^*)^T$.

Now the equilibria on the axes are given by $(\hat{x}, 0)^T = (\alpha, 0)^T$ and $(0, \hat{y})^T = (0, \beta)^T$. So $f(0, \hat{y}) = e^{\alpha - a\beta} > 1$ if $\beta < \frac{\alpha}{a}$ and $g(\hat{x}, 0) = e^{\beta - b\alpha} > 1$ if $\alpha < \frac{\beta}{b}$. Both inequalities hold if $a\beta < \alpha < \frac{\beta}{b}$. These conditions on the parameters are natural to ensure that all orbits are bounded above and stay away from the extinction equilibrium $(0, 0)^T$ as may be seen in Figure 3.30.

Hence, we conclude that system (3.79) is uniformly permanent.

One can show that conditions A_2 and A_4 also implies dissipativity (Problem 3).

Exercises 3.9

1. Show that the discrete quadratic model $x(t+1) = rx(t)(1 - x(t))$ is uniformly permanent if $r > 1$.

2. Show that the Ricker model
$$x(t+1) = x(t)e^{r\left(1 - \frac{x(t)}{K}\right)}$$
is always uniformly permanent.

3. Prove that conditions A_2 and A_4 imply that system (3.79) is dissipative.

4. Consider the structured model
$$\mathbf{x}(t+1) = P(\mathbf{x})x(t), \text{ where } P(\mathbf{x}) = \begin{pmatrix} 0 & \phi f(x_1, x_2) \\ \sigma_1 g(x_a, x_2) & 0 \end{pmatrix}.$$

   (a) Show that the system has 2-cycles c_2 of the form $(\overline{x_1}, 0)^T$, $(0, \overline{x_2})^T$.

   (b) Find $\overline{x_1}$ and $\overline{x_2}$.

5. Consider the following modified Ricker competition model of two species x and y.
$$x(t+1) = x(t) \exp(r_1(1 - (c_{11}x(t))^{\nu_{11}} - (c_{12}y(t))^{\nu_{12}})$$
$$y(t+1) = y(t) \exp(r_2(1 - (c_{21}x(t))^{\nu_{21}} - (c_{22}y(t))^{\nu_{22}}),$$
where $r_i > 0$, $c_{ij} > 0$, $\nu_{ij} > 0$, $i, j = 1, 2$.

(a) Show that this model satisfies conditions A_1, A_2, and A_4.

(b) Show that, if $\nu_{11} > \nu_{21}$, $\nu_{22} > \nu_{12}$, $c_{11} > c_{21}$ and $c_{22} > c_{12}$, then the system is uniformly permanent.

6. Consider the following predator–prey model

$$x(t+1) = x(t)\left[s_1 + b_1 \exp\left(-a_{11}x(t) - a_{12}y(t)\right)\right]$$
$$y(t+1) = y(t)\left[s_2 + b_2 \exp\left(-a_{21}x(t) - a_{22}y(t)\right)\right],$$

where $s_1 + b_1 > 1$, $0 < s_1 < 1$, $0 < s_2 + b_2 < 1$, $a_{ij} > 0$, $1 \leq i, j \leq 2$. Show that, if $\ln\left(\frac{b_1}{1-s_1}\right) > \frac{a_{11}}{a_{21}} \ln\left(\frac{1-s_2}{b_2}\right)$, then the system is uniformly permanent.

3.10 Bifurcation

In Chapter 1, we studied the bifurcation of single-species models represented by the one-parameter difference equation $x(t+1) = f(x(t), \mu)$, where $x \in \mathbb{R}^+$ and $\mu \in \mathbb{R}$. Let x^* be the equilibrium point associated with parameter value μ^*. If $f'(x^*, \mu^*) = -1$, then we have period-doubling bifurcation and if $f'(x^*, \mu) = 1$, then we have transcritical bifurcation or exchange of stability bifurcation.

In this section, we will extend this study to the bifurcation in multi-species models with multi-parameters represented by the difference equation

$$\mathbf{x}(t+1) = F(\mathbf{x}(t), \mu), \qquad (3.80)$$

where $x \in \mathbb{R}^n_+$ and $\mu \in \mathbb{R}^k$, for some positive integer k. Let $\mathbf{x}^*$ be the equilibrium point of (3.80) associated with the vector parameter value μ^*, that is $F(\mathbf{x}^*, \mu^*) = \mathbf{x}^*$. As we have seen in Section 3.3.2, the eigenvalues of the Jacobian matrix $JF(\mathbf{x}^*)$ play the same role in the dynamics of Equation (3.80) as played by derivative $f'(x^*)$ of the function f in single-species models. An important distinction in multi-species models is the fact in addition to period-doubling bifurcation and saddle-node bifurcation, we have one more type of bifurcation, the Neimark–Sacker bifurcation. Figure 3.31 shows a parameter map in the (det −tr) plane at which these bifurcations can occur. The three scenarios illustrated by the graph are:

(i) Saddle-node bifurcation occurs when one of the two eigenvalues of the Jacobian matrix JF at (x^*, μ^*) is equal to 1. In other words, if the point (tr , det) lies on the line det = tr − 1.

(ii) Period-doubling bifurcation occurs when one of the two eigenvalues of the Jacobian matrix JF at (x^*, μ^*) is equal to -1. In other words, if the point (tr , det) lies on the line det = −tr − 1.

(iii) Neimark–Sacker bifurcation occurs if there are complex conjugate eigenvalues with modulus 1. In other words, if the point (tr , det) lies on the line det = 1 and inside the parabola det = $\frac{1}{4}(\text{tr })^2$.

In the sequel, we will give a more detailed analysis of these bifurcations.

Remark 11 *It is noteworthy to know the differences and similarities between difference equations and differential equations in regard to bifurcations. For instance, in differential equations, there are only two types of bifurcations, namely, saddle-node bifurcation (when one of the eigenvalues of the Jacobian matrix is real and equal to zero) and Hopf-bifurcation (when there are pure imaginary conjugates eigenvalues of the Jacobian matrix, i.e., $\lambda_{1,2} = ib$).*

3.10. BIFURCATION

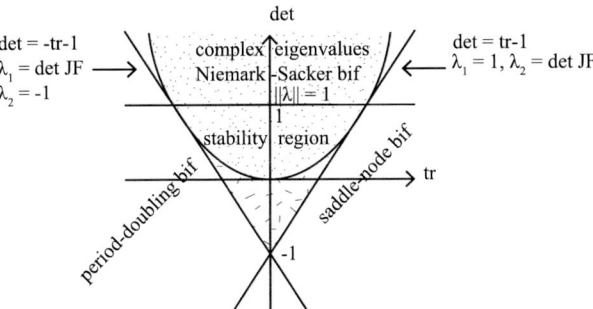

Figure 3.31: The det-tr figure showing the three main types of bifurcations on a two-dimensional map.

3.10.1 Period-Doubling Bifurcation

We now turn our attention to the phenomenon of period-doubling bifurcation in higher dimensional discrete systems. Recall that, in Example 54, we studied the stability of cycles of the Ricker competition model

$$x(t+1) = x(t)e^{\alpha - x(t) - ay(t)}$$
$$y(t+1) = y(t)e^{\beta - y(t) - by(t)}. \quad (3.81)$$

It was shown when the parameters α and β are increases, a locally asymptotically stable 2-cycle is born when the coexistence equilibrium loses its stability.

Numerical simulation shows that, in the symmetric case ($a = b$), as the pair (α, β) increases beyond the pair $(2,2)$, the Ricker model exhibits the phenomenon of *period-doubling route to chaos* and cycles of period $2, 4, 8, \ldots, 2^n, \ldots$ appear. The bifurcation pair here follows Table 1.1 in Chapter 1. The first bifurcation pair is $(\alpha_1, \beta_1) = (2,2)$. For $\alpha > 2$, $\beta > 2$, a 2-cycle is born and the positive equilibrium loses its stability, while the 2-cycle is asymptotically stable. The second bifurcation pair is $(\alpha_2, \beta_2) = (2.52646, 2.52646)$, beyond it the 2-cycle loses its stability an a new locally asymptotically stable 4-cycle is born. This 4-cycle loses its stability at the parameter $(\alpha_3, \beta_3) = (2.65635, 2.65635)$, and this periodic-doubling scenario continue until we reach $(\alpha_\infty, \beta_\infty) = (2.692, 2.692)$.

This is shown in the parameter-space bifurcation diagram (see Figure 3.32). In the black region if $(\alpha_1, \beta_1) \in S$, then the coexistence equilibrium point is asymptotically stable (Problem 1). If

$$(\alpha_1, \beta_1) \in R_1 = \{(\alpha_1, \alpha_2) : 0 < \alpha_1 \leq 2 \text{ and } \alpha_2 < b\alpha_1\},$$

then the exclusion equilibrium $(\alpha_1, 0)^T$ is asymptotically stable (Problem 2). And if

$$(\alpha_1, \alpha_2) \in Q_1 = \left\{(\alpha_1, \beta_1) : 0 < \beta_1 \leq 2 \text{ and } \beta_1 > \frac{\alpha_1}{a}\right\},$$

then the exclusion equilibrium $(0, \alpha_2)^T$ is asymptotically stable (Problem 3). In the blue region, these three equilibria lose their stability and a stable 2-cycle is born. Period-doubling continues in the green and red regions and beyond.

For the symmetric Ricker competition model with parameter values $a = 0.5$, $b = 0.7$, $\alpha = \beta$, Figures 3.33, 3.34 illustrate the phenomenon of period-doubling route to chaos. From these two figures we conclude that this dynamics is similar to the dynamics of one-dimensional Ricker in Chapter 1, Section 1.4.4.

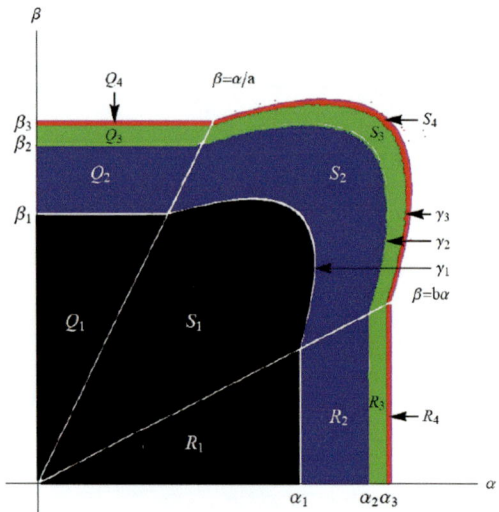

Figure 3.32: A parameter space (α, β) bifurcation diagram for system (3.81) which shows period-doubling in the three regions Q_1 (where species x goes to extinction), S_1 (where the coexistence equilibrium is asymptotically stable), and R_1 (where species y goes to extinction).

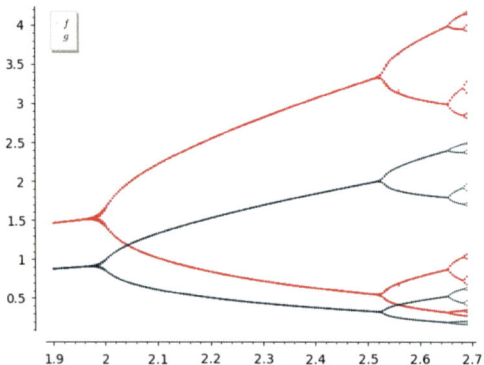

Figure 3.33: The Ricker map F is written as $F = (f, g)$. The figure shows a bifurcation diagram, on the x-axis we have the parameter α and on the y-axis we have the omega limit sets of both the x-components and the y-components of the orbits.

3.10. BIFURCATION

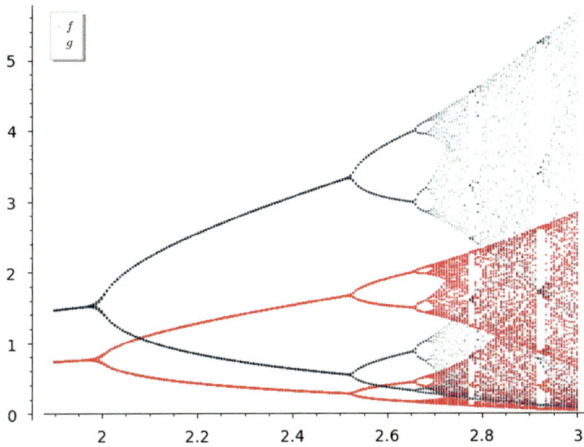

Figure 3.34: The figure illustrates the phenomenon of period-doubling route to chaos. The Ricker map F is written as $F = (f,g)^T$. The figure shows a bifurcation diagram, on the x-axis we have the parameter α and on the y-axis we have the omega limit sets of both the x-components and the y-components of the orbits.

It is an open question to prove that the two-dimensional Ricker Model exhibits period-doubling bifurcation. In addition, it is still an open problem to prove there is a universal Feigenbaum number for these types.

3.10.2 Neimark–Sacker Bifurcation

If the Jacobian of a system of n difference equations evaluated at an equilibrium has eigenvalues λ_i, then by the Linearization Principle the equilibrium is (locally asymptotically) stable if all eigenvalues lie inside the unit circle in the complex plane, i.e. satisty $|\lambda_i| < 1$. If a model parameter μ is changed, then in all likelihood so will the equilibrium and the associated eigenvalues. If one or more eigenvalues leave the unit circle as μ is increased (or decreased) through a value μ_0, then by the Linearization Principle the equilibrium loses its stability. We saw in Section 3.10.1 that if an eigenvalue leaves the unit circle through -1 then the result is generally the creation of 2-cycles, which bifurcate from the equilibrium as μ increases (or decreases) through μ_0. Thus, we associate a period doubling bifurcation with the eigenvalue -1. Another option is that the destabilization occurs because an eigenvalue leaves the unit circle instead through $+1$. We will see in Section 3.10.3 that this occurrence is generally associated with equilibrium bifurcations, ones that go by the names of transcritical, pitchfork, and saddle-node (and other equilibrium types of) bifurcations. In this section we consider the third possibility that can occur for systems of size $n \geq 2$, namely, that a complex conjugate pair of eigenvalues leaves the unit circle at a point $e^{i\theta}$, $0 < \theta < \pi$. In this case, a famous theorem due to Ju. I. Neimark and R. J. Sacker implies that in general a bifurcation occurs that creates an invariant loop.

An *invariant loop* is a closed curve in state space that consists of (usually infinitely many) orbits. According to the Neimark-Sacker Theorem[2] the result of destabilizing the equilibrium in this way results, under certain conditions, in the bifurcation of invariant loops emerging out of the equilibrium as μ increases (or decreases) through μ_0. An invariant loop is stable if all nearby orbits

---
[2]Ju. I. Neimark and R. J. Sacker published their work 5 years apart. Neimark published his theorem in 1967 in Russia [244], while Sacker published his work as part of his doctoral dissertation in 1974 at New York University [270].

approach the loop as $t \to \infty$ and is unstable if it is not stable. The bifurcation is sub-critical if the bifurcating loops exist before the equilibrium destabilizes and is super-critical if they exist when the equilibrium is unstable.

There are three technical conditions required for an application of the Neimark-Sacker Theorem. Suppose $\mathbf{x}(\mu)$ is an equilibrium associated with the parameter value μ. The conditions are that there exists an eigenvalue $\lambda(\mu) = r(\mu) e^{\theta(\mu)i}$ of the Jacobian at evaluated at $\mathbf{x}(\mu)$ and a value μ_0 such that

(i) $r(\mu_0) = 1$ and $r'(\mu_0) \neq 0$;

(ii) $e^{k\theta(\mu_0)i} \neq 1$ for $k = 1, 2, 3$ and 4;

(iii) a nondegeneracy condition holds (for more details, the reader is referred to Kuznetsov [202]).

Condition (i) is called the transversality condition; it ensures that an eigenvalue leaves the complex unit disk transversely without pausing. Condition (ii), that the critical eigenvalue is not a kth root of unity for $k = 1, 2, 3$ and 4, is called the nonresonance condition. (What kind of bifurcation occurs in these exceptional cases needs to be thoroughly understood.) The nondegeneracy condition (iii) requires the nonvanishing of a diagnostic quantity the formula for which, in practice, is usually very difficult the calculate, if not intractable. (See http://www.scholarpedia.org/article/Neimark-Sacker_bifurcation for a correct version of this formula, which is often in error in the literature.) If the nondegeneracy condition can be calculated its sign will indicate the direction of bifurcation and the stability properties of the invariant loop. In applications, once a complex pair crossing of the unit circle is determined, Neimark-Sacker bifurcations are usually explored by computer simulations.

Example 73 *We saw in Section 3.5.2 that the Beddington et al. predator-prey model*

$$H(t+1) = H(t) \exp\left(r\left(1 - \frac{H(t)}{K}\right) - \alpha P(t)\right)$$
$$P(t+1) = cH(t)\left(1 - e^{-\alpha P(t)}\right)$$

has a positive equilibrium that is a candidate for a Neimark-Sacker bifurcation in that it loses stability by a complex pair of eigenvalues leaving the complex unit circle; see Figure 3.18. As a specific example, take parameter values

$$r = 1.25, \quad \alpha = 1, \quad c = 1.2$$

and use $\mu = K$ as a bifurcation parameter. For $K = 2.5$ we numerically solve the equilibrium equations

$$H = H \exp\left(r\left(1 - \frac{H}{K}\right) - \alpha P\right)$$
$$P = cH\left(1 - e^{-\alpha P}\right)$$

for the positive equilibrium $(H, P)^T = (1.1468, 0.6766)^T$ (see Example 64). A straight forward calculation shows that the associated Jacobian has eigenvalue $0.98744 e^{0.9640i}$ which lies inside the complex unit circle, and therefore the positive equilibrium is stable. For $K = 2.7$ we numerically solve the equilibrium equations for the positive equilibrium $(H, P)^T = (1.1644, 0.71094)^T$. A straight forward calculation shows that the associated Jacobian has eigenvalue $1.0136 e^{0.96922i}$ which lies

3.10. BIFURCATION

outside the complex unit circle, and therefore the positive equilibrium is unstable. These calculations show that the complex eigenvalue associated with the equilibrium leaves the unit circle as K increases through a bifurcation value lying between 2.5 and 2.7. This suggests a Neimark-Sacker bifurcation occurs, which is corroborated by Figures 3.36, 3.37, 3.38 and 3.39.

Example 73 is an example of the paradox of enrichment principle in ecology, in which a predator-prey interaction destabilizes with increasing prey carrying capacity. For differential equation models, this principle occurs because of a Hopf bifurcation to a limit cycle. Example 73 illustrates the analogous phenomenon for a difference equation model by means of a Neimark-Sacker bifurcation (which is sometimes called a discrete Hopf bifurcation).

Figures 3.35, 3.36: With parameter values $r = 1.25$, $\alpha = 1$, and $c = 1.2$ in the Beddington et al. predator-prey model in Example 73, the orbit with initial conditions $(H(0), P(0))^T = (2, 1)^T$ yields an orbit that approaches an equilibria when $K = 2.5$. This is seen in both the time series plots of $H(t)$ and $P(t)$ and the orbit plotted in the phase plane.

Figures 3.37, 3.38: With parameter values $r = 1.25$, $\alpha = 1$, and $c = 1.2$ in the Beddington et al. predator-prey model in Example 73, the orbit with initial conditions $(H(0), P(0))^T = (2, 1)^T$ yields an orbit that approaches an invariant loop when $K = 2.7$, demonstrating the occurrence of a Neimark-Sacker bifurcation. This is seen in both the time series plots of $H(t)$ and $P(t)$ and the orbit plotted in the phase plane.

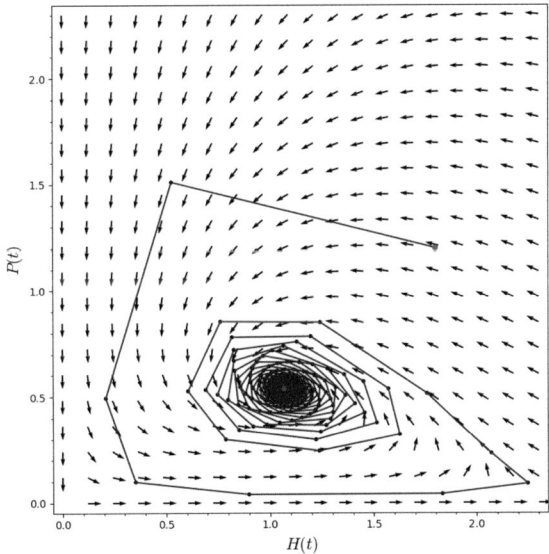

Figure 3.35: Phase space diagram for the parameters $r = 1.25$, $\alpha = 1$, $c = 1$, and $K = 2.5$.

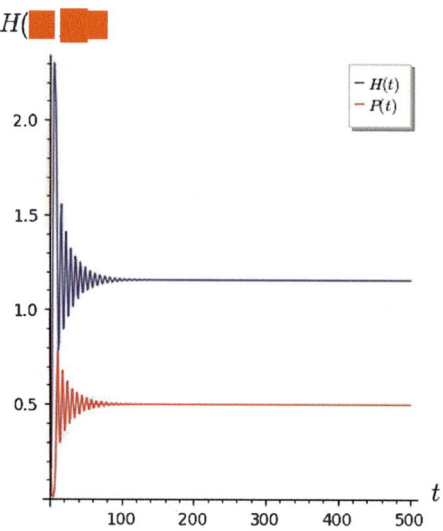

Figure 3.36: Time-series plots of $H(t)$ and $P(t)$ versus t, for the parameters $r = 1.25$, $\alpha = 1$, $c = 1$, and $K = 2.5$.

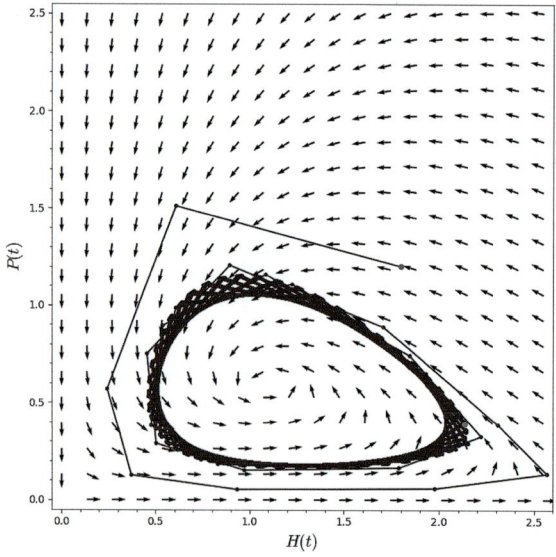

Figure 3.37: Phase space diagram for the parameters $r = 1.25$, $\alpha = 1$, $c = 1$, and $K = 2.7$.

3.10. BIFURCATION

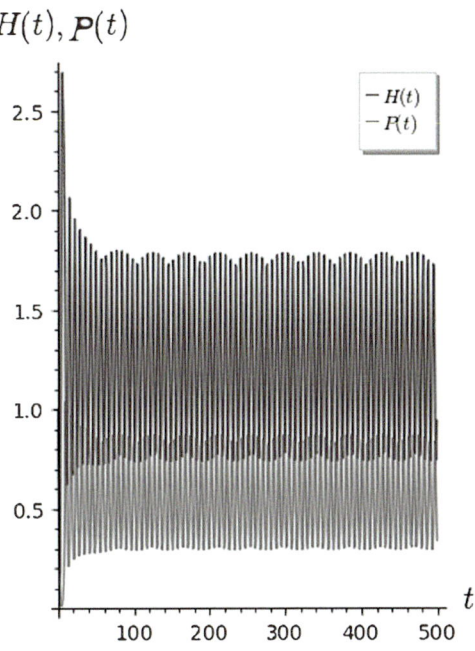

Figure 3.38: Time-series plots of $H(t)$ and $P(t)$ versus t, for the parameters $r = 1.25$, $\alpha = 1$, $c = 1$, and $K = 2.7$.

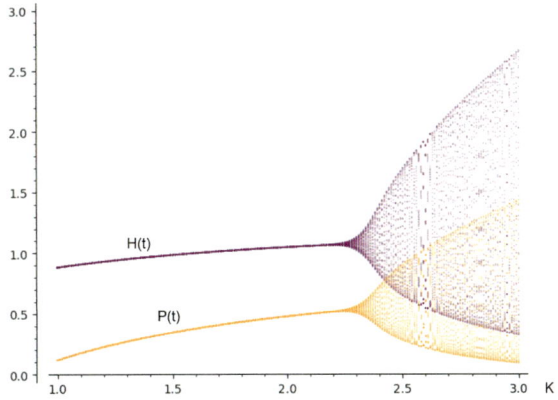

Figure 3.39: With parameter values $r = 1.25$, $\alpha = 1$, and $c = 1.2$ in the Beddington et al. predator-prey model in Example 73 the bifurcation diagram, using the bifurcation parameter K, shows a Neimark-Sacker bifurcation at approximately $K = 2.59$.

3.10.3 Saddle-Node Bifurcation

A +1 bifurcation of an equilibrium occurs when an eigenvalue of the Jacobian evaluated at the equilibrium passes through +1 as model a parameter changes. There are several types of +1 bifurcations, in transcritical, saddle-node, pitch-fork, and other less generic types. We first saw a transcritical bifurcation in Chapter 1 when studying the loss of stability of the extinction equilibrium. We will see this phenomenon again in Chapter 7 concerning the loss of stability of disease-free equilibria in epidemic models. What defines a transcritical bifurcation is the transversal intersection of two equilibrium branches. When this occurs there is typically an exchange of stability from one branch to the other (the exchange of stability property). The next example illustrates a transcritical bifurcation for a two species competition model.

Example 74 *[150] Consider the competition model*

$$x(t+1) = \frac{ax(t)(1-x(t))}{1+cy(t)}$$
$$y(t+1) = \frac{by(t)(1-y(t))}{1+dx(t)}, \quad (3.82)$$

where $a, b, c, d > 0$, $x, y \in [0,1] \times [0,1]$. A biologically feasible domain for this model is the unit square $[0,1] \times [0,1]$ when $a \leq 4$ and $b \leq 4$, which we assume throughout this example. There are four equilibrium points

$$E_1^* = (0,0)^T, \quad E_2^* = \left(\frac{a-1}{a}, 0\right)^T, \quad E_3^* = \left(0, \frac{b-1}{b}\right)^T, \quad E_4^* = (x^*, y^*)^T,$$

where

$$x^* = \frac{c(1-b) + b(a-1)}{ab - cd}, \quad y^* = \frac{d(1-a) + a(b-1)}{ab - cd}.$$

Let us write the model in a vector form

$$\mathbf{x}(t+1) = F(\mathbf{x}(t)).$$

The Jacobian matrix is given by

$$JF\begin{pmatrix} x \\ y \end{pmatrix} = \begin{pmatrix} \frac{a(1-x)-ax}{1+cy} & \frac{acx(1-x)}{(1+cy)^2} \\ \frac{-bdy(1-y)}{(1+dx)^2} & \frac{b(1-y)-by}{1+dx} \end{pmatrix}$$

Now

$$JF(E_1^*) = \begin{pmatrix} a & 0 \\ 0 & b \end{pmatrix}.$$

Hence, the extinction equilibrium point $E_1^* = (0,0)^T$ is locally asymptotically stable if $0 < a, b < 1$, and loses its stability if $a > 1$ and $b > 1$. In fact, E_1^* is globally asymptotically stable if $0 < b \leq 1$.

Lemma 5 *[150] If $0 < a \leq 1$ and $0 < b \leq 1$, the equilibrium point E_1^* is globally asymptotically stable on $\mathbb{R}_+^2$.*

3.10. BIFURCATION

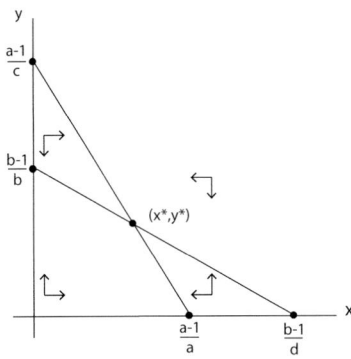

Figure 3.40: Isoclines: The coexistence equilibrium E_4^* exists if $\frac{a-1}{c} > \frac{b-1}{b}$ and $\frac{b-1}{d} > \frac{a-1}{a}$, $a > 1$, $b > 1$.

Proof. Note that on the forward invariant unit square, we have

$$0 \leq x(t+1) = \frac{ax(t)(1-x(t))}{1+cy(t)} \leq ax(t) - ax^2(t) \leq ax(t).$$

So if $0 < a < 1$, then $x(t) \leq x_0 a^t$, $t \in \mathbb{Z}^+$ and consequently, $\lim_{t \to \infty} x(t) = 0$. When $a = 1$, we have $x(t+1) < x(t)$, $t \in \mathbb{Z}^+$. Thus, $x(t)$ is a decreasing sequence bounded below by zero. This implies by the Monotone Convergence Theorem that $\lim_{t \to \infty} x(t) = L$, $0 \leq L < x(t)$. By Lemma 2 in Section 1.4.1, Chapter 1, L must be an equilibrium point. Since the only equilibrium point on the x-axis is zero, it follows that $\lim_{t \to \infty} x(t)$. Similarly, one may show that, if $0 < b \leq 1$, $\lim_{t \to \infty} y(t) = 0$. This completes the proof of the Lemma. ∎

We now turn our attention to the coexistence equilibrium $E_4^* = (x^*, y^*)^T$ which is positive if

(A). $a > 1, b > 1$ and $\quad \frac{a-1}{c} > \frac{b-1}{b}, \frac{b-1}{d} > \frac{a-1}{a}$

(See Figure 3.40). The Jacobian matrix at E_4^* is

$$JF(E_4^*) = \begin{pmatrix} \frac{1-2x^*}{1-x^*} & \frac{-cx^*}{a(1-x^*)} \\ \frac{-dy^*}{b(1-y^*)} & \frac{1-2y^*}{1-y^*} \end{pmatrix}.$$

Using the determinant-trace analysis, one can show that, if $1 < a, b < 3$, and assumption (A) holds, then the equilibrium point E_4^* is asymptotically stable (Problem 4). Hence, a transcritical bifurcation (exchange of stability) occurs at $(a, b) = (1, 1)$. Figure 3.41 shows a transcritical bifurcation for the symmetric case when $a = b$. The extinction equilibrium is asymptotically stable if $0 < a = b \leq 1$, and loses its stability for $a, b > 1$, and a new asymptotically stable equilibrium E_4^* is born. However, as a increases, a period-doubling bifurcation route to chaos occurs (Figure 3.42). At about $a \approx 4.2$, the simulation shows that both species go extinct (Figure 3.43)

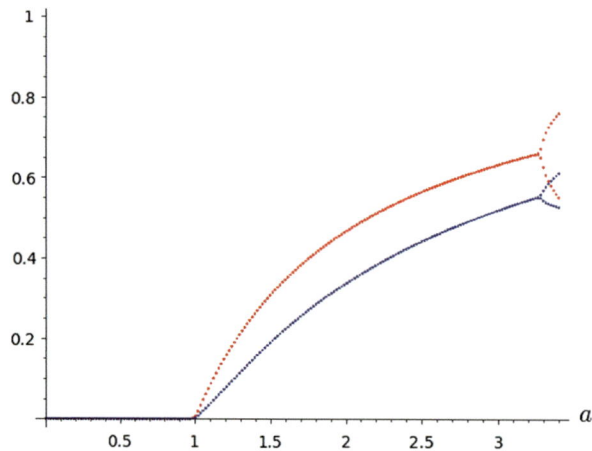

Figure 3.41: Shown is the bifurcation diagram for the symmetric version of the competition model (3.83). For $0 < a = b \leq 1$, the extinction equilibrium is asymptotically stable, and when a and b exceeds 1, a new asymptotically stable equilibrium point E_4^* is born.

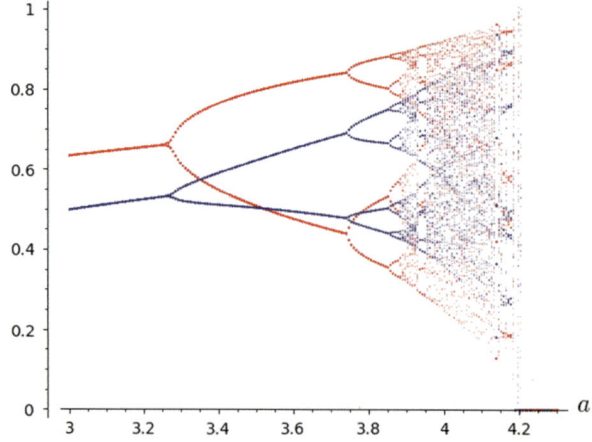

Figure 3.42: Period-doubling route to chaos is shown.

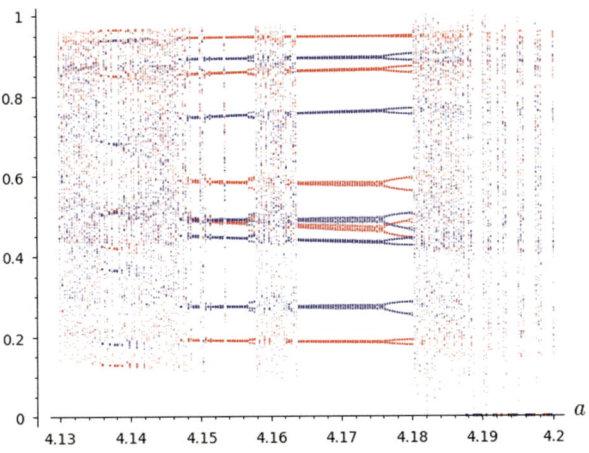

Figure 3.43: At $a \approx 4.165$ a 5-cycle appears and at $a \approx 4.2$, both species go the extinction.

Exercises 3.10

1. Consider the 2-species Ricker competition model (Example 70) and the parameter space $(\alpha - \beta)$ diagram (Figure 3.32).

    (a) Show that, if the parameters α, β in region $\mathcal{R}_2$, then there is a 2-cycle which is locally asymptotically stable.

    (b) Show that, if the parameters α, β in region $\mathcal{Q}_2$, then there is a 2-cycle which is locally asymptotically stable.

    (c) Show that, if the parameters α, β in region $\mathcal{S}_2$, then there is a 2-cycle which is locally asymptotically stable.

2.* Consider the following predator–prey model
$$x(t+1) = rx(t)e^{-y(t)}$$
$$y(t+1) = x(t)\left(1 - e^{-ay(t)}\right),$$
where x is the prey and y is the predator, $a > 0$. Investigate the Neimark–Sacker bifurcation. Let $\nu = r^a - 1$ and $r > 1$.

    (a) Show that when $0 < a < 2$, the coexistence equilibrium $(x^*, y^*)^T$ is unstable for all values of $\nu > 0$.

    (b) Show, by simulation, that a stable invariant closed curve appears in the phase space diagram for $a = 40$, $r = 3$. Show that the coexistence equilibrium is given by $(x^*, y^*)^T \approx (1.10, 1.10)^T$. Moreover, $(x^*, y^*)^T$ is unstable.

3. Consider the following discrete version of the predator–prey Lotka–Volterra model
$$x(t+1) = ax(t)(e^{-y(t)}(1-x(t)) - x(t)y(t)$$
$$y(t+1) = \frac{1}{b}x(t)y(t),$$
where x is the prey and y is the predator.

(a) Prove that the coexistence equilibrium undergoes a Neimark–Sacker bifurcation.

(b) Show that at $x \approx 1.64$, $b = 0.16$, the coexistence equilibrium (x^*, y^*) is given by $x^* \approx 1.64$, $y^* \approx 0.16$ and that it is unstable.

4. The population dynamics of Easter Island is given by
$$x(t+1) = x(t) + ax(t)\left(1 - \frac{x(t)}{R(t)}\right)$$
$$R(t+1) = R(t) + cR(t)\left(1 - \frac{R(t)}{K}\right) - hx(t),$$
where $P(t)$ is the size of population in Easter Island at time t, $R(t)$ is the amount of resources, say trees, animals, or crops, present at time t. The parameters, a, c, K, and h, represent the intrinsic growth rate of the population, the intrinsic growth rate of the resource, the resource carrying capacity, and the harvesting rate, respectively. For simplicity, assume $K = 1$.

(a) Determine conditions for asymptotic stability of the three equilibrium points.

(b) Show that, if $h = \frac{a+c-ac}{2a}$ and $a < c < a - 4 + \frac{8}{a}$, the interior equilibrium loses its stability.

(c) Determine the Neimark–Sacker bifurcation point and draw the resulting closed invariant curve if $h = 1.499$.

5. (Research Project) Consider the competition model
$$\begin{aligned} x(t+1) &= \frac{ax(t)(1-x(t))}{1+cy(t)} \\ y(t+1) &= \frac{by(t)(1-y(t))}{1+dx(t)}. \end{aligned} \quad (3.83)$$

(a) Under the Condition (A), show that the coexistence equilibrium point is asymptotically stable.

(b) Assume $c = d = 0.5$. Draw parameter space $(a - b)$ diagram similar to Figure 3.32.
    (i) Show that, if $(a, b) \in S_1$, then the coexistence equilibrium point is asymptotically stable.
    (ii) Show that, if $(a, b) \in S_2$, then we have a 2-cycle that is asymptotically stable.
    (iii) Repeat (i) and (ii) for R_1, R_2, Q_1, Q_2.

3.11 Appendix

LaSalle's Invariance Principle

In this appendix, we will present a second proof of LaSalle's Invariance Principle. To facilitate the proof, we need the following results on the properties of the omega limit set. Let us recall the definition of the omega limit set $\omega(\mathbf{x}_0)$ of a point $\mathbf{x}_0 \in \mathbb{R}^n$,

$$\omega(\mathbf{x}_0) = \{\mathbf{y} \in \mathbb{R}^n : \lim_{t \to \infty} \mathbf{x}(t) = \mathbf{y}\}$$

or

$$= \{\mathbf{y} \in \mathbb{R}^n : \lim_{t \to \infty} F^t(\mathbf{x}_0) = \mathbf{y}\},$$

where $F : \mathbb{R}^n \to \mathbb{R}^n$.

Lemma 6 (Elaydi [130]) *If the Orbit $O(X_0)$ is bounded, then the omega limit set $\omega(\mathbf{x}_0)$ is nonempty, closed bounded (compact), and positively invariant.*

Proof of LaSalle Invariance Principle. By Lemma 6, for $\mathbf{x}_0 \in G$, $\omega(\mathbf{x}_0) \subset \overline{G}$ since $\Delta V(\mathbf{x}(t)) = V(\mathbf{x}(t+1)) - V(\mathbf{x}(t)) \leq 0$, $V(\mathbf{x}(t+1)) \leq V(\mathbf{x}(t))$, it follows that $V(\mathbf{x}(t))$ is nonincreasing. Moreover, since $\overline{O(\mathbf{x}_0)}$ is closed and bounded (compact), $V(\mathbf{x}(t))$ is bounded below. Hence, $\lim_{t \to \infty} V(\mathbf{x}(t)) = c$. If $\mathbf{y} \in \omega(\mathbf{x})$, then $\lim_{n \to \infty} \mathbf{x}(t_n) = \mathbf{y}$. Consequently, $\lim_{i \to \infty} V(\mathbf{x}(t_{n_i})) = V(\mathbf{y}) = c$. Thus, $V(\omega(\mathbf{x}_0)) = c$, and thus $\omega(\mathbf{x}_0) \subset V^{-1}(c)$. Since $\omega(\mathbf{x}_0)$ is positively invariant, it follows that $\Delta V(\omega(\mathbf{x}_0)) = 0$. Let $E = \{\mathbf{x} \in \mathbb{R}^n_+ : \Delta V(\mathbf{x}) = 0\}$. Then $\omega(\mathbf{x}_0) \subset E \cap V^{-1}(c) \cap \overline{G}$. If M is the largest positively invariant set in E, then $\omega(\mathbf{x}_0) \subset M$. ∎

Chapter 4
Infectious Disease Models: Part I

4.1 Introduction

The theory of the dynamics of infectious diseases is one of the oldest branches of mathematical biology. In 1927, Kermack and McKendrick raised mathematical epidemiology to a new level when they introduced a continuous-time mathematical model of infectious diseases [187]. The seminal deterministic continuous-time infectious disease model framework of Kermack and McKendrick has now developed into a standard and well-established mathematical framework for constructing continuous-time and discrete-time infectious disease models.

Human and animal infectious diseases continue to cause harm and death, with new disease-causing pathogens emerging and old pathogens re-emerging or evolving. Infectious diseases may be transmitted directly from person to person by respiratory droplets (for example, smallpox, measles, and coronavirus caused diseases such as COVID-19 caused by SARS-CoV-2) or by body secretions (for example, chlamydia, HIV, and hepatitis B), indirect transmission by biting mosquitoes (for example, yellow fever, malaria, dengue, and zika) or by ingestion of pathogens in the environment (for example, anthrax, cholera, hepatitis A, and polio) [250], [251]. Mathematical models provide a framework for addressing questions about the time evolution of infectious diseases, and for quantifying possible infectious disease control and prevention strategies [9], [11], [19], [37], [44], [113], [143], [149], [167], [169], [187], [228], [251], [300].

4.2 Discrete-Time Kermack–McKendrick Type SIR Epidemic Model

One of the early uses of discrete-time epidemic models in the studies of the flow of an infectious disease in a population can be traced to the work of Daniel Bernoulli in 1760. Bernoulli used a discrete-time infectious disease model to analyze mortality due to smallpox infection [19], [143], [228]. Hamer, in 1906, used a discrete-time model of measles to study recurrent disease epidemics [153]. Discrete-time infectious disease models are especially appealing for the mathematical description of a disease epidemic process since such a process can be conceptualized as evolving through a set of discrete-time disease events. In addition, most infectious disease surveillance data are reported at discrete-time intervals, for example, daily, weekly, monthly, or yearly disease incidence or number of disease induced death [19], [143], [228], [250].

In this chapter, we begin with a study of discrete-time infectious disease models without demographic effects. Later, in the chapter, we include demographic effects in the disease models. As in continuous-time differential equations infectious disease models, we formulate our descriptions as compartmental models. That is, we partition the population under study into compartments with assumptions about the nature of inflow and outflow from each compartment during the discrete-time interval. As a result, our discrete-time epidemic models are formulated as *difference equations*.

Typically, introductions to mathematical models of infectious diseases are made through one of the first epidemic models proposed by Kermack and McKendrick (1927). Before introducing the discrete-time epidemic model, we state the classic Kermack and McKendrick *continuous-time* (ordinary differential equation) model.

$$\left.\begin{array}{rcl}\frac{dS}{dt} & = & -\beta SI \\ \frac{dI}{dt} & = & \beta SI - \alpha I \\ \frac{dR}{dt} & = & \alpha I,\end{array}\right\} \quad (4.1)$$

where

β is the transmission rate constant

and

α is the recovery rate constant.

In model (4.1),

$$\beta SI$$

is the density of the individuals of susceptible class S per unit of time who move to the infected class I. In general, the density of individuals who become infected per unit of time, in this case, βSI is called the *incidence*. The density of infected individuals per unit of time who recover from the infection is

$$\alpha I.$$

Model (4.1) is an SIR continuous-time model. In the next section, we introduce a discrete-time SIR epidemic model, and use it to introduce some important concepts in mathematical epidemiology.

4.2.1 Discrete-Time SIR Model

When an infectious disease invades an otherwise healthy population, it partitions the population into classes. As in the continuous-time Kermack-McKendrick model, we assume there are three such epidemiological classes:

1. Susceptible (S) class: This is the class of individuals who are healthy but capable of contracting infectious disease. These individuals are called *susceptibles*. At each discrete-time $t \in \mathbb{Z}_+$, the population density of individuals in the susceptible class is denoted by $S(t)$ or S.

2. Infected (I) class: This is the class of individuals who have contracted the infectious disease. As in the Kermack-McKendrick continuous-time model, we also assume that the infected individuals are *infectious*. These individuals are called *infecteds*. At each discrete-time $t \in \mathbb{Z}_+$, the population density of individuals in the infected class is denoted by $I(t)$ or I.

4.2. DISCRETE-TIME KERMACK–MCKENDRICK TYPE SIR EPIDEMIC MODEL

3. Recovered (R) class: This is the class of individuals who have recovered and cannot contract the infectious disease again. That is, they have acquired *immunity* from the infection. These individuals are called *recovered*. At each discrete-time $t \in \mathbb{Z}_+$, the population density of individuals in the recovered class is denoted by $R(t)$ or R.

Since infectious diseases partitioned the population under consideration into three classes, at each discrete-time $t \in \mathbb{Z}_+$, the total population, denoted by $N(t)$ or N, is given by

$$N(t) = S(t) + I(t) + R(t).$$

To formulate the model, as in the Kermack-McKendrick continuous-time model, we make the following simplifying assumption.

- Total population density remains constant for all time.

In the discrete-time model, the epidemiological classes are counted at discrete-time intervals. Thus, as in the previous chapters, our models consist of difference equations that describe how each class changes from time t to time $t+1$. When susceptible individuals are in close physical contact with infectious individuals, a fraction of the susceptible individuals become infected. This fraction of susceptibles moves into the infected class, while the remaining fraction who escaped the infection continues to stay in the susceptible class. Thus, in a unit of time, the number of susceptibles decreases by the number of susceptibles who have become infected. In the same unit of time, the number of infected increases by the number of newly infected individuals. In epidemiology, the number of individuals who become infected per unit of time is called the *incidence*. The remaining fraction of susceptibles at time $t+1$ is given by

$$S(t+1) = S(t)\phi(I),$$

where $\phi(I)$ is the fraction of susceptibles that escape the infection at time t. Consequently, in the same time interval, the fraction of susceptibles that did not escape from the infection is $1 - \phi(I)$, where $\phi(I)$ satisfies the following conditions unless otherwise stated, throughout the chapter.

(a) $\phi(I)$ is a continuous function of $I \geq 0$.

(b) $0 \leq \phi \leq 1$ for $I \geq 0$.

(c) If
$$\lim_{t \to \infty} \phi(z(t)) = 1$$
for some sequence $z(t) \geq 0$, then
$$\lim_{t \to \infty} z(t) = 0.$$

Remark 14 1. *Assumption (c) holds if*

$$\phi(0) = 1 \text{ and } \phi'(I) < 0 \text{ for } I \geq 0.$$

That is, (c) holds when all susceptible individuals remain in the susceptible class when there are no infectious individuals ($\phi(0) = 1$), and fewer susceptible individuals escape infection as the density of the infected class increases ($\phi' < 0$).

2. It follows from (c) that
$$\phi(0) = 1.$$

3. In general, numerous types of escape functions are utilized. In model (4.1), the force of infection is
$$\phi(I) = \beta I.$$
When the contact between susceptibles and infected individuals is modeled as Poisson processes, then we let
$$\phi(I) = e^{-c_\phi I},$$
where $c_\phi > 0$ is called the transmission coefficient. Another example of the escape function is
$$\phi(I) = \frac{1}{1 + c_\phi I}.$$

4. In some applications, the fraction ϕ is taken as a function of the proportion of individuals in the total population. That is, $\phi = \phi(\frac{I}{N})$, where N is the total population.

Those individuals who recover or die from the infection leave the infected class. We assume that, per each unit time interval, a constant fraction, $v \in (0, 1]$, of infected individuals recover while the remaining fraction, $1 - v \in [0, 1)$, do not recover and remain in the infected class. Consequently, the class of infectious individuals is
$$I(t+1) = S(t)(1 - \phi(I(t))) + (1 - v)I(t).$$

Infected individuals who recover leave the infectious class and move to the recovered class. Thus, the class of recovered individuals is
$$R(t+1) = vI(t) + R(t).$$

Therefore, our discrete-time SIR model is given by the following system of difference equations.
$$\begin{aligned} S(t+1) &= S(t)\phi(I(t)) \\ I(t+1) &= S(t)(1 - \phi(I(t))) + (1 - v)I(t) \\ R(t+1) &= vI(t) + R(t), \end{aligned} \tag{4.2}$$
together with the initial condition
$$S(0) \geq 0,\ I(0) \geq 0 \text{ and } R(0) \geq 0.$$
The right side of each equation in model (4.1) is the sum of nonnegative terms. Hence,
$$S(t) \geq 0,\ I(t) \geq 0 \text{ and } R(t) \geq 0$$
for each time $t \in \mathbb{Z}_+$. Furthermore, when we sum the three equations of model (4.2), we obtain that
$$N(t+1) = S(t+1) + I(t+1) + R(t+1) = N(t) = S(t) + I(t) + R(t).$$
That is,
$$N(t) = N(0) = N_0 \text{ for all } t \in \mathbb{Z}_+,$$
and the total population remains constant at its initial population density, N_0.

Model (4.2) is a discrete-time SIR model or SIR system. Its compartmental diagram is shown in Figure 4.1 (see Chapter 2).

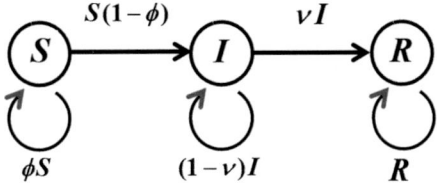

Figure 4.1: Flowchart for Discrete-time Kermack-McKendrick SIR model.

4.2.2 Some Properties of the SIR Model

As in the continuous-time Kermack–McKendrick model, model (4.2) has some distinctive dynamical behavior. The point
$$(S, I, R)^T = (N_0 - R^*, 0, R^*)^T$$
is an equilibrium point of model (4.2) for all values of $R^* \in [0, N_0]$. Moreover, these are the only equilibria (Problem 1).

Theorem 41 *In model (4.2), for any initial condition*
$$(S(0), I(0), R(0))^T \in \mathbb{R}_+^3,$$
the solution
$$(S(t), I(t), R(t))^T \in \mathbb{R}_+^3$$
equilibrates. That is,
$$\lim_{t \to \infty} (S(t), I(t), R(t))^T = (S^*, 0, R^*)^T \text{ for some } S^* \in [0, N_0],$$
where
$$R^* = N_0 - S^*.$$

Proof. (i) From the first difference equation of model (4.2), it follows that we have that for every $t \in \mathbb{Z}_+$,
$$0 \leq S(t+1) \leq 1 \cdot S(t),$$
and $S(t) \geq 0$ is a decreasing sequence, which must therefore converge to a point, $S^* \geq 0$.

(ii) $I(t) \geq 0$ is bounded for all $t \in \mathbb{Z}_+$. To see this, note that from the second difference equation of model (4.2),
$$0 \leq I(t+1) \leq S(0) + (1-v) I(t).$$
By mathematical induction,
$$0 \leq I(t) \leq x(t),$$
where $x(t)$ solves the (linear) difference equation
$$x(t+1) = S(0) + (1-v) x(t)$$
and $x(0) = I(0)$. Since $1 - v < 1$ it follows that
$$\lim_{t \to \infty} x(t) = S(0) v^{-1}.$$
Thus, $x(t)$ and hence $I(t)$, are bounded.

(iii) Suppose $S^* > 0$. Then from the first difference equation of model (4.2),
$$1 = \lim_{t \to \infty} \frac{S(t+1)}{S(t)} = \lim_{t \to \infty} \phi(I(t))$$
and by assumption (c) it follows that
$$\lim_{t \to \infty} I(t) = 0.$$

(iv) Suppose $S^* = 0$. Then from the second difference equation of model (4.2),
$$0 \leq I(t+1) \leq S(t) + (1-v)I(t).$$
Take the $\limsup_{t \to \infty}$ of both sides
$$0 \leq \limsup_{t \to \infty} I(t+1) \leq \limsup_{t \to \infty} [S(t) + (1-v)I(t)]$$
$$\leq \limsup_{t \to \infty} S(t) + (1-v) \limsup_{t \to \infty} I(t)$$
$$\leq (1-v) \limsup_{t \to \infty} I(t).$$
Hence, $1 - v < 1$ and
$$\limsup_{t \to \infty} I(t+1) \leq (1-v) \limsup_{t \to \infty} I(t)$$
imply
$$\limsup_{t \to \infty} I(t) = 0.$$
Since $I(t) \geq 0$ it follows that $\lim_{t \to \infty} I(t) = 0$.

(v) Taking the limit $t \to \infty$ on both sides of
$$N_0 = S(t) + I(t) + R(t)$$
we find that
$$N_0 - S^* = R^*,$$
where $\lim_{t \to \infty} S(t) = S^* \in [0, N_0]$. The quantity
$$S^* = N_0 - R^*$$
is called the **final size of the epidemic**. In model (4.2), the escape probability, $\phi(I) > 0$ for every $I \geq 0$. Consequently, some susceptible individuals *always* escape the disease infection, and the infection does not end because *all* the susceptibles have been infected and are now immune. In most real disease epidemics, $S^* > 0$.
∎

In model (4.2), the density of susceptible individuals decreases monotonically to S^* while that of recovered individuals increases monotonically to R^*. However, it is possible for the density of infected individuals to have non-monotonic dynamical behavior. The density of infected individuals can, under certain circumstances, initially increase to a maximum value, and then decrease to zero as predicted by Theorem 41.

4.2. DISCRETE-TIME KERMACK–MCKENDRICK TYPE SIR EPIDEMIC MODEL

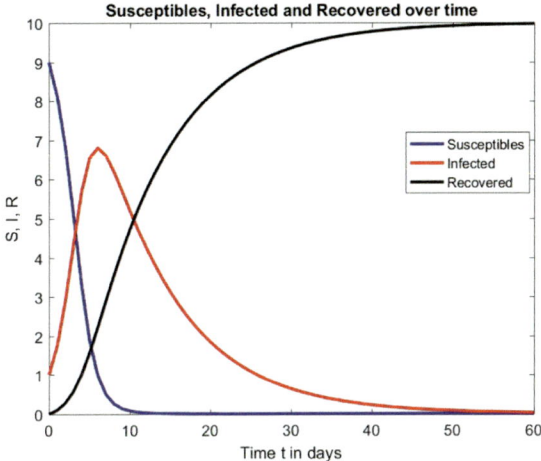

Figure 4.2: In model (4.2), the density of susceptibles decreases monotonically, density of recovered increases monotonically, while that of infected initially increases to a maximum value before decreasing to zero.

Example 75 *To illustrate the non-monotonic dynamical behavior of the infected individuals in model (4.2), we model contact between susceptibles and infected individuals as a Poisson process, and let*

$$\phi(I) = e^{-c_\phi I},$$

where

$$c_\phi = 0.1 \text{ and } v = 0.1.$$

With the initial condition

$$(S(0), I(0), R(0))^T = (9, 1, 0)^T.$$

Figure 4.2 shows the density of susceptibles decreasing monotonically and that of the recovered increasing monotonically, while the infected class initially increases to a maximum value before decreasing to zero.

On the other hand, just as the number of susceptibles, it is also possible for the density of infected individuals to decrease monotonically to zero. To illustrate this in model (4.2), we let

$$c_\phi = 0.008$$

and keep the other parameter values and initial conditions the same as in Figure 4.2. Figure 4.3 shows the population densities of susceptible and infected decreasing monotonically while that of the recovered increases monotonically.

In model (4.2), if $I(0)$ is small at the onset of the disease invasion, then

$$I(1) \approx (-S(0)\phi'(0) + 1 - v) I(0).$$

Thus, the infected class initially increases if

$$-S(0)\phi'(0) + 1 - v > 1.$$

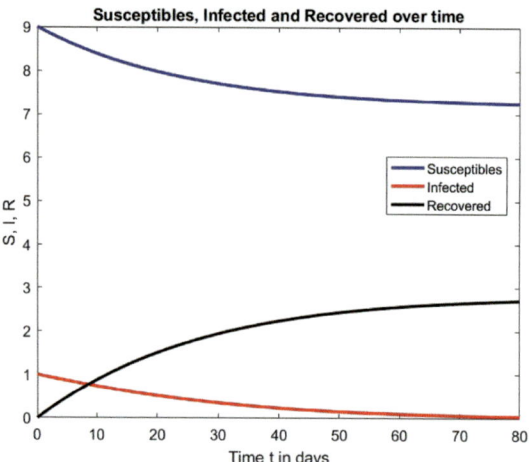

Figure 4.3: In model (4.2), the densities of susceptible and infected decrease monotonically while that of the recovered increases monotonically.

Hence, a sufficient condition for an initial increase in the density of the infected is

$$-S(0)\,\phi'(0) > v$$

or

$$\mathcal{R}_0 := \frac{-S(0)\,\phi'(0)}{v} > 1.$$

Whenever

$$I(1) > I(0)$$

we say that an **outbreak** occurs. That is, $\mathcal{R}_0 > 1$ implies an outbreak occurs, whereas $\mathcal{R}_0 < 1$ implies there is no disease outbreak. An initial increase in the density of infected individuals followed by its decline to zero is classical dynamical behavior of disease epidemics.

As an illustration, in Figure 4.2, $\mathcal{R}_0 = 9 > 1$, and a disease outbreak occurs before declining to zero. However, in Figure 4.3, $\mathcal{R}_0 = 0.72 < 1$, and the density of infected declines monotonically to zero (there is no disease outbreak). Threshold conditions for disease outbreaks, in this case defined using a diagnostic quantity $\mathcal{R}_0$, are common in epidemiology; we will discuss them later.

Model (4.2) is based on the following three assumptions.

- There are no births and deaths in the population.

- The population is closed. There are no immigrants into the population, and emigrants from the population.

- All recovered individuals have complete immunity and cannot get re-infected.

The above assumptions are not universal. However, **childhood diseases** like chickenpox, smallpox, rubella, and mumps can lead to permanent immunity and are well studied with SIR models.

4.2. DISCRETE-TIME KERMACK–MCKENDRICK TYPE SIR EPIDEMIC MODEL

Since
$$N_0 = S(t) + I(t) + R(t),$$
we obtain from the second equation of model (4.2) the inequality
$$I(t+1) \leq (N_0 - I(t))(1 - \phi(I(t))) + (1-v)I(t).$$
To obtain an estimate on the maximum density of infected individuals, we observe that when
$$\phi(I) = e^{-c_\phi I} \text{ or } \phi(I) = \frac{1}{1+c_\phi I}$$
then
$$\phi'(I) < 0 \text{ and } \phi''(I) > 0.$$
Thus, in addition to the monotonicity condition on ϕ, we assume throughout the chapter that
$$\phi''(I) > 0 \text{ for } I \geq 0$$
and obtain the following estimate on $1 - \phi(I)$.

Lemma 7 *If*
$$\phi(0) = 1, \text{ and for } I \geq 0, \, \phi'(I) < 0 \text{ and } \phi''(I) > 0,$$
then
$$1 - \phi(I) \leq -\phi'(0)I \text{ for } I \geq 0.$$

Proof. Since the result holds when $I = 0$, we assume that $I > 0$. Now, consider any closed interval $[0, I]$. By the Mean Value Theorem,
$$\frac{f(I) - f(0)}{I - 0} = f'(\xi) \text{ for some } \xi \in [0, I],$$
where
$$f(I) = 1 - \phi(I).$$
The result follows from the monotonicity and concavity conditions on ϕ. ∎

By Lemma 7,
$$I(t+1) \leq -(N_0 - I(t))\phi'(0)I(t) + (1-v)I(t).$$
By mathematical induction,
$$0 \leq I(t) \leq x(t),$$
where $x(t)$ solves the nonlinear equation
$$x(t+1) = -(N_0 - x(t))\phi'(0)x(t) + (1-v)x(t).$$
To obtain the maximum value of $x(t)$, we note that the quadratic polynomial in x,
$$h(x) = -(N_0 - x)\phi'(0)x + (1-v)x,$$
attains its maximum value for $x > 0$ at
$$x_{\max} = \frac{1}{2}\left(N_0 - \frac{(1-v)}{\phi'(0)}\right)$$

and
$$I_{\max} \leq h(x_{\max}),$$

where $I_{\max}$ is the maximum density of infected individuals reached in the SIR epidemic. That is, $I_{\max}$ signifies the maximum severity of the SIR infection. Typically, new infectious disease cases decline after attaining the maximum value, $I_{\max}$. Thus, estimates of $I_{\max}$ value for newly occurring infectious diseases can be used to determine the peak of the epidemic and the beginning of a decline in the density of new infections.

Final Size Equation

Assuming contact between susceptibles and infected individuals is a Poisson processes so that
$$\phi(I) = e^{-c_\phi I},$$
we now derive an equation for the final size S^* of the epidemic that is predicted by model (4.2). From first difference equation of (4.2) we see that for all $t \in Z_+$
$$S(t+1) = S(t) e^{-c_\phi I(t)}.$$
From which for $S(0) > 0$, using formula (11) in Chapter 1, we get that
$$S(t) = \left(\Pi_{k=0}^{k=t-1} e^{-c_\phi I(k)}\right) S(0) \tag{4.3}$$

Since $\lim_{t\to\infty} S(t) = S^*$ and $\lim_{t\to\infty} I(t) = 0$, taking the natural logarithm of both sides of this equation and the limit as $t \to \infty$ of the result, we find that the infinite series $\sum_{k=0}^{\infty} I(k)$ converges and
$$\sum_{k=0}^{\infty} I(k) = \frac{1}{c_\phi} \ln \frac{S(0)}{S^*}. \tag{4.4}$$

Another formula for this infinite series can be obtained as follows. By adding together the first two difference equations in (4.2) we get, after some algebraic rearrangement,
$$S(t+1) - S(t) + I(t+1) - I(t) = -vI(t) \tag{4.5}$$
for all $t \in Z_+$. Note that
$$\sum_{k=0}^{t} (S(k+1) - S(k)) = S(t) - S(0), \quad \sum_{k=0}^{t} (I(k+1) - I(k)) = I(t) - I(0) \tag{4.6}$$
for any $t \in Z_+$. When we replace t by k in Equation (4.5) and sum both sides from $k = 0$ to t,
$$S(t) - S(0) + I(t) - I(0) = -v \sum_{k=0}^{t} I(k)$$
whose limit as $t \to \infty$ gives
$$\sum_{k=0}^{\infty} I(k) = \frac{S(0) + I(0) - S^*}{v}.$$

This formula for the infinite series and the formula (4.4) yield the **final size equation**
$$\frac{1}{c_\phi} \ln \frac{S(0)}{S^*} = \frac{S(0) + I(0) - S^*}{v}$$
which S^* must satisfy. Recalling the definition of
$$\mathcal{R}_0 := \frac{S(0) c_\phi}{v}.$$
we can re-write this equation equivalently as
$$\ln \frac{S(0)}{S^*} = \left(1 + \frac{I(0)}{S(0)} - \frac{S^*}{S(0)}\right) \mathcal{R}_0. \tag{4.7}$$

While we cannot solve Equation (4.7) explicitly for the final size S^*, we can solve it numerically once $\mathcal{R}_0$ and the initial conditions $S(0)$ and $I(0)$ are known.

At the beginning of an emerging or re-emerging infectious disease epidemic, the population density of initial infected is small. Assuming $I(0) \approx 0$, the final size equation is approximately
$$\ln \frac{S(0)}{S^*} \approx \left(1 - \frac{S^*}{S(0)}\right) \mathcal{R}_0.$$

4.2.3 Calculation of the Mean Infectious Period

The mean time of the exposed period or the infectious period is usually measurable from observations. For the influenza disease, for example, the infectious period is about $3 - 7$ *days* with mean $4 - 5$ *days*. To see how this can help us estimate the recovery proportion, v, we assume that there is no movement of susceptibles into the infected class and at time $t = 0$, $I(0) = I_0 > 0$ infected individuals have been put into the infected class. The difference equation that describes the population dynamics of the class for all time $t \in \mathbb{Z}_+$ is
$$\begin{aligned} I(t+1) &= (1-v) I(t), \\ I(0) &= I_0. \end{aligned} \tag{4.8}$$

From the Chapter, we know that the solution of Equation (4.8) is
$$I(t) = (1-v)^t I_0 \text{ for time } t \in \mathbb{Z}_+.$$
Hence,
$$\frac{I(t)}{I_0} = (1-v)^t$$
gives the fraction of individuals who are still infectious at time $t \in \mathbb{Z}_+$. In probability language, this gives the probability of being still infectious at time t. Hence, the probability of being infectious in the unit time interval is
$$\frac{I(t-1)}{I_0} - \frac{I(t)}{I_0} = (1-v)^{t-1} v.$$

The average time spent in the infected class is given by the mean of the discrete random variable X, that denotes the time to exit the infected class [168]. That is, the mean of this geometric distribution is

$$\mu = E(X) = \sum_{t=1}^{\infty} t(1-v)^{t-1} v = (1)v + 2(1-v)v + 3(1-v)^2 v + \cdots .$$

Hence, the mean time spent in the infected class is

$$\mu = E(X) = \frac{1}{v} \text{ (see Problem 3)}.$$

For example, influenza infected individuals stay sick with the disease for $3 - 7$ days. If the mean time spent as infected with the disease is 5 days, then the proportion that recovers per unit time interval, measured in units of $[days]^{-1}$, is $\frac{1}{5}$.

4.3 A Case Study: The 1978 English Boarding School Influenza Outbreak

An outbreak of influenza infection occurred in a boarding school in North of England in January-February of 1978. The all boys boarding school had a total of 763 boys, and all of them were at risk of contracting the influenza infection. The students returned to the boarding school from their Christmas vacation, spent at various parts of the world, on January 10 when the school's spring term began. A student returning to the school from Hong Kong exhibited elevated temperature on 15–18 of January. On January 22, three students were sick with the disease. The following table gives the observed number of students ill on the nth day of the influenza outbreak, beginning from January 22 ($n = 1$) (Table 4.1).

Day	Number of Infected*	Day	Number of Infected
3	25	9	192
4	75	10	126
5	227	11	71
6	296	12	28
7	258	13	11
8	236	14	7

Table 4.1: Daily Number of Influenza Infected Students
*Data taken from "Influenza in a Boarding School", British Medical Journal, 4 March 1978.

The number of students who escaped the influenza infection was 19. The average time spent sick with the infection was 5–6 days. Since the infected students were isolated, perhaps they spent only about 2 days as infectious. A swab test taken from some of the sick students revealed that they were infected with the H1N1 influenza A virus. Most of the boarding school staff remained healthy, except only one staff member who displayed symptoms of the disease.

The given data give the values

$$S(3) = 738, \ I(3) = 25$$

and the number of students who escaped the influenza infection was $S^* = 19$. Using the finite size equation (4.7), we have that

$$\frac{c_\phi}{v} = \frac{\ln \frac{S(0)}{S^*}}{(S(0) + I(0) - S^*)},$$

and

$$\frac{c_\phi}{v} = \frac{\ln \frac{738}{19}}{763 - 19} = 0.0049168. \quad (4.9)$$

Since time is measured in days,

$$\text{January 21, 1978 is taken to be } t_0 = 0$$

and

$$\text{January 22, 1978 is taken to be } t = 1,$$

where

$$t_{end} = 14 \text{ is February 4, 1978.}$$

As in [230], we take the infective period to be 2.1 days. Consequently,

$$v = \frac{1}{2.1} = 0.476,$$

$$c_\phi = (0.0049168)(0.476) = 0.0023403968$$

and

$$\mathcal{R}_0 = \frac{S(0) c_\phi}{v} \approx 3.6285984 > 1.$$

Thus, because $\mathcal{R}_0 > 1$ the SIR model predicts an influenza outbreak (defined to be $I(1) > I(0)$), consistent with what occurred.

4.4 Discrete-Time SIRS Model

In model (4.2), once an individual has been infected and subsequently has recovered that individual cannot be reinfected with the SIR disease. Consequently, for all the model parameter values, the density of infectious individuals eventually tend to zero as $t \to \infty$, and the SIR infection vanishes in the population. In this section, we consider a slightly different model for infectious diseases that arises when we assume that the recovered individuals in model (4.2) may lose their immunity and become reinfected with the disease. Examples of this type of disease include *malaria* and *tuberculosis*. We assume that a fraction, $\mu \in (0,1)$, of the recovered individuals of model (4.2) return to the susceptible class. This leads to the following SIRS model (where the extra S in this acronym indicates that recovered individuals may re-enter the susceptible class).

$$\begin{aligned} S(t+1) &= S(t)\phi(I(t)) + \mu R(t) \\ I(t+1) &= S(t)(1 - \phi(I(t))) + (1-v)I(t) \\ R(t+1) &= vI(t) + (1-\mu)R(t), \end{aligned} \quad (4.10)$$

together with the initial condition

$$S(0) \geq 0, \; I(0) \geq 0 \text{ and } R(0) \geq 0.$$

As in the SIR model (4.2), we have in model (4.10),

$$N(t+1) = S(t+1) + I(t+1) + R(t+1) = N(t) = S(t) + I(t) + R(t),$$

and the total population is constant. That is,

$$N(t) = N(0) = N_0 \text{ for all } t \in \mathbb{Z}_+.$$

We may eliminate R from the SIRS system by setting

$$R(t) = N_0 - S(t) - I(t)$$

and obtaining the following reduced system.

$$\begin{aligned} S(t+1) &= S(t)\phi(I(t)) + \mu(N_0 - S(t) - I(t)) \\ I(t+1) &= S(t)(1 - \phi(I(t))) + (1-v)I(t) \end{aligned} \quad (4.11)$$

where $\mu, v \in (0,1)$ and $N_0 > 0$.

Unlike the SIR model (4.2), the SIRS model (4.11) can have two equilibria: the disease-free equilibrium,

$$(S^*, I^*)^T = (N_0, 0)^T,$$

and if

$$\frac{-N_0 \phi'(0)}{v} > 1,$$

the endemic equilibrium,

$$(S^*, I^*)^T = \left(\frac{vI^*}{1 - \phi(I^*)}, I^*\right)^T,$$

where $I^* > 0$ is the unique solution of the equation

$$1 - \phi(I) = \frac{v\mu I}{\mu N_0 - (v+\mu)I}.$$

We capture this in the following result.

Theorem 42 *For the SIRS model (4.11), define*

$$\mathcal{R}_0 = \frac{-N_0 \phi'(0)}{v}.$$

If $\mathcal{R}_0 < 1$, then the disease-free equilibrium point, $(N_0, 0)^T$, is globally asymptotically stable on the interior of $\mathbb{R}_+^2$, and there is no disease outbreak. However, if $\mathcal{R}_0 > 1$, then $(N_0, 0)^T$ is unstable, and model (4.11) has the unique endemic equilibrium point,

$$\begin{pmatrix} N_0 - \frac{v+\mu}{\mu}I^* \\ I^* \end{pmatrix}$$

where

$$I^* \in \left(0, \frac{\mu N_0}{v + \mu}\right)$$

is the unique positive solution of the equation

$$1 - \phi(I) = \frac{v\mu I}{\mu N_0 - (v+\mu)I}.$$

4.4. DISCRETE-TIME SIRS MODEL

Proof. The eigenvalues of the linearization of model (4.11) at

$$(N_0, 0)^T$$

are

$$\lambda_1 = 1 - \mu \text{ and } \lambda_2 = 1 + v(\mathcal{R}_0 - 1).$$

Clearly, $0 < \lambda_1 < 1$. Also, simple algebraic manipulations show that

$$0 < \lambda_2 < 1 \text{ whenever } \mathcal{R}_0 < 1.$$

and

$$\lambda_2 > 1 \text{ whenever } \mathcal{R}_0 > 1.$$

Thus, $(N_0, 0)^T$ is *locally* asymptotically when $\mathcal{R}_0 < 1$ and unstable when $\mathcal{R}_0 > 1$.

Next, we use the LaSalle Invariance Principle 29, to establish the global stability of the disease-free equilibrium point. Define the function

$$F : [0, \infty) \times [0, \infty) \to [0, \infty) \times [0, \infty)$$

by

$$F(S, I) := (S\phi(I) + \mu(N_0 - S - I), S(1 - \phi(I)) + (1 - v)I).$$

Then the set of iterates of the function of F is equivalent to the set of population density sequences generated by model (4.11). Now, define the function

$$V : [0, \infty) \times [0, \infty) \to [0, \infty)$$

by $V(S, I) = I$. Let

$$\Delta V(S, I) = V(F(S, I)) - V(S, I).$$

Then

$$\Delta V(S, I) = S(1 - \phi(I)) - vI.$$

Recall that

$$\phi(0) = 1, \text{ and for } I \geq 0, \phi'(I) < 0 \text{ and } \phi''(I) > 0.$$

Since $S \leq N_0$, using Lemma 7 we get

$$\Delta V(S, I) \leq (-N_0 \phi'(0) - v) I = v(\mathcal{R}_0 - 1) I.$$

Hence, $\mathcal{R}_0 < 1$ implies that

$$\Delta V(S, I) < 0 \text{ for all } I > 0.$$

Moreover,

$$V(S, I) \to \infty \text{ as } I \to \infty.$$

Consequently, V is a strict Liapunov function and by Theorem 3.9.2 we obtained that $(N_0, 0)^T$ is globally asymptotically on $\mathbb{R}_+^2$ whenever $\mathcal{R}_0 < 1$.

Recall that $(N_0, 0)^T$ is unstable when $\mathcal{R}_0 > 1$. To establish the existence of a unique endemic equilibrium we let

$$g(I) = 1 - \phi(I) \text{ and } h(I) = \frac{v\mu I}{\mu N_0 - (v + \mu) I}.$$

Then
$$g(0) = 0,\ g(N_0) = 1 - \phi(N_0),\ h(0) = 0 \text{ and } \lim_{I \to \left(\frac{\mu N_0}{v+\mu}\right)^-} h(I) \to \infty$$

Furthermore, for each
$$I \in \left[0, \frac{\mu N_0}{v+\mu}\right)$$

we have
$$g'(I) > 0,\ g''(I) < 0,\ h'(I) > 0 \text{ and } h''(I) > 0.$$

Hence, the graphs of the smooth functions g and h are both increasing in I on this interval. However, the graph of g is concave down while that of h is concave up. The slopes of the graphs g and h at $I = 0$ are, respectively,
$$g'(0) = -\phi'(0) \text{ and } h'(0) = \frac{v}{N_0}.$$

Consequently, $\mathcal{R}_0 > 1$ implies that $g'(0) > h'(0)$ and the graphs of g and h intersect at a unique point, $I^* > 0$, the endemic equilibrium, in the open interval, $\left(0, \frac{\mu N_0}{v+\mu}\right)$. Consequently, the unique endemic equilibrium point is
$$(S^*, I^*)^T = \left(\frac{vI^*}{1 - \phi(I^*)}, I^*\right)^T = \left(N_0 - \frac{(v+\mu)}{\mu}I^*, I^*\right)^T,$$

where I^* is the positive solution of the equation
$$1 - \phi(I) = \frac{v\mu I}{\mu N_0 - (v+\mu)I}, \text{ where } \mathcal{R}_0 > 1.$$

∎

By Theorem 42, the disease-free equilibrium point of model (4.10)
$$\begin{pmatrix} S^* \\ I^* \\ R^* \end{pmatrix} = \begin{pmatrix} N_0 \\ 0 \\ 0 \end{pmatrix}$$

is globally asymptotically stable on $\mathbb{R}_+^2$ whenever $\mathcal{R}_0 < 1$. However, when $\mathcal{R}_0 > 1$, then
$$(N_0, 0, 0)^T$$

is unstable and model (4.10) has a unique endemic point at
$$\begin{pmatrix} S^* \\ I^* \\ R^* \end{pmatrix} = \begin{pmatrix} \frac{vI^*}{1-\phi(I^*)} \\ I^* \\ N_0 - \left(\frac{vI^*}{1-\phi(I^*)} + I^*\right) \end{pmatrix} = \begin{pmatrix} N_0 - \frac{(v+\mu)}{\mu}I^* \\ I^* \\ \frac{(v+\mu)}{\mu}I^* - I^* \end{pmatrix}.$$

When the endemic equilibrium point is asymptotically stable, then the disease establishes in the population. We illustrate this in the following example.

4.5. HERD IMMUNITY

Example 76 *To illustrate the stable endemic equilibrium and stable disease-free equilibrium in model (4.10), we model contact between susceptibles and infected individuals as a Poisson process, and let*

$$\phi(I) = e^{-c_\phi I},$$

where

$$c_\phi = 0.1, \ v = 0.1 \ and \ \mu = 0.2.$$

With the initial condition

$$\begin{pmatrix} S(0) \\ I(0) \\ R(0) \end{pmatrix} = \begin{pmatrix} 9 \\ 1 \\ 0 \end{pmatrix}.$$

Then

$$\mathcal{R}_0 = \frac{N_0 c_\phi}{v} = 10 > 1$$

and as predicted by Theorem 42, Figure 4.4 suggests that the system has an asymptotically stable endemic equilibrium point (Problem 9) at

$$\begin{pmatrix} S^* \\ I^* \\ R^* \end{pmatrix} = \begin{pmatrix} 1.3172 \\ 5.7885 \\ 2.9843 \end{pmatrix}.$$

That is, the disease is established in the population when $\mathcal{R}_0 > 1$.

On the other hand, it is also possible for the disease-free equilibrium point

$$\begin{pmatrix} S^* \\ I^* \\ R^* \end{pmatrix} = \begin{pmatrix} 10 \\ 0 \\ 0 \end{pmatrix}$$

to be asymptotically stable. To illustrate this in model (4.10), we let

$$c_\phi = 0.002$$

and keep the other parameter values and initial conditions the same as in Figure 4.4. Then

$$\mathcal{R}_0 = \frac{N_0 c_\phi}{v} = 0.2 < 1$$

and as predicted by Theorem 42, Figure 4.5 shows the population densities of the infected and recovered decrease monotonically to zero while that of the susceptible population stabilizes at $N_0 = 10$. That is, the disease vanishes in the population when $\mathcal{R}_0 < 1$.

4.5 Herd Immunity

Herd immunity or *indirect protection* occurs when the spread of an infectious disease throughout a population is prevented because a large enough portion of the susceptible class is immune. For example, if 80% of a susceptible population is immune to an infectious virus or bacteria as a result of, say, a vaccination disease control measure, then four out of every five susceptibles who encounter

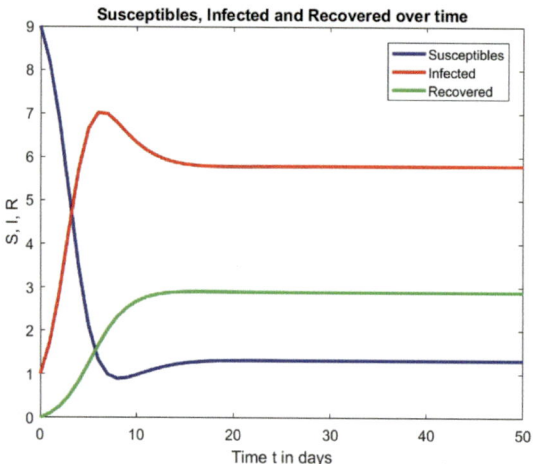

Figure 4.4: In model (4.10), $\mathcal{R}_0 = 10$, and there is an asymptotically stable endemic equilibrium point at $(S^*, I^*, R^*)^T = (1.3172, 5.7885, 2.9843)^T$.

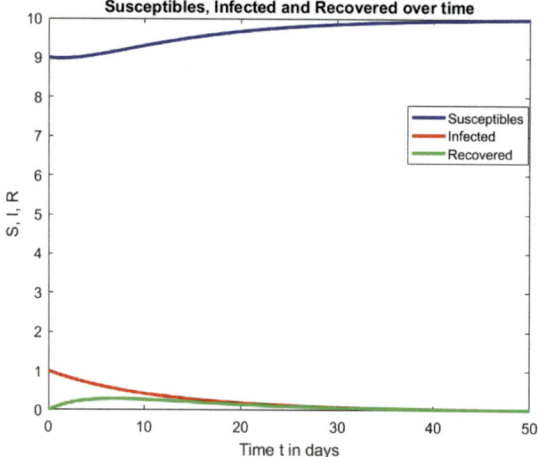

Figure 4.5: In model (4.10), $\mathcal{R}_0 = 0.2$, and the disease-free equilibrium point, $(S^*, I^*, R^*)^T = (10, 0, 0)^T$ is asymptotically stable.

4.5. HERD IMMUNITY

an infectious individual carrying the pathogen will not get infected and can therefore not transmit the disease to the other susceptibles in the population. This eventually decreases the density of infected individuals in the population.

Suppose a vaccine renders, at all times, a fraction p ($0 < p < 1$) of the susceptible population immune to the infection of the SIR model (4.2). Then the SIR model becomes the following model.

$$\begin{aligned} S(t+1) &= (1-p)\,S(t)\,\phi(I(t)) \\ I(t+1) &= (1-p)\,S(t)\,(1-\phi(I(t))) + (1-v)\,I(t) \\ R(t+1) &= v I(t) + p S(t) + R(t), \end{aligned} \quad (4.12)$$

Notice that model (4.12) reduces to model (4.2) when no susceptibles are vaccinated and $p = 0$.

Proceeding as we did with our analysis of model (4.2), we find that there will be no disease outbreak if

$$\frac{(1-p)\,S(0)\,c_\phi}{v} = (1-p)\,\mathcal{R}_0 < 1.$$

In other words, herd immunity against an outbreak is attained if the vaccinated fraction of susceptibles is sufficiently large, namely, if

$$p > 1 - \frac{1}{\mathcal{R}_0}. \quad (4.13)$$

Thus, the discrete-time SIR model predicts that diseases with large values of $\mathcal{R}_0$ require a large proportion of the susceptible population to be successfully vaccinated in order to prevent an outbreak of the disease [250]. For example, for the outbreak of influenza infection that occurred in the boarding school in North of England in January-February of 1978 (see Section 4.2),

$$\mathcal{R}_0 \approx 3.6285984.$$

Hence, if they had had a vaccine, then this model predicts herd immunity against the influenza outbreak would have been attained had at least 72.44% of the susceptible boarding school students have been vaccinated.

Exercises 4.1–4.5

1. Show that *all* the equilibrium points of model (4.2) are of the form

$$(S, I, R)^T = (N_0 - R^*, 0, R^*)^T$$

   for all values of $R^* \in [0, N_0]$.

2. In model (4.2), let

$$\phi(I) = \frac{1}{1 + c_\phi I},$$

   where

$$c_\phi = 0.1 \text{ and } v = 0.1.$$

   (a) Using a computer program and the initial condition

$$(S(0), I(0), R(0))^T = (9, 1, 0)^T,$$

graph the solution as in Figure 4.2 to show that the density of susceptibles decreases monotonically and that of the recovered increases monotonically, while the infected class initially increases to a maximum value before decreasing to zero.

(b) Repeat (a) with
$$c_\phi = 0.008 \text{ and } v = 0.1.$$

As shown in Figure 4.3 that population densities of susceptible and infected decrease monotonically while that of the recovered increases monotonically.

3. Consider the linear model
$$\begin{aligned} I(t+1) &= (1-v)I(t), \\ I(0) &= I_0. \end{aligned} \tag{4.14}$$

Show that the mean time spent in the infected class is
$$\mu = E(X) = \frac{1}{v}, \text{ where } 0 < v < 1.$$

4. Show that in model (4.12)
$$\mathcal{R}_0 = \frac{(1-p)S(0)c_\phi}{v}.$$

Use this parameter to predict the long-term disease dynamics for all initial population densities.

5. Show that
$$h(x) = -(N_0 - x)\phi'(0)x + (1-v)x,$$

attains its maximum value for $x > 0$ at
$$x_{\max} = \frac{1}{2}\left(N_0 - \frac{(1-v)}{\phi'(0)}\right)$$

and compute
$$h(x_{\max}).$$

6. Consider model (4.10) with
$$\mathcal{R}_0 > 1.$$

(a) Compute the Jacobian matrix, J, of the linearized equation at the endemic equilibrium point,
$$(S^*, I^*)^T = \left(N_0 - \frac{(v+\mu)}{\mu}I^*, I^*\right)^T,$$

where $I^* \in \left(0, \frac{\mu N_0}{v+\mu}\right)$ is the unique positive solution of the equation
$$1 - \phi(I) = \frac{v\mu I}{\mu N_0 - (v+\mu)I}.$$

(b) Using the Jury criteria,
$$|trace(J)| < \det J + 1 < 2,$$
establish conditions for the local asymptotic stability of the unique endemic equilibrium point.

7. Figure 4.4 shows that the system has an asymptotically stable endemic equilibrium point at
$$(S^*, I^*, R^*)^T = (1.3172, 5.7885, 2.9843)^T.$$
Use the Jury criteria to prove that
$$(S^*, I^*, R^*)^T = (1.3172, 5.7885, 2.9843)^T$$
is locally asymptotically stable.

8. Show that the total population of model (4.12) is constant. Then, find all the equilibrium points and use the Linearization principle to determine their stability.

9. Figure 4.4 shows that the system has an asymptotically stable endemic equilibrium point at
$$(S^*, I^*, R^*)^T = (1.3172, 5.7885, 2.9843)^T.$$
Use linear stability analysis to establish this.

10. In Section 4.5 the SIR model (4.2) includes a vaccination program that, at all times, vaccinates a fraction p of the susceptible individuals. In a similar way, modify the SIRS model to include such a vaccination program and show that the inequality (4.13) implies herd immunity in this model as well.

4.6 SIS Discrete-Time Epidemic Models With Birth and Death

Epidemic models are structured population dynamic models in which individuals are classified according to disease-related categories. As in Chapter 1, the discrete-time modeling methodology is based on a straightforward accounting for the individuals in each class found in a census at time $t+1$ according to whether they were present in the previous census at time t or are new arrivals after the previous census. New arrivals in a category can be due to any number of reasons: births, immigration, infection, recovery, etc. For example, let $S(t)$ denote the class of individuals in the population who are susceptible and not infected at time t. Then $S(t+1)$ consists of those susceptible individuals who were present at time t plus those who were not. The former consists of surviving susceptibles who avoided infection and the latter consists of new susceptible due to births, immigration, etc.

As a warm-up model, consider a disease in which all infected individuals are infectious (capable of transmitting the disease) and distinguish only two classes of individuals: the class I of infected individuals and the class S of uninfected but susceptible individuals. Assume the fraction of susceptible individuals that survive a time unit is σ_S ($0 \leq \sigma_S \leq 1$). Of those that survive assume a fraction ϕ do not become infected. Similarly, assume the proportion of infected individuals that

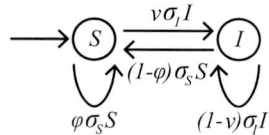

Figure 4.6: The compartmental diagram for the SI model (4.15).

survive a time unit is σ_I ($0 \leq \sigma_I \leq 1$). Of those that survive, assume a fraction v ($0 \leq v \leq 1$) do not gain immunity from the disease and hence become susceptible again. Then we have the equations

$$\begin{aligned} S(t+1) &= f + \phi \sigma_S S(t) + v\sigma_I I(t) \\ I(t+1) &= (1-\phi)\sigma_S S(t) + (1-v)\sigma_I I(t) \end{aligned} \quad (4.15)$$

where f is the recruitment of new susceptibles per unit of time (see Figure 4.6).

The disease is assumed transmitted by contact between infected and susceptible individuals and, in this example, we assume the probability ϕ that a susceptible individual escapes infection per unit time depends on the proportion of infected individuals in the population, i.e., $\phi = \phi(I/N)$ where

$$N = S + I$$

is the total population size. This probability equals 1 if there are no infected individuals and decreases as a function of I/N. In this warm-up example, we take

$$\phi\left(\frac{I}{N}\right) = \exp\left(-c_\phi \frac{I}{N}\right), \quad c_\phi > 0.$$

To complete the model, we need to specify a recruitment rate f. For this example we take f to be a constant over time, independent of population density. We have then the model equations

$$\begin{aligned} S(t+1) &= f + \exp\left(-c_\phi \frac{I(t)}{N(t)}\right)\sigma_S S(t) + v\sigma_I I(t) \\ I(t+1) &= \left(1 - \exp\left(-c_\phi \frac{I(t)}{N(t)}\right)\right)\sigma_S S(t) + (1-v)\sigma_I I(t) \end{aligned} \quad (4.16)$$

If $v = 1$, then all (surviving) infected individuals are susceptible and the model is for a disease with no acquired immunity (e.g., the common cold). Such a model is called an *SIS model*. If, on the other hand, $v = 0$ then no surviving infected individual who recovers is again susceptible (such as is the case for chicken pox and measles), and the model is an *SI model*. In this model, surviving recovered individuals leave the population in that they play no further role in the transmission of the disease or the recruitment of susceptibles.

In this model, the population's dynamics in the absence of the disease dynamics are governed by the linear difference equation

$$S(t+1) = f + \sigma_S S(t).$$

From the solution formula (1.14) in Section 1.2.2 of Chapter 1 we have, if $\sigma_S < 1$ (i.e., there is some susceptible mortality), that

$$S(t) = \sigma_S^t \left(S(0) - \frac{f}{1-\sigma_S}\right) + \frac{f}{1-\sigma_S}$$

4.6. SIS DISCRETE-TIME EPIDEMIC MODELS WITH BIRTH AND DEATH

for $t \geq 0$ and, as a result, we see that the disease-free population has a positive, globally asymptotically stable equilibrium

$$S^* := \frac{f}{1 - \sigma_S}.$$

In the presence of the disease, the model equations (4.16) have the *disease-free equilibrium* (denoted by DFE)

$$\begin{pmatrix} S \\ I \end{pmatrix} = \begin{pmatrix} S^* \\ 0 \end{pmatrix}.$$

To ascertain local stability properties of DFE, we apply the linearization principle (Chapter 3) by investigating the eigenvalues of the Jacobian associated with Equations (4.16) when evaluated at the equilibrium. This Jacobian is the triangular matrix

$$\begin{pmatrix} \sigma_S & v\sigma_I - \sigma_S c_\phi \\ 0 & (1-v)\sigma_I + \sigma_S c_\phi \end{pmatrix}$$

whose eigenvalues $\lambda_1 > 0$ and $\lambda_2 > 0$ are the diagonal entries. Since $0 < \lambda_1 = \sigma_S < 1$, the DFE is locally asymptotically stable if $\lambda_2 = (1-v)\sigma_I + \sigma_S c_\phi < 1$ and it is unstable, specifically a saddle, if $\lambda_2 > 0$. Re-writing these conclusions, we have that the DFE is locally asymptotically stable if

$$c_\phi \frac{\sigma_S}{1 - (1-v)\sigma_I} < 1$$

and is a saddle if the reverse inequality holds. This stability determining quantity has an important biological interpretation. According to the second equation for the dynamics of I in (4.16), the number of newly infected in individuals is the first term on the right-hand side and, hence, the number of new infections per infected individual in one unit of time is

$$\frac{\left(1 - \exp\left(-c_\phi \frac{I}{N}\right)\right) \sigma_S S}{I}$$

which, for a low infected population level, is approximately

$$\lim_{I \to 0} \frac{\left(1 - \exp\left(-c_\phi \frac{I}{N}\right)\right) \sigma_S S}{I} = c_\phi \sigma_S$$

(use l'Hopital's Rule). To obtain the expected number of new infections per infected individual by an infected individual, we multiple this number by the expected amount of time it spends in the infected class, which equals the geometric series

$$1 + (1-v)\sigma_I + ((1-v)\sigma_I)^2 + ((1-v)\sigma_I)^3 \cdots = \frac{1}{1 - (1-v)\sigma_I}.$$

Thus, we see that the stability determining quantity

$$\mathcal{R}_0 := c_\phi \sigma_S \frac{1}{1 - (1-v)\sigma_I}$$

has the interpretation of the expected number of infections (called secondary infections) caused by one infected individual. This quantity is called the *(inherent or intrinsic or basic) reproduction number*. We say inherent or intrinsic because it is an approximation to the expected number of secondary infections per infected individual when the population I of infected individuals is low.

4.7 The Reproduction Number $\mathcal{R}_0$

The inherent or basic reproduction number $\mathcal{R}_0$ is an important threshold parameter that provides insight when designing prevention and control strategies for established disease infections [11], [38]. In Chapter 1, $\mathcal{R}_0$ is defined for population models as the expected number of newborns per newborn per lifetime. In epidemiology, $\mathcal{R}_0$ is biologically defined to be *the expected number of infections produced by a single infectious individual* introduced into a totally susceptible population. Consequently, values of $\mathcal{R}_0 < 1$ are expected to imply that the number of infections will decrease over time and that the disease will eventually die out as the chain of transmission cannot be maintained. However, values of $\mathcal{R}_0 > 1$ imply that a low level of infection will increase infections in the population and a disease outbreak will occur. A specific example is measles which has an estimated $\mathcal{R}_0$ between 12 and 18 [250]. That is, each individual infected with measles (primary carrier) would, on average, transmit the measles infection to $12 - 18$ other individuals (secondary cases) in a totally susceptible population.

A next generation matrix method for defining and computing the basic reproduction number $\mathcal{R}_0$ for discrete-time compartmental epidemic models with a stable fixed point was developed by Allen and van den Driessche in [11]. This method is an extension of the method used in structured population dynamic models in [85] and [223] (see Theorem 18 in Chapter 2). The method considers discrete-time compartmental infectious disease models whose equations have the general mathematical form

$$\mathbf{x}(t+1) = \mathbf{g}(\mathbf{x}(t)) \tag{4.17}$$

with

$$\mathbf{x}(t) = (x_1(t), x_2(t), \cdots, x_n(t))^T.$$

We assume

> A1: $\mathbf{g}$ is continuously differentiable on an open set Ω in $\mathbb{R}^n$ that contains $\mathbb{R}_+^n \setminus \{\mathbf{0}\}$. Furthermore, assume $\mathbf{g}: \mathbb{R}_+^n \setminus \{\mathbf{0}\} \to \mathbb{R}_+^n \setminus \{\mathbf{0}\}$ and $\mathbf{g}: int(\mathbb{R}_+^n) \to int(\mathbb{R}_+^n)$.

Note that under assumption A1, $\mathbf{x}(0) \in \mathbb{R}_+^n \setminus \{\mathbf{0}\}$ (or $int(\mathbb{R}_+^n)$) implies $\mathbf{x}(t) \in \mathbb{R}_+^n \setminus \{\mathbf{0}\}$ (respectively, $int(\mathbb{R}_+^n)$) for all t. The reason the origin $\mathbf{0}$ is removed in assumption A1 is that some epidemic model equations are undefined at the origin; (4.16) is an example. For some epidemic models $\mathbb{R}_+^n \setminus \{\mathbf{0}\}$ can be replaced by $\mathbb{R}_+^n$.

We are interested in equilibrium solutions of Equation (4.17) and their stability properties. An equilibrium $\mathbf{x}^*$ is a solution $\mathbf{x} = \mathbf{x}^*$ of the algebraic equation

$$\mathbf{x} = \mathbf{g}(\mathbf{x}). \tag{4.18}$$

The linearization principle (Chapter 3) provides sufficient (but not necessary) conditions for the local asymptotic stability of $\mathbf{x}^*$ based on the eigenvalues of the Jacobian of $\mathbf{g}(\mathbf{x})$ evaluated at $\mathbf{x}^*$. In this section, we make use of the following notation for the Jacobian of a vector valued function $\mathbf{h}(\mathbf{y}) = (h_1(\mathbf{y}), \ldots, h_k(\mathbf{y}))^T \in \mathbb{R}^k$ of a $\mathbf{y} = (y_1, \ldots, y_k)^T \in \mathbb{R}^k$:

$$J_{\mathbf{y}}\mathbf{h}(\mathbf{y}) = (\partial_{y_i} h_j(\mathbf{y})) = \begin{pmatrix} \partial_{y_1} h_1(\mathbf{y}) & \partial_{y_2} h_1(\mathbf{y}) & \cdots & \partial_{y_k} h_1(\mathbf{y}) \\ \partial_{y_1} h_2(\mathbf{y}) & \partial_{y_2} h_2(\mathbf{y}) & \cdots & \partial_{y_k} h_2(\mathbf{y}) \\ \vdots & \vdots & & \vdots \\ \partial_{y_1} h_k(\mathbf{y}) & \partial_{y_2} h_k(\mathbf{y}) & \cdots & \partial_{y_k} h_k(\mathbf{y}) \end{pmatrix}.$$

4.7. THE REPRODUCTION NUMBER $\mathcal{R}_0$

When using equations of the form (4.17) to build a model an epidemic, we structure the population according to two classes of individuals: those who are infectious and those who are not. For our purposes, it is convenient to list the infectious classes first in the vector $\mathbf{x}$ and then decompose it as $\mathbf{x} = (\mathbf{x}_0, \mathbf{x}_1)^T$ where

$$\mathbf{x}_0 = (x_1, x_2, \cdots, x_m)^T$$

are m infectious states and

$$\mathbf{x}_1 = (x_{m+1}, x_{m+2}, \cdots, x_n)^T$$

are $n - m$ non-infectious states. Using this re-ordering of the variables, we can re-write the systems of Equation (4.17) as

$$\begin{aligned} \mathbf{x}_0(t+1) &= \mathbf{g}_0(\mathbf{x}_0(t), \mathbf{x}_1(t)) \\ \mathbf{x}_1(t+1) &= \mathbf{g}_1(\mathbf{x}_0(t), \mathbf{x}_1(t)) \end{aligned} \quad (4.19)$$

where we now distinguish the two components of $\mathbf{x}$ in argument lists:

$$\mathbf{g}(\mathbf{x}_0, \mathbf{x}_1) = (\mathbf{g}_0(\mathbf{x}_0, \mathbf{x}_1), \mathbf{g}_1(\mathbf{x}_0, \mathbf{x}_1))^T.$$

The component $\mathbf{g}_0$ describes the dynamics of the infectious states and $\mathbf{g}_1$ describes the dynamics of the non-infectious states. By A1 we have both $\mathbf{g}_i$ are continuously differentiable on $\Omega \setminus \{\mathbf{0}\}$ and

$$\begin{aligned} &\mathbf{g}_0 : \mathbb{R}_+^m \to \mathbb{R}_+^m && \text{and} && \mathbf{g}_0 : int(\mathbb{R}_+^m) \to int(\mathbb{R}_+^m) \\ &\mathbf{g}_1 \in C^1\left(\mathbb{R}_+^{n-m} \to \mathbb{R}_+^{n-m}\right) && \text{and} && \mathbf{g}_1 : int(\mathbb{R}_+^{n-m}) \to int(\mathbb{R}_+^{n-m}). \end{aligned} \quad (4.20)$$

The equilibrium equation (4.18) becomes the pair of equations

$$\begin{aligned} \mathbf{x}_0 &= \mathbf{g}_0(\mathbf{x}_0, \mathbf{x}_1) \\ \mathbf{x}_1 &= \mathbf{g}_1(\mathbf{x}_0, \mathbf{x}_1) \end{aligned}$$

whose Jacobian with respect to $\mathbf{x}$ is the 2×2 block matrix

$$J_{\mathbf{x}}\mathbf{g}(\mathbf{x}_0, \mathbf{x}_1) = \begin{pmatrix} J_{\mathbf{x}_0}\mathbf{g}_0(\mathbf{x}_0, \mathbf{x}_1) & J_{\mathbf{x}_1}\mathbf{g}_0(\mathbf{x}_0, \mathbf{x}_1) \\ J_{\mathbf{x}_0}\mathbf{g}_1(\mathbf{x}_0, \mathbf{x}_1) & J_{\mathbf{x}_1}\mathbf{g}_1(\mathbf{x}_0, \mathbf{x}_1) \end{pmatrix}. \quad (4.21)$$

In building an epidemic model on the basis of equation (4.17), we assume that the only source of infectious individuals is from the class of non-infected individuals, who become infected by contact with infectious individuals. That is to say, there is no immigration of infectious individuals into the population. So, if no infectious individuals are present at time t, then there will be no infectious individuals at time $t+1$. Mathematically, this means

$$\mathbf{g}_0(\mathbf{0}, \mathbf{x}_1) \equiv \mathbf{0} \text{ for all } \mathbf{x}_1 \in \mathbb{R}_+^{n-m}. \quad (4.22)$$

An equilibrium $(\mathbf{0}, \mathbf{x}_1^*)$ in which infectious individuals are absence is called a *disease-free equilibrium* (denoted by DFE). The component $\mathbf{x}_1^*$ of a DFE satisfies the algebraic equation

$$\mathbf{x}_1 = \mathbf{g}_1(\mathbf{0}, \mathbf{x}_1).$$

To study the stability properties of a DFE by means of the linearization Principle, we consider the Jacobian (4.21) evaluated at $(\mathbf{0}, \mathbf{x}_1^*)$, which is the block triangular matrix

$$J_{\mathbf{x}}\mathbf{g}(\mathbf{0}, \mathbf{x}_1^*) = \begin{pmatrix} J_{\mathbf{x}_0}\mathbf{g}_0(\mathbf{0}, \mathbf{x}_1^*) & O \\ J_{\mathbf{x}_0}\mathbf{g}_1(\mathbf{0}, \mathbf{x}_1^*) & J_{\mathbf{x}_1}\mathbf{g}_1(\mathbf{0}, \mathbf{x}_1^*) \end{pmatrix}. \quad (4.23)$$

The upper right corner O of this block matrix is the $m \times (n-m)$ matrix of zeros. This block is the zero matrix because of the identity (4.22). The eigenvalues of $J_\mathbf{x}\mathbf{g}(\mathbf{0}, \mathbf{x}_1^*)$ are those of the diagonal blocks $J_{\mathbf{x}_0}\mathbf{g}_0(\mathbf{0}, \mathbf{x}_1^*)$ and $J_{\mathbf{x}_1}\mathbf{g}_1(\mathbf{0}, \mathbf{x}_1^*)$, which we will investigate in turn.

Note that the component $\mathbf{x}_1^*$ of the DFE is an equilibrium of the difference equation

$$\mathbf{x}_1(t+1) = \mathbf{g}_1(\mathbf{0}, \mathbf{x}_1(t)). \tag{4.24}$$

This equation accounts for the disease-free dynamics of the population, i.e., the dynamics in the absence of any infectious individuals. To use the linearization principle to study the stability properties of the equilibrium $\mathbf{x}_1^*$ of the disease-free dynamic equation (4.24), we consider the eigenvalues of the Jacobian $J_{\mathbf{x}_1}\mathbf{g}_1(\mathbf{0}, \mathbf{x}_1^*)$, which is the same as the lower right block in the Jacobian (4.23) associated with the DFE of the epidemic model's equations (4.19).

If the spectral radius

$$\rho(J_{\mathbf{x}_1}\mathbf{g}_1(\mathbf{0}, \mathbf{x}_1^*)) > 1$$

then the population has an unstable equilibrium both in the absence and presence of infectious individuals. On the other hand, if

$$\rho(J_{\mathbf{x}_1}\mathbf{g}_1(\mathbf{0}, \mathbf{x}_1^*)) < 1 \tag{4.25}$$

then the population has a stability equilibrium in the absence of infectious individuals. In this case, to determine the local stability properties of the DFE $(\mathbf{0}, \mathbf{x}_1^*)$ of (4.19), we consider the eigenvalues of the upper left block $J_{\mathbf{x}_0}\mathbf{g}_0(\mathbf{0}, \mathbf{x}_1^*)$ in the Jacobian (4.23).

We summarize these results concerning a DFE in the following theorem, for which we make the assumption

A2: $\mathbf{g}_0(\mathbf{x}_0, \mathbf{x}_1)$ satisfies (4.22) and that $\mathbf{x}_1^* \in \mathbb{R}_+^{n-m}$ is a solution of the equation $\mathbf{x}_1 = \mathbf{g}_1(\mathbf{0}, \mathbf{x}_1)$ for which inequality (4.25) holds.

Theorem 43 *Assume A1 and A2. The disease-free equilibrium $(\mathbf{0}, \mathbf{x}_1^*)^T$ of model equations (4.19) is locally asymptotically stable if $\rho^* < 1$ and unstable if $\rho^* > 1$ where*

$$\rho^* := \rho(J_{\mathbf{x}_0}\mathbf{g}_0(\mathbf{0}, \mathbf{x}_1^*)).$$

Remark 15 *When $\rho^* < 1$ the DFE $(\mathbf{0}, \mathbf{x}_1^*)^T$ is locally asymptotically stable. According to the definition of local asymptotic stability, any initial condition $(\mathbf{x}_0(0), \mathbf{x}_1(0))^T$ sufficiently close to $(\mathbf{0}, \mathbf{x}_1^*)^T$ will give a solution of model equations (4.19) that asymptotically tends to $(\mathbf{0}, \mathbf{x}_1^*)^T$. In other words, the infectious class disappears in the long run and, in this sense, we can say that an epidemic does not occur. However, this statement requires that the population be initially close to its equilibrium and that the infectious class be small in number, i.e., the stability of a DFE implies that a low level of infection of a population near equilibrium cannot invade the population. Because Theorem 43 is based on the linearization principle, and hence only provides local stability, it tells us nothing about initial conditions lying outside a neighborhood of the DFE (for example, should a large initial infection occur). This would require a global analysis of the dynamics of the model equations.*

An application of Theorem 43 requires a calculation of ρ^*. The Jacobian $J_{\mathbf{x}_0}\mathbf{g}_0(\mathbf{0}, \mathbf{x}_1^*)$ is an $m \times m$ matrix and if m is large it is likely to be difficult, and even impossible, to find a general formula for ρ^* in terms of its entries. The reproduction number $\mathcal{R}_0$ is another important stability

4.7. THE REPRODUCTION NUMBER $\mathcal{R}_0$

determining diagnostic quantity that is often more analytically tractable. Our goal here is to show how to calculate $\mathcal{R}_0$ and to prove that it can replace ρ^* in Theorem 43. The method will be analogous to the next generation matrix method used for structured population models in Chapter 2.

We begin by noting that infectious individuals at time $t+1$ consist of *newly* infected individuals plus those infected individuals at time t who *survive* to time $t+1$. If we denote the former by $\mathbf{n}(\mathbf{x}_0, \mathbf{x}_1)$ and the latter by $\mathbf{s}(\mathbf{x}_0, \mathbf{x}_1)$, so that

$$\mathbf{g}_0(\mathbf{x}_0, \mathbf{x}_1) = \mathbf{n}(\mathbf{x}_0, \mathbf{x}_1) + \mathbf{s}(\mathbf{x}_0, \mathbf{x}_1)$$

in the model equations (4.19) then

$$J_{\mathbf{x}_0}\mathbf{g}_0(\mathbf{0}, \mathbf{x}_1^*) = F + T$$

where F is the Jacobian of $\mathbf{n}$ and T is the Jacobian of $\mathbf{s}$ evaluated at the DFE $(\mathbf{0}, \mathbf{x}_1^*)$, i.e.,

$$F := J_{x_0}\mathbf{n}(\mathbf{0}, \mathbf{x}_1^*), \quad T := J_{x_0}\mathbf{s}(\mathbf{0}, \mathbf{x}_1^*). \tag{4.26}$$

As is in Chapter 2, we have, under suitable conditions on F and T, that $\rho(F+T)$ and $\rho(F(I-T)^{-1})$ are both either less than or greater than or equal to 1.

A3: Assume F and T are nonnegative matrices and $\rho(T) < 1$.

As in Chapter 2, we mathematically define the *reproduction number* associated with a DFE as the spectral radius of the *next generation matrix* $F(I-T)^{-1}$:

$$\mathcal{R}_0 := \rho(F(I-T)^{-1}). \tag{4.27}$$

From Theorem 43 and 18 in Chapter 2, we obtain the following result, in which ρ^* is replaced by $\mathcal{R}_0$ in the stability criteria of Theorem 43 (see [11]).[1]

Theorem 44 *Assume A1, A2, and A3 and that $\mathcal{R}_0 > 0$. Then the disease-free equilibrium $(\mathbf{0}, \mathbf{x}_1^*)^T$ of model equations (4.19) is locally asymptotically stable if $\mathcal{R}_0 < 1$ and unstable if $\mathcal{R}_0 > 1$.*

The biological interpretation of $\mathcal{R}_0$ parallels that for the structured population models in Chapter 2 The linearization of model equations (4.19) at the DFE is the linear matrix equation

$$\begin{pmatrix} \mathbf{y}_0(t+1) \\ \mathbf{y}_1(t+1) \end{pmatrix} = \begin{pmatrix} F+T & O \\ J_{\mathbf{x}_0}\mathbf{g}_1(\mathbf{0}, \mathbf{x}_1^*) & J_{\mathbf{x}_1}\mathbf{g}_1(\mathbf{0}, \mathbf{x}_1^*) \end{pmatrix} \begin{pmatrix} \mathbf{y}_0(t) \\ \mathbf{y}_1(t) \end{pmatrix}$$

yields the linear matrix equation

$$\mathbf{y}_0(t+1) = (F+T)\mathbf{y}_0(t).$$

for $\mathbf{y}_0(t)$ which is, near the DFE, approximately the vector $\mathbf{x}_0(t)$ of infectious classes. The vector $F\mathbf{y}_0(t)$ is (approximately) the vector of newly infected individuals at time $t+1$ (in analogy to the newborns in the population models of Chapter 2) and $T\mathbf{y}_0(t)$ is (approximately) the vector of

---

[1] As pointed out in [223], the assumption that the column sums of the matrix T be less than or equal to 1 (which is true in population models in Chapter 2) can be dropped in Theorem 18 in Chapter 2. Therefore, we do not make that requirement here.

surviving infected individuals from time t. The inequality $\rho(T) < 1$ in A3 means that the expected time an individual spends in infectious classes during its lifetime is finite (as long as the population remains near the DFE).

In analogy with the population dynamic interpretation of $\mathcal{R}_0$, consider first the case when all newly infected individuals are placed into only one of the infectious classes in $\mathbf{x}_0$, say for simplicity the first class x_1. Thus, only the first row in F is nonzero and $\mathcal{R}_0$ is the first row, first column entry in $F(I-T)^{-1}$. In analog to linear, structured population models in Chapter 2 with a single newborn class, this definition of $\mathcal{R}_0$ has the biological interpretation of the *average number of new (secondary) infections produced by infected individuals over the time it spends in the infectious classes*. (For the more general case when newly infected individuals can be assigned to more than one possible infectious state in the model the interpretation of $\mathcal{R}_0$ follows analogously to the population model case when there are several classes of newborns; see Chapter 2.)

Example 77 *Consider the epidemic model* (4.16)

$$S(t+1) = f(S(t),I(t)) + v\sigma_I I(t) + \phi\left(\tfrac{I(t)}{N(t)}\right)\sigma_S S(t)$$
$$I(t+1) = \left(1 - \phi\left(\tfrac{I(t)}{N(t)}\right)\right)\sigma_S S(t) + (1-v)\sigma_I I(t)$$

with a general recruitment functional $f(S,I)$ and continuously differentiable escape function $\phi(z)$ that satisfies the conditions

$$0 \leq \phi(z) \leq 1 \text{ for all } 0 \leq z \leq 1 \text{ and } \phi(0)=1,\ \phi'(0) < 0. \tag{4.28}$$

For this $n=2$ dimensional model we have only $m=1$ infectious class. The infectious and susceptible class vectors are

$$\mathbf{x}_0 = I, \quad \mathbf{x}_1 = S,$$

respectively. The infectious and susceptible components of the model equations are, respectively,

$$\mathbf{g}_0(\mathbf{x}_0,\mathbf{x}_1) = \left(1 - \phi\left(\tfrac{I}{N}\right)\right)\sigma_S S + (1-v)\sigma_I I$$

$$\mathbf{g}_1(\mathbf{x}_0,\mathbf{x}_1) = f(S,I) + v\sigma_I I + \phi\left(\tfrac{I}{N}\right)\sigma_S S.$$

In order to fulfill assumptions A1 and A2, we assume the scalar difference equation (for the disease-free dynamics)

$$S(t+1) = f(S(t),0) + \sigma_S S(t) \tag{4.29}$$

has a positive equilibrium $S^* > 0$ which is locally asymptotically stable by linearization, i.e., there exists a positive root $S^* > 0$ of the equation the equilibrium equation

$$S^* = f(S^*,0) + \sigma_S S^*$$

that satisfies

$$\left|\frac{\partial h(S,0)}{\partial S} + \sigma_S\right|_{S=S^*} < 1. \tag{4.30}$$

To calculate $\mathcal{R}_0$ we need note that

$$\mathbf{n}_0(\mathbf{x}_0,\mathbf{x}_1) = \phi\left(\tfrac{I}{S+I}\right)\sigma_S S, \quad \mathbf{s}_0(\mathbf{x}_0,\mathbf{x}_1) = (1-v)\sigma_I I$$

4.7. THE REPRODUCTION NUMBER $\mathcal{R}_0$

and consequently the (1×1) Jacobians (4.26) are

$$F = \frac{\partial}{\partial I}\phi\left(\frac{I}{S+I}\right)\sigma_S S\bigg|_{S=S^*, I=0} = -\sigma_S \phi'(0)$$

$$T = \frac{\partial}{\partial I}(1-v)\sigma_I I\bigg|_{S=S^*, I=0} = (1-v)\sigma_I.$$

Note that A3 holds. Thus $\mathcal{R}_0 = F(I-T)^{-1}$ is

$$\mathcal{R}_0 = -\sigma_S \phi'(0) \frac{1}{1-(1-v)\sigma_I}. \tag{4.31}$$

It follows from Theorem 44 that if $\phi'(0) < 0$ (so that $\mathcal{R}_0 > 0$), then the DFE $(S, I)^T = (S^*, 0)^T$ is locally asymptotically stable if $\mathcal{R}_0 < 1$ and unstable if $\mathcal{R}_0 > 1$.

In the motivating example in Section 4.1, the recruitment function f was a constant. To include density dependence in the recruitment rate of susceptibles, one can use any of the numerous recruitment functions discussed in Chapter 1. Basic examples include density effects modeled by the discrete logistic (Beverton-Holt) and exponential (Ricker) models. How density affects recruitment determines how f depends on S and I. For example, modelers often assume recruitment is negatively affected by changes in total population size N, in which case one might take

$$f(S,I) = \begin{cases} be^{-cN} N & \text{Exponential recruitment} \\ \frac{b}{1+cN} N & \text{Discrete logistic recruitment.} \end{cases} \tag{4.32}$$

If, on the other hand, infected individuals are too ill in order to interfere with recruitment, then

$$f(S,I) = \begin{cases} be^{-cS} S & \text{Exponential recruitment} \\ b\frac{1}{1+cS} S & \text{Discrete logistic recruitment.} \end{cases}$$

For any of these examples, the disease-free dynamics is governed by the equations

$$S(t+1) = \begin{cases} be^{-cS(t)} S(t) + \sigma_S S(t) \\ b\frac{1}{1+cS(t)} S(t) + \sigma_S S(t). \end{cases} \tag{4.33}$$

It is left as a Problem 1 for the reader to show for these examples that

$$S^* = \begin{cases} \frac{1}{c}\ln\left(\frac{b}{1-\sigma_S}\right) & \text{for } 1 < \frac{b}{1-\sigma_S} < \exp\left(\frac{2}{1-\sigma_S}\right) \\ \frac{1}{c}\left(\frac{b}{1-\sigma_S} - 1\right) & \text{for } 1 < \frac{b}{1-\sigma_S} \end{cases} \tag{4.34}$$

are positive equilibria that are locally asymptotically stable by linearization, i.e., (4.30) holds for the indicated ranges of parameters. (Note that the ratio $b/(1-\sigma_S)$ in these examples is the reproduction number of the disease-free dynamics, as defined in Chapter 2.) For any of these examples used in Example 77, $\mathcal{R}_0$ is given by (4.31).

A commonly used infection escape function $\phi(z)$ that satisfies (4.28) in Example 77 is the exponential $\phi(z) = \exp(-cz)$, $c > 0$. This choice of ϕ can be given some biological underpinnings as follows. Divide the unit of time into small subintervals of length Δt and assume the probability a susceptible will encounter another individual, per unit time, is inversely proportional to the total population size N, i.e., is equal to $c\Delta t/N$. The probability of not encountering the individual during Δt is $1 - c\Delta t/N$ and over a unit interval consisting of $1/\Delta t$ increments is $(1 - c\Delta t/N)^{1/\Delta t}$. If there are I infected individuals and encountering them are independent events, then the probability of not encountering any infected interval during a unit of time is

$$\left((1 - c\Delta t/N)^{1/\Delta t}\right)^I.$$

As $\Delta t \to 0$ this expression approaches $\exp(-cI/N)$, which is the probability ϕ that an individual will not encounter any individual from the class I during a unit of time. Other circumstances that lead to different assumptions about the probability of interactions can lead to different expressions for ϕ. For example, if the probability of meeting and infected individual during a time interval is assumed simply to be proportional to the duration of the interval, then replacing $c\Delta t/N$ by $c\Delta t$ leads to $\phi = \exp(-cI)$.

Exercises 4.6

1. Derive the formulas (4.34) for the positive equilibria of Equations (4.33) and verify that they satisfy the local stability condition (4.30) for the indicated range of parameter values.

2. Consider the following equation for the susceptible population in its disease-free state

$$S(t+1) = \beta(S(t)) + \sigma_s S(t)$$

   where $S(0) \geq 0$ and $\sigma_s \in (0,1)$. Derive formulas for the positive equilibria in each case below. Also, determine (by using the Linearization Principle) the parameter values for which the equilibria are locally asymptotically stability. In all cases $\sigma_r \in (0,1)$ and $a, b, c > 0$.

   (a) $\beta(S) = b e^{-c\sigma_r S} \sigma_r S$ (b) $\beta(S) = b \dfrac{1}{(1 + c\sigma_r S)^a} \sigma_r S$

   (c) $\beta(S) = b e^{-cS(t)} \sigma_r S$ (d) $\beta(S) = b \dfrac{1}{(1 + cS)^a} \sigma_r S$

3. Consider the following SIS epidemic model with pre-birth mortality

$$\begin{aligned} S(t+1) &= h(\sigma_r S(t), I(t)) + v\sigma_I I(t) + \phi\left(\tfrac{I(t)}{N(t)}\right)\sigma_s S(t) \\ I(t+1) &= \left(1 - \phi\left(\tfrac{I(t)}{N(t)}\right)\right)\sigma_s S(t) + (1-v)\sigma_I I(t) \end{aligned}$$

   where $S(0) \geq 0$, $I(0) \geq 0$, $\sigma_r, \sigma_s, v \in (0,1)$. Compute the basic reproduction number $\mathcal{R}_0$

4. Consider the following SIS epidemic model with post-birth mortality

$$\begin{aligned} S(t+1) &= \sigma_r h(S(t), I(t)) + v\sigma_I I(t) + \phi\left(\tfrac{I(t)}{N(t)}\right)\sigma_s S(t) \\ I(t+1) &= \left(1 - \phi\left(\tfrac{I(t)}{N(t)}\right)\right)\sigma_s S(t) + (1-v)\sigma_I I(t) \end{aligned}$$

   where $S(0) \geq 0$, $I(0) \geq 0$, $\sigma_r, \sigma_s, v \in (0,1)$. Compute the basic reproduction number $\mathcal{R}_0$.

4.7. THE REPRODUCTION NUMBER $\mathcal{R}_0$

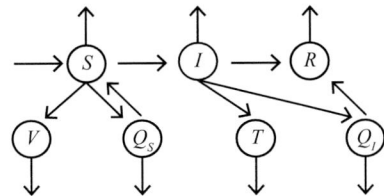

Figure 4.7: SIR Model with Vaccination and Quarantine.

5. From the accompanying Figure 4.7 diagram, construct a system of difference equations for the SIR model with vaccinated (V), quarantine (Q_S), treated (T) and isolated (Q_I) classes.

   (a) Assuming a constant birth function $h(S, I) = \Lambda > 0$, find the disease-free equilibrium point(s) and use the Linearization Principle to determine their stability.

   (b) Compute $\mathcal{R}_0$.

   (c) Use simulations to determine what happens to the disease in the long run.

   (d) By comparing the basic reproduction number $\mathcal{R}_0$ for this model with that of the SIR model without vaccination, quarantine, treatment and isolation, discuss the impacts that vaccination, quarantine, treatment and isolation have on $\mathcal{R}_0$ and on reducing the number of infections.

   (e) Assuming the birth function $h(S, I) = \tau S$, where $\tau > 0$, find the disease free equilibrium point(s) and use the Linearization Principle to determine the stability.

   (f) Compute $\mathcal{R}_0$ for the disease model in (f).

   (g) Use simulations to determine what happens to the disease in the long run for the model in (f).

6. Consider the basic reproduction number

$$\mathcal{R}_0 = c_\phi \sigma_S \frac{1}{(1 - (1 - \nu)\sigma_I)}$$

for the SI model (4.16). Here $c_\phi > 0$ is the transmission constant, $\gamma = \nu \sigma_I < 1$ is the fraction that recovers from the infection per unit time, and $\sigma = (1 - \nu)\sigma_I < 1$ is the fraction that stays infectious per unit time.

   (a) Show that $\mathcal{R}_0$ is an increasing function of the transmission constant. Consequently, high transmission among susceptible individuals can lead to disease invasion, while low transmissions lead to disease elimination.

   (b) Compute the relative sensitivity of $\mathcal{R}_0$ to the recovery and survival rates

$$\frac{\partial \mathcal{R}_0 / \partial \gamma}{\partial \mathcal{R}_0 / \partial \sigma}.$$

4.8 Global Stability of Equilibria

The stability of an equilibrium obtained by the Linearization Principle is local in that it describes the asymptotic dynamics of only those solutions that initially are sufficiently close to the equilibrium. Local asymptotic stability does not rule out the existence of other stable equilibria (or other types of attractors, such as k-cycles). In applications, it can be of importance to know whether an equilibrium point is globally asymptotically stable or is not. In this Section, we will introduce two methods for studying global asymptotic stability in epidemic models.

4.8.1 Liapunov Functions

For many disease models of the form Equation (4.17) in Section 4.7, the disease-free equilibrium

$$\mathbf{x}^* = \begin{pmatrix} 0 \\ \mathbf{x}_1^* \end{pmatrix} \in \mathbb{R}_+^{n+m}$$

is not only locally, but is globally asymptotically stable whenever $\mathcal{R}_0 < 1$. In this section, we use Liapunov functions to study the global asymptotic stability of the disease-free equilibrium; we follow the method developed in [251].

Using the notation $\mathbf{x} = (\mathbf{x}_0, \mathbf{x}_1)^T$ from Section 4.7 we define

$$\mathbf{f}(\mathbf{x}_0, \mathbf{x}_1) := (F + T)\mathbf{x}_0 - \mathbf{g}_0(\mathbf{x}_0, \mathbf{x}_1) \tag{4.35}$$

where F and T are defined by (4.26). The equation for the $\mathbf{x}_0(t)$ component in (4.19) can be re-written as

$$\mathbf{x}_0(t+1) = (F + T)\mathbf{x}_0(t) - \mathbf{f}(\mathbf{x}_0(t), \mathbf{x}_1(t)) \in \mathbb{R}_+^n. \tag{4.36}$$

Note that since $\mathbf{0} = \mathbf{g}_0(0, \mathbf{x}_1^*)$ it follows that $\mathbf{f}(0, \mathbf{x}_1^*) = \mathbf{0}$.

A4. In addition to A1, A2, and A3, assume
  (a) $(I - T)^{-1}$ is a nonnegative matrix and $(I - T)^{-1} F$ is irreducible.
  (b) $\mathbf{f}(\mathbf{x}_0, \mathbf{x}_1) \geq \mathbf{0}$ on $\mathbb{R}_+^{n-m}$

Since $(I - T)^{-1} F$ and $F(I - T)^{-1}$ have the same eigenvalues (see Problem 6) it follows that $\rho((I - T)^{-1} F) = \mathcal{R}_0$. Let $\omega \in int\left(\mathbb{R}_+^n\right)$ be a positive eigenvector of $(I - T)^{-1} F$ corresponding to $\mathcal{R}_0$ (whose existence is guaranteed by Perron-Frobenius theory). The real-valued function

$$V(\mathbf{x}_0, \mathbf{x}_1) = \omega^T (I - T)^{-1} \mathbf{x}_0 \tag{4.37}$$

is positive valued for all $(\mathbf{x}_0, \mathbf{x}_1)$ with $\mathbf{x}_0 \in \mathbb{R}_+^m \setminus \{\mathbf{0}\}$ and vanishes at $(\mathbf{x}_0, \mathbf{x}_1) = (\mathbf{0}, \mathbf{x}_1)$ for all $\mathbf{x}_1 \in \mathbb{R}_+^{n-m}$.

The assumption A2 implies the $\mathbf{x}_1^*$ is a locally asymptotically stable equilibrium of the disease-free equation (4.24). By Theorem 44 $\mathcal{R}_0 < 1$ implies the DFE $\mathbf{x}^* = (0, \mathbf{x}_1^*)^T$ is a locally asymptotically stable equilibrium of Equation (4.19). In the following theorem, we add the assumption that $\mathbf{x}_1^*$ is also a globally attracting equilibrium on $\mathbb{R}_+^{n-m}$ of the disease-free equation (4.24), i.e., that it is globally asymptotically stable on $\mathbb{R}_+^{n-m}$. The theorem gives conditions under which the DFE $\mathbf{x}^* = (0, \mathbf{x}_1^*)^T$ is globally asymptotically stable on $\mathbb{R}_+^{n-m}$ for (4.19).

Theorem 45 *[251] Assume A1, A2, A3, and A4 hold and that $\mathcal{R}_0 < 1$. Assume $\mathbf{x}_1^*$ is a globally asymptotically stable equilibrium of the disease-free equation (4.24) on $\mathbb{R}_+^{n-m}$. Then $\mathbf{x}^* = (0, \mathbf{x}_1^*)^T$ is a globally asymptotically stable equilibrium of (4.19) on $\mathbb{R}_+^n$.*

4.8. GLOBAL STABILITY OF EQUILIBRIA

Proof. The goal is to apply LaSalle's Invariance Principle [207] (see Theorem 28 in Chapter 3) by showing V defined by (4.37) is non-increasing along solutions $\mathbf{x}(t) = (\mathbf{x}_0(t), \mathbf{x}_1(t)) \in \mathbb{R}_+^n$ of (4.19) on $\mathbb{R}_+^n$. Define

$$\Delta V(\mathbf{x}_0, \mathbf{x}_1) := V(\mathbf{g}_0(\mathbf{x}_0, \mathbf{x}_1), \mathbf{g}_1(\mathbf{x}_0, \mathbf{x}_1)) - V(\mathbf{x}_0, \mathbf{x}_1).$$

for $(\mathbf{x}_0, \mathbf{x}_1) \in \mathbb{R}_+^n$. We need to show $\Delta V(\mathbf{x}_0(t), \mathbf{x}_1(t)) \leq 0$ for all t. Toward this end, we calculate

$$\begin{aligned}
\Delta V(\mathbf{x}_0(t), \mathbf{x}_1(t)) &= \omega^T (I-T)^{-1} \mathbf{x}_0(t+1) - \omega^T (I-T)^{-1} \mathbf{x}_0(t) \\
&= \omega^T (I-T)^{-1} (F+T) \mathbf{x}_0(t) \\
&\quad - \omega^T (I-T)^{-1} \mathbf{f}(\mathbf{x}_0(t), \mathbf{x}_1(t)) - \omega^T (I-T)^{-1} \mathbf{x}_0(t) \\
&= \omega^T (I-T)^{-1} (T - I + F + I) \mathbf{x}_0(t) \\
&\quad - \omega^T (I-T)^{-1} \mathbf{f}(\mathbf{x}_0(t), \mathbf{x}_1(t)) - \omega^T (I-T)^{-1} \mathbf{x}_0(t) \\
&= \omega^T (-1 + \mathcal{R}_0) \mathbf{x}_0(t) - \omega^T (I-T)^{-1} \mathbf{f}(\mathbf{x}_0(t), \mathbf{x}_1(t)).
\end{aligned}$$

The assumptions made imply

$$\omega^T (-1 + \mathcal{R}_0) \mathbf{x}_0(t) \leq 0 \quad \text{and} \quad \omega^T (I-T)^{-1} \mathbf{f}(\mathbf{x}_0(t), \mathbf{x}_1(t)) \geq 0$$

and consequently $\Delta V(\mathbf{x}_0(t), \mathbf{x}_1(t)) \leq 0$ for all t. LaSalle's Invariance Principle implies $\mathbf{x}(t)$ approaches the largest invariant subset of the set $M = \{\mathbf{x} \in \mathbb{R}_+^n | \Delta V(\mathbf{x}) = 0\}$ where, as calculated above,

$$\Delta V(\mathbf{x}) = \omega^T (-1 + \mathcal{R}_0) \mathbf{x}_0 - \omega^T (I-T)^{-1} \mathbf{f}(\mathbf{x}_0, \mathbf{x}_1).$$

It follows that $\mathbf{x} \in M$ if and only if $\mathbf{x}_0 = \mathbf{0}$ and hence $M = \{(\mathbf{0}, \mathbf{x}_1) | \mathbf{x}_1 \in \mathbb{R}_+^{n-m}\}$. Since $\mathbf{x}_1^*$ is a globally asymptotically stable equilibrium of the disease-free equation (4.24) on $\mathbb{R}_+^{n-m}$, it follows that the only invariant set in M is the disease-free equilibrium $(\mathbf{0}, \mathbf{x}_1^*)$ of (4.19). Thus, all solutions of (4.19) in $\mathbb{R}_+^n$ approach $(\mathbf{0}, \mathbf{x}_1^*)$ as $t \to +\infty$. This, together with the local asymptotic stability of $(\mathbf{0}, \mathbf{x}_1^*)$ which follows from Theorem 44 and $\mathcal{R}_0 < 1$, imply $(\mathbf{0}, \mathbf{x}_1^*)$ is globally asymptotically stable. ∎

Example 78 *For the SIS model*

$$\begin{aligned}
S(t+1) &= h(S(t), I(t)) + v\sigma_I I(t) + \phi\left(\frac{I(t)}{N(t)}\right) \sigma_S S(t) \\
I(t+1) &= \left(1 - \phi\left(\frac{I(t)}{N(t)}\right)\right) \sigma_S S(t) + (1-v)\sigma_I I(t)
\end{aligned} \quad (4.38)$$

we find from the calculations carried out there in Example 77 that assumptions A1, A2, A3 and A4(a) are satisfied. To apply Theorem 45 we need to consider only the assumption A4(b) that $\mathbf{f}(\mathbf{x}_0, \mathbf{x}_1) \geq \mathbf{0}$ *on* $\mathbb{R}_+^2$. *From Example 77 we have*

$$\mathbf{g}_0(\mathbf{x}_0, \mathbf{x}_1) = (1 - \phi\left(\frac{I}{S+I}\right))\sigma_S S + (1-v)\sigma_I I$$

$$F = -\sigma_S \phi'(0) \quad \text{and} \quad T = (1-v)\sigma_I$$

and from (4.35)

$$\mathbf{f}(\mathbf{x}_0, \mathbf{x}_1) = (-\sigma_S \phi'(0) + (1-v)\sigma_I) I - \left(\phi\left(\frac{I}{S+I}\right) \sigma_S S + (1-v)\sigma_I I \right)$$

$$= -\sigma_S \phi'(0) I - \phi\left(\frac{I}{S+I}\right) \sigma_S S$$

(note that $\omega = 1$). Making use of the Mean Value Theorem, we write

$$\phi\left(\frac{I}{S+I}\right) = 1 - \phi\left(\frac{I}{S+I}\right) = -\phi'(\xi) \frac{I}{S+I}$$

for some ξ satisfying $0 \leq \xi \leq \frac{I}{S+I}$ and obtain

$$\mathbf{f}(\mathbf{x}_0, \mathbf{x}_1) = \sigma_S I \left(-\phi'(0) + \phi'(\xi) \frac{S}{S+I} \right).$$

If we add to the conditions (4.28) the assumptions

$$\phi'(0) < 0 \text{ and } \phi''(z) \geq 0 \text{ for } 0 \leq z \leq 1$$

on the infection escape function (so that $\phi'(\xi) > \phi'(0)$), then

$$\mathbf{f}(\mathbf{x}_0, \mathbf{x}_1) \geq \sigma_S I \left(-\phi'(0) + \phi'(\xi) \right) \geq 0$$

and A4(b) is satisfied. Theorem 45 then implies that the DFE equilibrium $(S, I)^T = (S^*, 0)^T$ is globally asymptotically stable on $\mathbb{R}_+^2$ when

$$\mathcal{R}_0 = -\sigma_S \phi'(0) \frac{1}{1 - (1-v)\sigma_I} < 1.$$

Example 79 *For example, in the special case considered in Section 4.1 when susceptible recruitment is constant f and the escape function is $\phi(z) = \exp(-c_\phi z)$, the DFE equilibrium*

$$\begin{pmatrix} S^* \\ I^* \end{pmatrix} = \begin{pmatrix} \frac{f}{1-\sigma_S} \\ 0 \end{pmatrix}.$$

is globally asymptotically stable when

$$\mathcal{R}_0 = c_\phi \frac{\sigma_S}{1 - (1-v)\sigma_I} < 1.$$

4.8.2 Models with Constant or Asymptotically Constant Population Size

In many epidemic models, the total population size is constant or asymptotically constant. This fact provides a means for further analysis, which often includes results concerning the global dynamics of solutions.

The number of equations in an epidemic model whose population size $N(t)$ remains constant can, because of this fact, be reduced by one. This lowering of dimension often permits an analysis

4.8. GLOBAL STABILITY OF EQUILIBRIA

of the model, even when other methods such as those developed in Section 4.7 fail to apply. To illustrate this by means of an example, we return to the SI model (4.15) in Section 4.1 and make the following simplifying assumptions: no new susceptible individuals are added to the population (by birth or immigration) and that no deaths occur. (Such models are said to ignore "demography".) With $f = 0$ and $\sigma_I = \sigma_S = 0$, the model equations (4.16) become

$$\begin{aligned} S(t+1) &= \phi\left(\tfrac{I(t)}{N(t)}\right) S(t) + vI(t) \\ I(t+1) &= \left(1 - \phi\left(\tfrac{I(t)}{N(t)}\right)\right) S(t) + (1-v)I(t) \end{aligned} \quad (4.39)$$

with $0 < v < 1$. The analytic methodology and theorems developed in Section 4.7 concerning the stability of disease-free and endemic equilibria do not apply to this model because assumption A2 does not hold (i.e., there is no DFE that is locally asymptotically stable by linearization). Nonetheless, an analysis of this model can be performed by reducing it to a single difference equation because the total population size $N(t) = S(t) + I(t)$ remains constant for all t. To see this simply add both equations together to find that $S(t+1) + I(t+1) = S(t) + I(t)$. Using this, we can study whether an epidemic occurs or not by studying just the equation for $I(t)$, which we can write as the uncoupled equation

$$I(t+1) = \left(1 - \phi\left(\frac{I(t)}{N_0}\right)\right)(N_0 - I(t)) + (1-v)I(t) \quad (4.40)$$

where $N_0 := S(0) + I(0)$. A notational simplification can be made by dividing both sides by N_0 and defining $x(t) := I(t)/N_0$ (the proportion of infected individuals present at time t). This results in the difference equation

$$x(t+1) = (1 - \phi(x(t)))(1 - x(t)) + (1-v)x(t). \quad (4.41)$$

We have reduced the study of the dynamics of the two-dimensional SI model to the study of this one-dimensional difference equation for the fraction $x(t)$ of infected individuals in the population, for which the methods of Chapter 1 are available.

Example 80 *With the escape function*

$$\phi\left(\frac{I}{N}\right) = \frac{1}{1 + c_\phi \frac{I}{N}}, \quad c_\phi > 0$$

the difference equation (4.41) derived from the SI model (4.39) becomes

$$x(t+1) = \left(1 - \frac{1}{1 + c_\phi x(t)}\right)(1 - x(t)) + (1-v)x(t).$$

If $c_\phi < v$, then the equilibrium $x = 0$, which corresponds to the disease-free case $I = 0$ in the SI model, is locally asymptotically stable. This can be seen from the Linearization Principle by differentiating

$$h(x) = \left(1 - \frac{1}{1 + c_\phi x}\right)(1 - x) + (1-v)x \quad (4.42)$$

and evaluating the answer at $x = 0$ to find

$$0 < h'(0) = c_\phi + 1 - v < 1.$$

Moreover, for $0 \leq x \leq 1$ the inequalities

$$0 \leq h(x) = c_\phi x \frac{1-x}{1+c_\phi x} + (1-v)x$$
$$\leq c_\phi x + (1-v)x = (1 + c_\phi - v)x,$$

show that $0 \leq x(t+1) \leq (1 + c_\phi - v)x(t)$ which, since $1 + c_\phi - v < 1$, implies $x(t)$ is a decreasing sequence that satisfies $\lim_{t \to \infty} x(t) = 0$. Thus, when $c_\phi < v$ we see that $x = 0$ is both locally asymptotically stable and globally attracting on $0 \leq x \leq 1$, that is to say, it is globally asymptotically stable on the interval $0 \leq x \leq 1$.

What happens when $c_\phi > v$? In this case, there exists a unique equilibrium in the interval $0 < x < 1$, namely

$$x^* = \frac{c_\phi - v}{c_\phi + c_\phi v}.$$

A calculation shows

$$h'(x^*) = \frac{v - c_\phi v + v^2 + 1}{c_\phi + 1}$$

and straightforward manipulations with inequalities show

$$-1 < h'(x^*) < 1.$$

By the Linearization Principle, x^* is locally asymptotically stable. It is left as Problem 8 to show that (4.42) defines an S-map of the interval $0 \leq x \leq 1$ into itself (see Definition 4 in Chapter 1) and, hence, by the Allwright-Singer Theorem 6 (in Chapter 1) x^* is globally asymptotically stable on the interval $0 \leq x \leq 1$.

With respect to the SI model (4.39), these results show that $\lim_{t \to \infty} I(t) = 0$ (hence $\lim_{t \to \infty} S(t) = N_0$) for all initial conditions $(I(0), S(0))^T \in R_+^2 \setminus \{\mathbf{0}\}$ when $c_\phi < v$. This means that when the pathogen is not too infective, in the sense that c_ϕ is not too large, any initial infection of the susceptible population will ultimately die out and no epidemic will occur. But when the pathogen is more infectious pathogen, i.e., when $c_\phi > v$, an epidemic will occur in the sense that $\lim_{t \to \infty} I(t) = x^* N_0 > 0$ (hence $\lim_{t \to \infty} S(t) = (1 - x^*) N_0$) and the infected population will not disappear.

The SI model in Example 80 is a special case of the model (4.16) in Section 4.1. It is derived under the assumptions that no recruitment and no deaths occur, i.e., $f = 0$ and $\sigma_I = \sigma_S = 0$. It is these assumptions that result in a constant population size and allow for the global analysis in that special case. In general, epidemic models for which the population remains constant can, in a similar manner, be reduced in dimension by one which, in some cases at least, can lead to tractable analysis, as it did in Example 80. For models with recruitment and/or deaths, however, the population size will not in general remain constant. Nonetheless, in some models' total population size is asymptotically constant, a fact that can, with some added assumptions, lead to a global analysis of the dynamics. We illustrate this with an example that is again based on the SI model (4.16) in Section 4.6.

We assume recruitment is through the birth susceptible newborns and that the pathogen does not affect an individual's birth rate. Specifically, we take f to be a function of the total population size (see (4.32)):

$$h = b \frac{1}{1 + c_b N} N, \quad b, c_b > 0$$

4.8. GLOBAL STABILITY OF EQUILIBRIA

Secondly, we assume that the disease does not affect survival so that $\sigma_I = \sigma_S$. The SI model equations (4.16) are then

$$S(t+1) = b\frac{1}{1+c_bN(t)}N(t) + \phi\left(\frac{I(t)}{N(t)}\right)\sigma S(t) + \upsilon\sigma I(t)$$
$$I(t+1) = \left(1 - \phi\left(\frac{I(t)}{N(t)}\right)\right)\sigma S(t) + (1-\upsilon)\sigma I(t) \quad (4.43)$$

where $\sigma = \sigma_I = \sigma_S$. By adding these equations we get

$$N(t+1) = b\frac{1}{1+c_bN(t)}N(t) + \sigma N(t), \quad (4.44)$$

an uncoupled difference equation for the total population size $N(t) = S(t) + I(t)$. If $b + \sigma < 1$ then

$$0 \le N(t+1) \le bN(t) + \sigma N(t) = (b+\sigma)N(t)$$

which implies $0 \le \lim_{t\to\infty} N(t) \le \lim_{t\to\infty}(b+\sigma)^t = 0$. In this case the population goes extinct in the absence of the pathogen. If we assume $b + \sigma > 1$, Equation (4.44) has a unique positive equilibrium

$$N^* = \frac{1}{c_b}\left(\frac{b}{1-\sigma} - 1\right). \quad (4.45)$$

With

$$f(N) = b\frac{1}{1+c_bN}N + \sigma N$$

we calculate

$$0 < f'(N^*) = \frac{b\sigma + (1-\sigma)^2}{b} = 1 - \frac{1}{b}(1-\sigma)(b+\sigma-1) < 1$$

for $b+\sigma > 1$. By the Linearization Principle, N^* is locally asymptotically stable. It is left as Problem 11 to show, using Theorem 4 in Chapter 1, that N^* is in fact globally asymptotically stable on $int(\mathbb{R}_+)$ and, therefore, any positive initial condition $N(0) = N_0 > 0$ produces a unique positive solution $\bar{N}(t)$ of Equation (4.44) that approaches an equilibrium $N^* \ge 0$ as $t \to \infty$.

For any initial condition $(S(0,)I(0))^T \in \mathbb{R}_+^2 \setminus \{\mathbf{0}\}$ such that $S(0) + I(0) = N_0$, we can view SI model equations (4.43) or equivalently

$$S(t+1) = b\frac{1}{1+c_bN(t)}N(t) + \phi\left(\frac{N(t)-S(t)}{N(t)}\right)\sigma S(t) + \upsilon\sigma(N(t)-S(t))$$
$$I(t+1) = \left(1 - \phi\left(\frac{I(t)}{N(t)}\right)\right)\sigma(N(t)-I(t)) + (1-\upsilon)\sigma I(t) \quad (4.46)$$

as a nonautonomous system of difference equations

$$S(t+1) = b\frac{1}{1+c_b\bar{N}(t)}\bar{N}(t) + \phi\left(\frac{\bar{N}(t)-S(t)}{\bar{N}(t)}\right)\sigma S(t) + \upsilon\sigma(\bar{N}(t)-S(t))$$
$$I(t+1) = \left(1 - \phi\left(\frac{I(t)}{N(t)}\right)\right)\sigma(\bar{N}(t)-I(t)) + (1-\upsilon)\sigma I(t) \quad (4.47)$$

in which $\lim_{t\to\infty} \bar{N}(t) = N^*$. With $\bar{N}(t)$ replaced by N^*, the resulting associated *autonomous* system of difference equations

$$S(t+1) = b\frac{1}{1+c_bN^*}N^* + \phi\left(\frac{N^*-S(t)}{N^*}\right)\sigma S(t) + \upsilon\sigma(N^*-S(t))$$
$$I(t+1) = \left(1 - \phi\left(\frac{I(t)}{N^*}\right)\right)\sigma(N^*-I(t)) + (1-\upsilon)\sigma I(t) \quad (4.48)$$

is called the *limiting system* of the SI model (4.43). Since N^* is a globally attracting equilibrium of system (4.44), the system (4.47) has the same limiting equation (4.48) for all initial conditions $(S(0,) I(0))^T \in R_+^2 \setminus \{\mathbf{0}\}$. The question we ask is: what is it that we can learn about the dynamics of the epidemic model (4.43) by studying its limiting equation (4.48)? One way to obtain answers is by means of the following theorem.

Theorem 46 *[98], [241] Assume, for all $t \in Z_+$, that $\mathbf{g}$ and $\mathbf{g}_t$ are continuous on an open set Ω in $\mathbb{R}^n$ that contains $\mathbb{R}_+^n \setminus \{\mathbf{0}\}$. Assume $\mathbf{g} : \mathbb{R}_+^n \setminus \{\mathbf{0}\} \to \mathbb{R}_+^n \setminus \{\mathbf{0}\}$ and $int(\mathbb{R}_+^n) \to int(\mathbb{R}_+^n)$ and that $\mathbf{g}_t$ converges uniformly to $\mathbf{g}$ as $t \to \infty$. Suppose $\mathbf{x}^* \in \mathbb{R}_+^n \setminus \{\mathbf{0}\}$ is an equilibrium of the limiting equation $\mathbf{x}(t+1) = \mathbf{g}(\mathbf{x}(t))$.*

(i) If $\mathbf{x}^* \in \partial(\mathbb{R}_+^n) := \mathbb{R}_+^n \setminus \{int(\mathbb{R}_+^n)\}$ is a globally asymptotically stable equilibrium of the limiting equation on $int(\mathbb{R}_+^n)$, then all solutions of the nonautonomous equation $\mathbf{x}(t+1) = \mathbf{g}_t(\mathbf{x}(t))$ with $\mathbf{x}(0) \in int(\mathbb{R}_+^n)$ tend to $\mathbf{x}^*$ as $t \to \infty$.

(ii) If $\mathbf{x}^* \in int(\mathbb{R}_+^n)$ is a globally asymptotically stable equilibrium of the limiting equation on $int(\mathbb{R}_+^n)$, then all solutions of the nonautonomous equation $\mathbf{x}(t+1) = \mathbf{g}_t(\mathbf{x}(t))$ with $\mathbf{x}(0) \in int(\mathbb{R}_+^n)$ tend to $\mathbf{x}^*$ as $t \to \infty$.

Example 81 *The SIS model (4.46) with the escape function*

$$\phi\left(\frac{I}{N}\right) = \frac{1}{1 + c_\phi \frac{I}{N}}, \quad c_\phi > 0$$

is

$$\begin{aligned} S(t+1) &= b \frac{1}{1+c_b N(t)} N(t) + \frac{1}{1+c_\phi \frac{N(t)-S(t)}{N(t)}} \sigma S(t) + v\sigma(N(t) - S(t)) \\ I(t+1) &= \left(1 - \frac{1}{1+c_\phi \frac{I(t)}{N(t)}}\right) \sigma(N(t) - I(t)) + (1-v)\sigma I(t) \end{aligned} \quad (4.49)$$

As noted above, if $b + \sigma > 1$ then $\lim_{t \to \infty} N(t) = N^* > 0$ for all positive initial conditions $N(0) > 0$ and the limiting system is

$$\begin{aligned} S(t+1) &= b \frac{1}{1+c_b N^*} N^* + \frac{1}{1+c_\phi \frac{N^* - S(t)}{N^*}} \sigma S(t) + v\sigma(N^* - S(t)) \\ I(t+1) &= \left(1 - \frac{1}{1+c_\phi \frac{I(t)}{N^*}}\right) \sigma(N^* - I(t)) + (1-v)\sigma I(t) \end{aligned} \quad (4.50)$$

which has the disease-free equilibrium

$$\begin{pmatrix} S^* \\ I^* \end{pmatrix} = \begin{pmatrix} N^* \\ 0 \end{pmatrix} = \begin{pmatrix} \frac{1}{c_b}\left(\frac{b}{1-\sigma} - 1\right) \\ 0 \end{pmatrix} \in \partial(\mathbb{R}_+^2). \quad (4.51)$$

Our goal in this example is to determine conditions under which this equilibrium is globally asymptotically stable of the original SIS model (4.49). We will do this by showing that it is a globally asymptotically stable equilibrium of the limiting equation (4.48) and then invoking Theorem 46(i).

The Jacobian associated with (4.50), when evaluated at the disease-free equilibrium is the diagonal matrix

$$\begin{pmatrix} \sigma(c_\phi + 1 - v) & 0 \\ 0 & \sigma(c_\phi + 1 - v) \end{pmatrix}$$

whose (repeated) eigenvalue is $\sigma(c_\phi + 1 - v) > 0$. Thus, the equilibrium is locally asymptotically stable by the Linearization Principle if

$$c_\phi < \frac{1-\sigma}{\sigma} + v. \tag{4.52}$$

Furthermore, from the limiting equation for $I(t)$ in (4.50), we have

$$0 \leq I(t+1) = \frac{c_\phi I(t)}{N^* + c_\phi I(t)} \sigma(N^* - I(t)) + (1-v)\sigma I(t)$$

$$\leq \frac{c_\phi I(t)}{N^* + c_\phi I(t)} \sigma N^* + (1-v)\sigma I(t) = \frac{N^*}{N^* + c_\phi I(t)} \sigma c_\phi I(t) + (1-v)\sigma I(t)$$

$$\leq \sigma c_\phi I(t) + (1-v)\sigma I(t) = \sigma(c_\phi + (1-v)) I(t)$$

for all t. Since (4.52) implies $\sigma(c_\phi + (1-v)) < 1$, it follows that $\lim_{t \to \infty} I(t) = 0$ and $\lim_{t \to \infty} S(t) = \lim_{t \to \infty} (N(t) - I(t)) = N^*$, that is to say, the disease-free equilibrium is globally attracting on $\operatorname{int}(\mathbb{R}^2_+ \setminus \{\mathbf{0}\})$ and, hence, is globally asymptotically stable on $\operatorname{int}(\mathbb{R}^2_+ \setminus \{\mathbf{0}\})$.

Theorem 46(i) now implies that the disease-free equilibrium (4.51) is globally asymptotically stable on $\operatorname{int}(\mathbb{R}^2_+ \setminus \{\mathbf{0}\})$. Our conclusion is that if (4.52) holds, that is to say, if the pathogen is not highly infectious, then any initial infection will asymptotically die out and an epidemic will not occur.

Using formula (4.31) we derived in Example 77, we find that the reproduction number for the SI model in Example 81 is

$$\mathcal{R}_0 = \sigma c_\phi \frac{1}{1 - (1-v)\sigma}.$$

It is straightforward to show that the inequality (4.52) is equivalent to $\mathcal{R}_0 < 1$. Thus, in that example, we arrive at the familiar conclusion that the disease-free equilibrium is globally stable and no epidemic occurs if $\mathcal{R}_0 < 1$.

4.9 Disease Acquired Herd Immunity

Acquired immunity can come from a vaccine or exposure to an infection or disease. The acquired immunity is different from innate immunity, the immunity that one is born with. The innate immune system doesn't fight specific pathogens. Acquired immunity helps the immune system get stronger. Individuals with stronger immune system are less likely to get sick.

When a large portion of a population (herd) becomes immune to a disease and the spread of the disease from an individual to an individual is unlikely, then herd immunity occurs. In the previous section, we illustrated herd immunity as a result of a vaccination protocol. In this section, we use an SIRS model to illustrate the occurrence of herd immunity as a result of a large percentage of the population acquiring a disease immunity as a result of recovering from a SIRS infection. To illustrate this, we consider the following SIRS model with constant birth $f > 0$ per unit interval.

$$\begin{aligned} S(t+1) &= f + (1-\mu_S)S(t)\phi((1-\mu_I)I(t)) + v(1-\mu_R)R(t) \\ I(t+1) &= (1-\mu_S)S(t)(1-\phi((1-\mu_I)I(t))) + (1-\gamma)(1-\mu_I)I(t) \\ R(t+1) &= \gamma(1-\mu_I)I(t) + (1-v)(1-\mu_R)R(t) \end{aligned} \tag{4.53}$$

where per each unit interval, $\mu_S, \mu_I, \mu_R \in (0,1)$, respectively, denote the fractions of the population densities of susceptibles, infected and recovered that die. There is no permanent immunity and per each unit time $v \in (0,1)$ is the fraction of the recovered individuals that transition back to the susceptible class.

The disease-free equation of model (4.53) is

$$S(t+1) = f + (1-\mu_S)S(t).$$

Hence, in the absence of the disease,

$$\lim_{t\to\infty} S(t) = S^* = \frac{f}{\mu_S}.$$

The DFE of model (4.53) is

$$(S^*, I^*, R^*)^T = \left(\frac{f}{\mu_S}, 0, 0\right)^T.$$

Using linearization or the Next Generation Matrix Method at the DFE, we obtain

$$\mathcal{R}_0 = -\frac{(1-\mu_S)(1-\mu_I)f\phi'(0)}{\mu_S\left(1-(1-\gamma)(1-\mu_I)\right)}.$$

Hence, $\mathcal{R}_0 < 1$ implies the DFE is locally asymptotically and the SIRS disease dies out, whereas $\mathcal{R}_0 > 1$ implies the DFE is unstable and the disease is not eliminated from the population.

To include disease acquired permanent immunity in model (4.53), we let $\chi \in (0,1)$ denote the fraction of the infected population that have acquired permanent immunity. Then model (4.53) becomes the following system of equations.

$$\begin{aligned} S(t+1) &= f + (1-\mu_S)S(t)\phi((1-\chi)(1-\mu_I)I(t)) + (1-\chi)v(1-\mu_R)R(t) \\ I(t+1) &= (1-\mu_S)S(t)\left(1 - \phi((1-\chi)(1-\mu_I)I(t))\right) + (1-\gamma)(1-\mu_I)I(t) \\ R(t+1) &= (1-\chi)\gamma(1-\mu_I)I(t) + (1-v)(1-\mu_R)R(t). \end{aligned} \quad (4.54)$$

When there is no disease-acquired permanent immunity and $\chi = 0$, then model (4.54) reduces to model (4.53). Models (4.54) and model (4.53) share the same DFE

$$S^* = \frac{f}{\mu_S}.$$

Proceeding as we did with our analysis of model (4.2), we find that there will be no disease outbreak in model (4.54) if

$$(1-\chi)\mathcal{R}_0 < 1.$$

In other words, disease-acquired herd immunity against an outbreak is attained if the fraction of the population with the disease-acquired herd immunity is sufficiently large, namely, if

$$\chi > 1 - \frac{1}{\mathcal{R}_0}.$$

4.9. DISEASE ACQUIRED HERD IMMUNITY

Exercises 4.8–4.9

1. Consider the following SI disease model:
$$S(t+1) = b\frac{1}{(1+c\sigma_r S(t))^a}\sigma_r S(t) + \sigma_S S(t) e^{-\alpha I(t)}$$
$$I(t+1) = \beta \sigma_S S(t)\left(1 - e^{-\alpha I(t)}\right) + \mu I(t),$$
where $S(0) \geq 0$, $I(0) \geq 0$, $\sigma_r, \sigma_S \in (0,1)$, and $\alpha, \beta, a, b, c > 0$.

   (a) Under what conditions are $S(t)$, $I(t)$ bounded? (HINT: see Chapter 3 for proof of bounded orbits, $\lim_{t\to\infty} S(t) < \infty$ and $\lim_{t\to\infty} I(t) < \infty$)
   (b) Show that there is a unique disease-free equilibrium point.
   (c) Use the next generation matrix method to compute the basic reproduction number, $\mathcal{R}_0$.
   (d) What are the implications of $\mathcal{R}_0 > 1$ versus $\mathcal{R}_0 < 1$ on the disease dynamics?
   (e) How many endemic equilibria exist when $\mathcal{R}_0 > 1$?
   (f) What can you say about the stability of the endemic equilibrium point(s)?

2. Consider the following SI disease model:
$$S(t+1) = be^{-c\sigma_r S(t)}\sigma_r S(t) + \sigma_S S(t) e^{-\alpha I(t)}$$
$$I(t+1) = \beta \sigma_S S(t)\left(1 - e^{-\alpha I(t)}\right) + \mu I(t),$$
where $S(0) \geq 0$, $I(0) \geq 0$, $\sigma_r, \sigma_S \in (0,1)$, and $\alpha, \beta, b, c > 0$.

   (a) Under what conditions are $S(t)$, $I(t)$ bounded? (HINT: see Chapter 3.)
   (b) Show that there is a unique disease-free equilibrium point.
   (c) Use the next generation matrix method to compute the basic reproduction number $\mathcal{R}_0$.
   (d) What are the implications of $\mathcal{R}_0 > 1$ versus $\mathcal{R}_0 < 1$ on the disease dynamics?
   (e) How many endemic equilibria exist when $\mathcal{R}_0 > 1$?
   (f) What can you say about the stability of the endemic equilibrium point(s)?

3. Consider the following density-dependent discrete-time SI disease model with constant birth function (or new arrivals) $\Lambda > 0$ per unit time interval:
$$S(t+1) = \Lambda + aS(t) e^{-I(t)}$$
$$I(t+1) = \beta aS(t)\left(1 - e^{-I(t)}\right) + \mu I(t),$$
where $S(0) \geq 0$, $I(0) \geq 0$, $a \in (0,1)$, $b \in [0,1)$, and $\beta, \Lambda > 0$.

   (a) Under what conditions are $S(t)$, $I(t)$ bounded? (HINT: see Chapter 3.)
   (b) Show that there is a unique disease-free equilibrium point.
   (c) Use the Next Generation Matrix Method to compute the basic reproduction number, $\mathcal{R}_0$.
   (d) What are the implications of $\mathcal{R}_0 > 1$ versus $\mathcal{R}_0 < 1$?
   (e) How many endemic equilibria exist when $\mathcal{R}_0 > 1$?
   (f) What can you say about the stability of the endemic equilibrium point(s)?

4. Consider the following SI model:
$$S(t+1) = b\sigma_r S(t) \frac{1}{1+c\sigma_r S(t)} + \sigma_S S(t) \exp\left(-c_\varphi \frac{I(t)}{S(t)+I(t)}\right)$$
$$I(t+1) = \sigma_S S(t)\left(1 - \exp\left(-c_\varphi \frac{I(t)}{S(t)+I(t)}\right)\right) + \sigma_I I(t)$$

where $b, c, c_\varphi > 0$ and $0 < \sigma_I, \sigma_r, \sigma_S < 1$.

(a) Show that the disease-free equation is
$$S(t+1) = b\sigma_r S(t) \frac{1}{1+\sigma_r S(t)} + \sigma_S S(t).$$

(b) Give a biological interpretation of
$$\mathcal{R}_d = \frac{b\sigma_r}{1-\sigma_S}.$$

(c) In the absence of the SI disease, in the long-run what happens to the susceptible population when $\mathcal{R}_d < 1$?

(d) In the absence of the SI disease, in the long-run what happens to the susceptible population when $\mathcal{R}_d > 1$?

(e) Verify that Theorems 64 and 65 apply to the SI model. Calculate μ_0, κ, α, and $\mathcal{R}_0$.

5. Consider the following SI model:
$$S(t+1) = b\sigma_r S(t) \exp(-c\sigma_r S(t)) + \sigma_S S(t) \exp\left(-c_\varphi \frac{I(t)}{S(t)+I(t)}\right)$$
$$I(t+1) = \sigma_S S(t)\left(1 - \exp\left(-c_\varphi \frac{I(t)}{S(t)+I(t)}\right)\right) + \sigma_I I(t)$$

where $b, c, c_\varphi > 0$ and $0 < \sigma_I, \sigma_r, \sigma_S < 1$.

(a) Show that the disease-free equation is
$$S(t+1) = b\sigma_r S(t) \exp(-c\sigma_r S(t)) + \sigma_S S(t).$$

(b) Give a biological interpretation of
$$\mathcal{R}_d = \frac{b\sigma_r}{1-\sigma_S}.$$

(c) In the absence of the SI disease, in the long-run what happens to the susceptible population when $\mathcal{R}_d < 1$?

(d) In the absence of the SI disease, in the long-run what happens to the susceptible population when $\mathcal{R}_d > 1$?

(e) Verify that Theorems 64 and 65 apply to the SI model. Calculate μ_0, κ, α, and $\mathcal{R}_0$.

6. Show that $(I-T)^{-1} F$ and $F(I-T)^{-1}$ have the same eigenvalues.

7. (a) Let A be an $n \times n$ nonnegative matrix. Prove, by induction, the following comparison theorem. If $\mathbf{0} \leq \mathbf{x}(t+1) \leq A\mathbf{x}(t)$ for all $t = 0, 1, 2, \ldots$ and if $\mathbf{y}(t)$ solves the initial value problem $\mathbf{y}(t+1) = A\mathbf{y}(t)$, $\mathbf{y}(0) = \mathbf{x}(0)$, then $\mathbf{0} \leq \mathbf{x}(t) \leq \mathbf{y}(t)$ for all $t = 0, 1, 2, \ldots$.

4.9. DISEASE ACQUIRED HERD IMMUNITY

(b) As in Theorem 45 assume A1, A2, A3 and A4 hold, that $\mathcal{R}_0 < 1$, and that $\mathbf{x}_1^*$ is a globally asymptotically stable equilibrium of the disease-free equation (4.24) on $\mathbb{R}_+^{n-m}$. Use (a) to show that $\lim_{t\to\infty} \mathbf{x}_0(t) = \mathbf{0}$ for any solution $(\mathbf{x}_0(t), \mathbf{x}_1(t))$ of (4.19) with initial conditions in $\mathbb{R}_+^n$.

8. Show that (4.42) defines an S-map of the interval $0 \le x \le 1$ into itself.

9. In Example (80) when $c < v$ why is this the disease-free equilibrium $(S,0)^T = (N_0, 0)^T$ not locally (hence not globally) asymptotically stable as an equilibrium of the SI model (4.39)? Similarly, when $c > v$ why is the endemic equilibrium $(S,I)^T = ((1-x^*)N_0, x^*N_0)^T$ not locally (hence not globally) asymptotically stable on $int(R_+^2)$ as an equilibrium of the SI model (4.39)?

10. Repeat Example 80 using the escape function $\phi(I/n) = \exp(-cI/N)$.

11. Show when $b + \sigma > 1$ the solution of Equation (4.44) is globally asymptotically stable on $int(\mathbb{R}_+)$. (see Theorem 4 in Chapter 1.)

12. For this set of questions, we first introduce an anthrax epidemic model in a population of herbivores. We assume that at each time $t \in \{0, 1, 2, \ldots\}$, each live herbivore is either susceptible $(S(t))$ or exposed (have the disease and are mildly infectious, $E(t)$) or infectious (infected with anthrax disease, $I(t)$). That is, we let $S(t)$, $E(t)$, $I(t)$ and

$$N(t) = S(t) + E(t) + I(t),$$

respectively, denote the population size of susceptible, exposed, infectious, and total population of live herbivores at each time t. Furthermore, at each time $t \ge 0$, we let $B(t)$ denote the grams of anthrax spores in the environment, and let $C(t)$ denote the density of anthrax-infected carcasses. Sudden death of an anthrax-infected animal is by far the most common clinical sign of anthrax infection in animals, and we assume that anthrax-infected herbivores do not recover from the disease.

The herbivore's intrinsic birth rate per time interval is denoted by $r > 0$ and $d \in (0, 1)$ is the fraction of herbivores that die "naturally" per the time interval, and the fraction of animals that do not die naturally is $\widehat{d} = (1-d)$. We assume that anthrax-infected live herbivores die from anthrax with constant probability $\mu \in (0, 1)$ per time interval, and the fraction of infected animals $\widehat{\mu} = (1-\mu)$ do not die from the disease. The probability that an exposed herbivore progresses to the infectious class is the constant $\eta \in (0, 1)$ per time interval and remains in the exposed class with probability $\widehat{\eta} = (1 - \eta)$.

The total herbivore population dynamics are assumed to follow the discrete (Beverton-Holt) logistic model. Consequently, we assume all newborn herbivores are susceptibles, and

$$g(N) = r\frac{N}{1+bN}$$

are born into the susceptible class per time interval, where the scaling parameter $b > 0$. In the environment, anthrax spores grow on anthrax-infected carcasses and decay or are washed away. Therefore, at each time $t \ge 0$, we let $\beta > 0$ denote the constant per-capita spore growth rate per herbivore carcass, and $\alpha \in (0, 1)$ denote the fraction of spores that decay per the time interval. Carcasses are organic matter, and decay with probability $\kappa \in (0, 1)$ per time interval. During each time interval, susceptible live herbivores become exposed from grazing or inhaling anthrax spores in the environment with probability

$$\widehat{\phi}(B) = (1 - \exp(-\sigma B)),$$

and escape from inhalation anthrax with probability

$$\phi(B) = \exp(-\sigma B)$$

where $\sigma > 0$. During each time interval, we assume that the bacteria ($B.$ $anthracis$) are shed in the environment by the infectious herbivore at the constant rates $\delta \geq 0$. These assumptions and notation lead to the following discrete-time anthrax model in herbivore populations.

$$\begin{aligned}
S(t+1) &= r\frac{N(t)}{1+bN(t)} + \widehat{d}S(t)\exp(-\sigma B(t)) \\
E(t+1) &= \widehat{d}(S(t)(1-\exp(-\sigma B(t))) + \widehat{\eta}\widehat{\mu}E(t)) \\
I(t+1) &= \widehat{d}(\widehat{\eta}\mu E(t) + \widehat{\eta}I(t)) \\
B(t+1) &= (1-\alpha)B(t) + \beta C(t) + \delta I(t) \\
C(t+1) &= \left(d + \mu_A\widehat{d}\right)I(t) + (1-\kappa)C(t)
\end{aligned}$$

for time $t = 0, 1, 2, 3, \cdots$.

(a) By summing the equations for $S(t+1)$, $E(t+1)$ and $I(t+1)$ show that the total population of the herbivore population is bounded, and there is no unbounded population growth.

(b) Show that the disease-free equation is

$$S(t+1) = r\frac{S(t)}{1+bS(t)} + (1-d)S(t).$$

(c) Let $\mathcal{R}_d = r/d$. Give a biological interpretation of $\mathcal{R}_d$.

(d) In the absence of the anthrax disease, in the long-run what happens to the susceptible herbivore population when $\mathcal{R}_d < 1$?

(e) In the absence of the anthrax disease, in the long-run what happens to the susceptible herbivore population when $\mathcal{R}_d > 1$?

(f) What are the disease-free equilibrium points?

(g) Let

$$\Omega(\mathcal{R}_d \leq 1) = \{(\mathbf{x}, \mathbf{y}) \in \mathbb{R}_+^5 \mid 0 \leq N \leq N_0, 0 \leq B \leq B_0, 0 \leq C \leq C_0\}$$

and define the Liapunov function,

$$V : int(\Omega(\mathcal{R}_d \leq 1)) \to \mathbb{R}_+$$

by

$$V(S, E, I, B, C) = S + E + I + \nu B + \theta C$$

where

$$0 < \nu < \frac{\kappa}{\beta}\theta \text{ and } \nu\delta + \theta(d_A + \mu_A(1-d_A)) < \mu_A(1-d_A).$$

4.9. DISEASE ACQUIRED HERD IMMUNITY

Show that if $\mathcal{R}_d < 1$, then
$$V\left(S(t+1), E(t+1), I(t+1), B(t+1), C(t+1)\right)$$
$$< V\left(S(t), E(t), I(t), B(t), C(t)\right)$$
for all points
$$(S, E, I, B, C)^T \neq (0, 0, 0, 0, 0)^T \in int\left(\Omega\left(\mathcal{R}_d^A \leq 1\right)\right).$$

This shows that $(0,0,0,0,0)^T$ is globally asymptotically stable and a global catastrophic extinction of the herbivore population occurs when $\mathcal{R}_d < 1$.

(h) When $\mathcal{R}_d > 1$, assuming *Bacillus anthracis* shedding is a new infection, use the next generation matrix method at the non-trivial disease-free equilibrium
$$\left(\frac{1}{b}(\mathcal{R}_d - 1), 0, 0, 0, 0\right)^T$$
to compute the basic reproduction number $\mathcal{R}_0$ for the herbivores.

(i) Let $\mathcal{R}_d > 1$. Use Theorem 4 to show that if $\mathcal{R}_0 < 1$, then the nontrivial DFE
$$\left(\frac{1}{b}(\mathcal{R}_d - 1), 0, 0, 0, 0\right)^T$$
is globally asymptotically stable in the interior of
$$\Omega = [0, \infty) \times [0, \infty \times [0, \infty) \times [0, \infty),$$
and, hence, anthrax is eradicated in the herbivore population.

(j) What happens when $\mathcal{R}_d > 1$ and $\mathcal{R}_0 > 1$?

Chapter 5
Models with Multiple Attractors

5.1 Introduction

The main focus of this chapter is on the introduction of models that exhibit multiple attractors. Most single-species density-dependent population models without structure that are widely studied, such as the discrete logistic (Beverton-Holt) equation (1.25), the Leslie-Gower competition model (3.34), and the Ricker (exponential) competition model (3.55), are, in general, not capable of supporting multiple attractors [233]. However, we showed in Chapter 3 that the Leslie-Gower competition model can have multiple attractors on the boundary, in the case where there are no stable interior equilibrium points (3.34). In this chapter, we consider some single-species and multiple-species population models (with and without structure) that support multiple attractors. When multiple attractors occur, the long-term fate of the population depends on initial conditions. In such systems, one obviously cannot determine all possibilities for a population's ultimate fate from a single time series data set or initial condition. Alternative outcomes could result from other data sets or other initial conditions. This deterministic uncertainty is further compounded by the presence of noise, which can cause data points to jump between the basins of attraction of the different coexisting attractors. When the coexisting attractors consist of attracting population cycles or chaotic attractors and the system has sensitive dependence on initial conditions (see Chapter 1), then the basins of attraction of the coexisting attractors can be intermingled in complicated ways, making impossible the prediction of the ultimate fate of a population with certainty [5], [158], [301].

One example of a multiple attractor scenario that has received considerable attention is due to an *Allee effect*. This occurs when some animal and plant species do not show a monotonic decline in per-capita growth rate with increasing density. Instead per-capita growth rates increase with population density at low population levels and decline only at higher population levels. There are many biological reasons for this possibility [6], [7], [61]. For example, a species may require conspecifics for protection from predators or from climate extremes; other species may forage more effectively in groups than alone. In sexual species, individuals may have a difficult time finding mates at low densities, so mating rates increase with population density. The result of these types of effects is that it might not even be possible for the per-capita growth rate exceeds one until the population reaches some threshold density. If it is the case that a density threshold must be exceeded before survival is possible, then a *strong Allee effect* is said to occur. In this case, the population will have two attractors, the extinction state and a survival state. Typically, responsible for this situation is the presence of one or more Allee effects in the model. These are modeling components that imply an increase in an aspect of fitness (e.g., a birth rate, a survival rate or litter size) with an *increase* in low-level population density. This is to be contrasted with the more familiar modeling components that cause a decrease in fitness components when population density increases (such as those used in the discrete logistic (Beverton-Holt) and Ricker or exponential model equations). The presence of Allee effects in a population equation does not necessarily result in a strong Allee effect, however; they must be of significant magnitude to outweigh the negative effects of increased density in overall fitness, i.e., the per-capita population growth rate.

In this chapter, we study some unstructured single-species models that exhibit bistability and multistability via the strong Allee effect and also via the presence of hysteresis loops in their bifurcation diagrams [104], [127]. We also study some structured single-species and multiple-species models without the Allee effect that is capable of exhibiting multiple attractors. We illustrate bistability by a two patch (two-dimensional) dispersal linked single-species models, where the local patch pre-dispersal dynamics are described by density-dependent models without the Allee effect [158] and [301]. We will also look at some structured single-species models and multi-species models that are capable of generating multiple attractors (with or without the Allee effects) that include competition models [91], [89], [174], [14-16], [218], predator–prey models [144], age-structured models, and epidemic models [42].

5.2 Models with Multiple Attractors with no Allee Effects

In Section 5.2, we will study models with multiple attractors but with no Allee effects and in Section 5.3, we will study models with the Allee effects, where the existence of multiple attractors is one of the characteristics of the Allee effect.

We begin our presentation with a familiar model, the Ricker Competition model.

Example 82 *([90]) Multiple mixed-type attractors in which cycles are present may be exhibited by a Ricker competition model with survival rates. Such phenomenon has been observed in Tribolium experiments [90], [211], [239]. As an example, we consider the following exponential (Ricker) competition model of two species x and y with survival rates $0 \leq s_1, s_2 < 1$ and inherent birth rates $b_1, b_2 > 0$, and the density-dependent effects $c_{ij} > 0$ on newborn recruitment.*

$$x(t+1) = b_1 x(t) \exp(-c_{11} x(t) - c_{12} y(t)) + s_1 x(t)$$
$$y(t+1) = b_2 y(t) \exp(-c_{21} x(t) - c_{22} y(t)) + s_2 y(t)$$

The main objective here is to investigate the possible occurrence of mixed-type nonequilibrium attractors in the above model under symmetrically higher inter-specific competition as has been observed in more complicated models that include juvenile life-cycle stages. To simplify our analysis we let $c_{11} = c_{22} = 1$, and let $r = \frac{c_{21}}{c_{12}}$, and $c_{12} = c$. This yields the system

$$\begin{aligned} x(t+1) &= b_1 x(t) \exp(-x(t) - cy(t)) + s_1 x(t) \\ y(t+1) &= b_2 y(t) \exp(-rcx(t) - y(t)) + s_2 y(t) \end{aligned} \quad (5.1)$$

Let $b_1 = 8$, $b_2 = 10$, $s_1 = 0.65$, $s_2 = 0$, $r = 1.1$, $c = 1.9$. The equilibrium equations are

$$x = 8x \exp(-x - 1.9y) + 0.65x$$
$$y = 10y \exp(-2.09x - y).$$

5.2. MODELS WITH MULTIPLE ATTRACTORS WITH NO ALLEE EFFECTS

They have a solution $(x, y) = (x^*, 0)$, $x^* > 0$, if and only if x^* satisfies the equation $1 = 8 \exp(-x) + 0.65$. Thus

$$x^* = \ln\left(\frac{8}{0.35}\right) \approx 3.13.$$

To use the Linearization Principle to analyze the local stability of the equilibrium $(x^*, 0)$, define

$$f(x, y) := 8x \exp(-x - 1.9y) + 0.65x$$
$$g(x, y) := 10y \exp(-2.09x - y)$$
$$F(x, y) = (f(x, y), g(x, y))$$

and calculate the Jacobian

$$JF(x, y) = \begin{pmatrix} 8(1-x)e^{-x-1.9y} + 0.65 & -15.2xe^{-x-1.9y} \\ -20.9y \exp(-2.09x - y) & 10(1-y)\exp(-2.09x - y) \end{pmatrix}.$$

The eigenvalues $\lambda_1 = 0.0144$ and $\lambda_2 = -0.0952$ of

$$JF(x^*, 0) = JF\left(\ln\left(\frac{8}{0.35}\right), 0\right) \approx \begin{pmatrix} -0.0952 & -2.08 \\ 0 & 0.0144 \end{pmatrix}$$

satisfy $|\lambda_i| < 1$ and, as a result, the equilibrium $(x^*, 0)$ is locally asymptotically stable by the Linearization Principle.

The 2-cycles consist of two points $(x_1, y_1) \neq (x_2, y_2)$ where (x_1, y_1) is an equilibrium point of of the composite equations

$$f(f(x, y), g(x, y)) = x$$
$$g(f(x, y), g(x, y)) = y$$

and $(x_2, y_2) = (f(x_1, y_1), g(x_1, y_1))$. The left side of the first and second equations have factors of x and y, respectively, both of which can be canceled to get the equations

$$8[8 \exp(-x - 1.9y) + 0.65x][\exp(-f(x, y) - 1.9g(x, y)) + 0.65] = 1$$
$$10[10 \exp(-2.09x - y)]\exp(-2.09f(x, y) - g(x, y)) = 1$$

to be solved for positive x and y. A computer program can be used to obtain the solution $(\bar{x}_1, \bar{y}_1) \approx (0.109, 1.14)$. The second point on the 2-cycle is

$$(x_2, y_2) = (f(0.109, 1.14), g(0.109, 1.14)) \approx (0.160, 2.903).$$

The Jacobian of the composite evaluated at (x_1, y_1) is

$$JF(x_2, y_2) JF(x_1, y_1) = J(0.160, 2.903) JF(0.109, 1.14) \approx \begin{pmatrix} 0.981 & -0.112 \\ 1.240 & 0.672 \end{pmatrix}$$

whose complex eigenvalues $\lambda \approx 0.827 \pm 0.339i$ have absolute value is $|\lambda| \approx 0.894 < 1$. By the Linearization Principle the 2-cycle is locally asymptotically stable (see Figure 5.1).

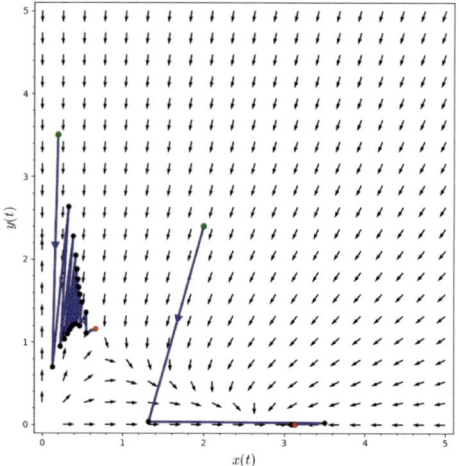

Figure 5.1: The graph shows two attractors, an equilibrium point on the x-axis, and a 2-cycle in the interior.

Example 83 *Another example is the bob white quail "hump with tail" model due to Errington [135], and Milton and Belair [239]*

$$x_i(t+1) = x_i(t)(k_i + (K_i/(1+x_i^{n_i}(t)))), \ i = 1, 2, \ldots, N \tag{5.2}$$

where $x_i(t)$ denotes the population size in N habitat patches. The parameter n_i measures the level and type of interspecific competition in the population, $k_i + K_i > 1$ represent the intrinsic growth rate, and $0 < k_i < 1$, $K_i > 0$. The dynamics of N patches, modeled by system (5.2), are coupled via the dispersal of individuals at a rate that is proportional to each local density. Reproduction is assumed to occur prior to dispersal within each generation and each patch. Hence, after reproduction, the fraction d_{ij} of the population disperses from patch i to j, $i \neq j$, $1 \leq i, j \leq N$. The dynamics of the population sizes in the N-patches are given by the following system [301]

$$x_i(t+1) = \left(1 - \sum_{\substack{j=1 \\ j \neq i}}^{N} d_{ji}\right) f_i(x_i(t)) + \sum_{\substack{j=1 \\ j \neq i}}^{N} d_{ji} f_j(x_j(t)) \ (1 \leq i \leq N) \tag{5.3}$$

where $d_{ji} > 0$, $0 < d_{ji} < 1$, $0 < \sum_{\substack{j=1 \\ j \neq i}}^{N} d_{ji} < 1$ and $d_{ii} = 0$. It is also assumed that the timing of reproduction and dispersal do not differ from patch to patch, that is, it models a multi-patch system with synchronous dispersal.

Let us consider the dynamics of two-patches

$$\begin{aligned} x(t+1) &= (1-d_{12})f_1(x(t)) + d_{21}f_2(y(t)) \\ y(t+1) &= (1-d_{21})f_2(y(t)) + d_{12}f_1(x(t)) \end{aligned} \tag{5.4}$$

5.2. MODELS WITH MULTIPLE ATTRACTORS WITH NO ALLEE EFFECTS

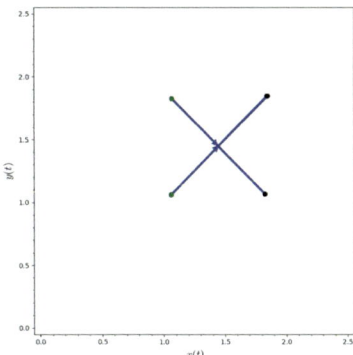

Figure 5.2: The graph shows two locally asymptotically stable 2-cycle, one in phase and one out of phase.

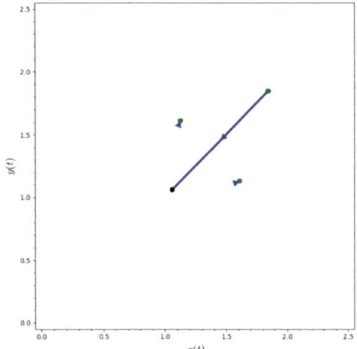

Figure 5.3: The graph shows one locally asymptotically 2-cycle and two asymptotically stable equilibrium points.

where $f_1(x) = x\left(k_1 + \frac{K_1}{1+x^{n_1}}\right)$, and $f_2(y) = y\left(k_2 + \frac{K_2}{1+y^{n_2}}\right)$ For the parameters $d_{12} = d_{21} = 0.01$, $k_1 = k_2 = 0.5$, $K_1 = K_2 = 3$ and $n_1 = n_2 = 6$, numerical simulation show the existence of two locally asymptotically 2-cycles given by $c_2^1 = \{(1.0595, 1.0595), (1.8461, 1.8461)\}$ and $c_2^2 = \{(1.0646, 1.8251), (1.8251, 1.0646)\}$. Note that these parameters were chosen in such a way that in the absence of dispersal, each local reproduction function f_i supports and asymptotically stable 2-cycle (scramble or over-compensatory dynamics). (see Figure 5.2) If k_i, K_i, and n_i are as above but we increase the symmetric dispersal rates $d_{12} = d_{21} = 0.9$, we obtain three attractors, two out of phase equilibrium points $(1.1275, 1.6108)$ and $(1.6108, 1.1275)$ and an in-phase 2-cycle $c_2^2 = \{(1.0646, 1.8251), (1.8251, 1.0646)\}$ (see Figure 5.3).

Exercises 5.2

1. Consider the Leslie-Gower model of two competing species x and y.

$$x_1(t+1) = \frac{b_1 x(t)}{1 + c_{11} x_1(t) + c_{12} y(t)}$$
$$x_2(t+1) = \frac{b_2 y(t)}{1 + c_{21} x(t) + c_{22} y(t)}$$
(5.5)

   where $b_1, b_2 > 1$ and $c_{ij} > 0$ for all $1 \leq i, j \leq 2$.

   (a) Find the four equilibrium points $E_1^* = (0,0)^T$, $E_2^* = (x_1, 0)$, $E_3^* = (0, x_2)^T$, and $E_4^* = (x^*, y^*)^T$

   (a) If $\frac{c_{22}}{c_{12}} < \frac{b_2 - 1}{b_1 - 1} < \frac{c_{21}}{c_{11}}$, show that E_2^* and E_3^* are locally asymptotically stable and that E_4^* is a saddle.

2. Consider the exponential (Ricker) competition model (5.1).

   (a) Show that the equilibrium $(x,y)^T = (0,0)^T$ is unstable if either one or both of the intrinsic reproduction numbers

   $$\mathcal{R}_0 = b_1 \frac{1}{1 - s_1}, \quad \mathcal{R}_0 = b_2 \frac{1}{1 - s_2}$$

   is greater than 1 and that it is locally asymptotically stable if both are less than 1.

   (b) In the absence of species y, the dynamics of species x are governed by the difference equation

   $$x(t+1) = b_1 x(t) \exp(-x(t)) + s_1 x(t).$$

   Show that $\mathcal{R}_0 < 1$ implies $x = 0$ is a locally asymptotically stable equilibrium of this equation. Show that if $\mathcal{R}_0 > 1$ then $x = 0$ is unstable and the equation has a positive equilibrium x^* which is locally asymptotically stable if

   $$\mathcal{R}_0 < \exp\left(2 \frac{1}{1 - s_1}\right)$$

   and unstable if this inequality is reversed.

   (c) Assume

   $$1 < \mathcal{R}_0 < \exp\left(2 \frac{1}{1 - s_1}\right)$$

   so that species x has a locally asymptotically stable equilibrium x^* in the absence of species y. Use the Linearization Principle to show that the exclusion equilibrium

   $$\begin{pmatrix} x \\ y \end{pmatrix} = \begin{pmatrix} x^* \\ 0 \end{pmatrix}, \quad x^* > 0$$

   of the competition model is locally asymptotically stable if

   $$\mathcal{R}_0 < e^{cr \ln \mathcal{R}_0}$$

   and unstable if this inequality is reversed.

(d) Let $b_1 = 8, b_2 = 10, s_1 = 0.65, s_2 = 0, r = 1.1$, and $c = 1.9$. Show that there exists an exclusion equilibrium $(x, y)^T = (x^*, 0)^T$, $x^* > 0$, and that it is locally asymptotically stable.

3. Consider the exponential (Ricker) competition model (5.1). Let $b_1 = 8, b_2 = 10, s_1 = 0.65$, $s_2 = 0$, and $r = 1.1$.

    (a) Show that if $c = 2.35$, the exclusion equilibrium $E_1 = (3.1292, 0)^T$ is asymptotically stable, and the 2-cycle $c_2 \approx \{(0, 1.045), (0, 3.5169)\}$ is also asymptotically stable.

    (b) Repeat what you did in (a) with $c = 1.9$ and show that in addition to the asymptotically stable exclusion equilibrium E_1, there is an asymptotically stable 2-cycle $c_2 \approx \{(0.1097, 1.1373), (0.1620, 2.8993)\}$.

4. Consider system (5.1), with $b_1 = 8$, $b_2 = 14$, $s_1 = 0.8$, $s_2 = 0$, $r = 0.8$. Show that if $c = 4.6$, then $E_1 \approx (3.688, 0)^T$ is asymptotically stable and the 4-cycle $c_4 \approx \{(6.832 \times 10^{-3}, 1.0206)^T, (8.5431 \times 10^{-3}, 3.9645)^T, (4.77 \times 10^{-3}, 0.4536)^T, (5.971 \times 10^{-3}, 5)^T\}$.

5. Consider system (5.1), with $b_1 = 8$, $b_2 = 14$, $s_1 = 0.8$, $s_2 = 0$, $r = 0.8$. Show that if $c = 3.5$, then there are two attractors: an equilibrium E_1 as in Problem 3(a) and a 2-cycle $c_2 \approx \{(8.876 \times 10^{-2}, 0.7959)^T, (0.1110, 3.9219)^T\}$.

6. Consider system (5.1), with $b_1 = 5$, $b_2 = 15$, $s_1 = 0.8$, $s_2 = 0$, $r = 0.8$. Show that if $c = 3.5$, then there are two attractors: an equilibrium $E_1 \approx (3.2188, 0)^T$ and a 2-cycle $c_2 \approx \{(5.735 \times 10^{-2}, 0.6716)^T, (7.168 \times 10^{-2}, 4.3831)^T\}$.

7. Consider system (5.1), with $b_1 = 8$, $b_2 = 14$, $s_1 = 0.8$, $s_2 = 0$, $r = 0.8$. Show that if $c = 6.5$, then there are two attractors: an equilibrium $E_1 = (3.6888, 0)^T$ and a 4-cycle $c_4 \approx \{(0, 5.1239)^T, (0, 1.1047)^T, (0, 3.900)^T, (0, 0.4270)^T\}$.

8. Consider system (5.4) with the parameters $k_1 = k_2 = 0.5$, $K_1 = K_2 = 3$, $n_1 = n_2 = 7$, $d_{12} = d_{21} = 0.9$.

    Use a numerical method to show the existence of a locally asymptotically 4-cycle $c_4^1 = \{(1.7648, 1.7648)^T, (0.9799, 0.9799)^T, (2.0642, 2.0642)^T, (1.0706, 1.0607)^T\}$, two locally asymptotically stable 4-cycle $c_4^2 = \{(1.8212, 2.0230)^T, (0.9922, 1.0543)^T, (2.0230, 1.8212)^T, (1.0543, 0.9922)^T\}$ and period-2 limit cycle.

    Illustrate your work with a phase space diagram.

9. Consider system (5.4) with the parameters $k_1 = k_2 = 0.5$, $K_1 = K_2 = 3$, $n_1 = n_2 = 6$, $d_{12} = d_{21} = 0.1$.

    Use a numerical method to show the existence of a locally asymptotically 2-cycle $c_2^1 = \{(1.8461, 1.8461^T), (1.0595, 1.0595)^T\}$, and an asymptotically stable 2-cycle $c_2^2 = \{(1.6107, 1.1274)^T, (1.1274, 1.6107)^T\}$.

    Illustrate your work with a phase space diagram.

10. Consider system (5.4) with the parameters $k_1 = k_2 = 0.5$, $K_1 = K_2 = 3$, $n_1 = n_2 = 20$, $d_{12} = d_{21} = 0.1$.

Use a numerical method to show the existence of a locally asymptotically 6-cycle $c_6 = \{(1.1680, 1.1680)^T, (2.3360, 2.3360)^T, (0.6676, 0.6676)^T, (1.2825, 1.2825)^T, (2.5650, 2.5650)^T, (0.7341, 0.7341)^T\}$, and an asymptotically stable 3-cycle $c_3 = \{(1.2990, 0.7765)^T, (2.7286, 1.4143)^T, (0.8735, 2.5000)^T\}$.

Illustrate your work with a phase space diagram.

5.3 Models with the Allee Effects

Recall that a (demographic) Allee effect is a phenomenon in biology characterized by a positive correlation between population density (size) and its per-capita growth rate at low densities [8]. It may be broadly defined as a decline in individual fitness at low population sizes or densities [60, 287]. In their book, Courchamp et al. [61] described the Allee effect in a straightforward manner: "The more the merrier". Explicitly, it is a notion of positive density dependence in which the "overall individual fitness" or one of its components, is positively related to population size or density. This is in contrast of the classical relationship between individual fitness and population density which is negative (fitness decreases with increasing density).

Stephens et al.[287] made the distinction between the component Allee effect, of particular interest to behaviorists, and the demographic Allee effect, of overriding concern to conservationists. Components Allee effects are at the level of components of individual fitness, for example, juvenile survival or litter size. On the other hand, demographic Allee effects are at the level of the overall mean individual fitness, by which is meant the effect on the per-capita population growth rate of the whole population. These effects can have profound consequences on the survival of a population. One example is the creation of a population density threshold that must be exceeded to avoid extinction, a scenario called a strong Allee effect. If the Allee effects do not create such a threshold, they are referred to as weak Allee effect [217, 287].

Figure 5.4 shows geometrically how a reduction in the per- capita population growth rate in small or sparse populations results in a (demographic) Allee effect. Strong Allee effects give rise to smaller than one per-capita population growth rates once the population size or density falls below the Allee threshold. On the other hand, a weak Allee effect does not give rise to an Allee threshold. While fewer red sea urchins give rise to worsening feeding conditions of their young and less protection from predation. Some mast flowering trees, such as *Spartina alterniflora*, with a low density have a lower probability of pollen grain finding stigma in wind-pollinated plants. The Allee effect may explain one of the most dramatic extinctions of modern times - that of the passenger pigeon *Estopistes migratorius* [287].

Allee et al. [8] provided experimental and field studies that confirmed the presence of the Allee effect among many species. The examples include bobwhite quails (*Colinus virginianus*) that huddle together to lower the surface presented to cold weather, and the disappearance of tsetse fly from an area in which the density of the flies fall below the threshold minimum density. Allee himself [8] considered the two types of Allee effect and observed the Allee effect caused by reduction in the number of mice, and the Allee effect caused by reduction in density of flour beetles, (*Tribolium confusum*). In [287], Stephens and Sutherland described several scenarios that cause the Allee effect in both animals and plants. For example, cod and many freshwater fish species have higher juvenile mortality when there are fewer adults.

Understanding the nature of the Allee effects is vital in preventing rare, declining, endangered or fragmented populations from extinction [29]. By providing predictions of the outcome of

5.4. MODEL DERIVATION FOR SINGLE SPECIES

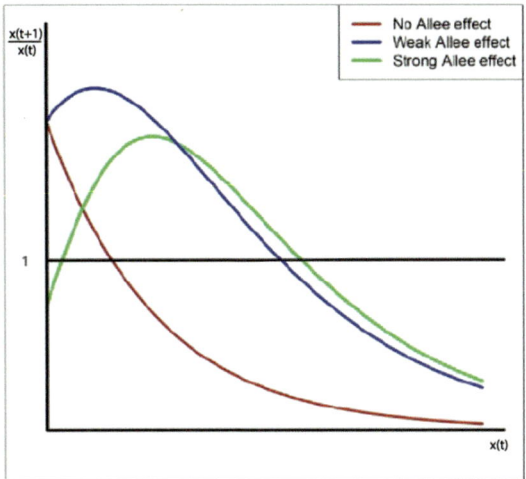

Figure 5.4: The graph shows the per-capita growth rate versus the size of the population. Strong Allee effect (solid), weak Allee effect (dashed), and no Allee effect (circles).

management actions in conservations (e.g., African wild dogs [60]), exploitation (e.g., red sea urchins [252]) or pest control (e.g., gypsy moth [177]), models save money where direct experimentation would be costly and allow decision-making where it would be unethical.

5.4 Model Derivation for Single Species

We start this section by formalizing the terms and informal definitions used in Section 5.3 [28].

Definition 25
- *An Allee threshold is a critical population size or density below which the per-capita growth rate is less than 1 and the population is threatened with extinct.*
- *A demographic Allee effect (or simply Allee effect) is a positive relationship between the overall individual fitness or the per-capita population growth rate and population size or density.*
- *A strong Allee effect is present when there exist (at least) two attractors, one of which is the extinction equilibrium and the other of which is a survival attractor.*
- *A weak Allee effect is an Allee effect that is not a strong Allee effect.*

Let $x(t)$ be the population density at time t, and $r(x(t))$ be the per-capita growth rate of the population. Then the dynamics of a population with synchronized generations are described by the difference equation

$$x(t+1) = r(x(t))x(t) = f(x(t)). \tag{5.6}$$

Schreiber [273] proposed that for species that possess the Allee effect, the per-capita growth rate function $r(x(t))$ can be viewed as the product of two functions $f(x(t)) = g(x(t))I(x(t))$, where $g(x(t))$ represents a negative density factor and $I(x(t))$ represents a positive density factor. In the absence of the Allee effect, the factor $I(x) \equiv 1$. Thus $g(x(t))$ is the per-capita growth rate of the population in the absence of the Allee effect and may be one of the various functions that describe the dynamics of single species, such as the exponential or Ricker model, the discrete logistic (Beverton-Holt) model, the Hassell model, etc. Two of numerous biological mechanisms of positive effects that can produce an Allee effect are predator saturation and mate limitation. We give examples motivated by these mechanisms below.

For the Allee effect due to predator saturation, we let

$$I(x) = \exp\left(-\frac{m}{1+sx}\right)$$

be the probability of escaping predation by a predator with a saturating functional response where m represents predation intensity and s is proportional to the handling time [156]. For instance, for the Ricker model, the difference equation (5.6) becomes

$$x(t+1) = x(t)\exp\left(\alpha\left(1 - \frac{x(t)}{k}\right)\right)\exp\left(-\frac{m}{1+sx(t)}\right). \tag{5.7}$$

The discrete logistic (Beverton-Holt) model becomes

$$x(t+1) = x(t)\frac{rk}{k+(r-1)x(t)}\exp\left(-\frac{m}{1+sx}\right) \tag{5.8}$$

and the Hassell model becomes

$$x(t+1) = x(t)\frac{r}{(1+kx(t))^b}\exp\left(-\frac{m}{1+sx}\right). \tag{5.9}$$

Allee effects can also be caused by the difficulty of finding mates at low densities. For instance, in a field experiment, Levitan et al. [213] found 0% of a small dispersed group of sea urchins *Strongylocentotus Franciscanus* were fertilized, while an 82.2% fertilization rate was reported in the center of a large aggregated group of sea urchins. To model mate limitation, we let

$$I(x) = \frac{sx}{1+sx}$$

be the probability of finding a mate where s is an individual's searching efficiency [104, 236, 272]. Hence, the Ricker model becomes

$$x(t+1) = x(t)\exp\left(\alpha\left(1 - \frac{x(t)}{k}\right)\right)\frac{sx(t)}{1+sx(t)}. \tag{5.10}$$

The discrete logistic (Beverton-Holt) model becomes

$$x(t+1) = x(t)\frac{rk}{k+(r-1)x(t)}\frac{sx(t)}{1+sx(t)}. \tag{5.11}$$

And the Hassell model becomes

$$x(t+1) = x(t)\frac{r}{(1+kx(t))^b} \cdot \frac{sx(t)}{1+sx(t)}. \tag{5.12}$$

5.4. MODEL DERIVATION FOR SINGLE SPECIES

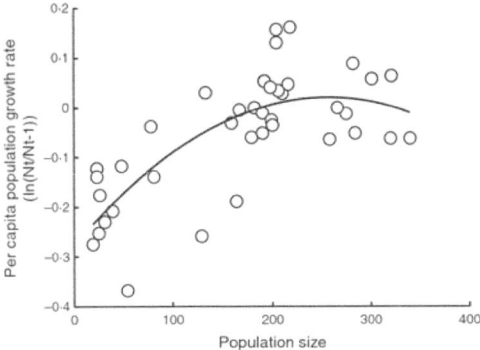

Figure 5.5: Annual counts of free-living V. I. marmots reveal strong inverse density dependence in *per capita* population growth from 1970 to 2007.

For a model that studies Allee effects may be caused by a combination of both predator saturation and mating limitation see [305].

Brashares et al. [36] identified the Vancouver Island Marmot as an endangered species that is on the brink of extinction. This large rodent is geographically restricted to Vancouver Island and evolved rapidly after its arrival after a glacial retreat 10,000 years ago. The authors provided evidence of the presence of the Allee effect in this population by exhibiting the graph of data, where the x-axis represents the population size $x(t)$ and the y-axis represents the logarithm of per-capita growth rate $\ln \frac{x(t+1)}{x(t)}$ (Fig. 5.5).

It is evident that the per-capita growth rate is below one, i.e.,

$$\ln\left(\frac{x(t+1)}{x(t)}\right) < 0$$

when the population size increases and reaches a threshold equilibrium size (called an Allee equilibrium, point A). The per-capita growth rate keeps increasing but at a certain critical size (around 200 marmots) [36] the curve turns downward, with a decreasing reproductive rate. The authors hypothesize that the marmots have increased their range due to a lack of mates nearby. Thus the very process of increasing their per-capita growth rates is hindered because they cannot find mates, and when they go looking, are more likely to be killed by predators or get lost in unfamiliar territory.

The following theorem provides conditions sufficient for a strong or weak Allee effect to occurs in the model equation (5.6).

Theorem 47 *[218] Equation (5.6) has a strong (weak) Allee effect if $r(x)$ is differentiable on $\mathbb{R}_+$ and the following conditions hold true:*

(i) $r'(x) > 0$ for $x \in (0, c)$ for some $c > 0$, and $r'(x) < 0$ for $x > c$

(ii) $r(0) < 1$ ($r(0) > 1$),

(iii) there exists a unique $K > 0$, such that $r(K) = 1$, $r'(K) < 0$, and $Kr^1(K) > -2$.

Proof. Now $f'(x) = xr'(x) + r(x)$. For the extinction equilibrium $f'(0) = r(0) < 1$, and thus it is locally asymptotically stable. For the equilibrium K we have $f'(K) = Kr'(K) + r(K) = Kr'(K) + 1 > -2$. Since $r'(K) < 0$, $f'(K) < 1$. Hence, K is asymptotically stable. ∎

Remark 16 *If in the above example $r'(K) = -2/K$, then there occurs a period-doubling bifurcation.*

Elaydi and Sacker [127] used the following rational function as the per-capita growth rate

$$r(x) = \frac{ax + r}{x^2 + cx + d}$$

to model the strong Allee effect and hence the model is given by

$$x(t+1) = \frac{ax(t)^2 + bx(t)}{(x(t)^2 + cx(t) + d)}. \tag{5.13}$$

This model is a variation of the discrete logistic (Beverton-Holt) model with an Allee effect caused by mate limitations as in [29]. Forcing the graph of $xr(x)$ to pass through the three equilibrium points shown in the cobweb diagram plotted in Figure 5.6, we obtain

$$r(x) = \frac{(A + K + c)x + \frac{r_0 AK}{1 - r_0}}{x^2 + cx + \frac{AK}{1 - r_0}},$$

where $r_0 = r(0)$. Using this $r(x)$ one may conclude that for large x, $xr(x)$ is asymptotic to $A+K+c$. The proof is left to the reader,

Figure 5.7 shows the two types of models with the strong Allee effect, the scramble competition such as the Ricker model (blue), and the contest competition (green) such as the discrete logistic (Beverton-Holt) model. In the next example, we revisit the Ricker model with the Allee effect due to predator saturation.

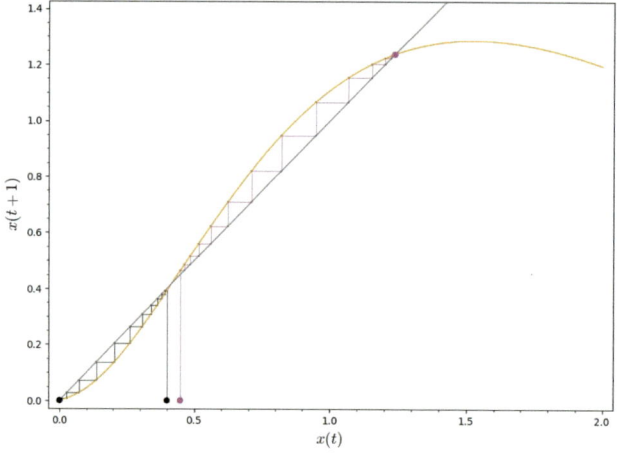

Figure 5.6: If the population density goes below the Allee threshold A, the the population will go extinct. The population may persist if its initial density is above the Allee threshold.

5.4. MODEL DERIVATION FOR SINGLE SPECIES

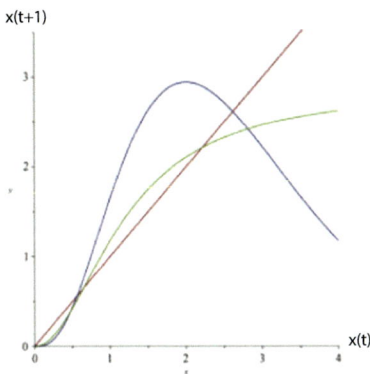

Figure 5.7: The green curve represents a contest competition model with the strong Allee effect while the blue curve represents a scramble competition model with the strong Allee effect.

Example 84 *Consider the Ricker model with the Allee effect due to predator saturation*

$$x(t+1) = x(t) \exp\left(\alpha(1-x(t)) - \frac{m}{1+sx(t)}\right). \tag{5.14}$$

Now

$$r(x) = \exp\left(\alpha(1-x)) - \frac{m}{1+sx}\right)$$

and the density-free per-capita growth rate $r(0) = \exp(\alpha - m)$. *Hence, if* $\alpha < m$, *the extinction equilibrium is asymptotically stable.*

Positive equilibria are roots of $r(x) = 1$, *i.e., of the equation*

$$\alpha(1-x) - \frac{m}{1+sx} = 0$$

or equivalently the quadratic equation $q(x) = 0$ *where*

$$q(x) := \alpha - m + \alpha(s-1)x - s\alpha x^2.$$

We can discover the positive root possibilities geometrically by noting the graph of $q(x)$ *is a concave down parabola with a vertical intercept of* $q(0) = \alpha - m$. *Thus, if* $\alpha - m > 0$ *there exists exactly one positive equilibrium. Suppose* $\alpha - m < 0$. *Then there exist no positive roots if the critical point*

$$x_c = \frac{s-1}{s}$$

of $q(x)$ *is negative, i.e.,* $s < 1$. *Suppose* $s > 1$ *then there exist exactly two positive roots if and only if the maximum of the parabola is positive, i.e.,* $q(x_c) > 0$. *A calculation shows* $q(x_c) = \alpha - m$. *In summary:*

(1) $\alpha - m > 0$ *implies there exists a unique positive equilibrium;*
(2) $\alpha - m < 0$ *and* $s < 1$ *implies there exist no positive equilibria;*
(3) $\alpha - m < 0$ *and* $s > 1$ *implies there exist exactly two positive equilibria.*

We are interested in case (3). First we note that $r'(0) = (ms - \alpha)\exp(\alpha - m) > 0$ *and, consequently, the model has a (demographic) Allee effect in this case. The two positive roots of* the quadratic $q(x)$ are in this case

$$A = \frac{\alpha(s-1) - \sqrt{\alpha^2(s+1)^2 - 4ms\alpha}}{2\alpha s}. \tag{5.15}$$

and

$$K = \frac{\alpha(s-1) + \sqrt{\alpha^2(s+1)^2 - 4ms\alpha}}{2\alpha s}. \tag{5.16}$$

It is left as Problem 5 to show, by applying the Linearization Principle with

$$f(x) = xr(x) = x\exp\left(\alpha(1-x) - \frac{m}{1+sx}\right),$$

that A is unstable and K is (locally asymptotically) stable. Thus, in case (3), there exist two stable positive equilibria. Since we showed above that $x = 0$ is stable when $\alpha - m < 0$, we conclude that there is a strong Allee effect in case (3).

Note that it is possible for $r'(0) > 0$, i.e., that there is an Allee effect in the other two cases as well. This shows that an Allee effect does not necessarily imply a strong Allee effect. In this case, it is said that there is a weak Allee effect.

5.5 Allee Effects and Hysteresis

In this Section we consider two scenarios in which multiple attractors occur. First, we consider backward bifurcations when the extinction equilibrium destabilizes, which typically give rise to strong Allee effects. Secondly, we look at hysteresis effects that involve multiple positive equilibria. Consider again the difference equation

$$x(t+1) = f(x(t)) = x(t)r(x(t)). \tag{5.17}$$

We make the following assumptions on $r(x)$.

A$_1$: $r(x)$ is C^2 and there exists a critical number $x_m > 0$ where $r(x)$ attains its global maximum.

A$_2$: There exists $z > x_m$ such that $r(z) < 1$.

There are two cases to consider.
Case (I): $r(x)$ is C^2 and there exists a critical number $x_m > 0$ where $r(x)$ attains its global maximum $0 < r(0) < 1$. This may be divided into three subcases.

(I_a): $r(x_m) < 1$. In this case, the only equilibrium point is the extinction equilibrium zero, since $r(x^*) = 1$ for positive equilibria x^*. Moreover, $f'(0) = r(0) < 1$ and $f(x) = xr(x) < x$. This implies that the extinction equilibrium is globally asymptotically stable.

(I_b): $r(x_m) = 1$. In this case, we have one positive equilibrium point, the Allee threshold $A = x_m$. Now $f'(A) = Ar'(A) + r(A) = r(A) = 1$. Thus A is nonhyperbolic. Since $f''(A) = Ar''(A) + 2r'(A) = Ar''(A) < 0$, it follows by Theorem 8 that A is semistable from the right. Hence, for $x_0 \in (A, f^{-1}(A))$, $\lim_{t\to\infty} x(t) = A$ and for $x_0 \in (0, A) \cup (f^{-1}(A), \infty)$, $\lim_{t\to\infty} X(t) = 0$. Therefore, in this subcase, the population possesses a strong Allee effect.

5.5. ALLEE EFFECTS AND HYSTERESIS

(I_c): $r(x_m) > 1$. Since $r(0) < 1$ and by Assumption A_2, there exists $z > x_m$ such that $r(z) < 1$, it follows that there are two positive equilibria, the Allee threshold $0 < A < x_m$ and the carrying capacity $x_m < K < \infty$. Since $f'(A) = Ar'(A) + r(A) > 1$, A is unstable (a repeller).

Note that $r'(K) = Kr'(x) + r(K)$, where $r'(K) < 0$ and $r(K) = 1$. Thus $|f'(K)| < 1$ if $Kr'(K) < -2$ and in this case K is locally asymptotically stable. Hence, the population case possesses a strong Allee effect.

Case (II): $r(0) > 1$. In this case, we have only one positive equilibrium point K by Assumption A_2. Furthermore, K is locally asymptotically stable if $Kr'(K) < -2$. In this case, the population possesses a weak Allee effect. The above scenarios are illustrated in Figure 5.8.

Remark 17 *Note that a backward bifurcation occurs when $r'(0) > 0$ and $r(0) = r_0 = 1$. A forward bifurcation occurs when $r'(0) < 0$ and $r(0) = r_0 = 1$.*

Figure 5.9 illustrates the strong Allee effect when we have backward bifurcation. The spectrum of positive equilibrium pairs is given by the set

$$G_+ = \{(r_0, x^*(r_0)) : r_0 \geq r_m\}$$

is the half line $r_0 \geq r_m$ and Equation (5.17) has

- no positive equilibria for $r_0 < r_m$;

- two positive equilibria for $r_m < r_0 < 1$;

- exactly one positive equilibrium for $r_0 > 1$.

The two positive equilibria coalesce at $r_0 = r_m$ and the point (r_m, x_m) is called a tangent (or blue-sky or saddle-node) bifurcation.

Example 85 *Consider again the Ricker model with the Allee effect due to predator saturation*

$$x(t+1) = x(t) \exp\left(\alpha - x(t) - \frac{m}{1+sx(t)}\right).$$

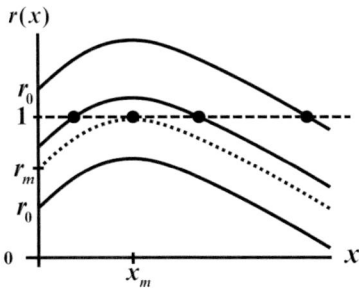

Figure 5.8: Plotting $r(x)$ versus x illustrates the two cases: (I) $r(0) < 1$, $r(x_m) < 1$, $r(x_m) = 1$, $f(x_m) > 1$ and (II) $r(0) > (1,0)$.

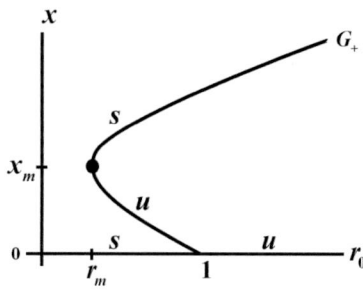

Figure 5.9: The backward bifurcation near the bifurcation point $(r_0, x) = 1$.

At positive equilibria

$$r(x) = \exp\left(\alpha - x - \frac{m}{1+sx}\right) = 1 \text{ or } r_0 = \alpha = x + \frac{m}{1+sx}.$$

Now

$$r'(x) = \left(-1 + \frac{m}{(1+sx)^2}\right)\exp\left(\alpha - x - \frac{m}{1+sx}\right) = 0$$

gives

$$x_m = \sqrt{\frac{m}{s}} - \frac{1}{x}$$

and

$$r_m = x_m + \frac{m}{1+sx_m}.$$

The set of positive equilibria is given by

$$G_+ = \left\{(r_0, x^*(r_0)) : r_0 = x^* + \frac{m}{1+sx^*}, r_0 \geq r_m\right\}.$$

Figure 5.9 shows the hallmark of the strong Allee effect, the backward bifurcation. The graph of G_+ can be easily obtained by reflecting through the line $r_0 = x$, the plot of

$$r_0 = x + \frac{m}{1+sx}$$

in the (x, r_0)-plane.

In the next example, we revisit the phenomenon of hysteresis which was introduced in Example 12 in Chapter 1.

Example 86 *In this example, we investigate the dynamics of a single species population model that exhibits hysteresis due to the presence of an Allee effect. It should be noted that the hysteresis is a commonly encountered phenomenon in ecology and epidemiology, where the observed equilibria of the system cannot be predicted solely based on environmental variables, but also required knowledge of the system's initial conditions. Examples include the theory of spruce budworm outbreaks and behavioral effects on disease transmission.*

5.5. ALLEE EFFECTS AND HYSTERESIS

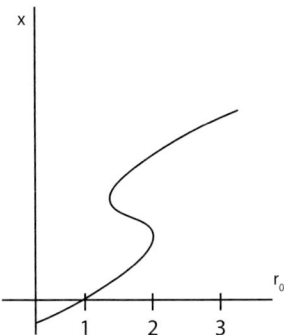

Figure 5.10: A forward bifurcation at $r_0 = 1$ with a hysteresis loop that creates the existence of multiple positive equilibria for an interval of r_0 values.

For the convenience of the reader, here is the difference equation that exhibits hysteresis due to the presence of a strong Allee effect.

$$x(t+1) = x(t)\left(r_0 + \frac{uvx^m}{v+x^m}\right)\left(\frac{1}{1+cx}\right) \quad (5.18)$$

where $c = 2$, $v = 4$, $u = 1.5$, $m = 4$. The bifurcation diagram in Figure 5.10 shows a distinctive S-shape indicative of a hysteresis effect. For an interval of r_0 values, between two saddle node bifurcations, there exist three positive equilibria. It is shown in [65] that, in general for one dimensional difference equation population models, the equilibria on decreasing segments of the bifurcation diagram are unstable and those on increasing segments near bifurcation points are locally asymptotically stable.

Exercises 5.4–5.5

1. For the equations (a)–(h) below

    (i) find the Allee threshold and the carrying capacity;

    (ii) give conditions for the presence of (i) strong Allee effect and (ii) weak Allee effect;

    (iii) state conditions for the stability of the carrying capacity.

    (iv) Sketch the bifurcation diagram and determine the direction of bifurcation.

    (a) $x(t+1) = \frac{\mu K x(t)}{K+(\mu-1)x(t)} \exp\left(\frac{-m}{1+sx(t)}\right)$

    (b) $x(t+1) = \frac{\mu K x(t)}{K+(\mu-1)x(t)} \frac{sx(t)}{1+sx(t)}$

    (c) $x(t+1) = \frac{rx(t)}{(1+ax(t))^b} \exp\left(\frac{-m}{1+sx(t)}\right)$ What are the effects of increasing b on the dynamics of the model?

    (d) $x(t+1) = \frac{rx(t)}{(1+ax(t))^b} \frac{sx(t)}{1+sx(t)}$ What are the effects of increasing b on the dynamics of the model?

    (e) $x(t+1) = \frac{ax^2(t)+bx(t)}{x^2(t)+cx(t)+d}$, $b \geq 0$, $a, c, d > 0$

(f) [235] $x(t+1) = \frac{rx(t)}{(1+(ax(t))^b)} \frac{sx(t)}{1+sx(t)}$

(g) [234] $x(t+1) = \frac{rx(t)}{(1+\exp(-\alpha(1-\frac{x(t)}{K})))}$, $\alpha, K > 0$.

(h) [155] $x(t+1) = \frac{rx(t)^c}{1+x(t)^c}$, $c > 1$

2. [127] Consider the b-Ricker model

$$x(t+1) = (x(t))^b \exp(\alpha(1-x(t))$$

where $b > 1$ introduces an Allee effect by causing some components of fitness to increase as x increases, so that fitness is maximum at intermediate population sizes. Draw a parameter-space diagram ($\alpha - b$) and draw some conclusions about the dynamics of the model. Here b introduces an Allee effect by causing some components of fitness (growth rate per capita) to increase as $x(t)$ increases, so that fitness is maximum at intermediate population sizes.

3. Consider the equation

$$x(t+1) = x(t)\exp\left(\alpha\left(1-\frac{x(t)}{K}\right)\left(\frac{x(t)}{A}-1\right)\right).$$

This exponential model was used to model the population of panda [43]. The panda population exhibits strong Allee effect due to the lack of mate selection. Assume that the carrying capacity of panda is 16,000, the intrinsic growth rate $\alpha = 1$, the initial population size is 1100, and the population size after 1 year is 1112.

(a) Find the Allee threshold.

(b) Determine the size of the population in 25 years.

(c) Find conditions under which the carrying capacity is asymptotically stable.

(d) Draw a bifurcation diagram ($\alpha-x$). What conclusions can you draw from the bifurcation diagram?

4. [225] Consider the following modified Ricker model with the strong Allee effect where the Holling-II type functional form is used.

$$x(t+1) = \frac{rx(t)^2}{\alpha + x(t)} \exp(-x(t)) = f(x(t)).$$

Let $x = c$ be the critical point of the map f and let A and K be the Allee threshold and the carrying capacity, respectively. Assume that $f(c) < f^{-1}(A)$.

(a) Find the Allee threshold A and the carrying capacity K.

(b) Show that the extinction equilibrium is locally asymptotically stable and find its basin of attraction.

(c) Find conditions under which the carrying capacity is asymptotically stable.

(d) Draw a bifurcation diagram ($\alpha-x$). What conclusions can you draw from the bifurcation diagram?

5. Using the Linearization Principle show, in case (3) of Example 84, that, the equilibrium A is unstable and the equilibrium K is (locally asymptotically) stable.

5.6. COMPETITION MODELS WITH THE ALLEE EFFECT

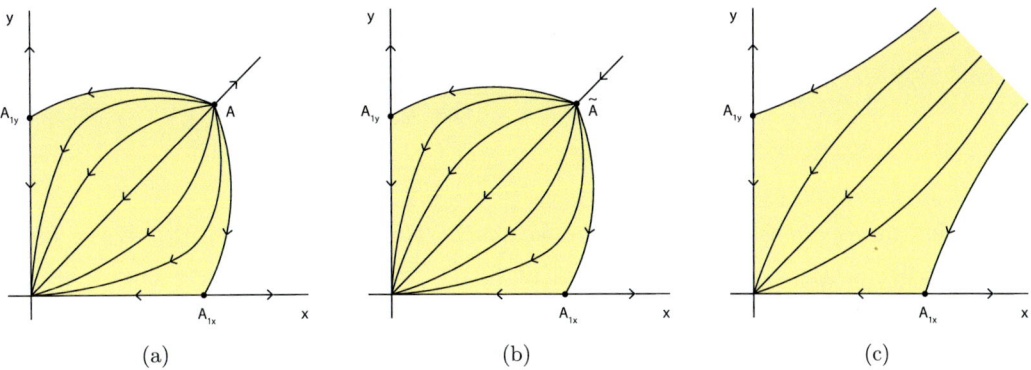

Figure 5.11: There are two possible rearrangements of the equilibrium along the Allee threshold curve (a,b) and one possibility of no interior equilibrium points, where the extinction region is unbounded (c).

5.6 Competition Models with the Allee Effect

In this section, we extend Definition 25 of an Allee effect from single-species models to multi-species competition models. Here, we have two cases to consider:
Case (I): All species possess the Allee effect.
Case (II): Not all species possess the Allee effect.

In both cases, we define a strong Allee effect as follows.

Definition 26 *Consider the difference equation*

$$\mathbf{x}(t+1) = F(\mathbf{x}(t)) \tag{5.19}$$

with $\mathbf{x}(t) \in \mathbb{R}^n$. Then Equation (5.19) is said to possess a strong Allee effect if there is an Allee threshold region $\mathbb{R}^n_+$ in which all species would go extinct, and a persistence region in which some or all the species survive. Weak Allee effect occurs if all the species possess Allee effects but there is no Allee threshold region.

In the strong Allee effect, the Allee effect region is, in general, bounded by an $(n-1)$-dimensional surface (or a curve in a two-dimensional system, called the Allee threshold curve). The only possible exception to this scenario is the case when there are no interior equilibrium points, where the extension region may be unbounded (Figure 5.11 a,b,c).

One may extend single-species model that we had in Section 5.4 to N-species in which each species possesses the strong Allee effect due to predator saturation.

Example 87 *The N-species Leslie-Gower competition model may be written as*

$$x_i(t+1) = \frac{a_i x_i^2(t)}{1 + x_i^2(t) + \sum_{j \neq i}^{N} b_i x_j}, \quad i = 1, 2, \ldots, N. \tag{5.20}$$

In this example, we will focus on the 2-species x and y model

$$x(t+1) = \frac{a_1 x^2(t)}{1 + x^2(t) + b_1 y(t)}$$
$$y(t+1) = \frac{a_2 y^2(t)}{1 + y^2(t) + b_1 x(t)}.$$
(5.21)

First, we find the equilibrium points on the boundary. They are $E^* = (0,0)^T$, $A_{1x} = (A_1, 0)^T$, $K_{1x} = (K_1, 0)^T$, $A_{2y} = (0, A_2)^T$, $K_{2y} = (0, K_2)^T$, where $A_1 = \frac{1}{2}\left(a_1 - \sqrt{a_1^2 - 4}\right)$, $K_1 = \frac{1}{2}\left(a_1 + \sqrt{a_1^2 - 4}\right)$, $A_2 = \frac{1}{2}\left(a_2 - \sqrt{a_2^2 - 4}\right)$, $K_2 = \frac{1}{2}\left(a_2 + \sqrt{a_2^2 - 4}\right)$. We assume that $a_1 > a_2$, and $a_2 > 2$ to ensure the existence of the boundary point. To find the interior equilibria, we find the intersections of the isoclines given by

$$1 + x^2 + b_1 y = a_1 x \tag{5.22}$$

and

$$1 + y^2 + b_2 x = a_2 y \tag{5.23}$$

which may be written as

$$y = -\frac{1}{b_1}\left(x - \frac{a_1}{2}\right)^2 + \frac{a_1^2 - 4}{4 b_1} \tag{5.24}$$

and

$$x = -\frac{1}{b_2}\left(y - \frac{a_2}{2}\right)^2 + \frac{a_2^2 - 4}{4 b_2}. \tag{5.25}$$

This leads to the fourth degree equation

$$x^4 - 2a_1 x^3 + (b_1 a_2 + a_1^2 + 2)x^2 + (b_1^2 b_2 - b_1 a_1 a_2 - 2a_1)x + (a + b_1 a_2 + b_1^2) = 0. \tag{5.26}$$

Once we solve for x, we compute y using Equation (5.22). Hence, there are from 0 to 4 interior equilibrium points. Geometrically, the solutions of the isoclines (5.22) and (5.23) are the intersections of two orthogonal parabolas shown in Figure 5.12.

There are seven scenarios where these intersections lead to 0-4 interior equilibrium point as shown in figure 5.13.

Since solving the fourth-degree equation (5.26) is a formidable task, we will consider here a special case, the symmetric case, where $a_1 = a_2 = a$ and $b_1 = b_2 = b$. Now subtracting (5.23) from (5.22) yields $(y-x)(y+x-b-a) = 0$. Hence, either (i) $x = y$ or (ii) $y + x = b + a$. Substituting $x = y$ in (5.22) yields the equation

$$x^2 + (b-a)x + 1 = 0. \tag{5.27}$$

Similarly, substitution $y + x = b + a$ in (5.23) yields

$$x^2 - (b-a)x + b(b+a) + 1 = 0. \tag{5.28}$$

Hence, the four possible interior points are given by

$$x_{1\pm} = \frac{1}{2}\left[(a-b) \pm \sqrt{(a-b)^2 - 4}\right], \quad y_{1\pm} = x_{1\pm}, \tag{5.29}$$

5.6. COMPETITION MODELS WITH THE ALLEE EFFECT

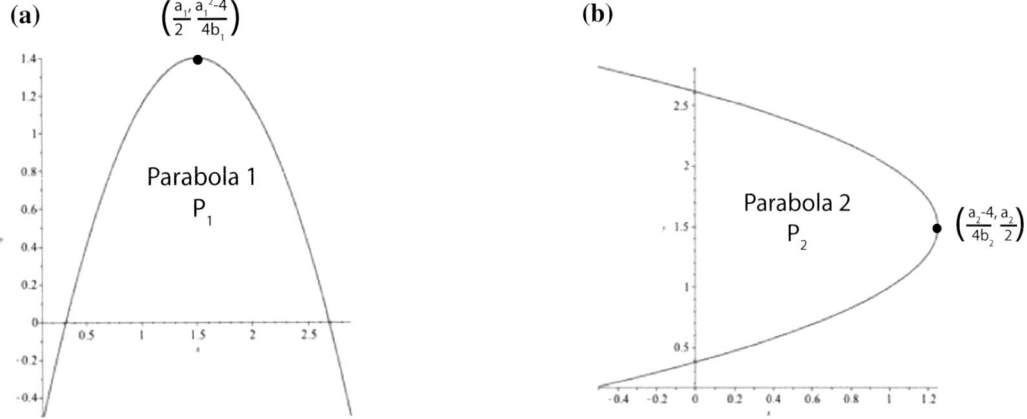

Figure 5.12: (a) The graph of isocline (5.21). (b) The graph of isocline (5.22).

and
$$x_{2\pm} = \frac{1}{2}\left[(a-b) \pm \sqrt{(a+b)(a-3b)-4}\right], \quad y_{2\pm} = b + a - x_{2\pm}. \tag{5.30}$$

As in Figure 5.13, we divide (x_{1-}, y_{1-}) by $\widetilde{A}$, (x_{1+}, y_{1+}) by $\widetilde{K}$, (x_{2-}, y_{2-}) by K_{2x}, and (x_{2+}, y_{2+}) by K_{2y}. We observe that
$$(a+b)(a-3b) = (a-b)^2 - 4b^2. \tag{5.31}$$

Hence, we have the following scenarios for the number of interior equilibrium points:

(i) if $(a-b) < 2$, there are no interior equilibria,

(ii) if $(a-b) = 2$, then we have one equilibrium point $\widetilde{A} = (1,1)$,

(iii) if $(a+b)(a-3b) < 4 < (a-b)^2$, then we have two equilibrium points $\widetilde{A}$ and $\widetilde{K}$ given by (5.29),

(iv) if $(a+b)(a-3b) = 4$, then we also have two equilibrium points $\widetilde{K} = \left(\frac{1}{2}(a+b), \frac{1}{2}(a+b)\right)$ and $\widetilde{A} = \left(\frac{1}{2}(a-3b), \frac{1}{2}(a-3b)\right)$,

(v) if $(a+b)(a-3b) > 4$, then we have four interior equilibrium points $\widetilde{A}$, $\widetilde{K}$, K_{2x} and K_{2y} given by formulas (5.29) and (5.30).

All the above scenarios are self-evident with the exception of (iv). To prove (iv), note that if $(a+b)(a-3b) = 4$, then from (5.31), $(a-b)^2 = 4 + 4b^2$. Thus
$$x_{2\pm} = \frac{1}{2}(a+b), \; y_{2\pm} = \frac{1}{2}(a+b), \; x_{1\pm} = \frac{1}{2}[a-b \pm 2b]$$

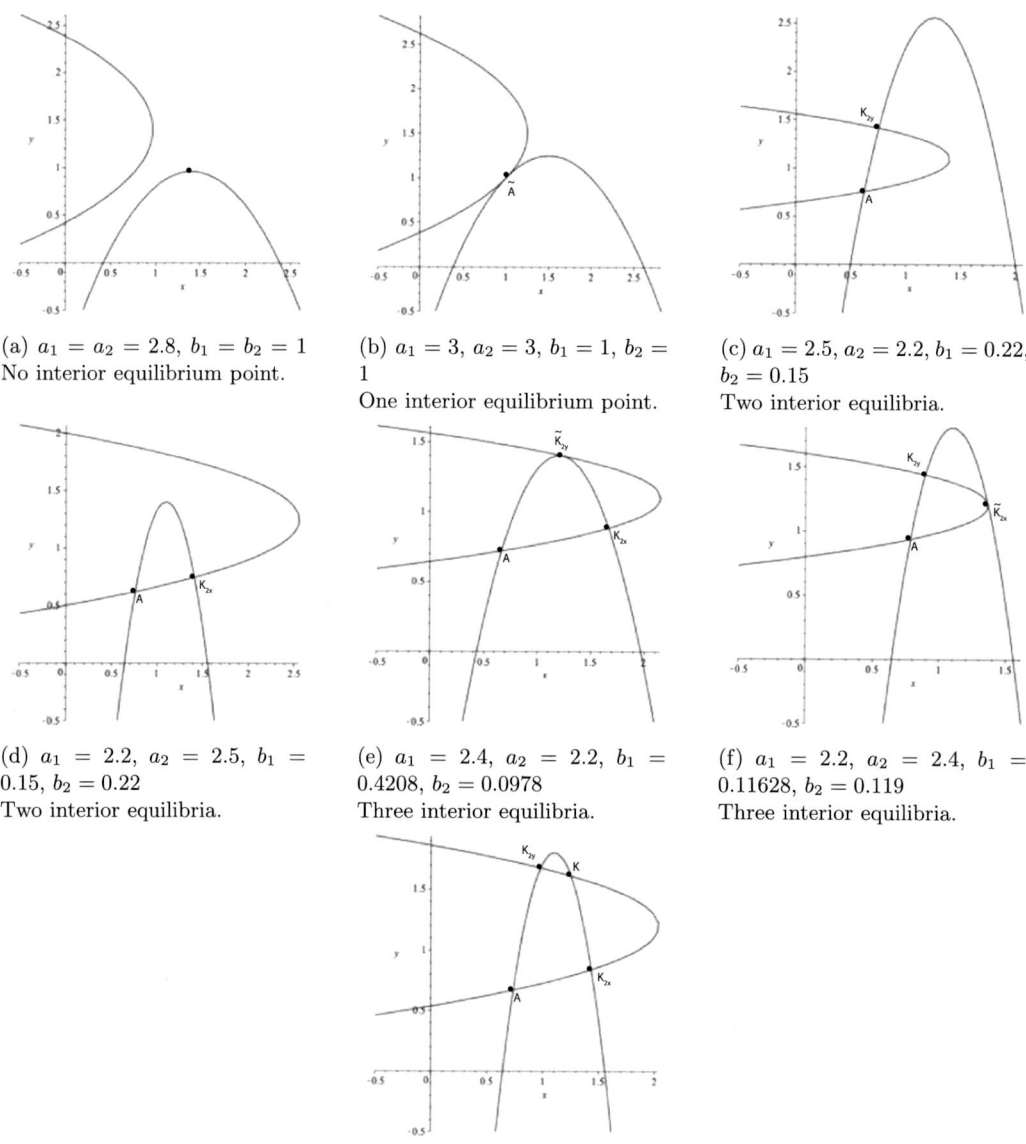

(a) $a_1 = a_2 = 2.8$, $b_1 = b_2 = 1$
No interior equilibrium point.

(b) $a_1 = 3$, $a_2 = 3$, $b_1 = 1$, $b_2 = 1$
One interior equilibrium point.

(c) $a_1 = 2.5$, $a_2 = 2.2$, $b_1 = 0.22$, $b_2 = 0.15$
Two interior equilibria.

(d) $a_1 = 2.2$, $a_2 = 2.5$, $b_1 = 0.15$, $b_2 = 0.22$
Two interior equilibria.

(e) $a_1 = 2.4$, $a_2 = 2.2$, $b_1 = 0.4208$, $b_2 = 0.0978$
Three interior equilibria.

(f) $a_1 = 2.2$, $a_2 = 2.4$, $b_1 = 0.11628$, $b_2 = 0.119$
Three interior equilibria.

(g) $a_1 = 2.2$, $a_2 = 2.46$, $b_1 = 0.11628$, $b_2 = 0.21628$
Four interior equilibria.

Figure 5.13: The seven scenarios of the intersection of the isocline (5.22) and (5.23) resulting from 0 to 4 interior equilibria.

or
$$x_{1+} = \frac{1}{2}(a+b), \ y_{1+} = \frac{1}{2}(a+b)$$
$$x_{1-} = \frac{1}{2}(a-3b), \ y_{1-} = \frac{1}{2}(a-3b).$$

Thus $x_{1\pm} = x_{2\pm}$, $y_{1\pm} = y_{2\pm}$ and hence we have only two interior equilibria.

Next, we investigate the local stability of the various equilibrium points. Now Equation (5.21) may be written as
$$\mathbf{x}(t+1) = F(\mathbf{x}(t)).$$

The Jacobian of the map is given by
$$JF(x,y) = \begin{pmatrix} \frac{2ax(by+1)}{(1+x^2+by)^2} & \frac{-abx^2}{(1+x^2+by)^2} \\ \frac{-aby^2}{(1+y^2+bx)^2} & \frac{2ay(bx+1)}{(1+y^2+bx)^2} \end{pmatrix}.$$

We will begin our analysis with the boundary equilibrium points A_{1x}, A_{1y}, K_{1x}, K_{1y}, and the extinction equilibrium $E^* = (0,0)^T$. For $E^* = (0,0)^T$,
$$JF(0,0) = \begin{pmatrix} 0 & 0 \\ 0 & 0 \end{pmatrix}$$

and the eigenvalues are $\lambda_1 = \lambda_2 = 0$. Hence, the extinction equilibrium is locally asymptotically stable. For $A_{1x} = (A_1, 0)^T$, $A_1 = \frac{1}{2}(a - \sqrt{a^2-4})$, $a > 2$. Using formula (5.22),
$$JF(A_{1x}) = \begin{pmatrix} \frac{2}{aA_1} & -\frac{b}{a} \\ 0 & 0 \end{pmatrix}.$$

Hence, the eigenvalues of $JF(A_{1x})$ are $\lambda_2 = 0$, and
$$\lambda_1 = \frac{4(a+\sqrt{a^2-4})}{4a} > 1.$$

Hence, A_{1x} is a saddle, where it is attracting on the interior and a repeller on the x-axis. Now for the equilibrium point $K_{1x} = (K_1, 0)^T$, we have $\lambda_2 = 0$ and
$$\lambda_1 = \frac{4(a-\sqrt{a^2-4})}{4a} < 1.$$

Hence, the equilibrium point K_{1x} is locally asymptotically stable.

Similarly, one may show that $A_{1y} = (0, A_2)$ is a saddle which is attracting in the interior and repelling on the y-axis and $K_{2y} = (0, K_2)$ is locally asymptotically stable (Problem 1).

Next, we study the stability of the interior equilibrium points $(x^*, y^*)^T$. Using the isocline equations (5.22) and (5.23), the Jacobian at any interior equilibrium point $(x^*, y^*)^T$ is given by
$$JF\begin{pmatrix} x^* \\ y^* \end{pmatrix} = \begin{pmatrix} \frac{2(by^*+1)}{ax^*} & \frac{-b}{a} \\ \frac{-b}{a} & \frac{2(bx^*+1)}{ay^*} \end{pmatrix}. \tag{5.32}$$

Theorem 48 *The following statements hold true.*

(i) If there is one interior equilibrium point, which is called $\widetilde{A}$, then $\widetilde{A}$ is unstable. More precisely, $\widetilde{A}$ has one unstable manifold connecting it with A_{1x} and A_{1y} and one semistable manifold.

(ii) If there are two interior equilibria, called $\widetilde{A}$ and $\widetilde{K}$, then $\widetilde{A}$ is a repeller (unstable) and $\widetilde{K}$ is locally asymptotically stable.

(iii) If there are four equilibrium points, then $\widetilde{A}$ is a repeller, K_{2x} and K_{2y} are saddle points and $\widetilde{K}$ is locally asymptotically stable.

Proof. Let us consider the above-mentioned three scenarios. Scenario (i) one interior equilibrium point that occurs when $(a - b) = 2$. We will call this point $\widetilde{A}$. From (5.29), $\widetilde{A} = (1,1)$. Moreover, $\widetilde{A}$ lies on the Allee threshold curve (Figure 5.14).

Here

$$JF(\widetilde{A}) = \begin{pmatrix} \frac{2b+2}{a} & \frac{-b}{a} \\ \frac{-b}{a} & \frac{2b+2}{a} \end{pmatrix} = \begin{pmatrix} \frac{2b+2}{b+2} & \frac{-b}{b+2} \\ \frac{-b}{b+2} & \frac{2b+2}{b+2} \end{pmatrix}$$

$$\det JF(\widetilde{A}) = \frac{4b^2 + 8b + 4}{b^2 + 4b + 4} - \frac{b^2}{b^2 + 4b + 4}$$

$$= \frac{3b^2 + 8b + 4}{b^2 + 4b + 4} > 1$$

$$\operatorname{tr} JF(\widetilde{A}) = \frac{4b + 4}{b + 2} = \det JF(\widetilde{A}) + 1.$$

It follows from Remark 3, Chapter 3 and Figure 3.4(a),(b) that the eigenvalues of $JF(\widetilde{A})$ are $\lambda_1 = 1$ and $\lambda_2 = \det JF(A_s) > 1$. Hence, $\widetilde{A}$ is unstable and more precisely, $\widetilde{A}$ has an unstable manifold connecting with A_{1x} and A_{1y} and a semi-stable manifold as shown in Figure 5.14(b).

Scenario (ii): Two equilibrium points that occur when $(a + b)(a - 3b) < 4 < (a - b)^2$. In this case, two equilibria $\widetilde{A} = (A, A)$ and $\widetilde{K} = (K, K)$,

$$\det JF(A) - 1 = \frac{4(bA + 1)^2}{a^2 A^2} - \frac{b^2}{a^2} - 1.$$

From formula (5.29), $A < \frac{a-b}{2}$ or $-b > 2A - a$. Moreover,

$$1 + A^2 + bA = aA$$

or

$$1 + bA = A(a - A).$$

Thus,

$$\det JF(\widetilde{A}) - 1 = \frac{4(a - A)^2}{a^2} - \frac{(2A - a)^2}{a^2} - 1 = 2\left(1 - \frac{2A}{a}\right) > 0$$

from (5.29). Thus, $\det JF(\widetilde{A}) > 1$. Next, we show that $\det JF(\widetilde{A}) > \operatorname{tr} JF(\widetilde{A}) - 1$. Note that

$$\det JF(A) = 1 + \frac{4(a - A)^2}{a^2} - \frac{(2A - a)^2}{a^2} - 1$$

$$= \frac{4(a - A)}{a} - 1$$

$$> \operatorname{tr} JF(A) - 1.$$

Hence, $\det JF(\widetilde{A}) > \operatorname{tr} JF(\widetilde{A}) - 1$. This implies that the equilibrium point $\widetilde{A}$ is a repeller.

5.6. COMPETITION MODELS WITH THE ALLEE EFFECT

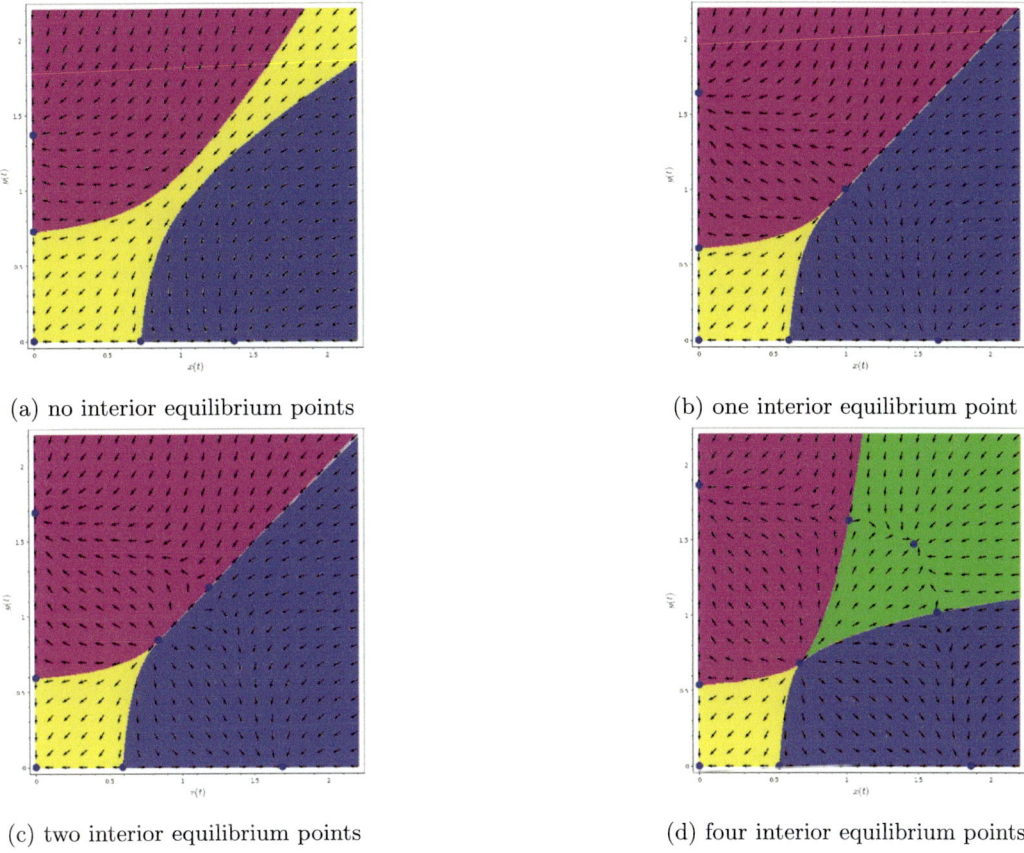

(a) no interior equilibrium points

(b) one interior equilibrium point

(c) two interior equilibrium points

(d) four interior equilibrium points

Figure 5.14: Phase space diagrams in which the yellow region is the extinction region, where both species go extinct.
(a) There are no interior equilibrium points and the extinction region is unbounded. The blue region is the basin of attraction of K_{1x}, and the magenta region is the basin of attraction of K_{1y}.
(b) There is one interior equilibrium point $\widetilde{A}$ which is semistable. The blue region is the basin of attraction of K_{1x}, and the magenta region is the basin of attraction of K_{1y}.
(c) There are two interior equilibria $\widetilde{A}$ which is a repeller, and $\widetilde{K}$, which is a saddle. The basin of attraction of K_{1x} and the magenta region is the basin of attraction of K_{1y}.
(d) There are four interior equilibria $\widetilde{A}$ which is a repeller, $\widetilde{K}$ which is local asymptotically stable with the green region as its basin of attraction, K_{2x} which is a saddle, and K_{2y} which is a saddle.

Next, we investigate the dynamics of the equilibrium point K. Note that from Figure 5.12, we see that
$$\frac{a}{2} < K < \frac{a^2 - 4}{4b}. \tag{5.33}$$

Now from (5.33)
$$\begin{aligned}
\det JF(\widetilde{K}) - 1 &= \frac{4(bK+1)^2}{a^2 K^2} - \frac{b^2}{a^2} - 1 \\
&< \frac{a^2}{4K^2} - \frac{b^2}{a^2} - 1 \\
&< 1 - \frac{b^2}{a^2} - 1 < 0.
\end{aligned}$$

Hence, $\det JF(\widetilde{K}) < 1$. Next, we will show that $\det JF(K) > \operatorname{tr} JF(K) - 1$.
$$\begin{aligned}
1 + \det JF(\widetilde{K}) - \operatorname{tr} JF(\widetilde{K}) &= 1 + \frac{4(a-K)^2}{a^2} - \frac{b^2}{a^2} - \frac{4(a-K)}{a} \\
&> 1 + \frac{4(a-K)^2}{a^2} - \frac{(a-2K)^2}{a^2} - \frac{4(a-K)}{a} = 0.
\end{aligned}$$

Hence, $\det JF(\widetilde{K}) > \operatorname{tr} JF(\widetilde{K}) - 1$. Therefore, the equilibrium point K is asymptotically stable. The proof of Scenario (iii) is similar to the proof of Scenario (ii). ∎

Figure 5.14 depicts the phase space diagrams illustrating the above theorem.

The dynamics of the symmetric system (5.21), as explained above, is illustrated by the phase space diagrams in Figure 5.14(a),(b),(c),(d).

Exercises 5.6

1. Consider Equation (5.21). Show that $A_{2y} = (0, A_2)$ is a saddle which is attracting in the interior and repelling on the y-axis that and $K_{2y} = (0, K_2)$ is locally asymptotically stable.

2. Consider the symmetric Leslie-Gower model (5.21). with the strong Allee effect. Prove the statement (iv) in Theorem 48.

In problems 3–13

(i) Find all the equilibrium points and determine the conditions for their existence.

(ii) Use the Linearization Principle to determine the local stability of the equilibrium points.

(iii) Draw the phase space diagrams for the possible scenarios.

3. Consider the Ricker competition model with the Allee effect between two species x and y where both are subject to predations by different generalist predators with saturating functional responses.

5.6. COMPETITION MODELS WITH THE ALLEE EFFECT

$$x(t+1) = x(t)\exp\left(\alpha(1-x(t)) - \frac{m_1}{1+s_1 x(t)} - ay(t)\right)$$

$$y(t+1) = y(t)\exp\left(\beta(1-y(t)) - \frac{m_2}{1+s_2 y(t)} - bx(t)\right)$$

4. Consider the following Ricker competition model between two species x and y where both are subject to mate limitation.

$$x(t+1) = x(t)\exp\left(\alpha(1-x(t)) - a_1 y(t)\left(\frac{b_1 x(t)}{1+b_1 x(t)}\right)\right)$$

$$y(t+1) = y(t)\exp\left(\beta(1-y(t)) - a_2 x(t)\left(\frac{b_2 y(t)}{1+b_2 y(t)}\right)\right)$$

5. Consider the following competition model between species x and y, where species x is subject to predation by a generalist predator with a saturating functional response and species y is subject to mate limitation.

$$x(t+1) = x(t)\exp\left(\alpha(1-x(t)) - \frac{m_1}{1+s_1 x(t)} - a_1 y(t)\right)$$

$$y(t+1) = y(t)\exp\left(\alpha(1-y(t)) - a_2 x(t)\left(\frac{by(t)}{1+by(t)}\right)\right)$$

6. Consider the following competition model of Leslie-Gower type with the Allee effect.

$$x(t+1) = \frac{a_1 x(t)}{1+x(t)+b_1 y(t)}\frac{x(t)}{m_1+x(t)}$$

$$y(t+1) = \frac{a_2 y(t)}{1+y(t)+b_2 x(t)}\frac{y(t)}{m_2+y(t)}$$

where the fraction $\frac{x(t)}{m_1+x(t)}$ is the probability of an individual successfully finds a mate or a cooperative individual when the population size is $x(t)$, and $\frac{1}{m_1}$ is the individual's search efficiency.

7.

$$x(t+1) = \frac{a_1 x(t)}{[1+x(t)+b_1 y(t)]^c}\frac{s_1 x(t)}{1+s_1 x(t)}$$

$$y(t+1) = \frac{a_2 y(t)}{[1+y(t)+b_2 x(t)]^c}\frac{s_2 y(t)}{1+s_2 y(t)}$$

where $c > 1$.

8.

$$x(t+1) = \frac{a_1 x^2(t) + b_1 x(t)}{1+x^2(t)+c_1 y(t)}$$

$$y(t+1) = \frac{a_2 y^2(t) + b_2 y(t)}{1+y^2(t)+c_2 x(t)}$$

9.
$$x(t+1) = x(t)^c \exp(\alpha(1-x(t)) - b_1 y(t))$$
$$y(t+1) = y(t)^c \exp(\beta(1-y(t)) - b_2 x(t))$$

where $c > 1$.

10.
$$x(t+1) = \frac{a_1 x(t)}{1 + x(t)^c + b_1 y(t)} \frac{s_1 x(t)}{1 + s_1 x(t)}$$
$$y(t+1) = \frac{a_2 y(t)}{1 + y(t)^b + b_2 x(t)} \frac{s_2 y(t)}{1 + s_2 y(t)}$$

where $b > 1$.

11.
$$x(t+1) = \frac{a_1 x(t)}{1 + \exp(-\alpha(1-x(t)) + b_1 y(t))}$$
$$y(t+1) = \frac{a_2 y(t)}{1 + \exp(-\beta(1-y(t)) - b_2 x(t))}$$

12. Consider the Sigmoid Leslie–Gower competition model with the Allee effect.
$$x(t+1) = \frac{a_1 x^\alpha(t)}{1 + x^\alpha(t) + b_1 y(t)}$$
$$y(t+1) = \frac{a_2 y^\alpha(t)}{1 + y^\alpha(t) + b_2 x(t)}$$

where $\alpha > 1$.

13. [225] Consider the following "extended" Ricker model with Allee effect.
$$x(t+1) = \frac{a_1 x^2(t)}{1 + x(t) + b_1 y(t)} e^{-x(t)}$$
$$y(t+1) = \frac{a_2 y^2(t)}{1 + y(t) + b_2 x(t)} e^{-y(t)}$$

5.7 Predator–Prey Models with Allee Effects

In Chapter 3, Section 3.5, we study predator–prey models where neither the prey nor the predator possesses the Allee effect. In this section, we extend our study of these models to the case when either one or both of the species possess the Allee effect.

Some of the conclusions we will find are the following. Allee effects in prey generally destabilize predator–prey dynamics. Firstly, strong Allee effects in prey can prevent predator–prey systems from exhibiting sustained cycles. For instance, if the prey's density falls below the Allee threshold, both prey and predator go extinct. Secondly, strong Allee effects in prey may destabilize predator–prey systems by causing the coexistence equilibrium to change from stable to unstable. Thirdly, strong Allee effects increase the vulnerability of the predator–prey system to collapse.

5.7. PREDATOR–PREY MODELS WITH ALLEE EFFECTS

One context in which Allee effects are important is the following. Herbivore-plan systems in which the plant population is subject to a demographic Allee effect may be considered from an optimal harvesting perspective. In arid or semi-arid areas where drought alternates with intense rainfall, positive feedback (Allee effect due to environmental conditioning), generating plant density (Allee threshold) below which the plant population collapses [259]. For this reason, a good strategy by a herdsman is to decrease his stock at the onset of the dry season and returns to a maximal sustainable yield after both plant and herbivore populations recover after the start of a new rainfall season.

As with strong Allee effects in prey, strong Allee effects in the predator growth rate may destabilize predator–prey systems by causing the coexistence equilibrium to change from stable to unstable [302]. Moreover, the strong Allee effect in the prey would increase the equilibrium prey density.

To illustrate these phenomena we consider a herbivore-plan for a deciduous plant and an herbivore such as the gypsy moth. Let $H(t)$ and $P(t)$ be the population densities of the plant (host/prey) and herbivore (parasite/predator) at generation t, respectively.

We will consider a host-parasitoid interaction with component Allee effects induced by predator saturation as follows [182].

$$\begin{aligned} H(t+1) &= H(t)\exp\left(r(1-H(t)) - \frac{m}{1+sH(t)}\right) - aP(t) \\ P(t+1) &= H(t)\exp\left(r(1-H(t)) - \frac{m}{1+sH(t)}\right)\left(1 - e^{-aP(t)}\right). \end{aligned} \quad (5.34)$$

This model is a modification of a model studied in [175] and [180], where the host is not subject to Allee effects. This model is given (by letting $m = 0$) by the system

$$\begin{aligned} H(t+1) &= H(t)\exp(r(1-H(t)) - aP(t)) \\ P(t+1) &= H(t)\exp(r(1-H(t)))(1 - \exp(-aP(t))). \end{aligned} \quad (5.35)$$

The next result summarizes the dynamics of system (5.35).

Theorem 49 *[180] System (5.35) is bounded and positively invariant in $\mathbb{R}^2_+$ and the following statements hold:*

- *If $0 < r < 2$, then system (5.35) is permanent in $\mathbb{R}^2_+$.*

- *If $a > 1$, system (5.35) has a unique interior equilibrium and if $a < 1$, system (5.35) may have two interior equilibria under certain range of r.*

- *System (5.35) has two attractors per certain parameter ranges when $r > 2$. In this case, the system may have a boundary attractor $(N*, 0)^T$ and an interior attractor.*

Proof. The proof is left as Problem 1 in Exercise 5.7. ∎

Next, we investigate the stability of the boundary equilibria of system (5.35). To find the equilibria, we write the isoclines of system (5.35):

$$r\left(1 - H - \frac{m}{1+sH}\right) - ap = 0 \quad (5.36)$$

$$H \exp\left(r\left(1 - H - \frac{m}{1+sH}\right)\right)(1 - \exp(-ap)) = p. \tag{5.37}$$

Clearly, the extinction equilibrium $E_1^* = (0,0)^T$ is present. Now letting $p = 0$ in (5.36) yields the equation

$$sH^2 - (s-1)H + m - 1 = 0.$$

Hence,

$$H_{\pm} = \frac{s - 1 \pm \sqrt{(s-1)^2 - 4s(m-1)}}{2s}. \tag{5.38}$$

It follows from (5.38) that in order for the host H to have a strong Allee effect we must assume

Assumption I: $m > 1$ and $1 < s < \frac{(s+1)^2}{4m}$,

With this assumption, there are two boundary equilibria, in addition to the extinction equilibrium E_1^*. These are $E_2^* = (A_H, 0)^T$ and $E_3^* = (K_H, 0)^T$, where $A_H = H_-$ and $K_H = H_+$ are the Allee threshold and the carrying capacity of the host in the absence of the herbivore. At the extinction equilibrium E_1^*, the Jacobian matrix of system (5.34) is given by

$$JF(0,0) = \begin{pmatrix} e^{r(1-m)} & 0 \\ 0 & 0 \end{pmatrix}$$

with eigenvalues $\lambda_1 = e^{r(1-m)}$ and $\lambda_2 = 0$. Since $m > 1$, $\lambda_1, \lambda_2 < 1$, and, consequently, E_1^* is asymptotically stable.

Next, we look at the other two boundary equilibria E_2^* and E_3^*. The Jacobian matrix of system (5.34) when $p = 0$ is given by

$$JF(H,0) = \begin{pmatrix} 1 + rH\left(\frac{ms}{(1+sH)^2} - 1\right) & -aH \\ 0 & aH \end{pmatrix}.$$

For the equilibrium point E_2^*, the eigenvalues of $JF(E_2^*)$ are given by $\lambda_1 = 1 + rA\left(\frac{ms}{(1+sA)^2} - a\right)$ and $\lambda_2 = aA$.

Note that from (5.36) and (5.38) $\frac{ms}{(1+sA_H)^2} = \frac{s(1-A_H)}{1+sA_H} > \frac{s\left(1 - \frac{s-1}{2s}\right)}{1+s\left(\frac{s-1}{2s}\right)} = 1$. Hence, $\lambda_1 > 1$. Since $\lambda_2 = aA_H$, $\lambda_2 > 1$ if $aA_H > 1$ or $\frac{1}{a} < A_H$, and thus E_2^* is a repeller (Figure 5.2(a)). On the other hand, if $aA_H < 1$ or $\frac{1}{a} > A_H$, then $\lambda_2 < 1$, and thus E_2^* is a saddle.

For the equilibrium point E_3^*, the eigenvalues are given by $\lambda_1 = 1 + rK_H\left(\frac{ms}{(1+sK_H)^2 - 1}\right)$ and $\lambda_2 = aK_H$. We will show that $|\lambda_1| < 1$. Note that $\lambda_1 < 1$ if $\frac{ms}{(1+sK_H)^2} < 1$. Observe that $\frac{ms}{(1+sK_H)^2} = \frac{s(1-K_H)}{1+sK_H} < \frac{s\left(1-\left(\frac{s-1}{2s}\right)\right)}{1+s\left(\frac{s-1}{2s}\right)} = 1$. Similarly, $\lambda_1 > -1$ if $rK_H\left(1 - \frac{ms}{(1+sK_H)^2}\right) < 2$. This is equivalent to $rK_H\left(1 - \frac{s(1-K_H)}{(+sK_H)}\right) < 2$ or $rsK_H^2 - K_H(r(s-1) + 2s) - 2 < 0$. This implies that $\left(\frac{s-1}{2} + \frac{s}{r}\right) - \frac{\sqrt{\left(\frac{s-1}{2} + \frac{s}{r}\right)^2 + 16rs}}{2s} < K_H < \frac{s-1}{2} + \frac{s}{r} + \frac{\sqrt{\left(\frac{s-1}{2} + \frac{s}{r}\right)^2 + 16rs}}{2s}$ which is clearly true. Thus, $|\lambda_1| < 1$. As for the value of λ_2, we have two cases. In Case 1, $\frac{1}{a} < K_H$ or $aK_H > 1$ (Figure 5.2a,c), and thus

$\lambda_1 > 1$ and E_3^* is a saddle, while in Case 2, $\frac{1}{a} > K_H$ or $aK_H < 1$ (Figure 5.2(b),(d)) and E_3^* is asymptotically stable.

Next, we turn our attention to the interior equilibria, their existence and stability. By substituting (5.36) into (5.37), the isocline equations may be written in the forms

$$P = \frac{r}{a}\left(1 - H - \frac{m}{1+sH}\right) \qquad (5.39)$$

$$H = \frac{P}{e^{aP} - 1}. \qquad (5.40)$$

Since formula (5.40) is transcendental, it is not possible to find exactly the interior equilibria. However, one may find conditions under which we have no interior equilibria, one interior equilibrium, and two interior equilibria. Figure 5.15 shows these three scenarios.

The following result summarizes what is exhibited in Figure 5.15.

Lemma 8 *Suppose that Assumption (I) holds true. Then the following statements hold.*

(1) *There are no interior equilibria if either $\frac{1}{a} < A_H$ (Figures 5.15a) or $\frac{1}{a} > K_H$ (Figure 5.15b).*

(2) *There is one interior equilibrium point if either $A_H < \frac{1}{a} < K_H$ (Figure 5.15c) or when the isocline (5.39) is tangent to the isocline (5.40).*

(3) *There are two equilibria if $f_1(H_c) > f_2^{-1}(H_c)$ and $K_H < \frac{1}{2}$, where $f_1(H) = \frac{r}{a}\left(1 - H - \frac{m}{1+sH}\right)$, $f_2(P) = \frac{P}{(e^{aP}-1)}$.*

Proof. The proof of (i) and (ii) can be easily obtained by easily concluded from Figure 5.15. Notice that $f_1'(H) = \frac{r}{a}\left(-1 + \frac{sm}{(1+sH)^2}\right)$. Hence, the critical number of f_1 is $H_c = \frac{1}{s}(\sqrt{ms} - 1)$. By Assumption (A) $H_c > 0$. Now $f_1''(H) = \frac{rms^2}{a(1+sH)^2} > 0$. Hence, f_1 is a convex function which is increasing on the interval $[0, H_c]$ and decreasing on the interval $[H_c, \infty)$. Consequently, if $f_1(H_c) > f_2^{-1}(H_c)$, then the curve of f_2 must intersect the curve of f_1 at a point (H^*, P^*), with $H^* < H_c$. Furthermore, the assumption that $\frac{1}{a} > K_H$ ensures the existence of the second interior point. ∎

Exercises 5.7

In problems 1–12

(a) Find all the equilibrium points and describe the conditions under which they exist and are nonnegative.

(b) Use the Linearization Principle to determine the stability of the equilibrium points.

(c) Draw the phase space diagrams for the different scenarios.

1.
$$H(t+1) = H(t)\exp\left(r - H(t) - \frac{m}{1+sH(t)} - aP(t)\right)$$
$$P(t+1) = bH(t)\left(1 - e^{-aP(t)}\right)$$

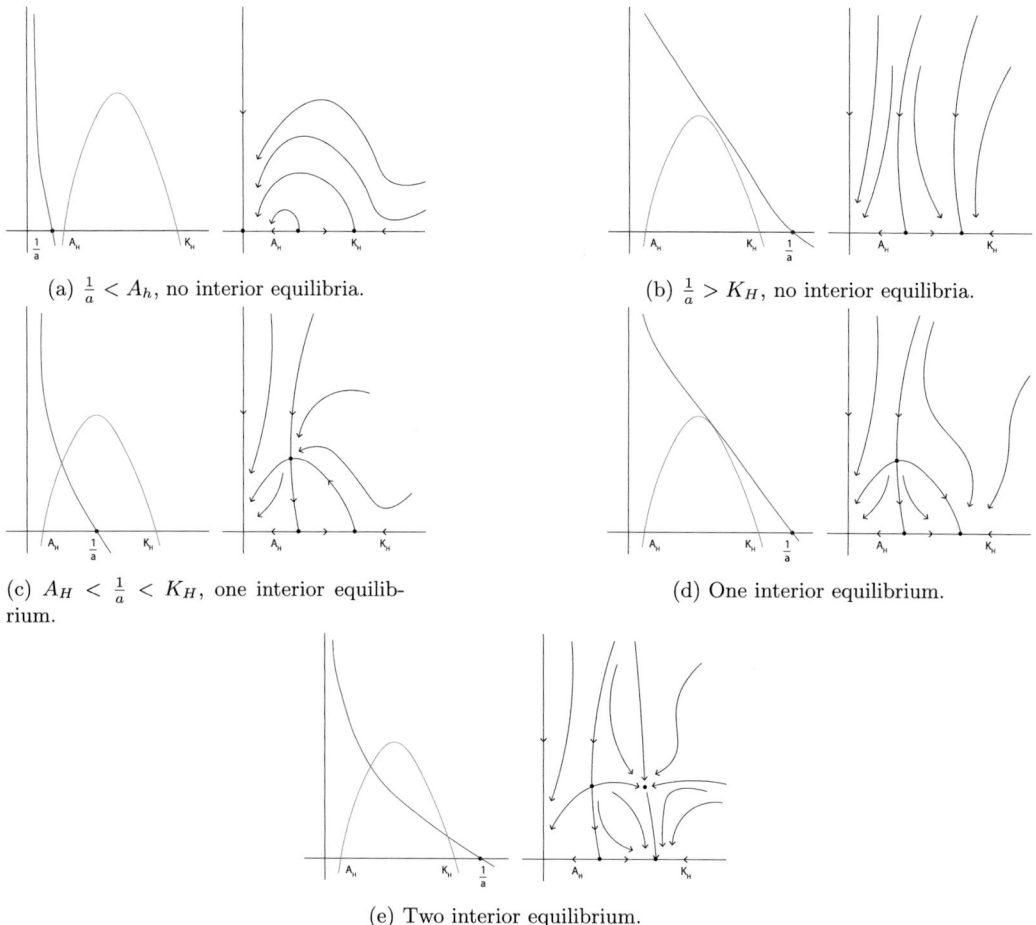

(a) $\frac{1}{a} < A_h$, no interior equilibria.

(b) $\frac{1}{a} > K_H$, no interior equilibria.

(c) $A_H < \frac{1}{a} < K_H$, one interior equilibrium.

(d) One interior equilibrium.

(e) Two interior equilibrium.

Figure 5.15: The five possible scenarios of the existence and stability of the interior equilibria. The interior equilibria are the intersection points of the two isoclines. Associated are the phase space diagrams depicting the global dynamics of the system.

2. In this problem, the predator (parasite) possesses an Allee effect caused by mate limitation.
$$H(t+1) = H(t)\exp\left(r\left(1-\frac{H(t)}{K}\right) - aP(t)\right)$$
$$P(t+1) = bH(t)\left(1-e^{-aP(t)}\right)\frac{sP(t)}{1+sP(t)}$$

3. In this problem, it is assumed that the host (prey) suffer from an Allee effects caused by mate limitations.
$$H(t+1) = H(t)\exp\left(r\left(1-\frac{H(t)}{K}\right) - aP(t)\right)\frac{sH}{1+sH}$$
$$P(t+1) = bH(t)\left(1-e^{-aP(t)}\right)$$

4. In this problem, it is assumed that both the host and the parasite suffer from Allee effects caused by mate limitations.
$$H(t+1) = H(t)\exp\left(r\left(1-\frac{H(t)}{K}\right) - aP(t)\right)\frac{s_1 H}{1+s_1 H}$$
$$P(t+1) = bH(t)\left(1-e^{-aP(t)}\right)\frac{s_2 P(t)}{1+s_2 P(t)}$$

5. Consider the Nicholson-Bailey model where the host population suffers from an Allee effect caused by predator saturation.
$$H(t+1) = H(t)\exp\left(-aP(t) - \frac{m}{1+sH(t)}\right)$$
$$P(t+1) = bH(t)\left(1-e^{-aP(t)}\right)$$

6. Consider the Nicholson-Bailey model where the parasite suffers from an Allee effect caused by mate limitation.
$$H(t+1) = H(t)\exp\left(-aP(t) - \frac{m}{1+sH(t)}\right)$$
$$P(t+1) = bH(t)\left(1-e^{-aP(t)}\right)\left(\frac{sP(t)}{1+sP(t)}\right)$$

7. Consider the following host-parasitoid model [232], where the host suffers from an Allee effect caused by mate limitation.
$$H(t+1) = H(t)\left(1+a\frac{P(t)}{k}\right)^{-k}\frac{sH}{1+sH}$$
$$P(t+1) = cH(t)\left[1-\left(1+a\frac{P(t)}{k}\right)^{-k}\right]$$

where $1 < k < \infty$.

8. Consider the model in Problem 7, where the parasite suffers from an Allee effect due to mate limitation.

9. Consider the model in Problem 7, where both host and parasite suffer from an Allee effect caused by mate limitation.

10. Consider the following host-parasite model where the host suffers from an Allee effect due to mate limitation.

$$H(t+1) = \frac{rH(t)}{1+aP(t)} \left(\frac{sH(t)}{1+sH(t)} \right)$$

$$P(t+1) = b \left(1 - \frac{1}{1+aP(t)} \right) H(t)$$

where $\frac{1}{1+aP(t)}$ is the fraction of the host population that survive from predation.

11. Consider the following revised Nicholson-Bailey host-parasitoid model

$$H(t+1) = rH(t) \exp\left(-aP(t) - \frac{m}{1+sH(t)} \right) + bH(t)$$

$$P(t+1) = rH(t) \left(1 - e^{-aP(t)} \right)$$

where the parasitoid suffers from a strong Allee effect caused by predator saturation and b is the fraction of hosts that survives to the next generation.

12. [292] The cinnabar moth *Tyria jacobacae* lives for one year on the perennial plant ragwort. At the end of the year, it lays eggs that hatch the following spring and then dies. Let $x(t)$ denote the total biomass of ragwort and $y(t)$ the number of insect eggs at the start of the year. We assume that the insect eggs suffer from the Allee effect due to mate limitation. This gives the model

$$x(t+1) = b \exp\left(\frac{-cy(t)}{x(t)} \right)$$

$$y(t+1) = ax(t) \left(\frac{sy(t)}{1+sy(t)} \right).$$

5.8 Global Dynamics of Population Models with the Allee Effect

In the previous sections, we focused on the local dynamics of models with the Allee effects. We needed numerical simulations to show the basins of attraction of equilibria using phase space diagrams. In this section, we are going to go one step further. We will show that the local dynamics may be extended to global dynamics for a two species hierarchical model with the Allee effects.

5.8. GLOBAL DYNAMICS OF POPULATION MODELS WITH THE ALLEE EFFECT

Consider the following two species where both species x and y possess the strong Allee effect [15].

$$\begin{aligned} x(t+1) &= x(t)\exp\left(\alpha - x(t) - \frac{m_1}{1+s_1 x(t)}\right) \\ y(t+1) &= y(t)\exp\left(\beta - y(t) - bx(t) - \frac{m_2}{1+s_2 y(t)}\right). \end{aligned} \quad (5.41)$$

This model may be written in a vector form $\mathbf{x}(t+1) = F(\mathbf{x}(t))$, where $F(x,y) = (f(x), g(x,y))$. Hence,

$$\begin{aligned} f(x) &= x\exp\left(\alpha - x - \frac{m_1}{1+s_1 x}\right) \\ g(x,y) &= y\exp\left(\beta - y - bx - \frac{m_2}{1+s_2 y}\right). \end{aligned} \quad (5.42)$$

This model is a modification of Example 64 in Chapter 3 in which the two species x and y do not possess the Allee effect. There are five boundary equilibrium points: $E^* = (0,0)^T$, $A_{1x} = (A_1, 0)^T$, $K_{1x} = (K_1, 0)^T$, $A_{1y} = (0, A_2)^T$, $K_{1y} = (0, K_2)^T$, where A_1 and A_2 are the single-species Allee threshold of species x and y, respectively, and K_1 and K_2 are their carry capacities, respectively. One may show that

$$A_1 = \frac{(\alpha s_1 - 1) - \sqrt{(\alpha s_1 - 1)^2 - 4s_1(m_1 - \alpha)}}{2s_1}$$

$$K_1 = \frac{(\alpha s_1 - 1) + \sqrt{(\alpha s_1 - 1)^2 - 4s_1(m_1 - \alpha)}}{2s_1}.$$

Furthermore,

$$A_2 = \frac{(\beta s_1 - 1) - \sqrt{(\beta s_2 - 1)^2 - 4s_2(m_2 - \beta)}}{2s_2}$$

$$K_1 = \frac{(\beta s_1 - 1) + \sqrt{(\beta s_2 - 1)^2 - 4s_2(m_2 - \beta)}}{2s_2}.$$

To ensure that both species x and y suffer from the strong Allee effect and consequently, A_1, A_2, K_1, K_2 are all positive, we make the following assumption.

(H_1) (a) $\alpha s_1 > 1$, $(\alpha s_1 + 1)^2 > 4 s_1 m_1$, $\alpha < m_1$

(b) $\beta s_2 > 1$, $(\beta s_2 + 1)^2 > 4 s_2 m_2$, $\alpha < m_1$.

To investigate the stability of equilibria, we compute the Jacobian

$$JF\begin{pmatrix}x\\y\end{pmatrix} = \begin{pmatrix} \left[1 - x\left(1 - \frac{m_1 s_1}{(1+s_1 x)^2}\right)\right]e^{\alpha - x - \frac{m_1}{1+s_1 x}} & 0 \\ -by e^{\beta - y - bx - \frac{m_2}{1+s_2 y}} & \left[1 - y\left(1 - \frac{m_2 s_2}{(1+s_2 y)^2}\right)\right]e^{\beta - y - \frac{m_2}{1+s_2 y}} \end{pmatrix}.$$

For $E^* = (0,0)^T$,

$$JF\begin{pmatrix}0\\0\end{pmatrix} = \begin{pmatrix} e^{\alpha - m_1} & 0 \\ 0 & e^{\beta - m_2} \end{pmatrix}.$$

From Assumption (H_1), $0 < \lambda_1, \lambda_2 < 1$ and thus E^* is locally asymptotically stable.

Next, we will investigate the stability of the other boundary equilibria.

$$JF\begin{pmatrix}A_1\\0\end{pmatrix} = \begin{pmatrix}1+A_1\left(\frac{m_1s_1}{(1+s_1A_1)^2}-1\right) & 0 \\ 0 & e^{\beta-m_2-bA_1}\end{pmatrix}.$$

Now $\lambda_2 = e^{\beta-m_2-bA_1} < 1$ (from Assumption H_b), $\lambda_1 > 1$ if $(1+s_1A_1)^2 < m_1s_1$. In this case, A_{1x} is a saddle.

Similarly, one may show that the equilibrium point K_{1x} is locally asymptotically stable if

$$H_{2a}: K_1\left(\frac{m_1s_1}{(1+s_1K_1)^2}-1\right) > -2 \text{ and } (1+s_1K_1)^2 > m_1s_1.$$

Similarly, one may show that A_{1y} is a saddle and K_{1y} is locally asymptotically stable if

$$H_{2b}: K_2\left(\frac{m_2s_2}{(1+s_2K_2)^2}-1\right) > -2 \text{ and } (1+s_2K_2)^2 > m_2s_2$$

(Exercise 1).

Next, we investigate the existence and the stability of the interior equilibrium points. Note that the interior equilibria are the intersection points of the isoclines $x = A_1$, $x = K_1$ and the curve

$$x = h(y) = \frac{1}{b}\left(\beta - y - \frac{m_2}{1+s_2y}\right). \tag{5.43}$$

Note that the existence of interior equilibria is determined by the critical point of the function $h(y)$. To find this critical point, we let

$$\frac{dx}{dy} = \frac{1}{b}\left(\frac{m_2s_2}{(1+s_2y)^2}-1\right) = 0.$$

The maximum value $x_{\max}$ of x is given by

$$x_{\max} = \frac{1}{b}\left(\beta - \frac{2\sqrt{m_2s_2}-1}{s_2}\right).$$

The next Lemma summarizes the above remarks, which may be seen in Figure 5.16.

Lemma 9 *The following statements hold true.*

(i) If $x_{\max} < A_1$, there are no interior equilibria.

(ii) If $x_{\max} = A_1$, then we have one interior equilibrium point.

(iii) If $A_1 < x_{\max} < K_1$, then we have two equilibria.

(iv) If $x_{\max} = K_1$, then we have three equilibria.

(v) If $x_{\max} > K_1$, then we have four interior equilibria.

5.8. GLOBAL DYNAMICS OF POPULATION MODELS WITH THE ALLEE EFFECT

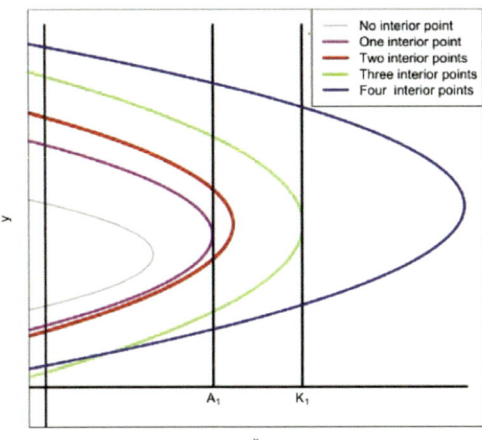

Figure 5.16: The five scenarios of the Allee effect. The graph depicts the various curves of the second isocline (parabola) The interior fixed points are the intersection of the second isocline with the isoclines $x = A_1$ and $x = K_1$. The five scenarios are (i) no interior fixed points, (ii) one interior fixed point, (iii) two interior fixed points, (iv) three interior fixed points, (v) four interior fixed points.

On the fiber $x = A_1$, we have

$$y \exp\left(\beta - y - bA_1 - m_2 \frac{1}{(1+s_2 y)}\right) < y$$

if and only if

$$s_2 y^2 - (\beta s_2 - 1 - b s_2 A_1) y + m_2 + b A_1 - \beta > 0 \tag{5.44}$$

and $y e^{\beta - y - b A_1 - m_2/(1+s_2 y)}$ if and only if

$$s_2 y^2 - (\beta s_2 - 1 - b s_2 A_1) y + m_2 + b A_1 - \beta < 0. \tag{5.45}$$

Now

$$s_2 y^2 - (\beta s_2 - 1 - b s_2 A_1) y + m_2 + b A_1 - \beta = 0. \tag{5.46}$$

If there are no interior points on the fiber $x = A_1$, then Equation (5.46) has no solutions. This implies that inequality (5.44) holds true and thus the orbit of any point (A_1, y) is monotonically decreasing. If there is only one interior fixed point, which will be denoted by (A_1, y_{0A_1}), on the fiber $x = A_1$, then Equation (5.46) has one repeated solution. Hence, every orbit on the fiber $x = A_1$, with the exception of the orbit of $(A_1, y_{0A_1})^T$ decreases monotonically. Hence, $(A_1, y_{0A_1})^T$ is semi-stable (i.e., repelling on one side and attracting on the other side) and its center manifold lies on the fiber $x = A_1$. Finally, if there are two interior fixed points, say $(A_1, y_{1A_1})^T$, $(A_1, y_{2A_2})^T$, then Equation (5.46) has two solutions. Hence, the orbits of points $(A_1, y)^T$ are increasing along the fiber $x = A_1$ if $y_{1A_1} < y < y_{2a_1}$, and decreasing if $y_{1A_1} > y$ or $y > y_{2A_1}$.

On the fiber $x = K_1$, we will have the following counterpart of Equation (5.46):

$$s_2 y^2 - (\beta s_2 - 1 - b s_2 K_1) y + m_2 + b K_1 - \beta = 0. \tag{5.47}$$

In this case, we have the same analysis as in the case of Equation (5.46), and thus the discussion will be omitted.

We will now investigate the stability of all the fixed points of the system in the following five scenarios, see Figure 5.16. Scenarios 2 and 4 will be discussed for completeness in Appendix 5.9 since they correspond to nonhyperbolic cases and are difficult to observe in nature.

To facilitate our analysis, we begin our exposition by the following general result.

Lemma 10 *Consider the continuous map $F : \mathbb{R}_+^N \to \mathbb{R}_+^N$. Let x^* be an asymptotically stable fixed point of the map F such that $\mathbf{x}^* \in \omega(\mathbf{x})$ for some point $\mathbf{x} \in \mathbb{R}_+^N$. Then, the orbit of x converges to x^*.*

Proof. Assume that $\mathbf{x}^* \in \omega(\mathbf{x})$. Since x^* is asymptotically stable, it follows that the immediate basin of attraction of x^* is an open ball $\mathcal{B}(x^*, \delta)$ around x^* [119]. Hence, $\mathcal{B}(x^*, \delta)$ must contain a point in the orbit of x. This implies that the orbit of x converges to x^*. This completes the proof of the lemma. ∎

We now resume our analysis of the various dynamical scenarios.

Scenario 1. There are no interior equilibrium points. This occurs when Equation (5.43) has no solutions or, equivalently, if $x_{\max} < A_1$. In this case, on the fibers $x = A_1$ and $x = K_1$ the function g Equation (5.42) is monotonically decreasing.

Let $(x_0, y_0) \in \mathbb{R}_+^2$. If $A_1 < x_0 < f^{-1}(A_1)$, then $\omega(x_0, y_0)$ lies on the fiber $x = K_1$. Since K_1 is asymptotically stable, it follows that $\lim f^n(x_0) = K_1$. Moreover, since the orbits are bounded, the omega limit set $\omega(x_0, y_0)$ is nonempty, closed, and invariant [110, 119]. Hence, there is a subsequence $F^{n_k}(x_0, y_0)$ of the orbit (x_0, y_0) such that $\lim_{n \to \infty} F^{n_k}(x_0, y_0) = (K_1, \hat{y})$ for some $\hat{y} \geq 0$. Note that $\lim_{n \to \infty} F^n(K_1, \hat{y}) = (K_1, 0)$. Thus, $(K_1, 0) \in \omega(x_0, y_0)$. Since $\omega(x_0, y_0)$ contains the asymptotically stable fixed point $(K_1, 0)$, it follows that $\omega(x_0, y_0) = \{(K - 1, 0)\}$. Hence, $\lim_{n \to \infty} F^n(x_0, y_0) = (K_1, 0)$. On the other hand if $0 < x_0 < A_1$ or $x_0 > f^{-1}(A_1)$, then $\omega(x_0, y_0)$ lies on the fiber $x = 0$. Let $(0, \hat{y}) \in \omega(x_0, y_0)$ and let $g_0(y) = g(0, y)$. If $A_2 < \hat{y} < g_0^{-1}(A_2)$, then $\lim_{n \to \infty} F^n(x_0, y_0) = (K_2, 0)$. Moreover, if $\hat{y} < A_2$ or $\hat{y} > g_0^{-1}(A_2)$, then $\lim_{n \to \infty} F^n(x_0, y_0) = (0, 0)$. Explicitly, if (x_0, y_0) or $F(x_0, y_0)$ lies between the global stable manifold $W^s(0, A_2)$ of $(0, A_2)$ and $F^{-1}(W^s(0, A_2))$, then (x_0, y_0) belongs to an exclusion region of x (orange), that is, $\lim_{n \to \infty} F^n(x_0, y_0) = (0, K_2)$. However, if (x_0, y_0) or $F(x_0, y_0)$ lies above $F^{-1}(W^s(0, A_2))$ or below $W^s(0, A_2)$, then (x_0, y_0) belongs to the extinction region ($\lim_{n \to \infty} F^n(x_0, y_0) = (0, 0)$) (yellow). Figure 5.17 shows an unbounded extinction region (yellow), where both species go to extinction, an exclusion region of y (magenta), and an exclusion region of x (orange).

Scenario 3. In scenario 3, there are two interior fixed points. This occurs when Equation (5.43) has two solutions or, equivalently, $A_1 < x_{\max} < K_1$. These fixed points lie on the fiber $x = A_1$, and will be denoted by (A_1, y_{1A_1}) and (A_1, y_{2A_1}). Now on the fiber $x = A_1$, the function g in Equation (5.42) is monotonically decreasing below the fixed point (A_1, y_{1A_1}) and above the fixed point (A_1, y_{2A_1}), and monotonically increasing between the fixed points (A_1, y_{1A_1}) and (A_1, y_{2A_1}). Moreover, the function g is monotonically decreasing on the fiber $x = K_1$. Hence, the fixed point (A_1, y_{1A_1}) is a repeller, and (A_1, y_{2A_1}) is a saddle. In this case, if $A_1 < x_0 < f^{-1}(A_1)$, then $\lim_{n \to \infty} f^n(x_0) = K_1$ and if $0 < x_0 < A_1$, or $x_0 > f^{-1}(A_1)$, then $\lim_{n \to \infty} f^n(x_0) = 0$. Thus, for $A_1 < x_0 < f^{-1}(A)$, the orbit of (x_0, y_0) converges to $(K_1 > 0)$. For $0 < x_0 < A_1$ or $x_0 > f^{-1}(A)$, the orbit of (x_0, y_0) either converges to $(0, 0)$ if (x_0, y_0) or $F(x_0, y_0)$ lies below the global stable manifold $W^s(0, A_2)$ of $(0, A_2)$, or above $F^{-1}(W^s(0, A_2))$ and converges to $(0, K_2)$ otherwise. Figure 5.18 depicts the regions of extinction (yellow), an exclusion region of x (orange), and an exclusion region of y (magenta).

5.8. GLOBAL DYNAMICS OF POPULATION MODELS WITH THE ALLEE EFFECT

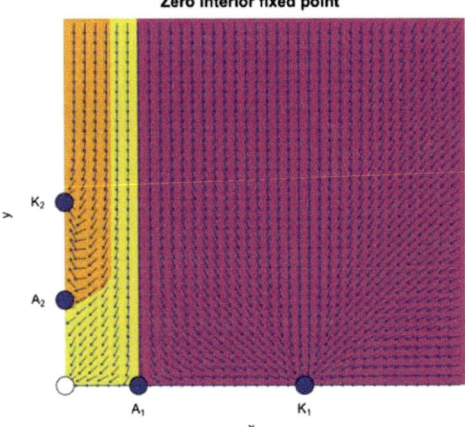

Figure 5.17: The phase space diagram in the case of no interior fixed points. The domain $\mathbb{R}_+^N$ is divided into three regions, an extinction region (yellow), an exclusion region of y (magenta), and an exclusion regions of x (orange). Note that the extinction region is unbounded.

Scenario 5. In scenario 5, there are four interior fixed points. This occurs when $x_{\max} > K_1$. Two fixed points lie on the fiber $x = A_1$, and will be denoted by (A_1, y_{1A_1}) and (A_1, y_{2A_1}) and two fixed points lie on the fiber $x = K_1$ and will be denoted by (K_1, y_{1K_1}) and (K_1, y_{2K_1}). Now on the fiber $x = A_1$, the function g in Equation (5.46) is monotonically decreasing below the fixed point (A_1, y_{1A_1}) and above the fixed point (A_1, y_{2A_1}), and monotonically increasing between the fixed points $(A_1, y_{1A_1})(A_1, y_{2A_1})$, and similarly on the fiber $x = K_1$. Hence, the fixed point (A_1, y_{1A_1}) is a repeller, (A_1, y_{2A_1}) is a saddle, (K_1, y_{1K_1}) is a saddle, and (K_1, y_{2K_1}) is asymptotically stable. In this case, for $A_1 < x_0 < f^{-1}(A_1)$, the orbit of (x_0, y_0) converges to $(K_1, 0)$ if (x_0, y_0) lies below the global stable manifold of (K_1, y_{1K_1}) and converges to (K_1, y_{2K_1}) if (x_0, y_0) lies above the global stable manifold of (K_1, y_{1K_1}). For $0 < x_0 < A_1$ or $x_0 > f^{-1}(A_1)$, the orbit of (x_0, y_0) either converges to $(0, 0)$ if (x_0, y_0) or $F(x_0, y_0)$ lies below the global stable manifold $W^s(0, A_2)$ of $(0, A_2)$, or above $F^{-1}(W^s(0, A_2))$ and converges to $(0, K_2)$ otherwise. Figure 5.19 depicts the regions of extinction (yellow), an exclusion region of x (orange), an exclusion region of y (magenta), and a coexistence region (green).

We now summarize the above analysis. Let

$$x_{\max} = \frac{1}{c}\left[\beta - \frac{2\sqrt{m_2}-1}{s_2}\right].$$

Theorem 50 *Under Assumptions (H_1) and (H_2), the following statements hold true.*

(i) *If $x_{\max} < A_1$, then there are no interior fixed points. Moreover, the positive quadrant in the phase space diagram is divided into three regions. The extinction region consists of three areas,*

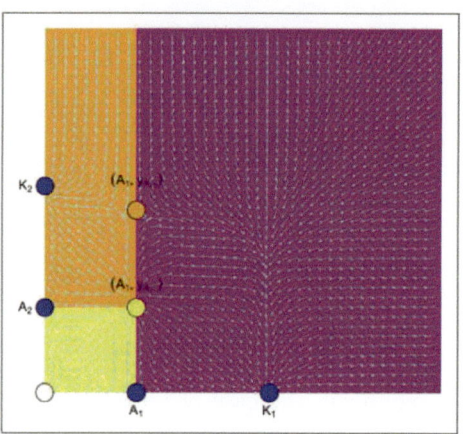

Figure 5.18: The phase space diagram in the case of two interior fixed points. The domain $\mathbb{R}_+^N$ is divided into three regions, an extinction region (yellow), an exclusion region of y (magenta), and an exclusion region of x (orange). Note that the extinction region is bounded.

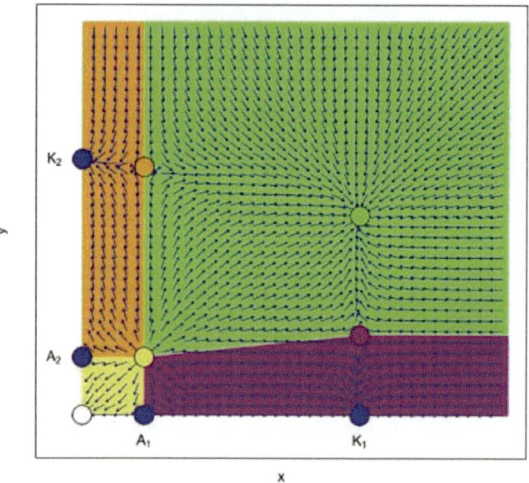

Figure 5.19: The phase space diagram in the case of four interior fixed points. The domain $\mathbb{R}_+^N$ is divided into four regions, an extinction region (yellow), an exclusion region of y (magenta), an exclusion region of x (orange), and a coexistence region (green).

the first is bounded by the fiber $x = A_1$ and the global stable manifold $W^s(0, A_2)$ of $(0, A_2)$, the second is bounded by the y-axis and $F^{-1}(W^s(0, A_2))$, and the third is part of the region $x > f^{-1}(A_1)$ for which $F(x_0, y_0)$ lies in the first two areas. An exclusion region of species x is bounded by the y-axis and the global stable manifold of $(0, A_2)$. An exclusion region of species y is bounded by the fibers $x = A_1$ and $x = f^{-1}(A_1)$ (Figure 5.17).

(ii) If $x_{\max} = A_1$, then we have one interior fixed point (A_1, y_{0A_1}), which is semi-stable from above. This is similar to Case 1 with the exception that the stable manifold of $(0, A_2)$ connects to (A_1, y_{0A_1}). Hence, the extinction region is a bounded set (Figure 5.20).

(iii) If $A_1 < x_{\max} < K_1$, then we have two interior fixed points, (A_1, y_{1A_1}) and (A_1, y_{2A_1}). (A_1, y_{1A_1}) is a repeller (unstable), and (A_1, y_{2A_1}) is a saddle with the stable manifold lying on the fiber $x = A_1$. Here, we have three regions as described in Case 2 (Figure 5.18).

(iv) If $x_{\max} = K_1$, then we have three interior fixed points, $(A_1, y_{1A_1})^T$, $(A_1, y_{2A_1})^T$, and $(K_1, y_{0K_1})^T$. The dynamics of $(A_1, y_{1A_1})^T$, $(A_1, y_{2A_1})^T$ are the same, while $(K_1, y_{0K_1})^T$ is semi-stable from above with the center manifold lying on the fiber $x = K_1$ and the stable manifold connects with $(K_2, 0)$. Here, we have a coexistence region bounded by $x = A_1$ and the stable manifold of (K_1, y_{0K_1}) (Figure 5.21).

(v) If $x_{\max} > K_1$, then we have four interior fixed points, $(A_1, y_{1A_1})^T$, $(A_1, y_{2A_1})^T$, $(K_1, y_{1K_1})^T$, $(K_1, y_{2K_1})^T$. Now $(K_1, y_{1K_1})^T$ is a repeller (unstable), and (K_1, y_{2K_1}) is asymptotically stable. A coexistence region is now bounded by $x = A_1$ and the stable manifold of (K_1, y_{1K_1}) (Figure 5.19).

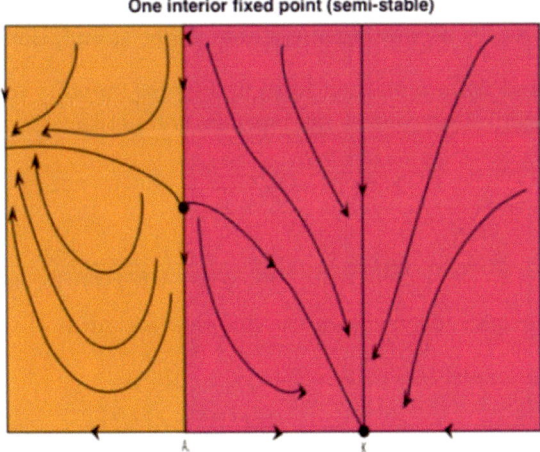

Figure 5.20: The phase space diagram in the case of one interior fixed point. The domain $\mathbb{R}_+^2$ is divided into three regions, an extinction region (yellow), an exclusion region of y (magenta), and an exclusion region of x (orange). Note that the extinction region is bounded.

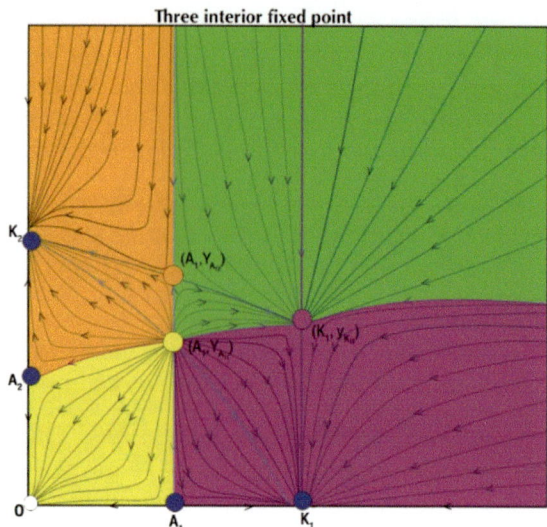

Figure 5.21: The phase space diagram in the case of three interior fixed points. The domain $\mathbb{R}_+^2$ is divided into four regions: an extinction region (yellow), an exclusion region of y (magenta), an exclusion region of x (orange), and a coexistence region (green). Note that this is the first instance when we have a coexistence region.

Exercises 5.8

1. Consider system (5.41). Use the Linearization Principle to show that the equilibrium point A_{1y} is a saddle and K_{1y} is asymptotically stable under assumptions H_1 and H_2.

2. Find the equilibria and determine their local and global stability of the following hierarchical competition model exhibiting contest interspecific competition

$$x(t+1) = \frac{r_1 x^2(t)}{1 + x^2(t)}$$

$$y(t+1) = \frac{r_2 y^2(t)}{1 + y^2(t) + b_2 x(t)}.$$

Then draw the phase space diagrams of the different scenarios of the system.

3. A modification of the model in Problem 2 is

$$x(t+1) = \frac{a_1 x^2(t) + r_1 x(t)}{1 + x^2(t)}$$

$$y(t+1) = \frac{a_2 y^2(t) + r_2 y(t)}{1 + y^2(t) + b_2 y(t)}.$$

IFind the equilibria and determine their local and global stability of the system and draw the phase space diagrams of various scenarios.

5.8. GLOBAL DYNAMICS OF POPULATION MODELS WITH THE ALLEE EFFECT

4.* Consider the following hierarchical competition model with the immigration and Allee effects.
$$x(t+1) = x(t)\exp\left(\alpha - x(t) - \frac{m_1}{1+s_1 x(t)}\right) - h_1 x(t)$$
$$y(t+1) = y(t)\exp\left(\beta - y(t) - bx(t) - \frac{m_2}{1+s_2 y(t)}\right) - h_1 y(t).$$

Find the equilibria and determine their local and global stability of the system and draw the corresponding phase space diagrams.

5.* Consider the following hierarchical competition model with the Allee and constant immigration
$$x(t+1) = x(t)\exp\left(\alpha - x(t) - \frac{m_1 s_1}{1+s_1 x(t)}\right) - h_1$$
$$y(t+1) = y(t)\exp\left(\beta - y(t) - bx(t) - \frac{m_2 s_2}{1+s_2 y(t)}\right) - h_1.$$

Find the equilibria and determine their local and global stability of the system and draw the corresponding phase space diagram.

6. Find the equilibria and determine their local and global stability of the hierarchical, quadratic competition model with the Allee effect caused by mate limitation. Assume $0 < a_1 < 4$ so that the unit square is mapped into itself.
$$x(t+1) = a_1 x(t)(1-x(t))\frac{s_1 x(t)}{1+s_1 x(t)}$$
$$y(t+1) = a_2 y(t)(1-x(t))\frac{s_2 y(t)}{1+s_2 y(t)}$$

7. Find the equilibria and determine their local and global stability of the following hierarchical model where the dormant species has an Allee effect caused by predator saturation and the other species suffers from an Allee effect caused by mate limitation.
$$x(t+1) = x(t)\exp\left(\alpha - x(t) - \frac{m}{1+s_1 x(t)}\right)$$
$$y(t+1) = \frac{ry(t)}{1+y(t)+bx(t)} \cdot \left(\frac{s_2 y(t)}{1+s_2 y(t)}\right).$$

8. (Research Project) Consider the following 2-species competition model with the Allee effect
$$x(t+1) = \frac{a_1 x^2(t)}{1+x^2(t)+b_1 y(t)}$$
$$y(t+1) = \frac{a_2 y^2(t)}{1+y^2(t)+b_2 x(t)}.$$

(a) Determine conditions under which the system is monotone.
(b) Investigate the global dynamics of the systems.

9. (Research Project) Consider the following 3-species hierarchical competition model with the Allee effect

$$x(t+1) = x(t)\exp\left(\alpha - x(t) - \frac{m_1}{1+s_1 x(t)}\right)$$

$$y(t+1) = y(t)\exp\left(\beta - y(t) - bx(t) - \frac{m_2}{1+s_2 y(t)}\right)$$

$$z(t+1) = z(t)\exp\left(\gamma - z(t) - cx(t) - dy(t) - \frac{m_3}{1+s_3 z(t)}\right).$$

5.9 Appendix

Hierarchical models Scenarios 2 and 4.

In this Appendix, we investigate scenarios 2 and 4 of Equation (5.41).

Scenario 2. In scenario 2, there is only one interior fixed point. This occurs when Equation (5.43) has one solution or, equivalently, $x_{\max} = A_1$. This point lies on the fiber $x = A_1$, and will be denoted by (A_1, y_{0A_1}). Now on the fiber $x = A_1$, the function g in Equation (5.42) is monotonically decreasing above and below the fixed point (A_1, y_{0A_1}). Moreover, the function g is monotonically decreasing on the fiber $x = K_1$. Hence, the fixed point (A_1, y_{0A_1}) has an unstable manifold and is semi-stable (from above), that is, its center manifold, which lies on the fiber $x = A_1$, is semi-stable. In this case, if $A_1 < x_0 < f^{-1}(A_1)$, then $\lim_{n\to\infty} f^n(x_0) = K_1$, and if $0 < x_0 < A_1$, or $x_0 > f^{-1}(A_1)$, then $\lim_{n\to\infty} f^n(x_0) = 0$. Thus, for $A_1 < x_0 < f^{-1}(A)$, the orbit of (x_0, y_0) converges to $(K_1, 0)$. For $0 < x_0 < A_1$ or $x_0 > f^{-1}(A)$, the orbit of (x_0, y_0) either converges to $(0,0)$ if (x_0, y_0) or $F(x_0, y_0)$ lies below the global stable manifold $W^s(0, A_2)$ of $(0, A_2)$, or above $F^{-1}(W^s(0, A_2)$ and converges to $(0, K_2)$ otherwise. Figure 5.20 depicts the regions of extinction (yellow), an exclusion region of x (orange), and an exclusion region of y (magenta).

Scenario 4. In scenario 4, there are three interior fixed points. This occurs when $x_{\max} = K_1$. Two fixed points lie on the fiber $x = A_1$, and will be denoted by (A_1, y_{1A_1}) and (A_1, y_{2A_1}) and one fixed point lies on the fiber $x = K_1$ and will be denoted by (K_1, y_{0K_1}). Now on the fiber $x = A_1$, the function g in Equation (5.42) is monotonically decreasing below the fixed point (A_1, y_{1A_1}) and above the fixed point (A_1, y_{2A_1}), and monotonically increasing between the fixed points (A_1, y_{1A_1}) (A_1, y_{2A_1}). Moreover, the function g is monotonically decreasing above and below the fixed point (K_1, y_{0K_1}), leading to its semi-stability. Hence, the fixed point (A_1, y_{1A_1}) is a repeller, (A_1, y_{2A_1}) is a saddle, and (K_1, y_{0K_1}) is semi-stable from above and has a stable manifold. In this case, if $A_1 < x_0 < f^{-1}(A_1)$, then $\lim_{n\to\infty} f^n(x_0) = K_1$, and if $0 < x_0 < A_1$, or $x_0 > f^{-1}(A)$, then $\lim_{n\to\infty} f^n(x_0) = 0$. Thus, for $A_1 < x_0 < f^{-1}(A_1)$, the orbit of (x_0, y_0) converges to $(K_1, 0)$ if (x_0, y_0) lies below the global stable manifold of $((K_1, y_{0K_1})$ and converges to (K_1, y_{0K_1}) if (x_0, y_0) lies above the global stable manifold of (K_1, y_{0K_1}). For $0 < x_0 < A_1$ or $x_0 > f^{-1}(A_1)$, the orbit of (x_0, y_0) either converges to $(0, 0)$ if (x_0, y_0) or $F(x_0, y_0)$ lies below the global stable manifold $W^s(0, A_2)$ of $(0, A_2)$, or above $F^{-1}(W^s(0, A_2)$ and converges to $(0, K_2)$ otherwise. Figure 5.21 depicts the regions of extinction (yellow), an exclusion region of x (orange), an exclusion region of y

(magenta), and a coexistence region (green). Note that this is the first instance where both species coexist if their densities lie in the green region. However, in any real ecological system the inevitable stochastic fluctuations would eventually perturb the populations of the orange equilibrium and into the magenta region, thus resulting in practical purposes in the exclusion of y.

Chapter 6
Nonlinear Structured Population Models

6.1 Introduction

In Chapter 2, we studied systems of linear difference equations of the form

$$\mathbf{x}(t+1) = A\mathbf{x}(t) \tag{6.1}$$

as a model for structured populations. In this equation, $A = (a_{ij})$ is a $k \times k$ nonnegative matrix, called the population projection matrix, and $\mathbf{x}(t) = (x_i(t))^T$ is a population demographic vector consisting of class-specific densities at time t. We refer to equations of the form (6.1) as a linear matrix equations [45]. In this chapter, we generally discuss matrix equations by simply specifying the projection matrix A, rather than writing an equation of the matrix form (6.1) or as a system of difference equations.

The Fundamental Theorem of Demography (Theorem 16 in Chapter 2) is a basic result concerning the long-term dynamics predicted by the linear model (6.1) and the fundamental biological question of extinction versus survival. Under the assumption that the population projection matrix A is irreducible and primitive, the spectral $r = \rho(A)$ is the strictly dominant eigenvalue of A (and is called the population growth rate). If $r < 1$ then the population goes extinct (the extinction equilibrium $\mathbf{x} = \mathbf{0}$ is globally asymptotically stable) whereas if $r > 1$ then the extinction equilibrium is unstable and population grows exponentially without bound. When $r = 1$ there exists a continuum of equilibria $\mathbf{x} = c\mathbf{w}_R$ where $\mathbf{w}_R$ is a positive, right eigenvector of A and c is an arbitrary constant.[1] Of course no biological population can grow without bound and consequently, in a linear matrix model (6.1), long-term population sustainability can only occur if $r = 1$ (which is a non-robust or non-generic requirement).

The Fundamental Theorem of Demography (Theorem 16 in Chapter 2) can be viewed as a transcritical bifurcation at $r = 1$ where the (horizontal) continuum of extinction equilibria that exists for all $r > 0$ intersects the (vertical) continuum of non-extinction equilibrium with $r = 1$.

---

[1] In order to avoid some notation conflicts, in this and subsequent chapters, we use $\mathbf{w}_R$ and $\mathbf{w}_L^T$ for right and left eigenvectors in place of $\mathbf{v}$ and $\mathbf{w}^T$ used in previous chapters.

The intersection or bifurcation point $r=1$ is the threshold at which the extinction equilibrium destabilizes. With regard to feasible equilibria for a population model, we note that the bifurcating continuum contains a subcontinuum of positive equilibria for arbitrary $c>0$.

The population growth rate r does not, in general, appear explicitly in the matrix equation (6.1). Instead r is a function of the parameters that do appear in the equation for which no formula in general is available. To express the bifurcation at $r=1$ in terms of a parameter μ that explicitly appears in the equation, we treat

$$A = A(\mu)$$

and $r = r(\mu)$ as functions of μ and locate a value μ_0 at which

$$r(\mu_0) = 1 \text{ and } \frac{dr(\mu_0)}{d\mu} \neq 0. \tag{6.2}$$

Then $r(\mu)$ increases through 1 as μ increases or decreases through μ_0 if this derivative is positive or negatively, respectively. If $a_{ij} = a_{ij}(\mu)$ then using the chain rule we have

$$r'(\mu) := \frac{dr(\mu)}{d\mu} = \sum_{i,j=1}^{k} \frac{dr}{da_{ij}} \frac{da_{ij}(\mu)}{d\mu}.$$

By (2.28) in Chapter 2, the sensitivity of r with respect to a_{ij} is

$$\frac{\partial r}{\partial a_{ij}} = \frac{w_{Li} w_{Rj}}{\mathbf{w}_L^T \mathbf{w}_R}$$

where $\mathbf{w}_L^T$ and $\mathbf{w}_R$ are positive left and right eigenvectors of $A(\mu_0)$ associated with eigenvalue $r(\mu_0) = 1$ and $\mathbf{w}_R = (w_{Ri})$ and $\mathbf{w}_L = (w_{Li})$. The formula

$$r'(\mu_0) = \frac{1}{\mathbf{w}_L^T \mathbf{w}_R} \sum_{i,j=1}^{k} w_{Li} w_{Rj} \frac{\partial a_{ij}(\mu_0)}{\partial \mu} = \frac{1}{\mathbf{w}_L^T \mathbf{w}_R} \mathbf{w}_L^T \frac{\partial A(\mu_0)}{\partial \mu} \mathbf{w}_R \tag{6.3}$$

relates the derivative and condition (6.2) to the sensitivities of the matrix entries $a_{ij}(\mu)$ to the parameter μ at μ_0. Note that if all these sensitivities are positive (or negative) then clearly the derivative $r'(\mu_0)$ is positive (or negative).

As an aid in locating a bifurcation point μ_0 one can make use of the reproduction number $\mathcal{R}_0(\mu)$, for which there often is, as seen in Chapter 2, an explicit formula in terms of the matrix entries $a_{ij}(\mu)$. One can also use the formula $\mathcal{R}_0(\mu)$ to validate the condition (6.2) since it is known that the two derivatives

$$r'(\mu_0) \text{ and } \mathcal{R}_0'(\mu_0) = \frac{d\mathcal{R}_0(\mu_0)}{d\mu}$$

vanish together or have the same signs (as also do the second derivatives); see Theorem 56 or [73].

Example 88 *Consider the linear Leslie matrix model for an age structured population studied in Chapter 2. Suppose the age-specific fertilities f_i in the population projection matrix*

$$\mathcal{L} = \begin{pmatrix} f_1 & f_2 & \cdots & f_{k-1} & f_k \\ s_1 & 0 & \cdots & 0 & 0 \\ 0 & s_2 & \cdots & 0 & 0 \\ \vdots & \vdots & & \vdots & \vdots \\ 0 & 0 & \cdots & s_{k-1} & s_k \end{pmatrix} \tag{6.4}$$

6.1. INTRODUCTION

are dependent on the amount of environmental food resource consumed, which in this example we assume is proportional to the amount of resource available as quantified by a parameter $\mu > 0$. If b_i is the number of newborns per unit time per adult per unit resource, then under this assumption

$$f_i = b_i c_i \mu$$

where c_i is a constant of proportionality. The reproduction number is

$$\mathcal{R}_0(\mu) = \sum_{i=1}^{k-1} f_i p_i + f_k p_k \frac{1}{1-s_k} = \mu \left(\sum_{i=1}^{k-1} b_i c_i p_i + b_k c_k p_k \frac{1}{1-s_k} \right)$$

where

$$p_i := s_0 s_1 s_2 \ldots s_{i-1}, \quad 1 \leq i \leq k$$

and $s_0 = 1$ (see (2.16) in Chapter 2). Then $\mathcal{R}_0(\mu_0) = 1$ for

$$\mu_0 = \left(\sum_{i=1}^{k-1} b_i c_i p_i + b_k c_k p_k \frac{1}{1-s_k} \right)^{-1}$$

and

$$\frac{d\mathcal{R}_0(\mu_0)}{d\mu} = \sum_{i=1}^{k-1} b_i c_i p_i + b_k c_k p_k \frac{1}{1-s_k} > 0.$$

Thus, the extinction equilibrium destabilizes as μ increases through μ_0 and this linear model predicts population extinction if the food resource availability μ is less than the threshold μ_0.

In this chapter, we consider nonlinear matrix models that arise when at least one of the projection matrix entries in a structured population model is density-dependent. That is, we consider matrix models

$$\mathbf{x}(t+1) = A(\mathbf{x}(t)) \mathbf{x}(t) \tag{6.5}$$

where $A(\mathbf{x}) = (a_{ij}(\mathbf{x}))$. The main goal is to explore in what way the bifurcation occurs in linear models, as provided by the Fundamental Theorem of Demography, extends to such nonlinear matrix equations.

Example 89 *The Leslie Logistic Model.* P. H. Leslie pioneered the use of matrix models to describe the dynamics of structured populations in two seminal papers [209], [210]. In the second of these papers, he addresses the shortcoming in a linear model when $r > 1$ and the model population grows without bounds. In [210] Leslie considers what he calls logistic growth by introducing nonlinear dependencies into the entries of an age-structured projection matrix (6.4). Specifically, Leslie assumes the survival of newborns and that of all individuals (regardless of age) are affected in the same negative way by an increase in total population size

$$\|\mathbf{x}\| = \sum_{i=1}^{k} |x_i|.$$

(Recall that f_i in (6.4) is the surviving newborns per i-class adult per unit time.) Leslie's logistic age-structured model has the projection matrix

$$\mathcal{L}(\mathbf{x}) = \frac{1}{1 + c\|\mathbf{x}\|} L \tag{6.6}$$

where $\mathcal{L}$ is the matrix (6.4). The resulting matrix equation

$$\mathbf{x}(t+1) = \frac{1}{1+c\,\|\mathbf{x}(t)\|}\mathcal{L}\mathbf{x}(t) \tag{6.7}$$

is an age-structured generalization of the discrete logistic (Beverton-Holt) equation for unstructured populations considered in Chapter 2.

Let $\mathbf{x}(t)$ be the solution of the nonlinear Leslie matrix model (6.7) with initial condition $\mathbf{x}(0) \in \mathbb{R}_+^k$, $\mathbf{x}(0) \neq \mathbf{0}$. Dividing both sides of the nonlinear Leslie equation (6.7) by $\|\mathbf{x}(t+1)\|$ we find

$$\frac{\mathbf{x}(t+1)}{\|\mathbf{x}(t+1)\|} = \frac{\frac{1}{1+c\|\mathbf{x}(t)\|}}{\left\|\frac{1}{1+c\|\mathbf{x}(t)\|}\mathcal{L}\mathbf{x}(t)\right\|}\mathcal{L}\mathbf{x}(t) = \frac{1}{\|\mathcal{L}\mathbf{x}(t)\|}\mathcal{L}\mathbf{x}(t)$$

$$= \frac{1}{\left\|\mathcal{L}\frac{\mathbf{x}(t)}{\|\mathbf{x}(t)\|}\right\|}\mathcal{L}\frac{\mathbf{x}(t)}{\|\mathbf{x}(t)\|}$$

and hence that the normalized age distribution vector

$$\mathbf{n}(t) = \frac{\mathbf{x}(t)}{\|\mathbf{x}(t)\|}$$

satisfies the equations

$$\begin{aligned}\mathbf{n}(t+1) &= \frac{1}{\|\mathcal{L}\mathbf{n}(t)\|}\mathcal{L}\mathbf{n}(t)\\ \mathbf{n}(0) &= \frac{\mathbf{x}(0)}{\|\mathbf{x}(0)\|}\\ \|\mathbf{n}(t)\| &= 1 \text{ for all } t.\end{aligned} \tag{6.8}$$

These equations have the (unique) solution $\mathbf{y}(t)/\|\mathbf{y}(t)\|$ where $\mathbf{y}(t)$ solves the linear Leslie equation $\mathbf{y}(t+1) = \mathcal{L}\mathbf{y}(t)$ with $\mathbf{y}(0) = \mathbf{x}(0)$. Thus, $\mathbf{n}(t) = \mathbf{y}(t)/\|\mathbf{y}(t)\|$ and, if $\mathcal{L}$ is primitive, it follows from the Fundamental Theorem of Demography (Theorem 16 in Chapter 2) that,

$$\lim_{t\to\infty}\mathbf{n}(t) = \frac{\mathbf{w}_R}{\|\mathbf{w}_R\|}$$

where $\mathbf{w}_R \in \mathring{\mathbb{R}}_+^k$ is a positive, right eigenvector associated with $r = \rho(\mathcal{L})$. That is to say, the Leslie logistic model has the same strong ergodic property that the linear Leslie model has.

But what is the long-term fate of the population, i.e., what happens to the total population size $\|\mathbf{x}(t)\|$ as $t \to \infty$? To answer this question we derive a difference equation for total population size $p(t) = \|\mathbf{x}(t)\|$ as follows:

$$p(t+1) = \|\mathbf{x}(t+1)\| = \frac{1}{1+c\,\|\mathbf{x}(t)\|}\|\mathcal{L}\mathbf{x}(t)\|$$

$$= \frac{1}{1+c\,\|\mathbf{x}(t)\|}\|\mathbf{x}(t)\|\left\|\mathcal{L}\frac{\mathbf{x}(t)}{\|\mathbf{x}(t)\|}\right\|$$

or

$$p(t+1) = \|\mathcal{L}\mathbf{n}(t)\|\frac{1}{1+cp(t)}p(t).$$

6.1. INTRODUCTION

This is a nonautonomous, scalar difference equation for $p(t)$ which, because by the strong ergodic property implies

$$\lim_{t \to \infty} \|\mathcal{L}\mathbf{n}(t)\| = \left\|\mathcal{L}\frac{\mathbf{w}_R}{\|\mathbf{w}_R\|}\right\| = r$$

is asymptotically autonomous with limiting equation

$$p(t+1) = r\frac{1}{1+cp(t)}p(t).$$

We recognize this equation as the discrete logistic (Beverton-Holt) equation (1.25) studied in Chapter 1 (with b replaced by the dominant eigenvalue r). Using Theorem 46 in Chapter 8 (also see [65]) we find that

$$\lim_{t \to \infty} p(t) = \lim_{t \to \infty} \|\mathbf{x}(t)\| = \begin{cases} 0 & \text{if } r < 1 \\ \frac{r-1}{c} & \text{if } r > 1 \end{cases}.$$

Note that, in sharp contrast to the linear Leslie model, this model has bounded solutions that equilibrate when $r > 1$. For more on nonlinear matrix models of this type and the strong ergodic property see [62], [65], [215].

The Leslie logistic model shares several features with linear matrix models: both have the strong ergodic property (a stabilizing normal age distribution); in both models, the extinction equilibrium destabilizes as r increases through 1; and both have a continuum of positive equilibria that (transcritically) bifurcates at $r = 1$. There is a significant difference, however: the Leslie logistic model has positive equilibria for all $r > 1$ and not just at the isolated point $r = 1$ (i.e., the bifurcation is no longer vertical) and it no longer predicts unbounded growth. The ergodic property of the normalized age distribution is a feature of the Leslie logistic model (due to the appearance of the nonlinear factor appearing as a multiple of the projection matrix) that nonlinear matrix models do not have in general, however. On the other hand, the bifurcation of positive equilibria when the extinction equilibrium destabilizes is a general feature of nonlinear matrix models, as we see in Theorem 54. Here is another example.

Example 90 *A Basic Juvenile-Adult Model.* The $k = 2$ dimensional Leslie model with projection matrix

$$\mathcal{L} = \begin{pmatrix} 0 & f_2 \\ s_1 & s_2 \end{pmatrix}$$

classifies individuals into a juvenile (non-reproducing) class x_1 and an adult (reproducing) class x_2 with the time unit taken equal to the juvenile maturation period. If we add adult density regulation of fertility to this linear model, then we obtain the projection matrix

$$\mathcal{L}(\mathbf{x}) = \begin{pmatrix} 0 & f_2\beta(x_2) \\ s_1 & s_2 \end{pmatrix}$$

where the factor $\beta(x_2)$ describes the effect of adult density on its fertility. We assume $\beta(0) = 1$ so that f_2 retains its biological interpretation of the inherent (intrinsic or density-free) fertility rate. If $\beta(x_2)$ is a decreasing function of $x_2 > 0$ then the model describes a negative density effect. Examples from Chapter 2 are

$$\beta(x_2) = \frac{1}{1+cx_2} \quad \text{and} \quad \exp(-cx_2) \quad \text{with } c > 0$$

which gives a structured population model that generalizes, respectively, the discrete logistic (Beverton-Holt) and Ricker equations in Chapter 2.

Consider the first case, which leads to the projection matrix

$$\mathcal{L}(\mathbf{x}) = \begin{pmatrix} 0 & f_2 \frac{1}{1+cx_2} \\ s_1 & s_2 \end{pmatrix}. \tag{6.9}$$

The equilibrium equation $\mathbf{x} = A(\mathbf{x})\mathbf{x}$ is algebraically tractable and yields, in addition to the extinction equilibrium $\mathbf{x} = \mathbf{0}$, the equilibrium

$$\mathbf{x} = \frac{1}{c}(\mathcal{R}_0 - 1)\begin{pmatrix} \frac{1-s_2}{s_1} \\ 1 \end{pmatrix}, \quad \mathcal{R}_0 = f_2 \frac{s_1}{1-s_2} \tag{6.10}$$

which is positive if and only if $\mathcal{R}_0 > 1$. Note that $\mathcal{R}_0$ is in fact the reproduction number associated with the matrix

$$\mathcal{L}(\mathbf{0}) = \begin{pmatrix} 0 & f_2 \\ s_1 & s_2 \end{pmatrix}$$

(see (2.16) in Chapter 2). We call $\mathcal{R}_0$ the **inherent** (or intrinsic or density-free) reproduction number associated with the nonlinear model and that $\mathcal{R}_0$ and the **inherent** population growth rate $r = \rho(A(\mathbf{0}))$ lie on the same side of 1.

Thus, although the matrix model with projection matrix (6.9) does not have the strong ergodic property, it does have the following features in common with the Fundamental Theorem of Demography:

- the extinction equilibrium loses stability as $\mathcal{R}_0$ (equivalently r) increases through 1
- a continuum of positive equilibria bifurcates from the extinction equilibrium at $\mathcal{R}_0 = r = 1$.

The significant difference between this nonlinear model and a linear matrix model is that it possesses positive equilibria for all $r > 1$ and not just the isolated value $r = 1$. Moreover, the positive equilibrium (6.10) is (locally asymptotically) stable for all $\mathcal{R}_0 > 1$, the proof of which is left as Problem 2.

6.2 Equilibria: The Fundamental Bifurcation Theorem

The equilibrium equation associated with the nonlinear matrix equation (6.5), i.e., the equation

$$\mathbf{x}(t+1) = A(\mathbf{x}(t))\mathbf{x}(t) \tag{6.11}$$

is the nonlinear algebraic equation

$$\mathbf{x} = A(\mathbf{x})\mathbf{x}. \tag{6.12}$$

The goal of this section is a study of the extinction equilibrium $\mathbf{x} = \mathbf{0}$ and the result of its destabilization. The point of view taken is that of the bifurcation implied by the Fundamental Theorem of Demography for linear matrix models, as illustrated by the examples in Section 6.1.

We assume the matrix $A(\mathbf{x})$ satisfies the following conditions. Recall that $\mathbb{R}_+^k$ is the set of nonnegative vectors in $\mathbb{R}^k$, $\mathring{\mathbb{R}}_+^k$ is the interior of $\mathbb{R}_+^k$ (the set of positive vectors in $\mathbb{R}^k$), and $\partial \mathbb{R}_+^k = \mathbb{R}_+^k \setminus \mathring{\mathbb{R}}_+^k$ is the boundary of $\mathbb{R}_+^k$ (the set of nonnegative vectors with at least on zero entry).

H1: The entries $a_{ij}(\mathbf{x})$ in the projection matrix $A(\mathbf{x}) = (a_{ij}(\mathbf{x}))$ satisfy $a_{ij} \in C^2(\Omega \to \mathbb{R}_+^1)$ where Ω is an open set in $\mathbb{R}^k$ containing $\mathbb{R}_+^k$. The nonnegative inherent projection matrix $A(\mathbf{0})$ is irreducible.

6.2. EQUILIBRIA: THE FUNDAMENTAL BIFURCATION THEOREM

Following the modeling methodology of Chapters 1 and 2 for a population closed to external immigration or seeding and loss to emigration or harvesting, the per capita contribution of a j-class individual, present at time t, to the ith class at the next census equals the surviving i-class newborns plus the j-class individuals who lie in the ith class at the next census time $t+1$ and so

$$a_{ij}(\mathbf{x}) = f_{ij}(\mathbf{x}) + \tau_{ij}(\mathbf{x})$$

where the terms $f_{ij}(\mathbf{x})$ and $\tau_{ij}(\mathbf{x})$ are the entries in the fertility and transition matrices

$$F(\mathbf{x}) = (f_{ij}(\mathbf{x})), \quad T(\mathbf{x}) = (\tau_{ij}(\mathbf{x})).$$

H2: The population projection matrix in H1 has the form $A(\mathbf{x}) = F(\mathbf{x}) + T(\mathbf{x})$ where the entries of $F(\mathbf{x}) = (f_{ij}(\mathbf{x}))$ and $T(\mathbf{x}) = (\tau_{ij}(\mathbf{x}))$ satisfy $f_{ij}, \tau_{ij} \in C^2(\Omega \to \mathbb{R}_+^k)$ with $0 \leq \tau_{ij}(\mathbf{x}) \leq 1$ and $\sum_{i=1}^k \tau_{ij}(\mathbf{x}) \leq 1$ for $\mathbf{x} \in \mathbb{R}_+^k$.

Under H2 the model equation (6.11) is

$$\mathbf{x}(t+1) = F(\mathbf{x}(t))\mathbf{x}(t) + T(\mathbf{x}(t))\mathbf{x}(t). \tag{6.13}$$

From the inherent population projection matrix $A(\mathbf{0}) = F(\mathbf{0}) + T(\mathbf{0})$ we obtain the (inherent) next-generation matrix $F(\mathbf{0})(I - T(\mathbf{0}))^{-1}$. We use this nonnegative matrix to define the inherent reproduction number

$$\mathcal{R}_0 := \rho\left(F(\mathbf{0})(I - T(\mathbf{0}))^{-1}\right)$$

provided

H3: $\rho(T(\mathbf{0})) < 1$.

If $\mathcal{R}_0 > 0$, then it follows from Theorem 18 in Chapter 2 that the inherent population growth rate

$$r := \rho(A(\mathbf{0}))$$

and $\mathcal{R}_0$ lie on the same side of 1. By the Perron-Frobenius Theorem 13 in Chapter 2, r is a simple positive eigenvalue of $P(\mathbf{0})$ and has associated positive right eigenvector $\mathbf{w}_R \in \mathring{\mathbb{R}}_+^k$ and left eigenvector $\mathbf{w}_L^T$, ($\mathbf{w}_L \in \mathring{\mathbb{R}}_+^k$).

Clearly $\mathbf{x} = \mathbf{0}$ is an equilibrium of the matrix equation (6.11). It is left as a Problem 6 to show that the Jacobian associated with equation (6.11), when evaluated at the extinction equilibrium $\mathbf{x} = \mathbf{0}$, equals the inherent projection matrix $P(\mathbf{0})$. The Linearization Principle (Theorem 22 in Chapter 3) gives the following result.

Theorem 51 *Local Stability of the Extinction Equilibrium. Assume H1. The extinction equilibrium $\mathbf{x} = \mathbf{0}$ of the nonlinear matrix equation (6.11) is (locally asymptotically) stable if $r < 1$ and unstable if $r > 1$.*

Under additional conditions on the projection matrix we can say more.

Theorem 52 *Global Stability of the Extinction Equilibrium. In addition to H1 assume*

$$A(\mathbf{x}) \leq A(\mathbf{0}) \tag{6.14}$$

for all $\mathbf{x} \in \mathbb{R}_+^k$. Then the extinction equilibrium $\mathbf{x} = \mathbf{0}$ of the nonlinear matrix equation (6.11) is globally asymptotically stable on $\mathbb{R}_+^k$ if $r < 1$.

Proof. By Theorem 51, $\mathbf{x} = \mathbf{0}$ is locally asymptotically stable so all we need to show is that $\mathbf{x} = \mathbf{0}$ is attracting on $\mathbb{R}_+^k$. From Equation (6.11) and (6.14) we get, for any initial condition $\mathbf{x}(0) \in \mathbb{R}_+^k$, that
$$0 \leq \mathbf{x}(t+1) \leq A(\mathbf{0})\mathbf{x}(t)$$
and hence $\mathbf{x}(t) \leq \mathbf{y}(t)$ for all $t = 0, 1, 2, 3, \ldots$ where $\mathbf{y}(t)$ is the solution of the linear matrix equation $\mathbf{y}(t+1) = A(\mathbf{0})\mathbf{y}(t)$. It follows that $r < 1$ implies $\mathbf{y}(t)$ and hence $\mathbf{x}(t)$ tend to $\mathbf{0}$ as $t \to \infty$. ∎

The condition (6.14) implies that all density dependent entries $a_{ij}(\mathbf{x})$ in the projection matrix can only decrease from the density-free values: $a_{ij}(\mathbf{x}) \leq a_{ij}(\mathbf{0})$. In this sense, (6.14) implies all density effects are negative. We see in Section 5.5 that an increase in an entry $a_{ij}(\mathbf{x}) > a_{ij}(\mathbf{0})$ above its density-free state (a so-called positive feedback or component Allee effect) can result in the existence of equilibria in $\mathbb{R}_+^k \setminus \{\mathbf{0}\}$ when $r < 1$, in which case $\mathbf{x} = \mathbf{0}$ cannot be globally stable. (Also see Chapter 1.)

To say more about the extinction equilibrium when $r > 1$ we need following concepts. (See Definition 21 in Chapter 3, Section 3.9.)

Definition 27 *(a) A set $D \subseteq \mathbb{R}^k$ is **forward invariant** with respect to the Equation (6.11) if $\mathbf{x}(0) \in D$ implies $\mathbf{x}(t) \in D$ for all $t = 1, 2, 3, \ldots$.*

*(b) Equation (6.11) is **dissipative** if there exists a real number $\beta > 0$ (independent of the initial condition $\mathbf{x}(0) \in \mathbb{R}_+^k$) and a time t^* (which can depend on the initial condition $\mathbf{x}(0) \in \mathbb{R}_+^k$) such that $\|\mathbf{x}(t)\| < \beta$ for all $t \geq t^*$ or equivalently such that $\limsup_{t \to \infty} \|\mathbf{x}(t)\| \leq \beta$.*

*(c) Equation (6.11) is **uniformly persistent** (with respect to $\mathbf{0}$) if there exists a real number $\eta > 0$ such that $\liminf_{t \to \infty} \|\mathbf{x}(t)\| \geq \eta$ for all $\mathbf{x}(0) \in \mathbb{R}_+^k \setminus \{\mathbf{0}\}$.*

*(d) Equation (6.11) is **permanent** (with respect to $\mathbf{0}$) if it is both dissipative and uniformly persistent.*

Remark 18 *Note that for each $\mathbf{x}(0) \in \mathbb{R}_+^k$ Equation (6.11) defines, under assumption H1, a unique solution sequence $\mathbf{x}(t) \in \mathbb{R}_+^k$, $t = 1, 2, 3, \ldots$, and hence that $\mathbb{R}_+^k$ is forward invariant. If Equation (6.11) is permanent, then all solutions with $\mathbf{x}(0) \in \mathbb{R}_+^k \setminus \{\mathbf{0}\}$ satisfy*
$$0 < \eta \leq \liminf_{t \to \infty} \|\mathbf{x}(t)\| \leq \limsup_{t \to \infty} \|\mathbf{x}(t)\| \leq \beta,$$
i.e., are bounded and do not tend to the extinction equilibrium.

Part (a) of the following theorem is proved in [199] (Theorem 3). Part (b) follows from part (a) and Theorem 18 in Chapter 2.

Theorem 53 Permanence. *Assume H1 holds, that $\mathbb{R}_+^k \setminus \{\mathbf{0}\}$ is forward invariant, and that Equation (6.11) is dissipative.*

(a) Then $r > 1$ implies Equation (6.11) is permanent.

(b) If, in addition, H2 and H3 hold, then $\mathcal{R}_0 > 1$ implies Equation (6.11) is permanent (with respect to $\mathbf{0}$).

6.2. EQUILIBRIA: THE FUNDAMENTAL BIFURCATION THEOREM

A sufficient criterion for the forward invariance of $\mathbb{R}_+^k \setminus \{\mathbf{0}\}$, satisfied by most population models is contained in the following lemma.

Lemma 11 *Assume H1 holds and that $A(\mathbf{x})$ is irreducible for all $\mathbf{x} \in \mathbb{R}_+^k$. Then $\mathbb{R}_+^k \setminus \{\mathbf{0}\}$ is forward invariant with respect to Equation (6.11).*

Proof. Let $\mathbf{x}(0) \in \mathbb{R}_+^k \setminus \{\mathbf{0}\}$. Then since $\mathbb{R}_+^k$ is forward invariant, we know $\mathbf{x}(t) \in \mathbb{R}_+^k$ for all $t = 1, 2, 3, \ldots$, and we only need to show $\mathbf{x}(t) \neq \mathbf{0}$ for all $t = 1, 2, 3, \ldots$. Suppose, for the purpose of contradiction, that this is not true and there exists a first time $t_0 > 1$ for which $\mathbf{x}(t_0) = \mathbf{0}$. Then by Equation (6.11)
$$A(\mathbf{x}(t_0 - 1))\mathbf{x}(t_0 - 1) = \mathbf{0}$$
which implies $\mathbf{x}(t_0 - 1)$ is a nonnegative eigenvector of the nonnegative irreducible matrix $A(\mathbf{x}(t_0 - 1))$ associated with the eigenvalue $\lambda = 0$. This contradicts the Perron-Frobenius Theorem 13 in Chapter 2, since only the dominant, positive eigenvalue of $A(\mathbf{x}(t_0 - 1))$ has a nonnegative eigenvector. ∎

The next lemma contains some conditions sufficient to guarantee that Equation (6.11) is dissipative. Recall that the operator norm of a matrix $M = (m_{ij})$ associated with the L^1 norm is $\|M\| = \max_{1 \leq j \leq k} \sum_{i=1}^{k} |m_{ij}|$.

Lemma 12 *Assume H1 and H2. If the fertility and transition matrices satisfy*
$$\|F(\mathbf{x})\mathbf{x}\| \leq \phi, \quad \|T(\mathbf{x})\| \leq \tau < 1 \tag{6.15}$$
for some constants ϕ and τ and all $\mathbf{x} \in \mathbb{R}_+^k$. Then Equation (6.13) is dissipative.

Proof. From Equation (6.13) we have
$$\|\mathbf{x}(t+1)\| \leq \|F(\mathbf{x}(t))\mathbf{x}(t)\| + \|T(\mathbf{x}(t))\mathbf{x}(t)\| \leq \phi + \|T(\mathbf{x}(t))\|\|\mathbf{x}(t)\|$$
and hence
$$\|\mathbf{x}(t+1)\| \leq \phi + \tau \|\mathbf{x}(t)\| \tag{6.16}$$
for all $t = 0, 1, 2, 3, \ldots$. Since $\tau < 1$ it follows that all solutions of the linear difference equation $y(t+1) = \phi + \tau y(t)$ satisfy
$$\lim_{t \to \infty} y(t) = \frac{\phi}{1-\tau} < \infty.$$
Using (6.16) and an induction argument, we find that $\|\mathbf{x}(t)\| \leq y(t)$ for all $t = 0, 1, 2, 3, \ldots$, provided we choose the initial condition $y(0) = \|\mathbf{x}(0)\|$ for $y(t)$. Then
$$\limsup_{t \to \infty} \|\mathbf{x}(t)\| \leq \limsup_{t \to \infty} y(t) = \frac{\phi}{1-\tau} = \beta.$$
∎

The condition on the fertility matrix $F(\mathbf{x})$ appearing in (6.15) is based on the assumption that there is an upper bound on the total number of newborns that can be produced (per unit time) by a population of any density $\mathbf{x}$. This is a reasonable biological assumption when limited resources and intraspecific competition constrain reproduction at high population densities. This condition is met, for example, by the commonly used rational and exponential type density factors

$$f_{ij}(\mathbf{x}) = \frac{1}{1 + c_{ij}\sum_{m=1}^{k} w_m x_m}, \quad f_{ij}(\mathbf{x}) = \exp\left(-c_{ij}\sum_{m=1}^{k} w_m x_m\right), \quad w_m, c_{ij} > 0$$

in which $\sum_{m=1}^{k} w_m x_m$ is a weighted total population size and c_{ij} is a competitive interaction coefficient that measures the effect of the j-class on the i-class.

The condition on the transition matrix $T(\mathbf{x})$ appearing in (6.15) means that in all age classes, there is some loss (e.g., due to death) per unit of time at all population densities.

Example 91 *The Levin-Goodyear Model Revisited. Motivated by the Ricker equation (Chapter 2) as a spawner-recruit model for a fish population, Levin and Goodyear [212] derive a nonlinear age-structured Leslie model with fertility and transition matrices of the form*

$$F(\mathbf{x}) = \begin{pmatrix} f_1(\mathbf{x}) & f_2(\mathbf{x}) & \cdots & f_{k-1}(\mathbf{x}) & f_k(\mathbf{x}) \\ 0 & 0 & \cdots & 0 & 0 \\ 0 & 0 & \cdots & 0 & 0 \\ \vdots & \vdots & & \vdots & \vdots \\ 0 & 0 & \cdots & 0 & 0 \end{pmatrix}, \quad T = \begin{pmatrix} 0 & 0 & \cdots & 0 & 0 \\ s_1 & 0 & \cdots & 0 & 0 \\ 0 & s_2 & \cdots & 0 & 0 \\ \vdots & \vdots & & \vdots & \vdots \\ 0 & 0 & \cdots & s_{k-1} & s_k \end{pmatrix}$$

with

$$f_j(\mathbf{x}) = b_j \exp\left(-c\sum_{i=1}^{k} w_i x_i\right), \quad b_j, c, w_i > 0, \quad 0 < s_i < 1$$

and apply it to Hudson River striped bass. The time unit is one year. By Corollary 3 in Chapter 2, the projection matrix $P(\mathbf{x}) = F(\mathbf{x}) + T$ is irreducible (in fact, primitive) for all $\mathbf{x} \in \mathbb{R}_+^k$ and both H1 and H2 hold with $\Omega = \mathbb{R}^k$. Moreover, H3 holds since $\rho(T) = s_k < 1$ and from (2.16) in Chapter 2 we have

$$\mathcal{R}_0 = \sum_{j=1}^{k-1} b_j p_j + b_k p_k \frac{1}{1-s_k} > 0$$

where

$$p_j := s_0 s_1 s_2 \ldots s_{j-1}, \quad 1 \leq j \leq k$$

($s_0 = 1$ for notational convenience). Clearly $f_j(\mathbf{x}) \leq b_j = f_j(\mathbf{0})$ and hence $A(\mathbf{x}) \leq A(\mathbf{0})$. From Theorem 52 we conclude that

$$\mathcal{R}_0 < 1 \text{ implies } \mathbf{x} = \mathbf{0} \text{ is globally asymptotically stable on } \mathbb{R}_+^k$$

and that $\mathcal{R}_0 > 1$ implies $\mathbf{x} = \mathbf{0}$ is unstable. Furthermore, Lemmas 11 and 12 together with Theorem 53 show

$$\mathcal{R}_0 > 1 \text{ implies permanence (with respect to } \mathbf{0}).$$

This is because the Leslie matrix $\mathcal{L}(\mathbf{x}) = F(\mathbf{x}) + T$ is irreducible for all $\mathbf{x} \in \mathbb{R}_+^k$ and because the inequalities

$$\|F(\mathbf{x})\mathbf{x}\| = \sum_{j=1}^{k} b_j x_j \exp\left(-c\sum_{i=1}^{k} w_i x_i\right) \leq \sum_{j=1}^{k} b_j x_j \exp(-c w_j x_j) \leq \sum_{j=1}^{k} b_j \frac{e^{-1}}{c w_j} = \phi$$

$$\|T(\mathbf{x})\| = \max_{1 \leq j \leq k} \sum_{i=1}^{k} |\tau_{ij}(\mathbf{x})| = \max_{1 \leq j \leq k} s_j = \tau < 1$$

show that inequalities (6.15) *hold.*

Upon destabilization of the extinction equilibria, we expect a bifurcation to occur with non-extinction equilibria and, in particular, with survival equilibria $\mathbf{x} \in \mathbb{R}_+^k$. This is illustrated in the $k = 1$ dimensional case in Chapter 1 and by the $k = 2$ dimensional matrix model in Example 6.9. In the latter case, note that as a function of $\mathcal{R}_0$ the equilibrium (6.10) constitutes a continuum branch of positive equilibria that coalesce with the extinction equilibrium when $\mathcal{R}_0 = 1$. (The entire branch of non-extinction equilibria (6.10) for all $\mathcal{R}_0$, positive, negative, and 1, transcritically bifurcates with the extinction equilibrium at $\mathcal{R}_0 = 1$, but of course the negative equilibria when $\mathcal{R}_0 < 1$ are not relevant for a biological model.) In this section, we show that this bifurcation phenomenon is quite general for nonlinear matrix models (at least locally near the bifurcation point).

As pointed out in Section 6.1, the basic quantities r and $\mathcal{R}_0$ are derived from model parameters and in general do not appear explicitly in the model equations. They are in general changed when a model parameter changes and if a model parameter of interest to the modeler is manipulated to as to cause r and $\mathcal{R}_0$ to increases through 1 then the extinction equilibrium will destabilize. To study the bifurcation that results from this, we let μ denote a model parameter of interest, include it in the argument list of the projection matrix so as to write

$$\mathbf{x}(t+1) = A(\mu, \mathbf{x}(t))\mathbf{x}(t) \quad (6.17)$$

and adjust the model assumptions H1 to include μ.

H4: The entries $a_{ij}(\mu, \mathbf{x})$ in the projection matrix $A(\mu, \mathbf{x}) = (a_{ij}(\mu, \mathbf{x}))$ satisfy $a_{ij} \in C^2\left(I_\mu \times \Omega \to \mathbb{R}_+^1\right)$ where I_μ denotes an open interval in $\mathbb{R}^1$ and Ω is an open set containing $\mathbb{R}_+^k$. The nonnegative inherent projection matrix $A(\mu, \mathbf{0})$ is irreducible for $\mu \in I_\mu$.

Under assumption H4 the inherent population growth rate $r(\mu)$ is a function of μ. We assume

H5: There exists $\mu_0 \in I_\mu$ such that $r(\mu_0) = 1$.

If $\mathbf{x}$ is an equilibrium of Equation (6.17) associated with parameter μ, i.e., if $\mathbf{x}$ satisfies the algebraic equation

$$\mathbf{x} = A(\mu, \mathbf{x})\mathbf{x}, \quad (6.18)$$

then we call $(\mu, \mathbf{x})$ an *equilibrium pair*. Note that $(\mu, \mathbf{0})$ is an *extinction equilibrium pair* for all $\mu \in I_\mu$. If $\mathbf{x} \in \mathbb{R}_+^k$ or $\mathring{\mathbb{R}}_+^k$ or $\partial \mathbb{R}_+^k$, then we call $(\mu, \mathbf{x})$ a *nonnegative* or *positive* or *boundary equilibrium pair*, respectively.

The next theorem concerns *positive equilibrium pairs* $(\mu, \mathbf{x})$, $\mathbf{x} \neq \mathbf{0}$. Define the scalar

$$\alpha := \mathbf{w}_L^T \frac{\partial A(\mu_0, \mathbf{0})}{\partial \mu} \mathbf{w}_R \quad (6.19)$$

where $\mathbf{w}_L^T$ and $\mathbf{w}_R$ are, respectively, left and right *positive* eigenvectors ($\mathbf{w}_L \in \mathring{\mathbb{R}}_+^k$ and $\mathbf{w}_R \in \mathring{\mathbb{R}}_+^k$) of the inherent projection matrix $A(\mu_0, \mathbf{0}) = (a_{ij}(\mu_0, \mathbf{0}))$ associated with the dominant eigenvalue $r(\mu_0) = 1$. Note that by (6.3)

$$r'(\mu_0) = \frac{\alpha}{\mathbf{w}_L^T \mathbf{w}_R} \quad (6.20)$$

so that $\alpha > 0$ implies $r(\mu)$ increases through 1, and hence the extinction equilibrium destabilizes, as μ increases through μ_0. If, on the other hand, $\alpha < 0$ then the extinction equilibrium destabilizes as μ decreases through μ_0.

A proof of the following theorem can be found in [**94**].

Theorem 54 Local Bifurcation. *Assume H4, H5 and $a \neq 0$. In an open neighborhood of $(\mu_0, \mathbf{0})$ in $I_\mu \times \mathbb{R}^k$ there exists a parameterized branch of equilibrium pairs $(\mu(\varepsilon), \mathbf{x}(\varepsilon))$ of Equation (6.17) for $\varepsilon \in I_\varepsilon$, where I_ε is an open interval in $\mathbb{R}^1$ containing $\varepsilon = 0$, and*

$$\mu(\varepsilon) = \mu_0 + \eta(\varepsilon), \quad \mathbf{x}(\varepsilon) = \varepsilon \mathbf{w}_R + \mathbf{z}(\varepsilon) \qquad (6.21)$$

where $\eta \in C^2(I_\varepsilon \to \mathbb{R}^1)$, $\mathbf{z} \in C^2(I_\varepsilon \to \mathbb{R}^k)$ with $\mathbf{z}(\varepsilon)^T \mathbf{w}_R = 0$ and where $\eta(0) = 0$ and $\mathbf{z}(0) = \mathbf{z}'(0) = \mathbf{0}$.

The expansions (6.21) describe a continuum (a closed and connected set) of nonnegative equilibrium pairs

$$\mathbb{C}_+ := \{(\mu(\varepsilon), \mathbf{x}(\varepsilon)) \mid 0 \leq \varepsilon \in I_\varepsilon\}$$

that contains (bifurcates from) the extinction equilibrium pair $(\mu_0, \mathbf{0})$. Since the eigenvector $\mathbf{w}_R \in \mathring{\mathbb{R}}_+^k$, it follows that the equilibria $\mathbf{x}(\varepsilon)$ in (6.21) also lie in $\mathring{\mathbb{R}}_+^k$ for small $\varepsilon > 0$, i.e., $\mathbb{C}_+ \setminus \{\mathbf{0}\}$ consists of positive equilibrium pairs for $\varepsilon > 0$ small. If, in a neighborhood of the bifurcation point $(\mu_0, \mathbf{0})$ (i.e., if for small $\varepsilon > 0$), it is true that $\mu(\varepsilon) > \mu_0$ for the equilibrium pairs $(\mu(\varepsilon), \mathbf{x}(\varepsilon)) \in \mathbb{C}_+$, then we say the bifurcation of positive equilibria if μ-forward. It is μ-backward if $\mu(\varepsilon) < \mu_0$ for small $\varepsilon > 0$.

Define

$$\kappa := -\mathbf{w}_L^T \left[\nabla_{\mathbf{x}} a_{ij}(\mu_0, \mathbf{0}) \mathbf{w}_R \right] \mathbf{w}_R \qquad (6.22)$$

where $\nabla_{\mathbf{x}}$ is the gradient (row) vector

$$\nabla_{\mathbf{x}} a_{ij}(\mu_0, \mathbf{0}) = \left(\begin{array}{cccc} \frac{\partial a_{ij}(\mu, \mathbf{x})}{\partial x_1} & \frac{\partial a_{ij}(\mu, \mathbf{x})}{\partial x_2} & \cdots & \frac{\partial a_{ij}(\mu, \mathbf{x})}{\partial x_k} \end{array} \right) \bigg|_{(\mu, \mathbf{x}) = (\mu_0, \mathbf{0})}.$$

Theorem 55 *[94]* **Direction of Bifurcation.** *Assume H4, H5, and $\alpha \neq 0$. Then*

$$\eta'(0) = \frac{\kappa}{\alpha}$$

and, as a result, the local bifurcation of positive equilibria in Theorem 54 is μ-forward if α and κ have the same sign and μ-backward if they have opposite signs.

The diagnostic quantities α and κ in Theorem 55 are linear combinations (with positive coefficients) of the partial derivatives of $a_{ij}(\mu, \mathbf{x})$ with respect to μ and the components x_i of $\mathbf{x}$. However, in order to apply Theorem 55, it is not necessarily required to calculate these derivatives. It is only required to ascertain the signs of α and κ and that can often be done in applications by inspection. For example, if the nonzero μ derivatives of all entries in the projection matrix $A(\mu, \mathbf{x})$ are positive, then we know $\alpha > 0$. In many applications, only negative density effects are present, which means all nonzero x_i partial derivatives of the entries $A(\mu, \mathbf{x})$ are negative and hence κ is positive. For such a case, the bifurcation is μ-forward.

Remark 19 *To locate the bifurcation point μ_0 in assumption H5, it is required to solve the equation $r(\mu_0) = 1$ for $\mu_0 \in I_\mu$. By Theorem 55 to determine the direction of bifurcation at μ_0 the sign of α needs to be determined or, by (6.20), the sign of $r'(\mu_0)$. Since a formula for $r(\mu)$ is rarely available or tractable (or even possible, for higher dimensions), it is often useful in applications to work instead with the reproduction number $\mathcal{R}_0(\mu)$. Theorem 18 in Chapter 2 and the theorem below shows that assumptions H5 and $\alpha \neq 0$ can be replaced in all theorems by the assumption:*

there exists $\mu_0 \in I_\mu$ such that $\mathcal{R}_0(\mu_0) = 1$ and $\mathcal{R}'(\mu_0) \neq 0$

and that the sign of α is the same as that of $\mathcal{R}'(\mu_0)$.

Theorem 56 *[73] Assume H4.*
(a) *Then for $\mu_0 \in I_\mu$*
$$r(\mu_0) = 1 \text{ if and only if } \mathcal{R}_0(\mu) = 1. \tag{6.23}$$

(b) *If (6.23) holds then there exists a positive constant $k > 0$ such that $r'(\mu_0) = k\mathcal{R}_0(\mu_0)$ and, as a result,*
$$r'(\mu_0) = 0 \text{ if and only if } \mathcal{R}'(\mu_0) = 0. \tag{6.24}$$

(c) *If (6.23) and (6.24) hold, then $r''(\mu_0) = k\mathcal{R}'(\mu_0)$.*

Before looking at examples, we consider the stability properties of the bifurcating positive equilibria in Theorem 55. It turns out that the primitivity of the inherent projection matrix is important in this regard, so this is the case that we consider first.

6.3 Stable Bifurcation: The Primitive Case

Definition 28 *If, in a neighborhood of the bifurcation point $(\mu, \mathbf{x}) = (\mu_0, \mathbf{0})$, the positive equilibria guaranteed by Theorem 54 are (locally asymptotically) stable or unstable, then we say that the bifurcation at $\mu = \mu_0$ is stable or unstable, respectively.*

To apply the Linearization Principle to study the local stability of the equilibria $(\mu(\varepsilon), \mathbf{x}(\varepsilon))$ given by (6.21) for small ε, we observe that the Jacobian $J(\varepsilon)$ associated with the Equation (6.17), when evaluated at the equilibria (6.21), becomes a function of ε as, therefore, do its eigenvalues $\lambda_i(\varepsilon)$ and associated eigenvectors $\mathbf{q}_i(\varepsilon)$, $i = 1, 2, \ldots, k$. Since eigenvalues are continuous functions of the entries in a matrix, we know that $\lambda_i(0)$ are the eigenvalues of the Jacobian at the bifurcation point, which is just the inherent projection matrix $A(\mu_0, 0)$. If we assume this matrix is primitive, then its strictly dominant eigenvalue, which we assume without loss in generality is the first eigenvalue in the list, is equal to 1. This means all other eigenvalues satisfy $|\lambda_i(0)| < 1$, $i = 2, \ldots, k$ which in turn implies
$$|\lambda_i(\varepsilon)| < 1, \ i = 2, \ldots, k \text{ for small } \varepsilon > 0.$$

As a result, the stability (by Linearization) of the bifurcating positive equilibria (6.21) depends on $\lambda_1(\varepsilon)$ for small $\varepsilon > 0$. Since $\lambda_1(0) = 1$ we see that $\lambda_1'(0) < 0$ implies $\lambda_1(\varepsilon) < 1$ for small $\varepsilon > 0$ and the equilibria are stable, whereas $\lambda_1'(0) > 0$ implies $\lambda_1(\varepsilon) > 1$ for small $\varepsilon > 0$ and the equilibria are unstable. To calculate $\lambda_1'(0)$ we let $\mathbf{q}_1(\varepsilon)$ be an eigenvector associated with $\lambda_1(\varepsilon)$ for which $\mathbf{q}_1(0) = \mathbf{w}_R$ and differentiate the equation
$$J(\varepsilon)\mathbf{q}_1(\varepsilon) = \lambda_1(\varepsilon)\mathbf{q}_1(\varepsilon)$$

with respect to ε, evaluate the result at $\varepsilon = 0$, and solve the resulting equation for $\lambda_1'(0)$. While straightforward, the details are tedious and we leave them for the reader (or refer the reader to [94]). The result is

$$\lambda_1'(0) = -\frac{\kappa}{\mathbf{w}_L^T \mathbf{w}_R} \tag{6.25}$$

from which, together with Theorem 55, we obtain at the following theorem.

Theorem 57 *Stability of the Local Bifurcation. Assume H4, H5 and $\alpha \neq 0$ hold. Assume further that $A(\mu_0, \mathbf{0})$ is primitive. Then the local bifurcation of positive equilibria at $\mu = \mu_0$ is stable if $\kappa > 0$ and unstable if $\kappa < 0$.*

Putting this result together with Theorem 55 we get the following result.

Corollary 7 *Stability and the Direction of Bifurcation. Assume H4, H5 and $\alpha \neq 0$ hold. Assume further that $A(\mu_0, \mathbf{0})$ is primitive.*
(a) If $\alpha > 0$ then the local bifurcation of positive equilibria at $\mu = \mu_0$ is stable if it is μ-forward and unstable if it is μ-backward.
(b) If $\alpha < 0$ then the bifurcation at $\mu = \mu_0$ is stable if it is μ-backward and unstable if it is μ-forward.

This corollary relates the direction of bifurcation to the stability properties of the bifurcation, all of which can be determined from the signs of α and κ without the need of a stability analysis.

It is important to remember that the existence and stability of results of the positive equilibria, created by bifurcation upon destabilization of the extinction equilibrium, as given in the theorems above, are local. First of all, they are local in nature because the theorems guarantee the existence of positive equilibria in only a neighborhood of the bifurcation point and, secondly, because the stability properties are local (having been attained by the Linearization Principle). The strength of these theorems lies in their generality and their ease of use (with a minimum amount of analysis required). They apply to the vast majority of applications found in the literature. However, they only constitute the basic first step in the analysis of a model in that equilibria (and/or other attractors) lying outside the neighborhood of the bifurcation point are also important in applications.

A more global accounting of the existence, number, and stability properties of equilibria of a nonlinear matrix model depend on the specific details of the nonlinear density effects described in the entries $a_{ij}(\mathbf{x})$ and usually requires an ad hoc study of the equation of interest. However, with regard to the existence of positive equilibria outside the neighborhood of the bifurcation point, there are some general results concerning the global extent of the branch $\mathbb{C}_+$ of positive equilibrium pairs. These results derive from the celebrated global bifurcation theorem of Rabinowitz [189], [253], but require some specialized assumptions concerning the appearance of the bifurcation parameter μ in the projection matrix. We consider just one example here.

H6. Assume the inherent projection matrix has the form $A(\mu, \mathbf{0}) = A_1 + \mu A_2$ where A_1, A_2 are nonnegative matrices where $I - A_1$ is invertible and A_2 is not the zero matrix.

Since H6 implies at least one entry in $A(\mu, \mathbf{0})$ is strictly increasing with μ, it follows that $r(\mu)$ is strictly increasing in μ and that μ_0 is therefore the only value of μ for which $r(\mu) = 1$.

6.3. STABLE BIFURCATION: THE PRIMITIVE CASE

Theorem 58 Global Bifurcation. *Assume H4, H5 holds and $\alpha \neq 0$. In addition, assume H6 holds and that $A(\mu, \mathbf{x})$ is primitive for all $(\mu, \mathbf{x}) \in \mathbb{R}_+^1 \times \mathbb{R}_+^k$. The locally bifurcating continuum $\mathbb{C}_+$ in Theorem 54 is contained in an unbounded continuum $\mathbb{C}$ of equilibrium pairs $(\mu, \mathbf{x})$ for which $\mathbb{C} \setminus \{(\mu_0, \mathbf{0})\} \subset \mathbb{R}^1 \times \mathring{\mathbb{R}}_+^k$.*

Proof. Theorem 54 implies the existence of the bifurcating branch $\mathbb{C}_+$ of positive equilibrium pairs from the point $(\mu_0, \mathbf{0})$. These pairs satisfy the equilibrium equation $\mathbf{x} = A(\mu, \mathbf{x})\mathbf{x}$, Performing a Taylor expansion with the remainder to the right side, we re-write the equilibrium equation as

$$\mathbf{x} = (A_1 + \mu A_2)\mathbf{x} + \mathbf{g}(\mu, \mathbf{x})$$

where $\mathbf{g}(\mu, \mathbf{x}) = o\left(\|\mathbf{x}\|^2\right)$ for $\mathbf{x}$ near $\mathbf{0}$, uniformly on bounded μ intervals, or using H6 as

$$\mathbf{x} = \mu L \mathbf{x} + \mathbf{h}(\mu, \mathbf{x}) \tag{6.26}$$

where $L := (I - A_1)^{-1} A_2$ and

$$\mathbf{h}(\mu, \mathbf{x}) := (I - A_1)^{-1} \mathbf{g}(\mu, \mathbf{x}) = o\left(\|\mathbf{x}\|^2\right)$$

for $\mathbf{x}$ near $\mathbf{0}$, uniformly on bounded μ intervals. By H5 and Perron-Frobenius theory, 1 is the (geometrically) simple dominant eigenvalue of $A(\mu_0, \mathbf{0}) = A_1 + \mu_0 A_2$, all of whose eigenvectors of satisfy $\mathbf{w}_R = A(\mu_0, \mathbf{0})\mathbf{w}_R$ and hence $\mathbf{w}_R = \mu_0 L \mathbf{w}_R$. It follows that μ_0 is a characteristic value of L with multiplicity 1 and the celebrated bifurcation theorem of Rabinowitz [189], [253] implies that the local continuum $\mathbb{C}_+$ of positive equilibrium pairs has a global continuum extension $\mathbb{C}$ that

   (i) either meets ∞ (is unbounded) in $\mathbb{R}^1 \times \mathbb{R}^k$
   (ii) or meets a point $(\tilde{\mu}, \mathbf{0})$ where $\tilde{\mu} \neq \mu_0$ is a characteristic value of L.

Can $\mathbb{C}$ contain a boundary equilibrium other than $(\mu_0, \mathbf{0})$? For purposes of contradiction, assume it does. This would imply the existence of a sequence of positive equilibrium pairs that approach a boundary equilibrium pair $(\mu^*, \mathbf{x}^*)$ other than the bifurcation point $(\mu_0, \mathbf{0})$:

$$\lim_{i \to \infty} (\mu_i, \mathbf{x}_i) = (\mu^*, \mathbf{x}^*) \neq (\mu_0, \mathbf{0}), \quad \mathbf{x}_i \in \mathring{\mathbb{R}}_+^k, \quad \mathbf{x}^* \in \partial \mathbb{R}_+^k$$

$(\mu_i, \mathbf{x}_i) \in \mathbb{C}$ implies $\mathbf{x}_i = \mu_i L \mathbf{x}_i + \mathbf{h}(\mu_i, \mathbf{x}_i)$ and hence

$$\mathbf{x}_i = A(\mu_i, \mathbf{x}_i)\mathbf{x}_i \tag{6.27}$$

Case 1: $\mathbf{x}^* \neq \mathbf{0}$. By passing to the limit in Equation (6.27) we obtain $\mathbf{x}^* = A(\mu^*, \mathbf{x}^*)\mathbf{x}^*$. Because $A(\mu^*, \mathbf{x}^*)$ is primitive and $\mathbf{x}^* \in \partial \mathbb{R}_+^k$ is not positive we have a contradiction to Perron-Frobenius theory that says the only nonnegative eigenvector is the positive Perron eigenvector.

Case 2: $\mathbf{x}^* = \mathbf{0}$. By choosing a subsequence if necessary we can assume that the sequence of positive unit vectors $\mathbf{x}_i / \|\mathbf{x}_i\| \in \mathring{\mathbb{R}}_+^k$ converges to a nonnegative unit vector $\mathbf{v}^* \in \mathbb{R}_+^k$. Equation (6.27) gives

$$\lim_{i \to \infty} \frac{\mathbf{x}}{\|\mathbf{x}_i\|} = \lim_{i \to \infty} A(\mu_i, \mathbf{x}_i) \frac{\mathbf{x}}{\|\mathbf{x}_i\|}$$

which implies $\mathbf{v}^* = A(\mu^*, \mathbf{0})\mathbf{v}^*$ and hence that $\mathbf{v}^*$ is a nonnegative eigenvector associated with eigenvalue 1 of $A(\mu^*, \mathbf{0})$. Unless $\mathbf{v}^*$ is positive this contradicts Perron-Frobenius theory. If $\mathbf{v}^*$ is positive, then implies $r(\mu^*) = 1$, which in turn implies the contradiction that $\mu^* = \mu_0$.

We conclude that $\mathbb{C}$ contains no equilibrium pair $(\mu, \mathbf{x})$, other than $(\mu_0, \mathbf{0})$, with $\mathbf{x}$ on the boundary $\partial \mathbb{R}_+^k$. This rules out alternative (ii) and implies all equilibrium pairs in $\mathbb{C} \setminus \{(\mu_0, \mathbf{0})\}$ are positive. ∎

Of interest in applications is the range of parameter values μ for which there exists (at least one) positive equilibrium. The set
$$\Lambda := \{\mu \mid (\mu, \mathbf{x}) \in \mathbb{C} \setminus (\mu_0, \mathbf{0})\}$$
(called the *spectrum* of $\mathbb{C}$) is a connected interval of parameter values in $\mathbb{R}^1$ for which there is at least one positive equilibrium.[2] The local bifurcation Theorem 54 implies that μ_0 lies in the closure $\bar{\Lambda}$ of Λ.

We give two examples to illustrate the local and global bifurcation results above. The first example involves a forward bifurcation for which Λ is half-life $\mu > \mu_0$. The second application involves a backward bifurcation and multiple equilibria (and a strong Allee effect) for some $\mu < \mu_0$.

Example 92 *Suppose that the age-specific, inherent birth rate b_j in in the Levin-Goodyear model in Example 91 is proportional to the amount of environmental food resources μ consumed by a j-class individual per unit time, so that*
$$b_j = \beta_j \varphi_j \mu$$
where β_j is the number of newborns produced per unit resource consumed. The population projection matrix of the Levin-Goodyear model is the $A(\mu, \mathbf{x}) = \mu \Phi(\mathbf{x}) + T(\mathbf{x})$ where

$$\Phi(\mathbf{x}) = \begin{pmatrix} \varphi_1(\mathbf{x}) & \varphi_2(\mathbf{x}) & \cdots & \varphi_{k-1}(\mathbf{x}) & \varphi_k(\mathbf{x}) \\ 0 & 0 & \cdots & 0 & 0 \\ 0 & 0 & \cdots & 0 & 0 \\ \vdots & \vdots & & \vdots & \vdots \\ 0 & 0 & \cdots & 0 & 0 \end{pmatrix}, \quad T = \begin{pmatrix} 0 & 0 & \cdots & 0 & 0 \\ s_1 & 0 & \cdots & 0 & 0 \\ 0 & s_2 & \cdots & 0 & 0 \\ \vdots & \vdots & & \vdots & \vdots \\ 0 & 0 & \cdots & s_{k-1} & s_k \end{pmatrix}$$

with
$$\varphi_j(\mathbf{x}) = \beta_j \varphi_j \exp\left(-c \sum_{i=1}^k w_i x_i\right), \quad b_i, c, w_i > 0, \quad 0 < s_i < 1.$$

We saw in Example 91 that
$$\mathcal{R}_0 = \mu \left(\sum_{j=1}^{k-1} \beta_j \varphi_j p_j + \beta_k \varphi_k \frac{1}{1 - s_k} \right) < 1$$

implies that the extinction equilibrium is globally asymptotically stable on $\mathbb{R}_+^k$ and $\mathcal{R}_0 > 1$ implies permanence (with respect to $\mathbf{0}$).

It is straightforward to check that all the assumptions required for the application of the local bifurcation Theorem 54 and Corollary 7 hold for this matrix model with $\Omega = \mathbb{R}^k$, $I_\mu = \mathbb{R}_+^1$, and
$$\mu_0 = \left(\sum_{j=1}^{k-1} \beta_j \varphi_j p_j + \beta_k \varphi_k \frac{1}{1 - s_k} \right)^{-1}.$$

---

[2] There is no general guarantee that all equilibria lie on the bifurcating continuum $\mathbb{C}$.

Since the entries in $A(\mu,\mathbf{x})$ that depend on μ are increasing in μ, it follows that $\alpha > 0$ and since the entries that depend on $\mathbf{x}$ are decreasing in x_i it follows that $\kappa > 0$. Thus, the local bifurcation of positive equilibria at $\mu = \mu_0$ is μ-forward and stable.

The Leslie population projection matrix $\mathcal{L}(\mu,\mathbf{x})$ is primitive for all $(\mu,\mathbf{x}) \in \mathbb{R}^k \times \mathbb{R}^k$ and H6 holds with $A_1 = T$ and $A_2 = \Phi(\mathbf{0})$. The global bifurcation Theorem 58 implies the extension of the local bifurcating positive equilibrium pairs to a continuum of positive equilibrium pairs that are unbounded in $\mathbb{R}^1 \times \mathring{\mathbb{R}}_+^k$. The spectrum Λ contains no $\mu < \mu_0$ because then $\mathcal{R}_0(\mu) < 1$ and $\mathbf{x} = \mathbf{0}$ is globally asymptotically stable, implying the existence of no positive equilibrium. Thus $\mathbb{C}$ is unbounded in $\mathring{\mathbb{R}}^1 \times \mathring{\mathbb{R}}_+^k$. In fact, the spectrum Λ is unbounded in $\mathring{\mathbb{R}}^1$, as the following argument shows. For any positive equilibrium pair in $\mathbb{C}$ we saw in Example 91

$$\|\mathbf{x}\| \leq \mu \|\Phi(\mathbf{x})\mathbf{x}\| + \|T\|\|\mathbf{x}\| \leq \mu\phi + \tau\|\mathbf{x}\|.$$

From this, it follows that if Λ were bounded, so would $\mathbb{C}$ be bounded, a contradiction. Thus Λ is an unbounded interval in $\mathring{\mathbb{R}}_+^k$ (a half line) with μ_0 in its closure. In other words, Λ is the half-line $\mu > \mu_0$ and thus The Levin-Goodyear model has (at least one) positive equilibrium for each $\mu > \mu_0$ (or equivalently for each reproduction number $\mathcal{R}_0(\mu_0) > 1$).

The generality of the bifurcation theorems above is illustrated by noticing that the analysis of the Levin-Goodyear model in Example 92 can be carried out with the Ricker nonlinearities replaced by any $\varphi_j(\mathbf{x}) \in C^2(\Omega \to \mathbb{R}_+^1)$ satisfying the conditions

$$\begin{aligned}\frac{\partial \varphi_j(\mathbf{x})}{\partial x_i} &\leq 0 \text{ for all } 1 \leq i,j \leq k \text{ (with at least one strict inequality)} \\ 0 &\leq \varphi_j(\mathbf{x})x_j \leq \phi_j < 0 \text{ for all } 1 \leq j \leq k\end{aligned} \quad (6.28)$$

for all $\mathbf{x} \in \mathbb{R}_+^k$. See Problem 9.

The global bifurcation Theorem 58 concerns the existence of positive equilibria but does not address the stability properties of the equilibria from the continuum $\mathbb{C}$. While the local bifurcation theorem tells use the stability properties of the equilibrium from $\mathbb{C}$ near the bifurcation point, these properties do not necessarily persist everywhere along the continuum $\mathbb{C}$. We know this, for example, from the Ricker equation (which is a special case of the Levin-Goodyear model with $k = 1$ and semelparity $s_1 = 0$.) for which, as we saw in Chapter 1, Section 1.4.5, the positive equilibria resulting from a forward and stable bifurcation eventually destabilize and follow a period-doubling cascade to chaos. Therefore, we would expect the same for the general Levin-Goodyear model, a study of which appears in [212]. See Section 6.5. Similarly, the unstable equilibria from $\mathbb{C}$ outside the neighborhood of a backward bifurcation can be stable, which typically results in a strong Allee effect. This is illustrated in the following example.

Example 93 *A Juvenile-Adult Model with Allee Effect.* Consider a juvenile-adult structured model with

$$\mathbf{x}(t) = \begin{pmatrix} \text{juveniles} \\ \text{adults} \end{pmatrix}$$

and population projection matrix $A(\mathbf{x}) = F(\mathbf{x}) + T(\mathbf{x})$ with fertility and transition matrices

$$F(\mathbf{x}) = \begin{pmatrix} 0 & b\varphi(x_2) \\ 0 & 0 \end{pmatrix}, \quad T(\mathbf{x}) = \begin{pmatrix} 0 & 0 \\ s_1\alpha(x_2) & s_2 \end{pmatrix}.$$

The unit of time is the maturation period. The inherent parameters are the fertility rate $b > 0$, the juvenile survival rate $0 < s_1 < 1$ and the adult survival rate $0 < s_2 < 1$ per unit time. We take the fertility density factor $\varphi \in C^2\left(\Omega \to \mathbb{R}_+^2\right)$, $\varphi(0) = 1$, to be a decreasing function of adult population density and the juvenile survival density factor $\alpha \in C^2\left(\Omega \to \mathbb{R}_+^2\right)$, $\alpha(0) = 1$, to be an increasing function of adult population density (an effect attributed, say, to adult herding projection and nurturing of juveniles):

$$\varphi'(x) < 0, \quad \alpha'(x) > 0.$$

In addition, the survival rate $s_1\alpha(x)$ must remain less than 1 for all x and therefore we assume

$$\lim_{x \to \infty} s_1\alpha(x) := \bar{s}_1 < 1$$

where $\bar{s}_1$ is the maximal possible juvenile survival rate. We also assume an upper bound on fertility

$$\varphi(x) x \leq \phi < \infty$$

for all x.

Under these conditions, Theorems 51 and 53 and Lemmas 11 and 12 imply that the extinction equilibrium is stable for $r < 1$ and the model is permanent (with respect to $\mathbf{x} = \mathbf{0}$) if $r > 1$ and that these assertions remain valid with r replaced by the reproduction number.

$$\mathcal{R}_0 = b\frac{s_1}{1 - s_2}.$$

If we use $\mu = \mathcal{R}_0$ as a bifurcation parameter then we can write the projection matrix as

$$A(\mathcal{R}_0, \mathbf{x}) = \begin{pmatrix} 0 & \mathcal{R}_0 \frac{1-s_2}{s_1}\varphi(x_2) \\ s_1\alpha(x_2) & s_2 \end{pmatrix}.$$

Note that $\alpha > 0$ since the only entry in

$$A(\mathcal{R}_0, \mathbf{0}) = \begin{pmatrix} 0 & \mathcal{R}_0 \frac{1-s_2}{s_1} \\ s_1 & s_2 \end{pmatrix}$$

that involves $\mathcal{R}_0$ is an increasing function of $\mathcal{R}_0$. Theorem 54 implies the local bifurcation of positive equilibrium pairs $(\mathcal{R}_0, \mathbf{x})$ from the extinction equilibrium pair $(1, \mathbf{0})$. Since $A(\mathcal{R}_0, \mathbf{x})$ is primitive for all $\mathcal{R}_0$ and $\mathbf{x}$, the direction of bifurcation (and hence the stability) of the bifurcation depends on the sign of κ. Using formula (6.22) and the left and right eigenvectors

$$\mathbf{w}_L = \begin{pmatrix} s_1 \\ 1 \end{pmatrix}, \quad \mathbf{w}_R = \begin{pmatrix} 1 - s_2 \\ s_1 \end{pmatrix}$$

of

$$A(1, \mathbf{0}) = \begin{pmatrix} 0 & \frac{1-s_2}{s_1} \\ s_1 & s_2 \end{pmatrix}$$

we calculate

$$\kappa = -s_1^2(1 - s_2)(\varphi'(0) + \alpha'(0)).$$

Thus, the bifurcation can be either forward and stable or backward and unstable depending on the relative absolute values of $\varphi'(0) < 0$ and $\alpha'(0) > 0$.

6.3. STABLE BIFURCATION: THE PRIMITIVE CASE

We conclude that the bifurcation is $\mathcal{R}_0$-forward and stable if the negative density effects on fertility (due to increases in low-level adult density) is stronger than the positive effect described by the factor φ:

$$\varphi'(0) + \alpha'(0) < 0 \Rightarrow \text{ bifurcation is } \mathcal{R}_0\text{-forward and stable.}$$

whereas if the positive effect on juvenile survival is sufficient strong then

$$\varphi'(0) + \alpha'(0) > 0 \Rightarrow \text{ bifurcation is } \mathcal{R}_0\text{-backward and unstable.}$$

Noting that H6 holds with

$$A_1 = \begin{pmatrix} 0 & 0 \\ s_1 & s_2 \end{pmatrix}, \quad A_2 = \begin{pmatrix} 0 & \frac{1-s_2}{s_1} \\ 0 & 0 \end{pmatrix}$$

we can apply the global bifurcation Theorem 58 to conclude that the locally bifurcating continuum of the positive equilibrium pairs $(\mathcal{R}_0, \mathbf{x})$ extends globally and is unbounded in $\mathbb{R}^1 \times \mathring{\mathbb{R}}^2_+$. We can say more about the spectrum Λ of $\mathcal{R}_0$ values from the positive equilibrium pairs on the continuum. First of all, $\Lambda \subset \mathring{\mathbb{R}}^1$. If this were not true, then the continuum would contain positive pair $(\mathcal{R}_0, \mathbf{x}) = (0, \mathbf{x})$ with positive $\mathbf{x} \in \mathring{\mathbb{R}}^2_+$, which the equilibrium equations $\mathbf{x} = \mathbf{A}(\mathcal{R}_0, \mathbf{x})\mathbf{x}$ or

$$x_1 = \mathcal{R}_0 \frac{1-s_2}{s_1} \varphi(x_2) x_2$$

$$x_2 = s_1 \alpha(x_2) x_1 + s_2 x_2$$

which shows this is impossible. Secondly, $\Lambda \subset \mathring{\mathbb{R}}^1$ is unbounded. This follows from the inequalities

$$0 \leq x_1 \leq \mathcal{R}_0 \frac{1-s_2}{s_1} \varphi$$

$$0 \leq x_2 \leq s_1 x_1 + s_2 x_2$$

which imply

$$0 \leq x_1 + x_2 \leq \mathcal{R}_0 \frac{1-s_2}{s_1} \varphi + s_m (x_1 + x_2)$$

and hence

$$0 \leq x_1 + x_2 \leq \mathcal{R}_0 \frac{1}{1-s_m} \frac{1-s_2}{s_1} \varphi$$

where $s_m := \max\{s_1, s_2\} < 1$. If Λ were bounded, then the equilibria $\mathbf{x}$ from the continuum would also be bounded, which contradicts that the continuum is unbounded.

It follows that the spectrum Λ is a half line of positive real numbers that contains 1 in its closure. Thus, there exists at least one positive equilibrium for each $\mathcal{R}_0 > 1$.

If the bifurcation is backward ($\kappa < 0$) then Λ also contains points $\mathcal{R}_0 < 1$ and since the continuum outside a neighborhood of the bifurcation point is unbounded is must "turn around" and produce positive equilibrium pairs outside the neighborhood. This multiple-positive equilibrium scenario usually leads to a strong Allee effect, by which we mean scenarios in which both a positive equilibrium and the extinction equilibrium are stable and survival is initial condition dependent. To verify this in this example, we need to know that there are stable positive equilibria for $\mathcal{R}_0 < 1$.

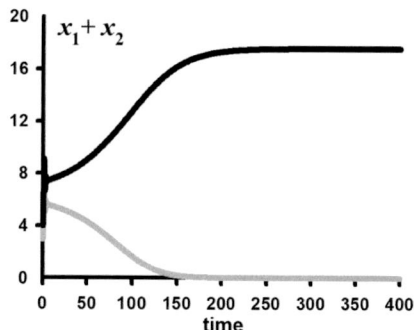

Figure 6.1: The total population sizes $x_1(t) + x_2(t)$ of two solutions of the juvenile-adult model in Example 93 with density factors (6.29) with parameter values $\mathcal{R}_0 = 0.85$, $s_1 = 0.2$, $s_{1m} = 0.4$, $s_2 = 0.5$, $c_1 = 0.05$, $c_2 = 1$, which yield $\kappa = -0.003$. The solution with initial condition $(x_1(0), x_2(0))^T = (4,0)^T$ (shown in black) approaches the positive equilibrium $(10.773, 6.791)^T$ while that with initial condition $(x_1(0), x_2(0))^T = (3,0)^T$ (shown in gray) approaches the extinction equilibrium $(0,0)^T$.

We have no general proof of this, but illustrate the phenomenon with sample simulations shown in Figure 6.1 using the density factors

$$\varphi(x_2) = \frac{1}{1 + c_1 x_2}, \quad \alpha(x_2) = \frac{1 + c_2 s_{1m} x_2}{1 + c_2 s_1 x_2} \qquad (6.29)$$

in the model with $c_i > 0$ and $s_1 < s_{1m} < 1$. The Allee factor $\alpha(x_2)$ increases juvenile survival from s_1 to a maximum of s_{1m} as adult density increases. For this case

$$\kappa = s_1^2 (1 - s_2)(c_1 - c_2(s_1^* - s_1))$$

is negative if the Allee coefficient c_2 is sufficiently large, specifically if $c_2 > c_1/(s_1^* - s_1)$. Figure 6.1 shows sample solutions for a selection of parameter values that imply a backward bifurcation occurs. One solution tends to a positive equilibrium while the other tends to the extinction equilibrium, suggesting that a strong Allee effect is present.

Example 93 illustrates how a backward bifurcation can produce a strong Allee effect, which is a common occurrence in population models. The reason is that the positive density effect (component Allee effect) at low population densities that causes the backward bifurcation is overridden, in virtually all population models, by negative density effects at high densities. For more on backward bifurcations and strong Allee effectssee [76]. Strong Allee effects, can, however, also occur when the local bifurcation is forward and stable and a hysteresis effect is results in stable positive equilibria for $\mathcal{R}_0 < 1$.

6.4 Stable Bifurcation: The Imprimitive Case

The local bifurcation Theorem 54 concerning the existence of positive equilibria does not require a primitivity assumption on the population projection matrix. The local stability Theorem 57 and

Corollary 7, on the other hand, assume the primitivity of the inherent population projection matrix. This added assumption is required for the validity of these results, which relate the direction of bifurcation to the stability of the bifurcating positive equilibria (in a neighborhood of the bifurcation point). If the inherent the projection matrix is imprimitive, then in general it is no longer true that the stability of the bifurcation can be determined from its direction. This issue is not just of academic interest, since there are notable applications in which the projection matrix is not primitive (e.g., see [40], [115], [116], [297]). In this case, the nature of the bifurcation that occurs when the extinction equilibrium destabilizes is not thoroughly understood, except in low-dimensional or specialized cases.

A type of model with imprimitive projection matrix that has found considerable application in population dynamics is the Leslie age-structured model with projection matrix

$$\mathcal{L} = \begin{pmatrix} 0 & 0 & \cdots & 0 & f_k \\ s_1 & 0 & \cdots & 0 & 0 \\ 0 & s_2 & \cdots & 0 & 0 \\ \vdots & \vdots & & \vdots & \vdots \\ 0 & 0 & \cdots & s_{k-1} & 0 \end{pmatrix}, \quad f_k > 0,\ 0 < s_i < 1. \tag{6.30}$$

In this model, the individuals in the first $k-1$ age classes are juveniles (non-reproducing individuals) and only those in the kth class are adults (reproducing individuals), which accounts for the zeros in the first row of $\mathcal{L}$. The zero in the lower right corner implies that adults do not survive a unit of time and so this model is often used to study the dynamics of a semelparous population.[3] The k eigenvalues of $\mathcal{L}$ are the kth the roots of the the reproduction number

$$\mathcal{R}_0 = f_k \Pi_{k=1}^{k-1} s_i$$

and hence have equal absolute values (Problem 8). Since the dominant eigenvalue (spectral radius) $r = \rho(0) = \mathcal{R}_0$ is not strictly dominant, this Leslie matrix is imprimitive. Notice that when $\mathcal{R}_0$ increases through 1, all eigenvalues simultaneously leave the unit circle in the complex plane.

A nonlinear semelparous Leslie model has fertility and survival density factors $\beta(\mathbf{x})$ and $\sigma_i(\mathbf{x})$ in the projection matrix

$$\mathcal{L}(\mathbf{x}) = \begin{pmatrix} 0 & 0 & \cdots & 0 & f_k\beta(\mathbf{x}) \\ s_1\sigma_1(\mathbf{x}) & 0 & \cdots & 0 & 0 \\ 0 & s_2\sigma_2(\mathbf{x}) & \cdots & 0 & 0 \\ \vdots & \vdots & & \vdots & \vdots \\ 0 & 0 & \cdots & s_{k-1}\sigma_{k-1}(\mathbf{x}) & 0 \end{pmatrix}, \quad f_k > 0, 0 < s_i < 1. \tag{6.31}$$

that satisfy H1 and

$$\beta(\mathbf{0}) = \sigma_i(\mathbf{0}) = 0$$

so that f_k and s_i retain their interpretation as the density-free fertility and survival rates. Perhaps the most well-known application of a model of this time is the population dynamics of the

---

[3] A semelparous population is one in which individuals have only one reproductive episode in their lifetime, after which they usually die. Annual plants are the prototypical example, but many animal species are semelparous as well (including many insects, such as the famous periodic cicadas; some mollusks, such as octopi and squids, and fish, such as some species of salmon; and even a few reptiles, amphibians, and mammals.)

famous periodical cicadas (*Cicadidae Magicicada* spp.), the longest-lived insects, which emerge every $k = 17$ years to reproduce [40]. There are also thousands of other semelparous species spread throughout many taxa, including numerous insert, arachnid, mollusk, fish, and of course plant species. This has stimulated numerous mathematical studies of nonlinear semelparous Leslie models [69], [71], [74], [77], [92], [100], [101], [102], [112], [194], [195], [200], [296].

We first note that Theorems 54 and 55 apply to the nonlinear semelparous Leslie model, since it meets all the requirements of those theorems for which the primitivity of $\mathcal{L}(\mathbf{0})$ is not required. Thus, we know that the extinction equilibrium destabilizes as $\mathcal{R}_0$ increases through 1 with the result that a branch of positive equilibria bifurcate from the extinction equilibrium with the direction of bifurcation is determined by the sign of κ given by (6.22). What we do not know is whether the direction of bifurcation determines the stability of the bifurcating positive equilibria, i.e., whether a forward bifurcation ($\kappa > 0$) is stable and a backward bifurcation ($\kappa < 0$) is unstable. This is because Theorem 54 and Corollary 7 require the primitivity of $\mathcal{L}(\mathbf{0})$.

6.4.1 The Case $k = 2$

The following example in fact shows that, in the absence of primitivity, the direction of bifurcation does *not* in general determine the stability of the bifurcating positive equilibrium.

Example 94 *Consider the $k = 2$ dimensional model $\mathbf{x}(t+1) = \mathcal{L}(\mathbf{x}(t))\mathbf{x}(t)$ with the Leslie matrix* (6.31) *and*

$$\beta(\mathbf{x}) = \frac{1}{1 + c_{21}x_1 + c_{22}x_2}, \quad \sigma_1(\mathbf{x}) \equiv 1. \tag{6.32}$$

In Problem 1, the reader is asked to show that the only equilibrium, other than the extinction equilibrium, is

$$\mathbf{x}^* = \begin{pmatrix} \frac{1}{c_{21}+sc_{22}}(\mathcal{R}_0 - 1) \\ s\frac{1}{c_{21}+sc_{22}}(\mathcal{R}_0 - 1) \end{pmatrix} \tag{6.33}$$

for $\mathcal{R}_0 = bs \neq 1$. Thus, we see that in this example the local bifurcation of positive equilibria at $\mathcal{R}_0 = 1$ guaranteed by Theorem 54 extends globally with spectrum $\mathcal{R}_0 > 1$. Also from Problem 1 we have that the Jacobian of the model equation, when evaluated at this equilibrium, is

$$J(\mathbf{x}^*) = \begin{pmatrix} -\frac{1}{\mathcal{R}_0}c_{21}\frac{\mathcal{R}_0-1}{c_{21}+sc_{22}} & \frac{1}{\mathcal{R}_0}\frac{1}{s}\frac{\mathcal{R}_0 c_{21}+sc_{22}}{c_{21}+sc_{22}} \\ s & 0 \end{pmatrix}. \tag{6.34}$$

Recall the Jury conditions that guarantee the eigenvalues to satisfy $|\lambda| < 1$ are

$$|\operatorname{tr} J(\mathbf{x}^*)| < 1 + \det J(\mathbf{x}^*) \quad \text{and} \quad \det J(\mathbf{x}^*) < 1$$

(Chapter 3, Section 3.2.2). The second inequality is satisfied because

$$\det J(\mathbf{x}^*) = -\frac{1}{\mathcal{R}_0}\frac{\mathcal{R}_0 c_{21}+sc_{22}}{c_{21}+sc_{22}} < 0.$$

Thus, $\mathbf{x}^*$ is stable if the first inequality

$$\frac{1}{\mathcal{R}_0}c_{21}\frac{\mathcal{R}_0-1}{c_{21}+sc_{22}} < 1 - \frac{1}{\mathcal{R}_0}\frac{\mathcal{R}_0 c_{21}+sc_{22}}{c_{21}+sc_{22}}$$

6.4. STABLE BIFURCATION: THE IMPRIMITIVE CASE

is satisfied. Some algebraic manipulations show this condition is equivalent to $c_{21} < sc_{22}$. The Jury test also tells us that the equilibrium is unstable if the inequality is reversed. We have then that

$\frac{c_{21}}{c_{22}} < s$ implies the positive equilibrium $\mathbf{x}^*$ is (locally asymptotically) stable
$\frac{c_{21}}{c_{22}} > s$ implies the positive equilibrium $\mathbf{x}^*$ is unstable.

Biologically, we conclude that weak intra-class competition relative to within class competition (as measured by the ratio c_{21}/c_{22}) implies the equilibrium is stable, whereas strong intra-class competition relative to within class competition implies the equilibrium is unstable.

Example 94 shows that a forward bifurcation guaranteed by the local bifurcation Theorem 54 is not necessarily stable and that the assumption of primitivity cannot be dropped in Corollary 7.

In Example 94 the destabilization of the extinction equilibrium at $\mathcal{R}_0 = 1$ is due not only because an eigenvalue of the inherent projection matrix (i.e., the Jacobian evaluated at the extinction equilibrium) leaves the unit circle in the complex plane through $+1$ but also because the other eigenvalue leaves simultaneously through -1, as can be seen by the Jacobian evaluated at the bifurcation point

$$J(\mathbf{0}) = \begin{pmatrix} 0 & \frac{1}{s} \\ s & 0 \end{pmatrix}$$

whose eigenvalues are ± 1. The first case is associated with the transcritical bifurcation of equilibria at $\mathcal{R}_0 = 1$. The second case is generally associated with a period-doubling bifurcation, which leads us to suspect that 2-cycles play a role in Example 94. That this is the case is seen in the next example.

Example 95 *Using the projection matrix*

$$\mathcal{L}(\mathbf{x}) = \begin{pmatrix} 0 & b\frac{1}{1+c_{21}x_1+c_{22}x_2} \\ s & 0 \end{pmatrix} \tag{6.35}$$

from Example 94, we notice that the boundary $\partial \mathbb{R}_+^2$ is forward invariant because the positive x_1-axis maps to the positive x_2-axis and vice versa. For example, starting from an initial condition of the positive x_1-axis, we get a solution of the form

$$\begin{pmatrix} x_1(0) \\ 0 \end{pmatrix} \to \begin{pmatrix} 0 \\ x_2(1) \end{pmatrix} \to \begin{pmatrix} x_1(3) \\ 0 \end{pmatrix} \to \begin{pmatrix} 0 \\ x_2(4) \end{pmatrix} \to \begin{pmatrix} x_1(5) \\ 0 \end{pmatrix} \to \begin{pmatrix} 0 \\ x_2(6) \end{pmatrix} \to \cdots.$$

*These synchronized orbits have temporally separated juvenile and adult classes. If such a sequence is periodic with period 2, then we call the orbit a **synchronous or a single class 2-cycle**. Single class 2-cycles are fixed points of the composite map and are found by solving the composite map's equilibrium equation*

$$\begin{pmatrix} x_1 \\ 0 \end{pmatrix} = \begin{pmatrix} 0 & b \\ s & 0 \end{pmatrix} \begin{pmatrix} 0 \\ sx_1 \end{pmatrix}$$

for

$$x_1 = \frac{1}{c_{21} + sc_{22}} (\mathcal{R}_0 - 1).$$

Thus, in addition to a positive equilibrium for each $\mathcal{R}_0 > 1$, this model has a single class 2-cycle consisting of the two points

$$\mathbf{x}_1 = \begin{pmatrix} \frac{1}{sc_{22}}(\mathcal{R}_0 - 1) \\ 0 \end{pmatrix}, \quad \mathbf{x}_2 = \begin{pmatrix} 0 \\ \frac{1}{c_{22}}(\mathcal{R}_0 - 1) \end{pmatrix} \quad \text{for } \mathcal{R}_0 > 1. \tag{6.36}$$

Note that these single class 2-cycles also bifurcate from the extinction equilibrium at $\mathcal{R}_0 = 1$. Their stability can be determined by the Linearization Principle applied to the composite map (Chapter 3, Section 3.5). The Jacobian of the composite evaluated at the 2-cycle is the product of the Jacobians along the cycle, which is

$$J(\mathbf{x}_2)J(\mathbf{x}_1) = \begin{pmatrix} \frac{1}{\mathcal{R}_0} & \frac{1}{s}c_{21}\frac{\mathcal{R}_0 - 1}{c_{21} - sc_{22} - \mathcal{R}_0 c_{21}} \\ 0 & \mathcal{R}_0 \frac{sc_{22}}{(\mathcal{R}_0 - 1)c_{21} + sc_{22}} \end{pmatrix}$$

Note that one eigenvalue $\lambda_1 = 1/\mathcal{R}_0$ is positive and less than 1. The second eigenvalue is positive and satisfies

$$\lambda_2 = \mathcal{R}_0 \frac{sc_{22}}{(\mathcal{R}_0 - 1)c_{21} + sc_{22}} > 1$$

if $c_{21} < sc_{22}$ and $\lambda_2 < 1$ if $c_{21} > sc_{22}$. We conclude that

$\frac{c_{21}}{c_{22}} < s$ implies the synchronous 2-cycle is unstable
$\frac{c_{21}}{c_{22}} > s$ implies the synchronous 2-cycle (locally asymptotically) stable.

Combining the results in Examples 94 and 95, we see that in the model equation with the imprimitive projection matrix (6.35) there occurs the simultaneous, forward bifurcation of positive equilibria and single class 2-cycles at $\mathcal{R}_0 = 1$ and that they exhibit a *dynamic dichotomy*, by which is meant that one is stable and the other is unstable, depending on the ratio c_{21}/c_{22}. See Figure 6.2 for a numerical simulation that illustrates this dichotomy.

The simultaneous bifurcation of positive equilibria in $\mathring{\mathbb{R}}_+^k$ and single class cycles on the boundary $\partial \mathbb{R}_+^k$ is a general feature of nonlinear semelparous Leslie models. A single-class k-cycle is a periodic orbit consisting of the points

$$\mathbf{x}_1 = \begin{pmatrix} x_1 \\ 0 \\ 0 \\ \vdots \\ 0 \end{pmatrix}, \mathbf{x}_2 = \begin{pmatrix} 0 \\ x_2 \\ 0 \\ \vdots \\ 0 \end{pmatrix}, \mathbf{x}_3 = \begin{pmatrix} 0 \\ 0 \\ x_3 \\ \vdots \\ 0 \end{pmatrix}, \ldots, \mathbf{x}_k = \begin{pmatrix} 0 \\ 0 \\ 0 \\ \vdots \\ x_k \end{pmatrix}$$

on the boundary $\partial \mathbb{R}_+^k$. The next theorem concerns single class k-cycles and the semelparous Leslie model.

Theorem 59 *Consider the semelparous Leslie model with projection matrix (6.31) with density factors that satisfy H1 and $\beta(\mathbf{0}) = \sigma_i(\mathbf{0}) = 0$. The extinction equilibrium destabilizes as $\mathcal{R}_0 = f_k \Pi_{k=1}^{k-1} s_i$ increases through 1.*
(a) *If*

$$\kappa := -\sum_{i=1}^{k-1}\sum_{j=1}^{k} p_j \frac{\partial \sigma_i(\mathbf{0})}{\partial x_j} - \sum_{j=1}^{k} p_j \frac{\partial \beta(\mathbf{0})}{\partial x_j} \neq 0 \quad (6.37)$$

where

$$p_j = \begin{cases} 1 & \text{for } j = 1 \\ \Pi_{n=1}^{j-1} s_n & \text{for } j = 2, 3, \ldots, k \end{cases},$$

then a branch of positive equilibria bifurcates from the extinction equilibrium at $\mathcal{R}_0 = 1$. The bifurcation is forward if $\kappa > 0$ and backward if $\kappa < 0$.

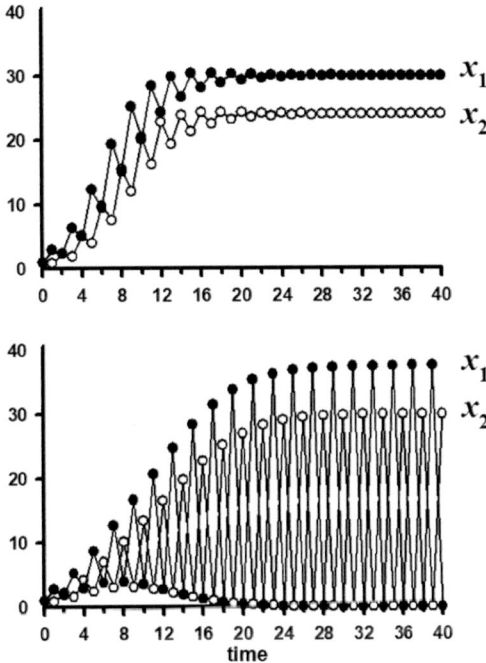

Figure 6.2: A sample solution, with initial condition $(x_1(0), x_2(0))^T = (1,1)^T$ of the model equation in Examples 94 and 95 with coefficients $\mathcal{R}_0 = 2.5$, $s = 0.5$ and $c_{22} = 0.01$, is shown. In the upper plot, $c_{21} = 0.01$ so that $c_{21}/c_{22} = 0.1 < s$ and the positive equilibrium is stable (and the synchronous 2-cycle is unstable). In the lower plot, $c_{21} = 0.1$ so that $c_{21}/c_{22} = 1 > s$ and the 2-cycle is stable (and the positive equilibrium is unstable).

(b) If
$$\delta := -\sum_{i=1}^{k-1} \frac{\partial \sigma_i(\mathbf{0})}{\partial x_i} p_i - \frac{\partial \beta(\mathbf{0})}{\partial x_k} p_k \neq 0$$
then a branch of single-class k-cycles bifurcates from the extinction equilibrium at $\mathcal{R}_0 = 1$. The bifurcation is forward if $\delta > 0$ and backward if $\delta < 0$.

Proof. (a) These assertions follow from Theorem 54 and Corollary 7.

(b) As pointed out above, the positive x_1-axis is forward invariant and single-class k-cycles are fixed points of the kth composite. For notational convenience in this proof, we write $\beta(\mathbf{x}) = \beta(x_1,\ldots,x_k)$ and $\sigma_i(\mathbf{x}) = \sigma_i(x_1,\ldots,x_k)$ and define

$$\bar{\beta}(x_k) := \beta(0,\ldots,0,x_k) \text{ and } \bar{\sigma}_i(x_i) := \sigma(0,\ldots,0,x_i,0,\ldots,0).$$

Note that
$$\bar{\beta}(0) = \bar{\sigma}_i(0) = 1$$
$$\bar{\beta}'(0) = \frac{\partial \beta(\mathbf{0})}{\partial x_k} \text{ and } \bar{\sigma}'_i(0) = \frac{\partial \sigma_i(\mathbf{0})}{\partial x_i}. \tag{6.38}$$

An initial point on the x_1-axis generates the orbit

$$\begin{pmatrix} x_1 \\ 0 \\ 0 \\ \vdots \\ 0 \end{pmatrix} \to \begin{pmatrix} 0 \\ x_2(x_1) \\ 0 \\ \vdots \\ 0 \end{pmatrix} \to \begin{pmatrix} 0 \\ 0 \\ x_3(x_1) \\ \vdots \\ 0 \end{pmatrix} \to \cdots \to \begin{pmatrix} 0 \\ 0 \\ 0 \\ \vdots \\ x_k(x_1) \end{pmatrix}$$

where the sequence $x_i(x_1)$ is defined recursively by

$$x_1(x_1) = x_1$$
$$x_2(x_1) = s_1 \bar{\sigma}_1(x_1) x_1$$
$$x_3(x_1) = s_2 \bar{\sigma}_2(x_2(x_1)) x_2(x_1)$$
$$= s_2 s_1 \bar{\sigma}_2(x_2(x_1)) \bar{\sigma}_1(x_1) x_1$$
$$x_4(x_1) = s_3 \bar{\sigma}_3(x_3(x_1)) x_3(x_1)$$
$$= s_3 s_2 s_1 \bar{\sigma}_3(x_3(x_1)) \bar{\sigma}_2(x_2(x_1)) \bar{\sigma}_1(x_1) x_1$$
$$\vdots$$
$$x_k(x_1) = s_{k-1} \bar{\sigma}_{k-1}(x_{k-1}(x_1)) x_{k-1}(x_1)$$
$$= \left(\Pi_{i=1}^{k-1} s_i\right) \Pi_{i=1}^{k-1} \bar{\sigma}_i(x_i(x_1)) x_1$$

Note $x_i(0) = 0$ for all i. The dynamics on the x_1-axis is described by the map

$$\begin{pmatrix} x_1 \\ 0 \\ 0 \\ \vdots \\ 0 \end{pmatrix} \to \begin{pmatrix} f_k \bar{\beta}(x_k(x_1)) x_k(x_1) \\ 0 \\ 0 \\ \vdots \\ 0 \end{pmatrix} = \begin{pmatrix} \mathcal{R}_0 \bar{\beta}(x_k(x_1)) \Pi_{i=1}^{k-1} \bar{\sigma}_i(x_i(x_1)) x_1 \\ 0 \\ 0 \\ \vdots \\ 0 \end{pmatrix}$$

6.4. STABLE BIFURCATION: THE IMPRIMITIVE CASE

or simply by the (one dimensional) difference equation

$$x_1(t+1) = a(x_1(t))x_1(t)$$

where

$$a(x_1) := \mathcal{R}_0 \bar{\beta}(x_k(x_1)) \Pi_{i=1}^{k-1} \bar{\sigma}_i(x_i(x_1))$$

to which we can apply Theorems 54 and 55 (with $k = 1$ and bifurcation parameter $\mathcal{R}_0$). The equilibrium $x_1 = 0$ destabilizes as $\mathcal{R}_0$ increases through 1 resulting in the bifurcation of positive equilibria, which correspond to single class k-cycles of the semelparous Leslie model. The direction of bifurcation is determined by κ for this one-dimensional case, which is equal to $da(x_1)/dx_1$ evaluated at the bifurcation point $x_1 = 0$ and $\mathcal{R}_0 = 1$. An application of the chain rule and the use of (6.38) shows that this derivative equals $-\delta$. ∎

Theorem 59 establishes the dual bifurcation of positive equilibria and single class k-cycles for the general semelparous Leslie model, but it does not address the stability properties of either. A thorough understanding of the bifurcation at $\mathcal{R}_0 = 1$ is currently available only for dimension $k = 2$ with projection matrix

$$\mathcal{L}(\mathbf{x}) = \begin{pmatrix} 0 & f\beta(\mathbf{x}) \\ s\sigma(\mathbf{x}) & 0 \end{pmatrix}, \quad f > 0, \ 0 < s < 1$$

(unneeded subscripts have been dropped). This matrix equation is known as (the generalized) Ebenman model [83], [115], [116]. The density factors β and σ are assumed to satisfy H1 and

$$\beta(\mathbf{0}) = \sigma(\mathbf{0}) = 1.$$

Using the formula $\mathcal{R}_0 = fs$ for the (inherent) reproduction number, we can introduce $\mathcal{R}_0$ explicitly into the projection matrix

$$\mathcal{L}(\mathbf{x}) = \begin{pmatrix} 0 & \mathcal{R}_0 \frac{1}{s} \beta(\mathbf{x}) \\ s\sigma(\mathbf{x}) & 0 \end{pmatrix} \tag{6.39}$$

for use as the bifurcation parameter μ in Theorem 54. When $\mathcal{R}_0 = 1$ we get

$$\mathcal{L}(\mathbf{0}) = \begin{pmatrix} 0 & \frac{1}{s} \\ s & 0 \end{pmatrix}$$

and right and left positive eigenvectors (associated with eigenvalue 1)

$$\mathbf{w}_R = \begin{pmatrix} 1 \\ s \end{pmatrix}, \quad \mathbf{w}_L = \begin{pmatrix} 1 \\ \frac{1}{s} \end{pmatrix}. \tag{6.40}$$

From formulas (6.19) and (6.22) with bifurcation parameter $\mu = \mathcal{R}_0$ we calculate

$$\alpha = 1, \quad \kappa = -(\delta_1 + \delta_2) \tag{6.41}$$

where

$$\delta_1 := s \frac{\partial \beta(0)}{\partial x_2} + \frac{\partial \sigma(0)}{\partial x_1}, \quad \delta_2 := s \frac{\partial \sigma(0)}{\partial x_2} + \frac{\partial \beta(0)}{\partial x_1}. \tag{6.42}$$

Note that δ_1 measures the within class density effects while δ_2 measures the between class density effects. From Theorem 54 we obtain the (local) bifurcation of a continuum $\mathbb{C}_+$ of positive equilibrium pairs $(\mathcal{R}_0(\varepsilon), \mathbf{x}(\varepsilon))$ with the parameterization

$$\begin{aligned} \mathcal{R}_0(\varepsilon) &= 1 + \kappa\varepsilon + O(\varepsilon^2) \\ \mathbf{x}(\varepsilon) &= \mathbf{w}_R \varepsilon + O(\varepsilon^2) \end{aligned} \tag{6.43}$$

for small $\varepsilon > 0$.

Theorem 60 *Consider the $k=2$ nonlinear semelparous Leslie model (6.39) under the assumptions H1 and $\beta(\mathbf{0}) = \sigma(\mathbf{0}) = 1$. Assume $\delta_1 \pm \delta_2 \neq 0$ and $\delta_1 \neq 0$ where δ_1 and δ_2 are defined by (6.42). Then a continuum of positive equilibria $\mathbb{C}_+$ and a continuum $\mathbb{C}_2$ (nonnegative) synchronous, single class 2-cycles both bifurcate from the extinction equilibrium at $\mathcal{R}_0 = 1$.*

(a) *$\mathbb{C}_+$ bifurcates forward if $\delta_1 + \delta_2 < 0$ and backward if $\delta_1 + \delta_2 > 0$ and $\mathbb{C}_2$ bifurcates forward if $\delta_1 < 0$ and backward if $\delta_1 > 0$.*

(b) *If both $\mathbb{C}_+$ and $\mathbb{C}_2$ both bifurcate forward, then a dynamic dichotomy occurs between them, i.e., one bifurcation is stable and the other is unstable. Specifically, $\mathbb{C}_+$ is a stable bifurcation if $\delta_1 - \delta_2 < 0$ and $\mathbb{C}_+$ is an unstable bifurcation if $\delta_1 - \delta_2 > 0$.*

(c) *If either $\mathbb{C}_+$ or $\mathbb{C}_2$ bifurcates backward, then both bifurcations are unstable.*

Proof. The existence of a continuum $\mathbb{C}_+$ of equilibrium pairs (6.43)

$$\mathcal{R}_0(\varepsilon) = 1 - (\delta_1 + \delta_2)\varepsilon + O(\varepsilon^2)$$

$$\mathbf{x}(\varepsilon) = \begin{pmatrix} 1 \\ s \end{pmatrix} \varepsilon + O(\varepsilon^2)$$

is guaranteed by Theorem 54. To study the stability of these equilibria by means of the Linearization Principle, we evaluate the Jacobian of $\mathcal{L}(\mathbf{x})\mathbf{x}$ at the equilibrium pair (6.43). The resulting Jacobian matrix is a function of ε, which we denote by $J(\varepsilon)$, and hence so are its two eigenvalues $\lambda_i(\varepsilon)$, $i = 1$ and 2. Since the eigenvalues of $J(0) = \mathcal{L}(0)$ are ± 1 we have $\lambda_1(0) = 1$ and $\lambda_2(0) = -1$. We can determine whether these eigenvalues satisfy $|\lambda_i(\varepsilon)| < 1$ (or > 1) for small $\varepsilon > 0$ from the derivatives $\lambda_i'(0)$. From (6.25) and (6.41) we have

$$\lambda_1'(0) = \frac{\delta_1 + \delta_2}{2}.$$

To calculate $\lambda_2'(0)$ let $\mathbf{w}_{R2}(\varepsilon)$ and $\mathbf{w}_{L2}(\varepsilon)$ denote right and left eigenvectors of the Jacobian $J(\varepsilon)$ associated with $\lambda_2(\varepsilon)$. Since $J(0) = \mathcal{L}(0)$, $\mathbf{w}_{R2}(0)$ and $\mathbf{w}_{L2}(0)$ are right and left eigenvectors of $\mathcal{L}(0)$ associated with eigenvalue -1, which we take as

$$\mathbf{w}_{R2}(0) = \begin{pmatrix} 1 \\ -s \end{pmatrix}, \quad \mathbf{w}_{L2}(0) = \begin{pmatrix} 1 \\ -\frac{1}{s} \end{pmatrix}.$$

By differentiating $J(\varepsilon)\mathbf{w}_{R2}(\varepsilon) = \lambda_2(\varepsilon)\mathbf{w}_{R2}(\varepsilon)$ with respect to ε and evaluate the answer at $\varepsilon = 0$ we obtain the equation

$$(J(0) - \lambda_2(0)I)\mathbf{w}_{R2}'(0) = (J'(0) - \lambda_2'(0))\mathbf{w}_{R2}(0)$$

which is solvable for $\mathbf{w}_{R2}'(0)$ if and only if (by the Fredholm Alternative) the right side is orthogonal to $\mathbf{w}_{L2}(0)$. This leads to

$$\lambda_2'(0) = \frac{\mathbf{w}_{L2}(0)^T J'(0) \mathbf{w}_{R2}(0)}{\mathbf{w}_{L2}(0)^T \mathbf{w}_{R2}(0)}.$$

where

$$J(\varepsilon) = \begin{pmatrix} \mathcal{R}_0(\varepsilon)\frac{1}{s}\frac{\partial \beta(\mathbf{x}(\varepsilon))}{\partial x_1}x_2 & \mathcal{R}_0(\varepsilon)\frac{1}{s}\frac{\partial \beta(\mathbf{x}(\varepsilon))}{\partial x_2}x_2 + \mathcal{R}_0(\varepsilon)\frac{1}{s}\beta(\mathbf{x}(\varepsilon)) \\ s\frac{\partial \sigma(\mathbf{x}(\varepsilon))}{\partial x_1}x_1 + s\sigma(\mathbf{x}(\varepsilon)) & s\frac{\partial \sigma(\mathbf{x}(\varepsilon))}{\partial x_2}x_1 \end{pmatrix}$$

6.4. STABLE BIFURCATION: THE IMPRIMITIVE CASE

and hence

$$J'(0) = \begin{pmatrix} \frac{1}{s}\frac{\partial \beta(0)}{\partial x_1}s & \frac{1}{s}\frac{\partial \beta(0)}{\partial x_2}s + \frac{1}{s}\left(\frac{\partial \beta(0)}{\partial x_1} + \frac{\partial \beta(0)}{\partial x_2}s\right) + \kappa\frac{1}{s} \\ s\frac{\partial \sigma(0)}{\partial x_1} + s\left(\frac{\partial \sigma(0)}{\partial x_1} + \frac{\partial \sigma(0)}{\partial x_2}s\right) & s\frac{\partial \sigma(0)}{\partial x_2} \end{pmatrix}.$$

Thus

$$\lambda_2'(0) = -\frac{1}{2}(\delta_1 - \delta_2)$$

and we have the expansions

$$\begin{aligned}\lambda_1(\varepsilon) &= 1 + \tfrac{1}{2}(\delta_1 + \delta_2)\varepsilon + O(\varepsilon^2) \\ \lambda_2(\varepsilon) &= -1 - \tfrac{1}{2}(\delta_1 - \delta_2)\varepsilon + O(\varepsilon^2)\end{aligned} \qquad (6.44)$$

for the eigenvalues of the Jacobian.

With regard to synchronous 2-cycles, we saw in the proof of Theorem 59(b), that they consist of two points

$$\mathbf{x}_1 = \begin{pmatrix} x_1 \\ 0 \end{pmatrix}, \quad \mathbf{x}_2 = \begin{pmatrix} 0 \\ x_2 \end{pmatrix}$$

where $x_1 > 0$ is a fixed point of the $k = 1$ dimensional (composite) difference equation

$$x_1(t+1) = \mathcal{R}_0 a(x_1(t)) x_1(t) \qquad (6.45)$$

with

$$a(x_1) := \beta(0, s\sigma(x_1, 0) x_1) \sigma(x_1, 0)$$

and the notation $\beta(\mathbf{x}) = \beta(x_1, x_2)$ and $\sigma(\mathbf{x}) = \sigma(x_1, x_2)$. Note that

$$a(0) = 1, \quad a'(0) = \delta_1.$$

Theorem 54 can be applied to this ($k = 1$ dimensional matrix) equation with the result that a continuum of positive equilibrium pairs of the parameterized form

$$\begin{aligned}\mathcal{R}_0(\varepsilon) &= 1 + \kappa\varepsilon + O(\varepsilon^2) \\ x_1(\varepsilon) &= \varepsilon + O(\varepsilon^2)\end{aligned}$$

bifurcates from $x_1 = 0$ at $\mathcal{R}_0 = 1$ for small $\varepsilon > 0$. In this case, by the chain rule, we have

$$\kappa = -a'(0) = -\left(\frac{\partial \sigma(0)}{\partial x_1} + \frac{\partial \beta(0)}{\partial x_2}s\right) = -\delta_1.$$

The component of the second point in the 2-cycle has the expansion

$$x_2(\varepsilon) = s\sigma(x_1(\varepsilon), 0) x_1(\varepsilon) = s\varepsilon + O(\varepsilon^2)$$

This proves the existence of the bifurcating synchronous 2-cycle pairs

$$\mathcal{R}_0(\varepsilon) = 1 - \delta_1\varepsilon + O(\varepsilon^2)$$

$$\mathbf{x}_1(\varepsilon) = \begin{pmatrix} 1 \\ 0 \end{pmatrix}\varepsilon + O(\varepsilon^2), \quad \mathbf{x}_2(\varepsilon) = \begin{pmatrix} 0 \\ s \end{pmatrix}\varepsilon + O(\varepsilon^2)$$

for small $\varepsilon > 0$. The stability of these 2-cycles can by determined by the Linearization Principle applied to the composite matrix model, which is the product $P(\varepsilon)$ of the Jacobian of $\mathcal{L}(\mathbf{x})\mathbf{x}$ evaluated at $\mathbf{x}_1(\varepsilon)$ and $\mathbf{x}_2(\varepsilon)$ (see Chapter 3, Section 3.5). It turns out that this Jacobian is a lower triangular matrix with eigenvalues

$$\begin{aligned}\mu_1 &= 1 - (\delta_1 - \delta_2)\varepsilon + O(\varepsilon^2) \\ \mu_2 &= 1 + \delta_1 \varepsilon + O(\varepsilon^2)\end{aligned} \tag{6.46}$$

for small $\varepsilon > 0$.

The conclusions (a)–(c) follow from a study of the eigenvalue expansions (6.44) and (6.46) and when they are, or are not, less than 1 in absolute value (for small $\varepsilon > 0$). ∎

Theorem 60(b) shows that the dynamic dichotomy seen in Examples 94 and 95 occurs generally when both the positive equilibria and the synchronous 2-cycles forward bifurcate at $\mathcal{R}_0 = 1$ in a $k = 2$ dimensional semelparous Leslie model. As discussed in Section 6.3 and in Section 5.5 of Chapter 5, a backward bifurcation at $\mathcal{R}_0 = 1$ in a population model, as in Theorem 60(c), usually results (due to negative density feedback at high densities) in multiple attractor scenario for $\mathcal{R}_0 < 1$ when the extinction equilibrium is stable (i.e., a strong Allee effect). One possibility in Theorem 60(c) is that both bifurcations are backward and that, as a result, there can occur a strong Allee effect with three attractors.

Example 96 *A strong Allee effect with three attractors. Consider the $k = 2$ nonlinear semelparous Leslie model* (6.39) *with density factors*

$$\beta(x_1, x_2) = \frac{1}{1 + c_{21}x_1 + c_{22}x_2}, \quad \sigma(x_1, x_2) = \frac{1 + s_m(c_{11}x_1 + c_{21}x_2)}{1 + s(c_{11}x_1 + c_{21}x_2)} \tag{6.47}$$

$$c_{ij} \geq 0, \quad 0 < s < s_m \leq 1.$$

In this population model, adult fertility is subject to negative density effects while juvenile survival is subject to positive density effects (Allee effects). Juvenile survival $s\sigma(x_1, x_2)$ is an increasing function of the weighted population size $c_{11}x_1 + c_{21}x_2$ that increases from s to a maximum of s_m. Calculations using formulas (6.42) *show*

$$\delta_1 = (s_m - s)c_{11} - sc_{22}, \quad \delta_2 = -(1 - (s_m - s)s)c_{21}.$$

According to Theorem 60 both branches $\mathbb{C}_+$ and $\mathbb{C}_2$ bifurcate backward if $\delta_1 + \delta_2 > 0$ and $\delta_1 > 0$ which, since $\delta_2 < 0$, occurs if only the first inequality holds. Thus, if the density coefficients c_{ij} satisfy

$$(s_m - s)c_{11} - sc_{22} - (1 - (s_m - s)s)c_{21} > 0 \tag{6.48}$$

we can expect strong Allee effects from both the positive equilibrium and synchronous 2-cycle branches for $\mathcal{R}_0 \lesssim 1$, i.e., a three attractor scenario in which one attractor is a positive equilibrium, one is a synchronous 2-cycle, and the third is the extinction equilibrium. While we have no rigorous proof of this, Figure 6.3 supports this conjecture by showing, for selected parameter values that satisfy inequality (6.48), *three initial conditions whose orbits approach these three different attractors.*

6.4. STABLE BIFURCATION: THE IMPRIMITIVE CASE

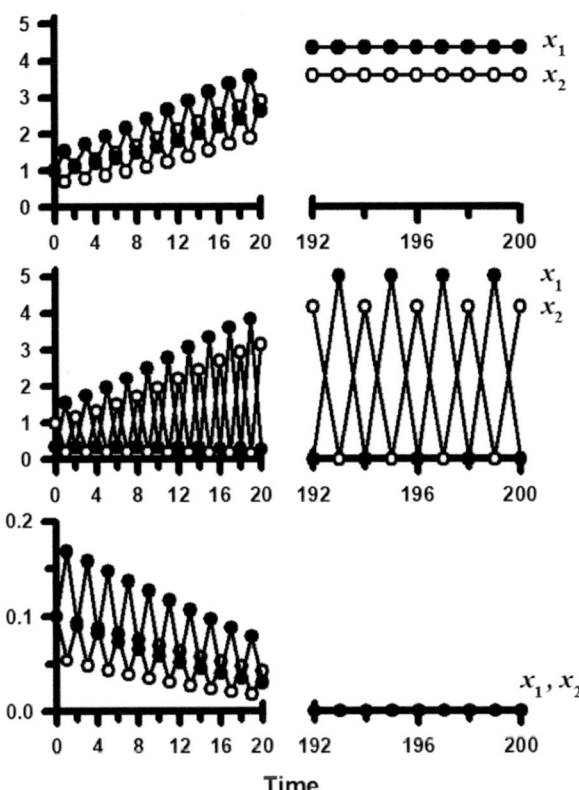

Figure 6.3: Strong Allee effect with three attractors. Shown are three sample orbits of the $k = 2$ dimensional semelparous Leslie model ((6.39)) with density factors (6.47) and parameter values $\mathcal{R}_0 = 0.85$, $s = 0.5$, $s_m = 0.9$, $c_{11} = 2$, $c_{12} = 0.1$, $c_{21} = 0.01$, $c_{22} = 0.1$. The inequality (6.48) is satisfied and hence Theorem 60 implies that the branch $\mathbb{C}_+$ of positive equilibria and the branch $\mathbb{C}_2$ of synchronous 2-cycles both bifurcate backwards at $\mathcal{R}_0 = 1$. The top plot shows that the orbit with initial condition $(x_1(0), x_2(0))^T = (1,1)^T$ approaches a positive equilibrium. The middle plot shows that the orbit with initial condition $(x_1(0), x_2(0))^T = (0.35, 1)^T$ approaches a synchronous 2-cycle. The middle plot shows that the orbit with initial condition $(x_1(0), x_2(0))^T = (0.1, 0.1)^T$ approaches the extinction equilibrium.

6.4.2 The Case $k=3$

Theorem 60 illustrates the complexity, even at the low dimension $k=2$, that can occur in the basic bifurcation at $\mathcal{R}_0 = 1$ in matrix models when the projection matrix is imprimitive. As the dimension k increases, this complexity significantly increases. To illustrate this we take a brief look at a general class of $k=3$ dimensional models for which the bifurcation at $\mathcal{R}_0 = 1$ is well understood. Specially, we consider the semelparous Leslie model (6.31) when $k=3$

$$\mathcal{L}(\mathbf{x}) = \begin{pmatrix} 0 & 0 & f_3\beta(w_3) \\ s_1\sigma_1(w_1) & 0 & 0 \\ 0 & s_2\sigma_2(w_2) & 0 \end{pmatrix}, \quad f_k > 0,\ 0 < s_i < 1 \qquad (6.49)$$

where

$$w_i := \sum_{j=1}^{3} c_{ij} x_j, \quad i=1,2,3$$

are weighted population sizes determined by the entries in the competition coefficient matrix

$$C = (c_{ij}), \quad c_{ij} \in \mathbb{R}_+^1 \text{ and at least one } c_{ii} > 0.$$

The density factors in $\mathcal{L}(\mathbf{x})$ satisfy

H7: $\sigma_i \in C^2(\Omega \to \mathbb{R}_+^1)$, $\sigma_i(0) = \beta(0) = 1$, $\sigma_i'(0) = \beta'(0) = -1$, $\sigma_i'(w)$ and $\beta'(w) < 0$ for $w > 0$.

In addition we assume

H8: $\sigma_i(w)w$ and $\beta(w)w$ are bounded and (strictly) monotonically increasing for $w \geq 0$.

A basic example density factor that satisfies H7 and H8 (with $\Omega = (-1, +\infty)$) is the rational function

$$\frac{1}{1+w} \qquad (6.50)$$

used in Leslie's logistic model (6.6).

In the $k=2$ dimensional case, when both branches of positive equilibria and synchronous 2-cycles forward bifurcate, Theorem 60 describes a dynamic dichotomy between them: one continuum is stable and other is not. Theorem 61 below also describes a dynamic dichotomy in the $k=3$ dimensional case, not between the bifurcating positive equilibria and synchronous 3-cycles, but between the positive equilibria and the boundary $\partial \mathbb{R}_+^3$ of the cone $\mathbb{R}_+^3$. Note that the three nonnegative coordinate axes are invariant, as are three nonnegative coordinate planes. Orbits on the coordinate axis are single class orbits (i.e., are orbits which have exactly one zero component) while those in the coordinate plane are two class orbits (i.e., are orbits which have exactly one zero component). Thus, $\partial \mathbb{R}_+^3$ is invariant.

We say the boundary $\partial \mathbb{R}_+^3$ is an *attractor* if there exists a neighborhood such that the omega limit set of all orbits with initial conditions in the neighborhood lies in $\partial \mathbb{R}_+^3$. We say the boundary $\partial \mathbb{R}_+^3$ is a *repeller* if there is a neighborhood N of $\partial \mathbb{R}_+^3$ such that for all $x(0) \notin \partial \mathbb{R}_+^3$ there exists a time $T(x(0)) > 0$ such that $x(t) \notin N$ for all $t \geq T$. [200].

Define the quantities

$$\rho_1 := \frac{c_{21} + sc_{32} + s_1 s_2 c_{13}}{c_{11} + s_1 c_{22} + s_1 s_2 c_{33}}, \quad \rho_2 := \frac{c_{31} + s_1 c_{12} + s_1 s_2 c_{23}}{c_{11} + s_1 c_{22} + s_1 s_2 c_{33}}.$$

6.4. STABLE BIFURCATION: THE IMPRIMITIVE CASE

Theorem 61 *[71], [92] Assume the $k = 3$ dimensional semelparous Leslie model, with the projection matrix (6.49) satisfies H7 and H8.*

(a) *If $\mathcal{R}_0 = f_3 s_1 s_2 < 1$ then the extinction equilibrium $\mathbf{x} = \mathbf{0} \in \mathbb{R}^3$ is globally asymptotically stable on $\mathbb{R}_+^3$. If $\mathcal{R}_0 > 1$ then the model is permanent with respect to $\mathbf{0}$.*

(b) *A continuum $\mathbb{C}_+$ of positive equilibria forward bifurcates from the extinction equilibrium at $\mathcal{R}_0 = 1$ and there exists a positive equilibrium for each $\mathcal{R}_0 > 1$. A continuum $\mathbb{C}_3$ of (nonnegative) single class 3-cycle forward bifurcates from the extinction equilibrium at $\mathcal{R}_0 = 1$ and there exists a single class 3-cycle for each $\mathcal{R}_0 > 1$.*

(c) *(Dynamic Dichotomy) For $\mathcal{R}_0 \gtrsim 1$ the bifurcating positive equilibria are stable and $\partial \mathbb{R}_+^3$ is a repeller if $\rho_1 + \rho_2 < 2$. For $\mathcal{R}_0 \gtrsim 1$ the bifurcating positive equilibria are unstable and $\partial \mathbb{R}_+^3$ is an attractor if $\rho_1 + \rho_2 > 2$.*

Proof.

(a) The first sentence follows from Theorem 52 and the second sentence from Theorem 53 together with Lemmas 11 and 12.

(b) Noticing that $f_3 = \mathcal{R}_0/s_1 s_2$ we see that $\mathcal{R}_0$ appears linearly in the Leslie projection matrix. Theorem 58 implies that the continuum $\mathbb{C}_+$ of locally bifurcating positive equilibria pairs $(\mathcal{R}_0, \mathbf{x})$ guaranteed by Theorem 59 is unbounded in $\mathring{\mathbb{R}}_+^1 \times \mathring{\mathbb{R}}_+^3$. That the spectrum of $\mathcal{R}_0$ values on the continuum remains in $\mathring{\mathbb{R}}_+^1$ follows by observing that should the continuum contain a negative $\mathcal{R}_0$ value, it would (by continuity) have to contain $\mathcal{R}_0 = 0$, i.e., there would be a positive equilibrium when $\mathcal{R}_0 = 0$. That this is impossible is easily seen from the equilibrium equations

$$x_1 = \mathcal{R}_0 \frac{1}{s_1 s_2} \beta(w_3) x_3$$
$$x_2 = s_1 \sigma_1 (w_1) x_1$$
$$x_3 = s_2 \sigma_2 (w_2) x_2.$$

By assumption at least one $c_{ii} > 0$. Suppose $c_{33} > 0$ (the cases $c_{11} > 0$ or $c_{22} > 0$ are handled similarly). From the equilibrium equations we also have, for any positive equilibrium, that

$$0 < x_1 = \mathcal{R}_0 \frac{1}{s_1 s_2} \beta(w_3) x_3 \leq \mathcal{R}_0 \frac{1}{s_1 s_2} \frac{1}{c_{33}} \beta(c_{33} x_3) c_{33} x_3 \leq \mathcal{R}_0 \frac{1}{s_1 s_2} \frac{1}{c_{33}} \beta_m$$
$$0 < x_2 = s_1 \sigma_1(w_1) x_1 \leq s_1 x_1$$
$$0 < x_3 = s_2 \sigma_2(w_2) x_2 \leq s_2 x_2$$

for a bound $\beta_m > 0$ on $\beta(z) z$. These inequalities in turn imply

$$0 < x_1 \leq \mathcal{R}_0 \frac{1}{s_1 s_2} \frac{1}{c_{33}} \beta_m$$
$$0 < x_2 \leq \mathcal{R}_0 \frac{1}{s_2} \frac{1}{c_{33}} \beta_m$$
$$0 < x_3 \leq \mathcal{R}_0 \frac{1}{c_{33}} \beta_m.$$

It follows that if the spectrum of $\mathcal{R}_0$ from the continuum of positive solution pairs is bounded, then so is the range of equilibria from the continuum. This contradicts that the continuum is unbounded. A similar argument applies to the equilibrium equations of the third composite and hence the bifurcating $\mathbb{C}_3$ of single class 3-cycles.

(c) This dynamic dichotomy is proved in [71] (also see [92]) using both the Linearization Principle and average Liapunov functions. ∎

The biological interpretation of the inequalities involving $\rho_1 + \rho_2$ in Theorem 61(c) can be ascertained by noting that the numerators in ρ_i measure the intensity of *inter*-class density effects (competition) whereas the denominator measures the intensity of *intra*-class density effects (competition). Thus, the case $\rho_1 + \rho_2 > 2$ when the boundary $\partial \mathbb{R}^3_+$ is an attractor occurs when inter-class effects are high (relative to intra-class effects), whereas the case $\rho_1 + \rho_2 < 2$ when the positive equilibrium is stable occurs when inter-class effects are low (relative to intra-class class effects). In the first case $\rho_1 + \rho_2 > 2$ when the boundary $\partial \mathbb{R}^3_+$ is an attractor, it is not clear what the asymptotic dynamics of orbits are. To determine this, we need to study the dynamics on $\partial \mathbb{R}^3_+$ to which orbits are attracted.

The extinction equilibrium destabilizes at the bifurcation point $\mathcal{R}_0 = 1$ because, as the Linearization Principle shows, eigenvalues of the Jacobian

$$\mathcal{L}(\mathbf{0}) = \begin{pmatrix} 0 & 0 & \frac{\mathcal{R}_0}{s_1 s_2} \\ s_1 & 0 & 0 \\ 0 & s_2 & 0 \end{pmatrix},$$

which are $\sqrt[3]{\mathcal{R}_0} u_i$ where u_i are the three roots of unit, all leave the complex unit circle simultaneously. When $R_0 = 1$ the eigenvalue $\lambda_1 = 1$ is associated with the bifurcation of the positive equilibria. The other two complex conjugate eigenvalues λ_2 and λ_3 would normally be associated with a Neimark-Sacker bifurcation of an invariant loop (discrete Hopf bifurcation) (Chapter 1, Section 3.10). That famous theorem does not apply here, however, because this case falls into one of the exception resonance cases that must be avoided for application of the theorem.

That invariant loops do indeed bifurcate at $\mathcal{R}_0 = 1$, however, is proved in [71]. The main tools are, first of all, the invariance of the coordinate axes on which the dynamics can be studied by the composite map, which is a one-dimensional difference equation of the type studied in Chapter 3. Secondly, the coordinate planes are invariant and the dynamics on them can be studied by the composite map which is a two-dimensional (planar) map that is, by assumptions H7 and H8, monotone. This permits the use of powerful theorems from the theory of monotone maps. The results from this analysis show that an invariant loop, lying on the boundary $\partial \mathbb{R}^3_+$, bifurcates from the extinction equilibrium at $\mathcal{R}_0 = 1$, along with the positive equilibria in $\mathring{\mathbb{R}}^3_+$. The loop is, depending on parameter values, one of the three types shown in Figure 6.4. For the composite map, these loops consist of equilibrium points connected by heteroclinic orbits lying in the coordinate planes. For the original Leslie model, they consist of a single class 3-cycle and, in some classes, an additional two class 3-cycle connected by heteroclinic orbits in the coordinate planes.

Returning to the case $\rho_1 + \rho_2 > 2$, we have the following theorem, obtained from the analysis in [71], that provides more detail about the dynamic dichotomy in Theorem 61 when $\partial \mathbb{R}^3_+$ is the attractor.

6.4. STABLE BIFURCATION: THE IMPRIMITIVE CASE

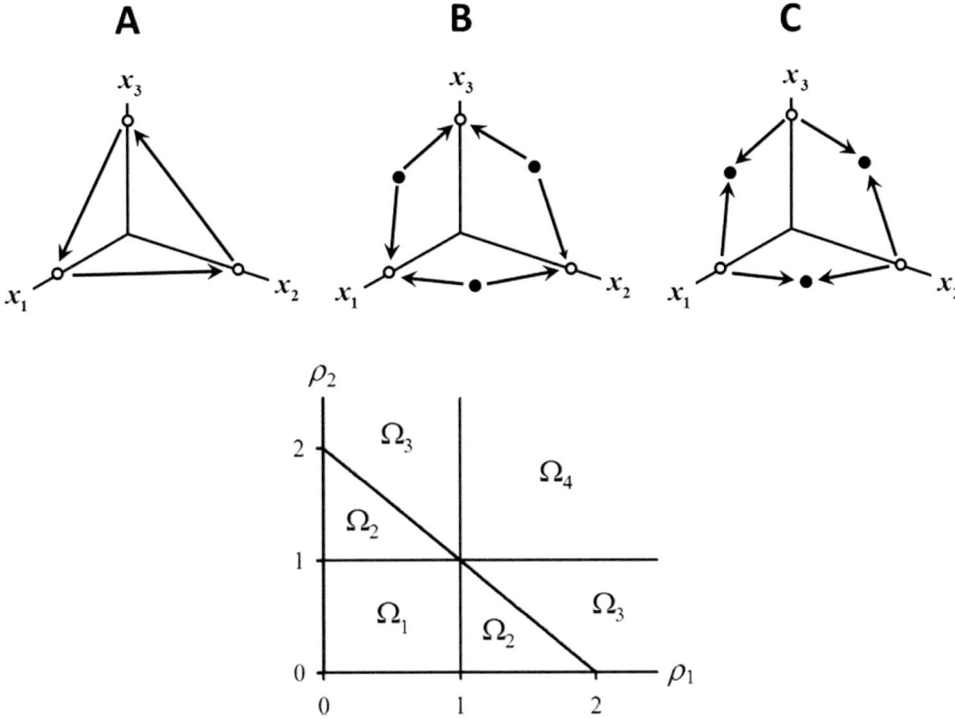

Figure 6.4: For $\mathcal{R}_0 \gtrapprox 1$ in the $k = 3$ dimensional semelparous Leslie model with projection matrix (6.49), there exist an invariant loop of one of the three geometries shown. In the ρ_1 and ρ_2 parameter region $\Omega_2 \cup \Omega_3$, the loop is of type A. In region Ω_4, the loop is of type B and in region Ω_1 it is of type C. The open regions Ω_i in the (ρ_1, ρ_2)-plane are defined in the lower graph. The open circles on the coordinate axes are the fixed points of the composite map and correspond to the three different phases of a single-class 3-cycle. In types B and C, the solid circles interior to the coordinate planes are also fixed points of the composite map, but correspond to the three phases of a two-class 3-cycle. The oriented connecting lines are heteroclinic orbits that connect the phase shifts of these 3-cycles. The temporal motion is counter-clockwise, visiting the coordinate planes sequentially. Figure reproduced with permission from [71].

Theorem 62 *Suppose in Theorem 61 that $\rho_1 + \rho_2 > 2$ so that the bifurcating positive equilibria are unstable and the boundary $\partial \mathbb{R}_+^3$ is an attractor.*

> *(a) If both $\rho_1 > 1$ and $\rho_2 > 1$ then the bifurcating single class 3-cycles are (locally asymptotically) stable. These 3-cycles lie on a bifurcating invariant loop of type B on the boundary $\partial \mathbb{R}_+^3$ as shown in Figure 6.4.*
>
> *(b) If either $\rho_1 < 1$ or $\rho_2 < 1$ then the bifurcating invariant loops on the boundary $\partial \mathbb{R}_+^3$ are of type A in Figure 6.4 and are attracting.*

In Theorem 62(b), the time series dynamics of orbits that approach an invariant loop on the boundary $\partial \mathbb{R}_+^3$ are distinctively interesting. For the composite map, the orbits spiral outward and approach the loop and, in doing so, recurrently visit (for longer and longer episodes) the three saddle fixed points on the coordinate axes. These three fixed points correspond to the three different phases of the single class 3-cycle. Thus, orbits of the Leslie model spiral outward toward the boundary $\partial \mathbb{R}_+^3$ recurrently visiting the three phases of the single class 3-cycle, spending longer and longer episodes near each phase in turn and separated by quick transitions in between. See Figure 6.5.

For evidence of synchronous 3-cycles found in laboratory experiments with flour beetles, see [77].

6.4.3 The Case $k \geq 4$

Our goal in looking at the $k = 2$ and $k = 3$ semelparous Leslie models in some detail was primarily to demonstrate the complications that arise at the bifurcation point $\mathcal{R}_0 = 1$ when the (inherent) population projection matrix is imprimitive. For dimensions $k \geq 4$, these complications significantly increase in general. Since the characteristic equation of the inherent projection Leslie matrix $\mathcal{L}(\mathbf{0})$ (the Jacobian at $\mathbf{x} = \mathbf{0}$) is $\lambda^k - \mathcal{R}_0 = 0$, the eigenvalues are $\lambda_i = \sqrt[k]{\mathcal{R}_0} u_i$ where u_i are the kth roots of unity. Thus, all k eigenvalues leave the complex unit circle simultaneously as $\mathcal{R}_0$ increases through 1, creating the possibility for a complicated array of invariant loop bifurcations in subsets of the boundary $\partial \mathbb{R}_+^3$. It is interesting to note that Bulmer, in his study of the case $k = 17$ for the famous periodical cicadas, numerically observed time series like those in Figure 6.5 in which there were phase shifts of single class orbits of period 17. Little is rigorously known, however, about the nature of the bifurcation at $\mathcal{R}_0 = 1$ for dimensions $k \geq 4$. Some results concerning the stability properties of the bifurcating positive equilibria are given in [92], but whether there is in general a dynamic dichotomy between them and the boundary $\partial \mathbb{R}_+^3$ when $k \geq 4$ is unknown. The dichotomy has been established for some special types of projection matrices, for example, for those in which the density effects are hierarchical [74] or for those possessing what is called the beta property [295].

For more on in Leslie models see [27], [100], [101], [102], [114], [194], [195], [197], [200], [196], [240], and the references cited therein.

6.5 Secondary Bifurcations

Sections 6.2, 6.3, and 6.4 are concerned with the primary bifurcation of equilibria and cycles that occurs when a parameter change causes the extinction equilibrium to destabilize. These general theorems are local in two senses. First, they describe the nature of the bifurcation only in a neighborhood of the bifurcation point. Secondly, the stability assertions are (with the exception of Theorem 52) local stability assertions, being derived from the Linearization Principle. The strength of these results is their generality and relative ease of application. Their weakness is that they do not

6.5. SECONDARY BIFURCATIONS

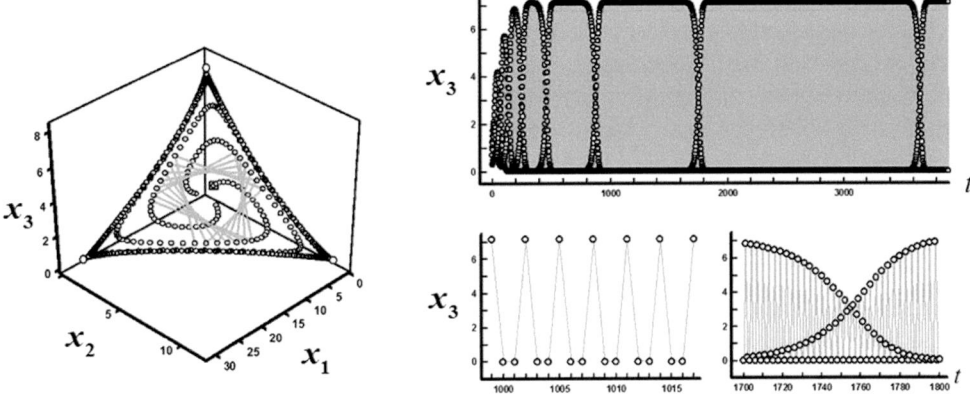

Figure 6.5: These graphs show a sample orbit and time series for the $k = 3$ dimensional semelparous Leslie model (6.49) with density factors $\beta(w) = \sigma_i(w) = 1/(1+w)$, competition coefficient matrix
$C = \begin{pmatrix} 0.01 & 0 & 0 \\ 0.03 & 0.01 & 0 \\ 0.01 & 0.02 & 0.01 \end{pmatrix}$, and parameters $b = 4$, $s_1 = 0.5$, $s_2 = 0.75$, which imply $R_0 = 1.5$
and $(\rho_1, \rho_2) = (2.133, 0.5333) \in \Omega_3$. In the three-dimensional plot on the left, the orbit with initial condition $\mathbf{x}(0) = (1,1,1)$ in the interior $\mathbb{R}^3_+$ (the open square) is seen tending, in an outward spiraling fashion, to a heteroclinic cycle of type A lying on the boundary ∂R^3_+. The top plot on the right shows the time series for the adult component x_3 of the orbit. The lower two graphs show, left to right, respectively, an episode of synchronous 3-cycle dynamics and an episode showing a transitional shift from one phase to another phase of the 3-cycle. Figures reproduced with permission from [71].

describe the dynamics outside a neighborhood of the bifurcation point. An exception is Theorem 58 which establishes the global existence of the bifurcating continuum of positive equilibria. Even with this in hand, however, the stability properties of the bifurcating equilibria near the bifurcation point might not persist global throughout the continuum. This is well known to us from the $k = 1$ dimensional case of the Ricker model equation in Chapter 1, Section 1.4.5, where the continuum of positive equilibria ultimately lose stability through a period-doubling route to chaos. Such secondary bifurcations do not in general occur, however; they are highly dependent on the specifics of the model equations. We see this, for example, in the $k = 1$ discrete logistic (Beverton-Holt) equation for which the bifurcating positive equilibria never destabilize. This is also true for the logistic Leslie model in Example 89 and for the $k = 2$ dimensional juvenile-adult model in Example 90 (see Problem 2). A change in the density factor in the latter example, from a discrete logistic factor to a exponential factor such as is used in the Ricker equation can, as is no surprise, result in secondary bifurcations. This model is a special, low dimensional, case of the general Levin-Goodyear model in Examples 91 and 92 which is known to have secondary bifurcations; this was the principal focus of the study of this model in [212]. In the next example, we illustrate secondary bifurcations by looking at the $k = 2$ juvenile-adult model.

Example 97 *Consider the $k = 2$ dimensional matrix model with the primitive, Leslie projection matrix*

$$\mathcal{L}(\mathbf{x}) = \begin{pmatrix} 0 & f_2 \exp(-cx_2) \\ s_1 & s_2 \end{pmatrix} \quad (6.51)$$
$$0 < s_1, s_2 < 1, \quad f_2, c > 0$$

in which the only density effects are adult density effects on adult fertility. Solving the equilibrium equations of the associated matrix equation

$$x_1 = f_2 \exp(-cx_2) x_2$$
$$x_2 = s_1 x_1 + s_2 x_2$$

we obtain a positive equilibrium

$$\mathbf{x} = \begin{pmatrix} \frac{1-s_2}{s_1} \frac{1}{c} \ln \mathcal{R}_0 \\ \frac{1}{c} \ln \mathcal{R}_0 \end{pmatrix}$$

if and only if

$$\mathcal{R}_0 = f_2 \frac{s_1}{1 - s_2} > 1$$

(in which case it is unique). The Jacobian evaluated at this equilibrium is

$$\begin{pmatrix} 0 & -(\ln \mathcal{R}_0 - 1) \frac{1-s_2}{s_1} \\ s_1 & s_2 \end{pmatrix}.$$

Using the Jury (trace-determinant) criteria (Chapter 3, Section 3.2.3), we find that the equilibrium is (locally asymptotically) stable if

$$\mathcal{R}_0 < \mathcal{R}_0^* := \exp\left(\frac{2 - s_2}{1 - s_2}\right)$$

and unstable if $\mathcal{R}_0 > \mathcal{R}_0^*$. Thus, the forward and stable bifurcating positive equilibria destabilize as $\mathcal{R}_0$ increases through $\mathcal{R}_0^*$. Moreover, the Jury conditions imply that this loss of stability is due to a complex pair of eigenvalues leaving the complex unit circle. From the characteristic equation of the Jacobian, we find that the two eigenvalues at $R_0 = 1$ are

$$\frac{1}{2}s_2 \pm i\sqrt{1 - \frac{1}{4}s_2^2} \neq 1$$

which are not equal to square, cubic or quartic roots of unity. Thus, a Neimark-Sacker bifurcation occurs at $\mathcal{R}_0 = \mathcal{R}_0^*$ (Chapter 3, Section 3.10). See Figure 6.6 for an illustrative example.

We can view the structured model with matrix (6.51) as an extension of the exponential difference equation model (1.29) for an unstructured population introduced in Chapter 1, namely

$$x(t+1) = bx(t)\exp\left(-cx(t)\right). \tag{6.52}$$

Mathematically, this example shows that one consequence of including a juvenile maturation period in the model is to change the result of the destabilization of the positive equilibrium from a period-doubling bifurcation to an invariant loop bifurcation, a quite different dynamic. Another biological punch line, obtained from this example, is that the structured model predicts the onset of nonequilibrium dynamics occurs at a larger critical value of $\mathcal{R}_0$. This is because the onset of the period-doubling cascade in the Equation (6.52) occurs at $\mathcal{R}_0 = b = \exp(2)$ and because $\mathcal{R}_0^* > \exp(2)$. (In fact, $\mathcal{R}_0^*$ is significantly larger than $\exp(2)$ for adults survival probability s_2 near 1). In this sense, the maturation period has a stabilizing influence on the population dynamics and works against the onset of chaotic dynamics, in comparison to the unstructured model.

6.6 A Case Study: The LPA Model

Throughout the twentieth century, since the "golden age of theoretical ecology" [275], mathematical models played a significant role in the formulation of the most basic principles in population biology and ecology, including density dependence and self-regulation, competitive exclusion, limiting similarity and ecological niche, predator-prey cycles, and so on. Moreover, models can clarify hypotheses, expose gaps in knowledge, help in experimental and observation design, and organize thinking about a population's dynamics and role in an ecosystem. There was, however, a lack of successful use of models to make quantitatively accurate predictions, i.e., accurate forecasts of population numbers or densities made by a model parameterized using independent data (e.g., see [1], [118]). A concerted, two decade long project to address this issue was carried out at the turn of the century, using cultures of the flour beetle (*Tribolium castaneum*) in a laboratory setting and matrix models of the type studied in this chapter.

In addition to the goal of interfacing models with data, this project addressed a large variety of dynamical phenomenon predicted by the models and conducted controlled and replicated experiments that validated the predictions. The centerpiece of this multifaceted project was a bifurcation diagram that includes numerous types of attractors, ranging from equilibria to chaotic strange attractors, as a model parameter is changed. This approach in studying a complex biological system is in agreement with E. O. Wilson who writes that the "best procedure, as in the rest of science, is first to simplify the system, then to hold it more or less constant while varying the important parameters one or two at a time to see what happens" [299]. Here we focus on only a few highlights of this project; for more the reader can consult [59], [86], [88] and the many references therein.

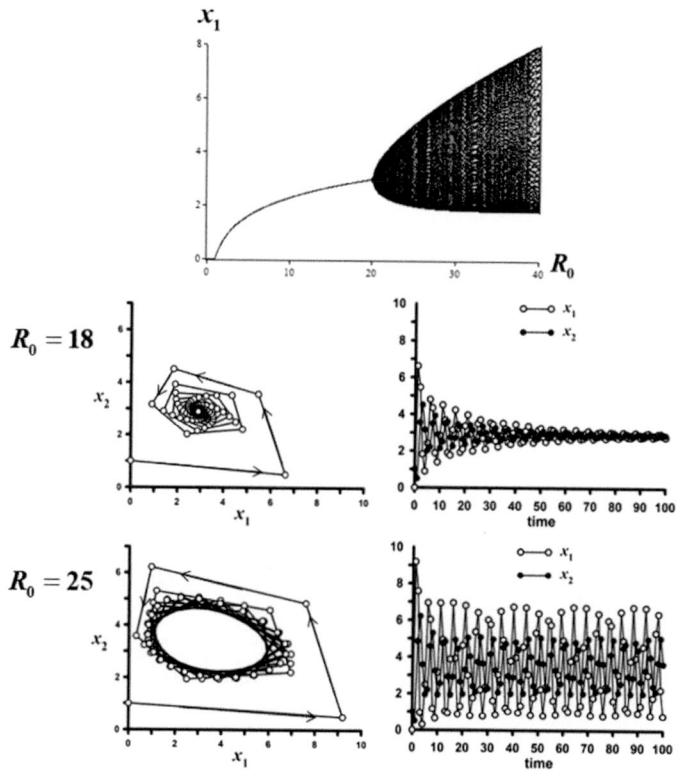

Figure 6.6: The bifurcation diagram for the $k = 2$ dimensional matrix model with the Leslie projection matrix (6.51) with $f_3 = \mathcal{R}_0\left(1 - s_2\right)/s_1$ and $s_1 = s_2 = 0.5$ and $c = 1$ show a Neimark-Sacker bifurcation to an invariant loop at $\mathcal{R}_0^* = e^3 \approx 20.09$. A sample orbit with initial condition $\mathbf{x}\left(0\right) = \left(0, 1\right)^T$ is displayed in the phase plane and in time series for a value of $\mathcal{R}_0 = 18$ before the bifurcation and $\mathcal{R}_0 = 25$ after the bifurcation.

6.6.1 Model Derivation

The LPA model classifies a flour beetle population into three classes. The L-stage consists of feeding larvae, the P-stage consists of nonfeeding larvae, pupae, and callow adults and the A class consists of sexually mature adults

$$\mathbf{x} = \begin{pmatrix} x_1 \\ x_2 \\ x_3 \end{pmatrix} = \begin{pmatrix} L \\ P \\ A \end{pmatrix}.$$

The unit of time is 2 weeks, which is the approximate average time spent in both the L-stage and the P-stage under the experimental conditions used. During the experiments, the beetle populations were kept in constant environmental conditions in culture jars and were counted and fed every 2 weeks so that they suffered no food resource shortage. The density free projection matrix is the Leslie matrix form $\mathcal{L} = F + T$ where

$$F = \begin{pmatrix} 0 & 0 & b \\ 0 & 0 & 0 \\ 0 & 0 & 0 \end{pmatrix}, \quad T = \begin{pmatrix} 0 & 0 & 0 \\ 1-\mu_l & 0 & 0 \\ 0 & 1 & 1-\mu_a \end{pmatrix}.$$

Here $b > 0$ is the average number of larvae produced per adult and μ_l and μ_a are the larval and adult probabilities of dying per unit time $(0 < \mu_l, \mu_a < 1)$. Note that the model assumes there is no inherent P-stage mortality. The main density induced mortality is due to cannibalism. Specifically, eggs are cannibalized by both L and A-stage individuals. (Note the egg stage is a relatively short 4 days and is ignored as a state variable in this model.) In addition, P-stage individuals are cannibalized by A-stage individuals.

To derive a density factor that describes cannibalism mortality, consider cannibalism of P-stage individuals by A-stage individuals. What is the probability that a P-stage individual avoids death by cannibalism during a unit of time? Consider a short interval of time Δt and the $n = 1/\Delta t$ steps it takes for a unit of time. There is no evidence that an adult searches for victims to cannibalize, but instead simply cannibalizes a P-stage individual if it encounters one at random. In the presence of a single A-stage individual, assume a P-stage individual will encounter that adult individual (and hence be cannibalized) during a time period Δt with a probability that is proportional to Δt, i.e., $c_{pa}\Delta t$ where c_{pa} is the constant of proportionality. Thus, the probability it escapes cannibalism is approximately $1 - c_{pa}\Delta t$. Assuming the independence of these events during the n time steps, we have that the probability of escaping cannibalism during a time unit in the presence of a single A-stage individual is approximately $(1 - c_{pa}\Delta t)^{1/\Delta t}$. If we also assume that cannibalism by different A-stage individuals are independent events, then the probability a P-stage individual escapes cannibalism per unit time in the presence of x_3 individuals in the A-stage is approximately $(1 - c_{pa}\Delta t)^{x_3/\Delta t}$. Passing $\Delta t \to 0$ we obtain the per capita A-stage probability of escaping cannibalism by the A-stage population x_3:

$$\lim_{\Delta t \to 0} (1 - c_{pa}\Delta t)^{x_3/\Delta t} = \exp(-c_{pa}x_e).$$

Similar arguments produce similar exponential type density factors for egg cannibalism by L and A-stage individuals, which then lead to the LPA projection matrix

$$\mathcal{L}(\mathbf{x}) = \begin{pmatrix} 0 & 0 & b\exp(-c_{el}x_1 - c_{ea}x_3) \\ 1-\mu_l & 0 & 0 \\ 0 & \exp(-c_{pa}x_3) & 1-\mu_a \end{pmatrix} \quad (6.53)$$

with coefficients $0 < \mu_l, \mu_a < 1$, $b, c_{el}, c_{ea}, c_{pa} > 0$.

It is left as Problem 14 to show that H2 and H3 hold with $\Omega = \mathbb{R}^3$, that the inherent net reproduction number is

$$\mathcal{R}_0 = b \frac{1 - \mu_l}{\mu_a}, \tag{6.54}$$

and that $\mathcal{L}(\mathbf{x})$ is primitive for all $\mathbf{x} \in \mathbb{R}_+^3$.

6.6.2 Basic Analysis of the LPA Model

Consider first the extinction equilibrium $\mathbf{x} = \mathbf{0}$. From the exponential density factors in the LPA projection matrix (6.53), it is straightforward to see that (6.14) holds and hence the extinction equilibrium of the LPA model is globally asymptotically stable when $\mathcal{R}_0 < 1$. Similarly, for the same reason, conditions (6.15) hold and hence Lemmas 12 and 11 and Theorem 52 imply the model is permanent with respect to $\mathbf{x} = \mathbf{0}$ when $\mathcal{R}_0 > 1$.

Next, we turn our attention to positive equilibria. To apply Theorem 54 we need to validate assumptions H4 and H5. Although various model parameters were used as bifurcation parameters in the beetle project, here we use $\mu = b$. Note that the exponential density factors in (6.53) imply H4 holds with $I_\mu = \mathbb{R}_+^1$ and $\Omega = \mathbb{R}^3$. The remaining assumption H5 requires that there exists a value b_0 of b for which $r(b_0) = 1$, $r'(b_0) \neq 0$. It is convenient here to make use of Remark 19 and instead use the equivalent requirements that $\mathcal{R}_0(b_0) = 1$, $\mathcal{R}_0'(b_0) \neq 0$. From formula (6.54), we find that

$$b_0 = \frac{\mu_a}{1 - \mu_l} \tag{6.55}$$

and

$$\mathcal{R}_0'(b_0) = \frac{1 - \mu_l}{\mu_a} > 0.$$

Theorem 54 now tells us that, as b increases through b_0, a branch of positive equilibria bifurcates from the extinction equilibrium. Moreover, the exponential density factors are all decreasing functions of x_i and it follows from formula (6.22) that $\kappa > 0$. Since $\alpha > 0$ in (6.19) is clearly positive when bifurcation parameter $\mu = b$ is used in the projection matrix (6.53), it follows from Corollary 7 that the bifurcation of positive equilibria at $b = b_0$ is forward and stable. Finally, from $\mathcal{L}(\mu, \mathbf{0}) = A_1 + \mu A_2$ where

$$A_1 = \begin{pmatrix} 0 & 0 & 0 \\ 1 - \mu_l & 0 & 0 \\ 0 & 1 & 1 - \mu_a \end{pmatrix}, \quad A_2 = \begin{pmatrix} 0 & 0 & 1 \\ 0 & 0 & 0 \\ 0 & 0 & 0 \end{pmatrix}$$

we find that assumption H6 holds. Theorem 58 guarantees that the locally bifurcating branch of positive equilibrium at $b = b_0$ has a global existence, i.e., is contained in an unbounded continuum $\mathbb{C}$ of equilibrium pairs $(b, \mathbf{x})$ for which $\mathbb{C} \setminus \{(b_0, \mathbf{0})\} \subset \mathbb{R}^1 \times \mathring{\mathbb{R}}_+^3$. Furthermore, using an argument similar to that found in the proof of Theorem 61(b), one can show that the spectrum of b values from positive equilibrium pairs in $\mathbb{C} \setminus \{(b_0, \mathbf{0})\}$ is the half line $b > b_0$ (Problem 14).

In the following theorem, we summarize the results about the LPA model we have obtained from the theorems in Sections 6.2 and 6.3.

Theorem 63 *Consider the LPA model with projection matrix* (6.53).

(a) The extinction $\mathbf{x} = \mathbf{0}$ equilibrium is globally asymptotically stable on $\mathbb{R}_+^3$ when $b < b_0 = \mu_a/(1-\mu_l)$. The model is permanent (with respect to $\mathbf{x} = \mathbf{0}$) when $b > b_0$.

(b) A forward and stable bifurcation of positive equilibria from the extinction equilibrium occurs at $b = b_0$. For each $b > b_0$ there exists at least one positive equilibrium.

We have seen (e.g., see Example 97) that when the primary bifurcation is forward and stable in a nonlinear matrix model, it is not necessarily true that the bifurcating positive equilibria outside a neighborhood of the bifurcation point $(b, \mathbf{x}) = (b_0, \mathbf{0})$ are stable (nor, if they are stable, that they are globally stable). For some models it can be the case that all positive equilibria are stable (e.g., see for the juvenile-adult model in Example 90), whereas for other models they lose stability as $b > b_0$ increases. For example, the latter case occurs for the juvenile-adult model in Example 97, which has an exponential density factor as does the LPA projection matrix. This leads us to suspect that the positive equilibria in the LPA model might lose stability as $b > b_0$ increases. That this is true is left as a problem (Problem 14).

For $b > b_0$ close to b_0 the bifurcating positive equilibria are stable, but it is an open question whether or not they are globally asymptotically stable. Only in some special cases has it been proved that they are globally stable, namely when $c_{el} = 0$ [201] or when $c_{pa} = 0$ [173]. In general, however, local stability does not imply global stability in the LPA model; we see an example of this in the following section.

6.6.3 Secondary Bifurcations in the LPA Model

We do not have rigorous proofs concerning the occurrence or type of secondary bifurcations, except in a few special cases. For example, the case when there is no larvae cannibalism of eggs $c_{el} = 0$ is analyzed in [106] where period-doubling bifurcations to 2-cycles and Neimark-Sacker bifurcations to invariant loops, depending on parameter values, are shown to exist. Figures 6.7 and 6.8 illustrate both types of bifurcation (but in an example with $c_{el} > 0$).

The sample bifurcation diagrams shown in Figures 6.7 and 6.8 involve forward-stable bifurcations when the positive equilibrium eventually loses stability as b_0 increases. It is also possible, however, that this secondary bifurcation can be backward-unstable. Figure 6.9 illustrates a case when a 2-cycle bifurcation is backward and unstable. Note, however, that in this case, the bifurcating branch of 2-cycles "turns around" (at a saddle-node bifurcation point) and stabilizes. This leads to an interval of b values where there exist both a stable positive equilibrium and 2-cycle. (This multi-attractor scenario is reminiscent of a backward bifurcation when the extinction equilibrium loses stability and creates a multi-attractor scenario involving the extinction equilibrium, i.e., a strong Allee effect.)

In addition to the secondary bifurcation caused by the destabilization of the positive equilibrium, there can be (and typically are) further bifurcations as attractors lose stability with increases in b. The sequence of bifurcations and resulting new attractors can be extraordinarily complex, often involving in chaotic and strange attractors. Such a bifurcation sequence was the focal point of a sequence of controlled and replicated laboratory experiments involving flour beetles that were carried out during the 1990s and early 2000s. Using a bifurcation diagram predicted by a parameterized LPA model—one that had been validated using historical and additional preliminary experiments—this project manipulated replicated cultures of beetles placed along the bifurcation diagram at selected locations predicted to have distinctly different attractors. The data obtained

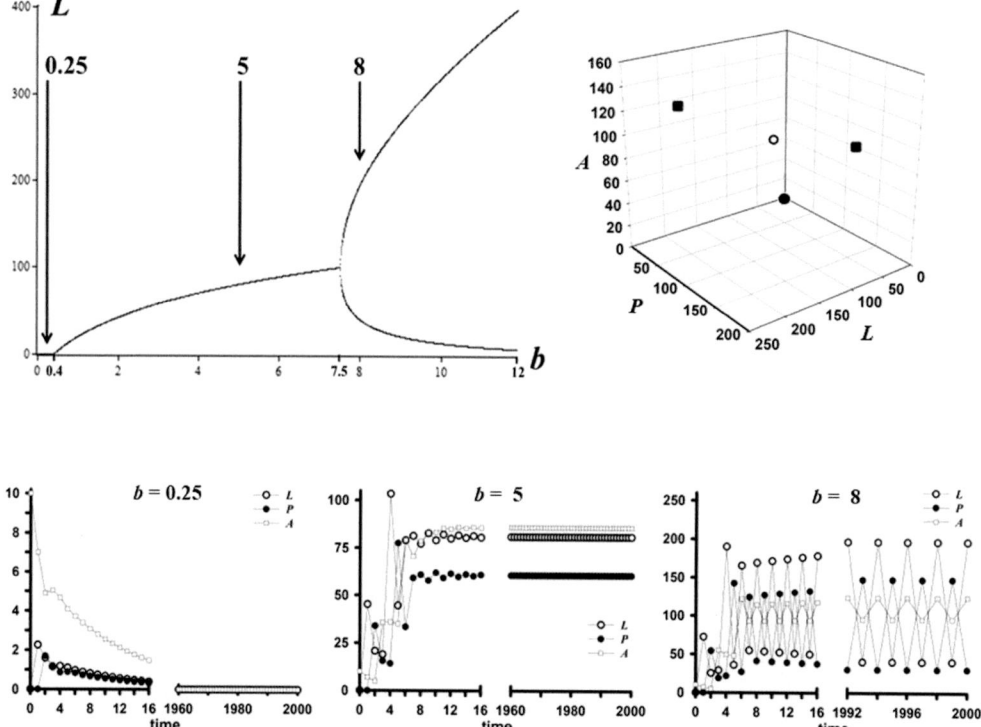

Figure 6.7: Shown is the bifurcation diagram for the LPA model with projection matrix (6.53), using b as the bifurcation parameter, and plotting the L component of the attractor. Parameter values are $c_{el} = c_{eq} = c_{pa} = 0.01$, $\mu_l = 0.25$, and $\mu_a = 0.3$. The forward, stable bifurcation of positive equilibria occurs at $b_0 = 0.4$ and a 2-cycle bifurcation occurs at $b \approx 7.5$. The upper right plot shows three sample attractors in phase space: the extinction equilibrium (solid dot) when $b = 0.25$; a positive equilibrium (open dot) when $b = 5$; and 2-cycle (open squares) when $b = 8$. Sample solution time series (with initial condition $\mathbf{x}(0) = (L, P, A)^T = (0, 0, 10)$) are shown for each of these three cases in the bottom row.

6.6. A CASE STUDY: THE LPA MODEL

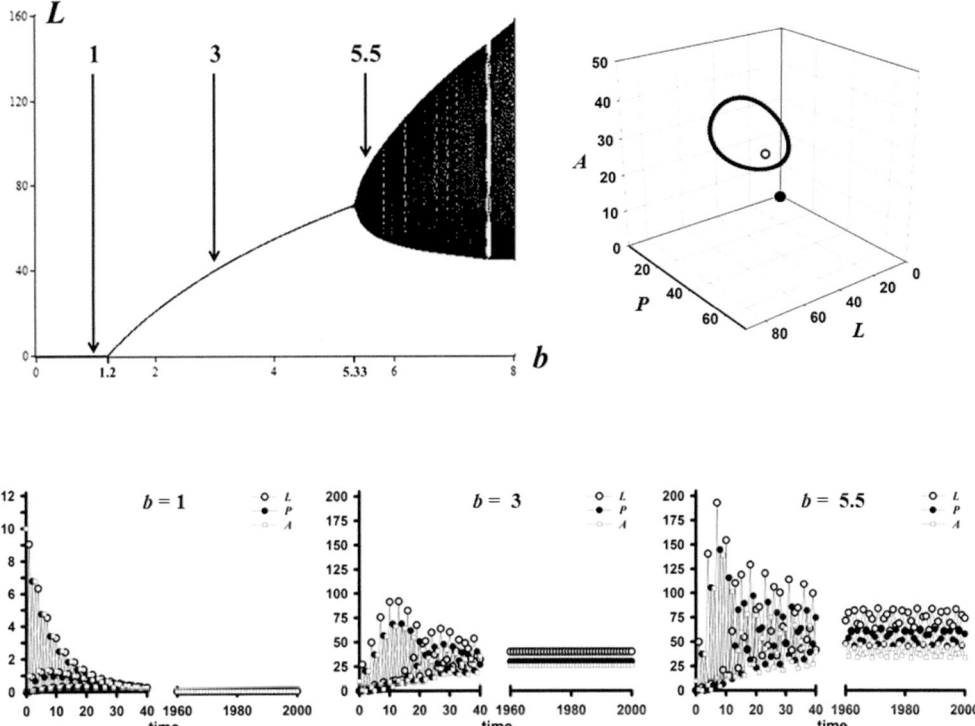

Figure 6.8: Shown is the bifurcation diagram for the LPA model with projection matrix (6.53), using b as the bifurcation parameter, and plotting the L component of the attractor. Parameter values are $c_{el} = c_{eq} = c_{pa} = 0.01$, $\mu_l = 0.25$, and $\mu_a = 0.9$. The forward, stable bifurcation of positive equilibria occurs at $b_0 = 1.2$ and a Neimark-Sacker bifurcation to an invariant loop occurs at $b \approx 5.33$. The upper right plot shows three sample attractors in phase space: the extinction equilibrium (solid dot) when $b = 1$; a positive equilibrium (open dot) when $b = 3$; and an invariant loop when $b = 5.5$. Sample solution time series (with initial condition $\mathbf{x}(0) = (L, P, A)^T = (0, 0, 10)$) are shown for each of these three cases in the bottom row.

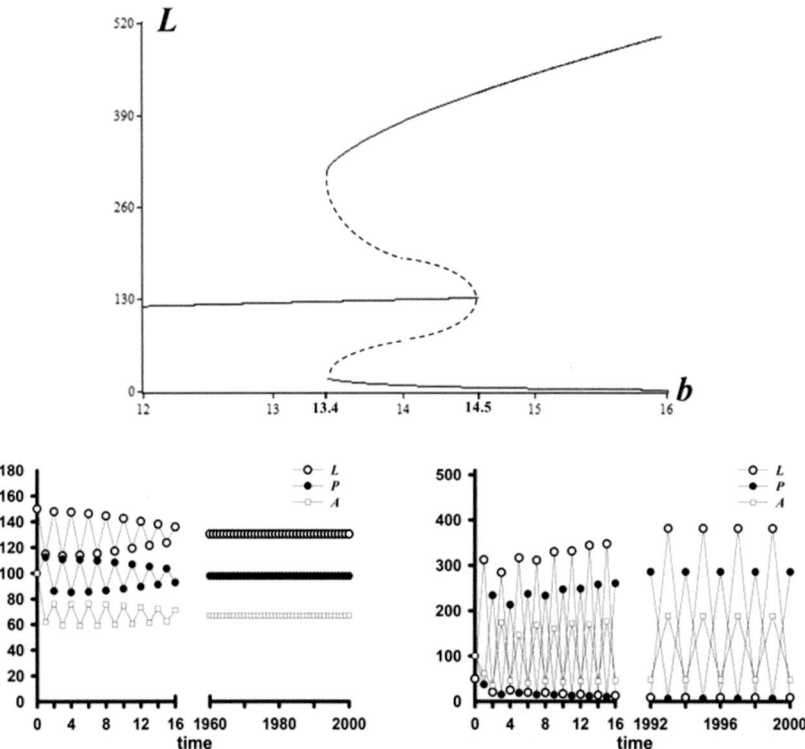

Figure 6.9: Shown is the bifurcation diagram for the LPA model with projection matrix (6.53), using b as the bifurcation parameter, and plotting the L component of the attractor. Parameter values are $c_{el} = c_{eq} = c_{pa} = 0.01$, $\mu_l = 0.25$, and $\mu_a = 0.75$. The forward, stable bifurcation of positive equilibria occurs at $b_0 = 1$ (not shown) and a backward-unstable 2-cycle bifurcation occurs at $b \approx 14.5$ (the dotted lines). A saddle-node bifurcation occurs at $b \approx 13.4$, giving rise to stable 2-cycles. For b in the interval $13.4 < b < 14.5$, there exist a stable positive equilibrium and a stable 2-cycle. This is illustrated by the two time series plots at $b = 14$, one of which (with initial condition $\mathbf{x}(0) = (150, 100, 100)^T$) tends to the equilibrium and the other of which (with initial condition $\mathbf{x}(0) = (50, 100, 100)^T$) tends to a 2-cycle.

6.6. A CASE STUDY: THE LPA MODEL

from the experiments remarkably followed the predicted sequence of attractors can be seen in the reports on the project in [57], [88], [105].

Figures 6.10 and 6.11 show two illustrative example treatments taken from the experiment, one of which involves a periodic 3-cycle attractor and the other of which a chaotic attractor. Figure 6.10(D) shows the L-stage component of a 3-cycle oscillation predicted in advance of the experiment and (E) shows the L-stage obtained from one of the thrice replicated experimental treatment over 40 weeks. Figure 6.11(D) shows the L-stage component of a predicted chaotic attractor and (E) shows the L-stage obtained from one of the thrice replicated experimental treatment; in order to study the complex dynamics in this case, the experiment was performed without interruption for 424 weeks (over 8 years or 112 generations!). A detailed study of this chaotic treatment, its attractor and the data from the experiments can be found in [88] and [190].

6.6.4 Discussion

The LPA model and its role in experimental studies involving flour beetles demonstrate the powerful role that discrete time models can play in both describing and predicting the dynamics of biological populations.

In addition to its predictions of attractor bifurcations in these experiments described in Section 6.6.3, the LPA model has shown remarkable success in describing numerous other dynamic features observed in flour beetle population data in numerous experiments. One example is the frequently seen evidence of model predicted unstable equilibria and the geometric features of their stable and unstable manifolds. This is due to the inevitable presence of stochastic perturbations in the beetle population data that sometimes place the demographic vector $\mathbf{x}(t)$ near a model predicted unstable equilibrium point with the result that the data ultimately moves away in the model predicted equilibrium in phase space. An example is seen in Figure 6.11(E) from week 350 to week 375 where the L-stage component in that plot is placed near what turns out to be the parameterized LPA model's predicted unstable equilibrium; see [190] for details. For other examples of this phenomenon see [87].

Other examples include: subtle cyclic patterns from periodic cycles embedded in a chaotic attractor [190]; the control of chaos by making use of model predicted sensitivity locations (as measured by Liapunov exponents) in phase space [108]; resonance due to periodically fluctuating environments [58], [160]; multiple attractors in a periodic environment (which were counter-intuitive predictions subsequently born out by controlled experiments) [162], [161], [164]; cycles due to lattice effects in phase space (integer valued population numbers) [88], [163].

Extensions of the LPA model have been used in various studies as well. Competitively coupled LPA models challenge the classic competitive exclusion principle (which is based on equilibrium dynamics) [117]; Darwinian dynamic versions of the LPA model were used in evolutionary changes in flour beetle populations in [254], [255]; spatial patterns observed in laboratory cultures were studied using a integrodifference equation version of the LPA model in [260], [261], [262]; and stochastic versions (in addition to being central to the parameterization and validation studies) have been used to demonstrate how both attractor stability and instability are required to understand time-series data [166] and to provide experimental support of the scaling rule for demographic stochasticity [109].

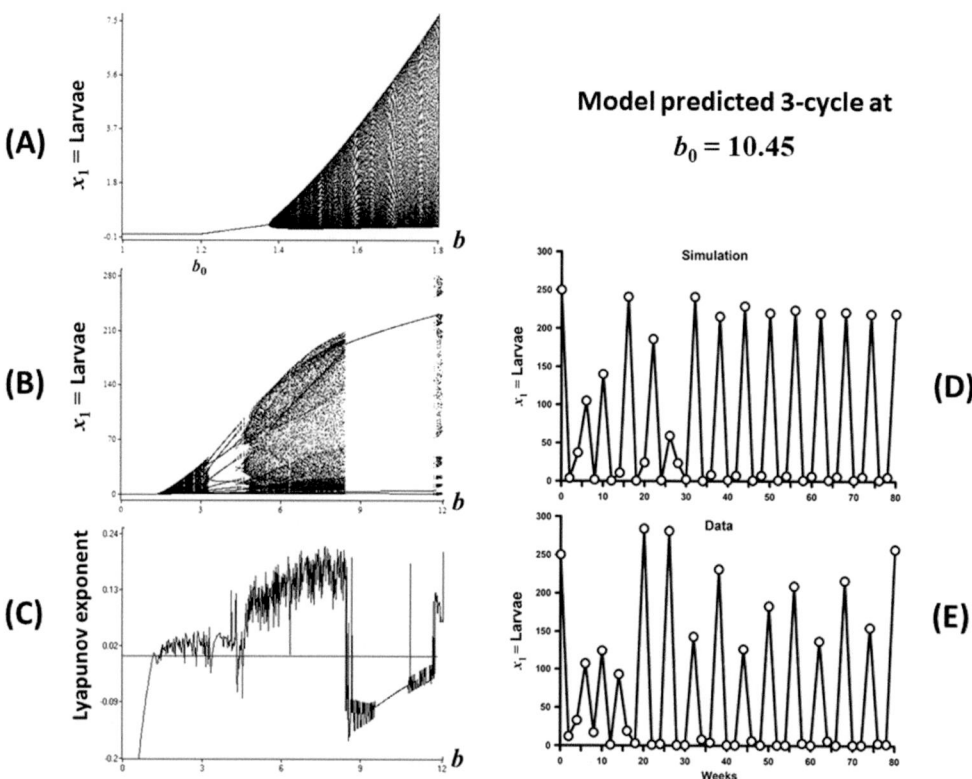

Figure 6.10: Conditioned least square estimates for the LPA model parameters derived in [105] (Table 1) (also see Table 4.1 in [88]) are $c_{el} = 0.01731$, $c_{ea} = 0.01310$, and $\mu_l = 0.2000$. (A)–(B) The bifurcation diagram for the LPA model is shown using the experimentally set parameter values $\mu_a = 0.96$ and $c_{pa} = 0.5$ and the dominant Lyapunov exponent is shown in (C). A Neimark-Sacker bifurcation occurs at approximately $b \approx 1.376 > b_0 = 1.2$. The conditioned least square estimate for b given in [105] is $b = 10.45$, which the diagram predicts, will result in a periodic 3-cycle attractor. In (D), the model predicted time series for the L-stage over 40 weeks, starting from the experimental initial condition $\mathbf{x}(0) = (250, 5, 100)^T$, is seen to approach a 3-cycle oscillation, after about 30 weeks of transients. Plot (E) shows the L-stage data from one of three replicated cultures in the experiment.

6.6. A CASE STUDY: THE LPA MODEL

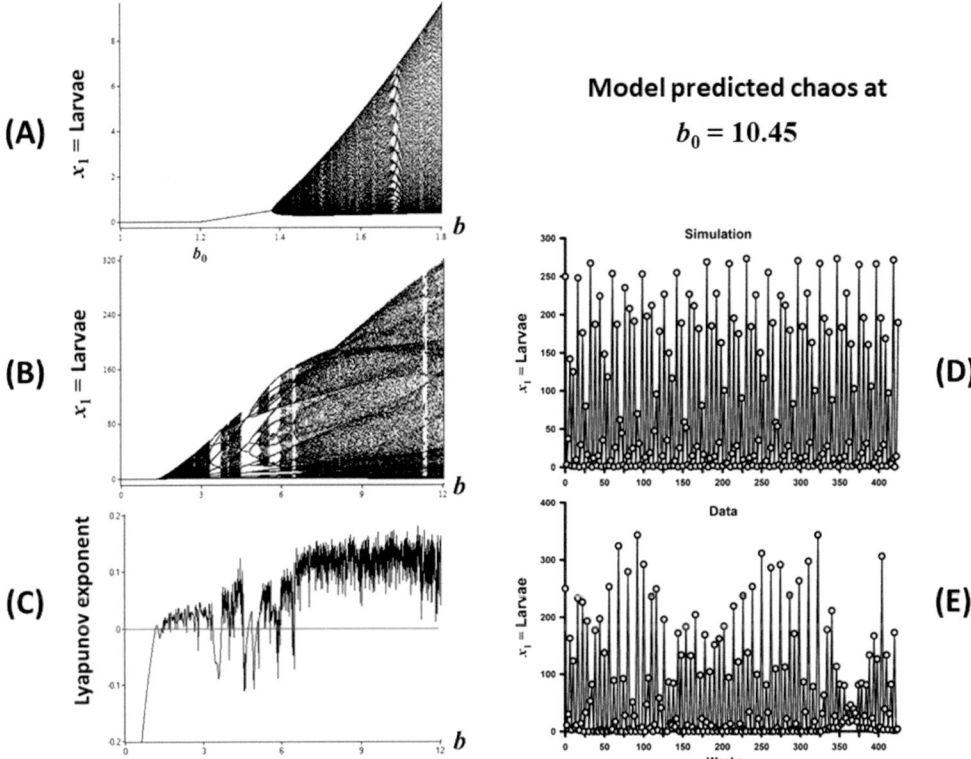

Figure 6.11: The same plots are shown as in Figure 6.10 but with c_{pa} changed from 0.5 to 0.35. At $b = 10.45$ the predicted attractor is now chaotic.

Exercises 6

1. For the $k = 2$ dimensional model $\mathbf{x}(t+1) = \mathcal{L}(\mathbf{x}(t))\mathbf{x}(t)$ with the Leslie matrix (6.31) and
$$\beta(\mathbf{x}) = \frac{1}{1 + c_{21}x_1 + c_{22}x_2}, \quad \sigma_1(\mathbf{x}) \equiv 1$$
in Example 94 show that the only equilibrium, other than the extinction equilibrium, is $\mathbf{x}^*$ given by (6.33) and that the Jacobian evaluated at $\mathbf{x}^*$ is (6.34).

2. Use the Linearization Principle to prove that the equilibrium (6.10) of the Juvenile-Adult model in Example 90 is locally asymptotically stable for all $\mathcal{R}_0 > 1$.

   (a) Do this by calculating the eigenvalues of the Jacobian evaluated at the equilibrium.

   (b) Do this by using the Jury conditions (the Trace-Determinant conditions).

3. The population projection matrix
$$\mathcal{L}(\mathbf{x}) = \begin{pmatrix} 0 & f_2 \\ s_1 \frac{1}{1+cx_1} & s_2 \end{pmatrix}, \quad f_2, c > 0, \, 0 < s_i < 1$$
defines a population model structured on juvenile and adult stages (as in Example 90) in which juvenile survival is dependent on juvenile density x_1.

   (a) Calculate the inherent reproduction number $\mathcal{R}_0$.

   (b) Use the Linearization Principle to show that the extinction equilibrium is locally asymptotically stable if $\mathcal{R}_0 < 1$ and unstable if $\mathcal{R}_0 > 1$.

   (c) Use Theorem 52 to show that if $\mathcal{R}_0 < 1$ then the extinction equilibrium is globally asymptotically stable on $\mathbb{R}_+^k$.

   (d) By algebraically solving the equilibrium equations, show that for $\mathcal{R}_0 > 1$ there exists a unique positive equilibrium.

   (e) Use the equilibrium formulas calculated in (d) to show, by using the Linearization Principle, that for $\mathcal{R}_0 > 1$ the positive equilibrium is locally asymptotically stable. Do this both by calculating the eigenvalues of the Jacobian and by using the Jury conditions (the Trace-Determinant conditions).

4. The population projection matrix
$$\mathcal{L}(\mathbf{x}) = \begin{pmatrix} 0 & f_2 \\ s_1 & s_2 \frac{1}{1+cx_2} \end{pmatrix}, \quad f_2, c > 0, \, 0 < s_i < 1$$
defines a population model structured on juvenile and adult stages (as in Example 90) in which adult survival is dependent on juvenile density x_1.

   (a) Calculate the inherent reproduction number $\mathcal{R}_0$.

   (b) Use the Linearization Principle to show that the extinction equilibrium is locally asymptotically stable if $\mathcal{R}_0 < 1$ and unstable if $\mathcal{R}_0 > 1$.

(c) Use Theorem 52 to show that if $\mathcal{R}_0 < 1$ then the extinction equilibrium is globally asymptotically stable on $\mathbb{R}_+^2$.

(d) By algebraically solving the equilibrium equations show that there exists a positive equilibrium if and only if $1 < \mathcal{R}_0 < (1-s_2)^{-1}$ (and that it is unique).

(e) Use the equilibrium formula calculated in (d) to show, by using the Linearization Principle, that when it exists the positive equilibrium is locally asymptotically stable. Do this both by calculating the eigenvalues of the Jacobian and by using the Jury conditions (the Trace-Determinant conditions).

(f) Show that if $\mathcal{R}_0 > (1-s_2)^{-1}$ then $\mathbf{0} \neq \mathbf{x}(0) \in \mathbb{R}_+^2$ implies $\lim_{t \to \infty} x_i(t) = +\infty$. In this case, the density dependence in this model is not sufficient to control unlimited population growth. (Hint: show $x_2(t+1) > x_2(t) + k$ for $t \geq 1$ and some positive constant $k > 0$.)

5. Repeat the analysis of the Juvenile-Adult model in Example 90 using $\beta(x_2) = e^{-cx}$, $c > 0$. Unlike the model in Example 90 show that the positive equilibria are not stable for all $\mathcal{R}_0 > 1$ but are unstable for $\mathcal{R}_0$ greater than a critical value. Calculate this critical value. What type of bifurcation from the positive equilibria occurs at this critical value?

6. Show that the Jacobian of (6.5) evaluated at $\mathbf{x} = \mathbf{0}$ equals the inherent projection matrix $P(\mathbf{0})$.

7. Analyze the Levin-Goodyear model as is done in Examples 92 and 91 but with the Ricker nonlinearity replaced by
$$f_j(\mathbf{x}) = b_j \frac{1}{1 + c\sum_{i=1}^k w_i x_i}.$$

8. Show that the eigenvalues of the inherent semelparous Leslie matrix (6.30) are $\lambda_i = \mathcal{R}_0^{1/2} u_i$ where u_i are the kth roots of unity and $\mathcal{R}_0 = f_k \Pi_{k=1}^{k-1} s_i$.

9. Repeat the analysis in Example 92 with nonlinearities satisfying (6.28).

10. Consider the projection matrices below that model various juvenile and adult class scenarios. All coefficients are positive, all survival coefficients s_i and s_{ij} are also less than 1, and $a > 1$.

1 juvenile and 1 adult: $P(\mathbf{x}) = \begin{pmatrix} 0 & f_2 \frac{1}{1+c_2 x_2} \\ s_1 \frac{1}{1+c_1 x_1} & s_2 \end{pmatrix}$

1 juvenile and 1 adult: $P(\mathbf{x}) = \begin{pmatrix} 0 & f_2 \frac{1}{1+cx_2} \frac{1+ax_2}{1+x_2} \\ s_1 & s_2 \end{pmatrix}$

1 juvenile and 2 adult: $P(\mathbf{x}) = \begin{pmatrix} 0 & f_2 \exp(-c||\mathbf{x}||) & f_3 \exp(-c||\mathbf{x}||) \\ s_{21} & s_{22} & 0 \\ s_{31} & 0 & s_{33} \end{pmatrix}$

2 juvenile and 1 adult: $P(\mathbf{x}) = \begin{pmatrix} 0 & 0 & f_2 \frac{1}{1+c_1 x_3} \\ 0 & 0 & f_3 \frac{1}{1+c_2 x_3} \\ s_{31} \frac{1+ax_1}{1+x_1} & s_{32} \frac{1+ax_2}{1+x_2} & s_{33} \end{pmatrix}$

(a) Show that the inherent projection matrix $P(\mathbf{0})$ is primitive.

(b) Calculate a formula for the inherent reproduction number $\mathcal{R}_0$.

(c) Apply the Theorems in Section 6.2, when applicable, to determine the stability properties of the extinction equilibrium.

(d) Using the bifurcation parameter $\mu = f_2$ find the bifurcation value μ_0 and apply the Theorems in Section 6.4 to determine the properties of the bifurcating positive equilibria at $\mu = \mu_0$.

11. Consider the $k = 4$ dimensional, nonlinear semelparous Leslie model with the projection matrix
$$P(\mathbf{x}) = \begin{pmatrix} 0 & 0 & 0 & b\frac{1}{1+x_4} \\ s_1 & 0 & 0 & 0 \\ 0 & s_2 & 0 & 0 \\ 0 & 0 & s_3 & 0 \end{pmatrix}, \quad b > 0,\ 0 < s_i < 1.$$

(a) Apply the Theorems in Section 6.2, when possible, to determine the stability properties of the extinction equilibrium and to show that a forward (local) bifurcation of positive equilibria occurs at $\mathcal{R}_0 = 1$.

(b) Algebraically solve the equilibrium equations to show that there exists a unique positive equilibrium for every $\mathcal{R}_0 > 1$.

(c) Use your answer in (b) and the Linearization Principle to show that the positive equilibrium is locally asymptotically stable for every $\mathcal{R}_0 > 1$.

(d) Show that for each $\mathcal{R}_0 > 1$ there exists a unique single class 4-cycle and that it is unstable.

12. Consider the $k = 4$ dimensional, nonlinear semelparous Leslie model with the projection matrix
$$P(\mathbf{x}) = \begin{pmatrix} 0 & 0 & 0 & b\frac{1}{1+cx_3+x_4} \\ s_1 & 0 & 0 & 0 \\ 0 & s_2 & 0 & 0 \\ 0 & 0 & s_3 & 0 \end{pmatrix}, \quad b, c > 0,\ 0 < s_i < 1.$$

(a) Apply the Theorems in Section 6.2, when possible, to determine the stability properties of the extinction equilibrium and to show that a forward (local) bifurcation of positive equilibria occurs at $\mathcal{R}_0 = 1$.

(b) Algebraically solve the equilibrium equations to show that there exists a unique positive equilibrium for every $\mathcal{R}_0 > 1$.

(c) Use your answer in (b) and the Linearization Principle to show that if $c > s_3$, then the positive equilibrium is unstable for every $\mathcal{R}_0 > 1$. (Hint: Show that the fourth order characteristic polynomial $p(\lambda)$ of the Jacobian evaluated at the positive equilibrium satisfies $p(-1) < 0$ and $\lim_{\lambda \to -\infty} p(\lambda) = +\infty$).

13. Show that the LPA model is dissipative, with a constant of the form $\beta = m\mathcal{R}_0 + n$ with $m, n > 0$. Show that H4 (with $\Omega = \mathbb{R}^3$), H5 and H5 hold. Calculate a formula for α and show it is positive. Show that the spectrum of $\mathcal{R}_0$ values from the unbounded, bifurcating continuum of positive equilibrium pairs is an unbounded interval in $\mathring{\mathbb{R}}^3_+$ that contains $\mathcal{R}_0 = 1$ in its closure.

14. The following problems concern the LPA model with projection matrix $\mathcal{L}(\mathbf{x})$ given by formula (6.53).

   (a) Show that assumptions H2 and H3 hold with $\Omega = \mathbb{R}^3$.

   (b) Show that the inherent net reproduction number $\mathcal{R}_0$ is given by the formula (6.54).

   (c) Prove that the LPA projection matrix $\mathcal{L}(\mathbf{x})$ is primitive for all $\mathbf{x} \in \mathbb{R}^3_+$.

   (d) Show that the spectrum of b values corresponding to positive equilibria on the continuum $\mathbb{C}\setminus\{(b_0, \mathbf{0})\}$ is the half line $b > b_0$. (Hint: see the proof of Theorem 61(b).)

   (e) Treat the positive equilibrium $\mathbf{x}(b)$ as a function of b. Show that $\mathbf{x}(b)$ is unstable for large $b > b_0$. (Hint: Use the (three) equilibrium equations for the equilibrium components $x_i(b)$ to show that all three are strictly increasing functions of b and that $\lim_{t \to \infty} x_i(b) = +\infty$. The Jacobian of the LPA model, when evaluated at the equilibrium $\mathbf{x}(b)$, is a function $J(b)$ of b. Show for sufficiently large b that the eigenvalues of $J(b)$ are real, two of which approach $-\infty$ and the third of which approaches 0 as $b \to \infty$.)

Chapter 7
Infectious Disease Models Part: II

7.1 Introduction

Infectious diseases are among the leading causes of mortality worldwide, presenting complex challenges in public health, economic stability, and social structures. The continuous emergence and re-emergence of pathogens, driven by factors such as global mobility, environmental changes, and microbial adaptation, necessitate dynamic and robust responses. Mathematical models provide critical insights into disease dynamics and serve as essential tools in the design and evaluation of intervention strategies.

Infectious agents spread through diverse and complex pathways. Direct transmission occurs through close contact, involving bodily fluids or skin contact. Diseases like HIV and influenza exemplify this mechanism. The mathematical modeling of direct transmission focuses on contact rates and the probability of transmission per contact.

Vectors such as mosquitoes and ticks play a critical role in the transmission of diseases like malaria and Lyme disease. Models in this area often incorporate climatic variables which influence vector populations and disease transmission rates.

Diseases like cholera are spread through contaminated water sources. Modeling these diseases requires the integration of environmental science, hydrology, and human behavior to understand and predict their spread.

Mathematical models are indispensable in dissecting the complexities of infectious diseases. They range from simple SIR models to complex simulations incorporating genetics and behavior.

Using difference equations, we consider time in discrete steps and are suited to diseases with clear incubation periods or reporting cycles. They are crucial for outbreak response and vaccine impact studies. Moreover, a discrete-time view of disease progression is vital for chronic diseases or those with gradual intervention rollout. Examples include the SEIR model and models incorporating age structure or spatial dynamics. Furthermore, models have helped optimize the allocation of resources and strategize vaccination campaigns, contributing significantly to the near-eradication of polio.

Modeling has been central in predicting disease spread, evaluating lockdown measures, and guiding vaccination strategies, demonstrating the power of real-time data integration and simulation in managing global health crises. Despite their successes, epidemiological models face challenges such as data quality, assumptions validity, and integrating multidisciplinary approaches. Future directions include the use of more real-time data, machine learning techniques, and enhanced collaboration across scientific disciplines. The strategic application of mathematical models in epidemiology has substantially shaped public health responses to infectious diseases. Continued advancements in computational techniques and interdisciplinary research are essential to enhance the effectiveness and precision of these models.

In this chapter, we lay the foundation for modeling infectious diseases. We will introduce some of the popular terminology such as herd immunity, net reproduction number, disease-free equilibrium, and endemic equilibrium.

7.2 Endemic Equilibria

Theorems 43 and 44 concern the stability and instability of a disease-free equilibrium of an epidemic model formulated by Equations (4.19). The stability determining spectral radius ρ^* in Theorem 43 and reproduction number $\mathcal{R}_0$ in Theorem 44 are quantities that depend on the properties of $\mathbf{g}_0$ from the first equation in (4.19). If a parameter in $\mathbf{g}_0$ is manipulated so that ρ^* (equivalently $\mathcal{R}_0$) increases through 1, then the disease-free equilibrium $(\mathbf{0}_m, \mathbf{x}_1^*)$ destabilizes. When a disease-free equilibrium destabilizes in this way, we expect in general that another branch of equilibria will intersect the disease-free equilibrium to form what is called a transcritical bifurcation of equilibria. In this section, we will investigate the arising, in this way, of *endemic equilibria*, i.e., equilibria of the form $(\mathbf{x}_0, \mathbf{x}_1)$ where $\mathbf{x}_0$ is positive (which means the disease is present at equilibrium).

For this approach, a modeler must identify a model parameter that is to be manipulated so as to cause the destabilization of the disease-free equilibrium. We denote the parameter by μ and include it in our general model equations

$$\begin{aligned} \mathbf{x}_0(t+1) &= \mathbf{g}_0(\mu, \mathbf{x}_0(t), \mathbf{x}_1(t)) \\ \mathbf{x}_1(t+1) &= \mathbf{g}_1(\mu, \mathbf{x}_0(t), \mathbf{x}_1(t)) \end{aligned} \tag{7.1}$$

where $\mathbf{g}_0$ and $\mathbf{g}_1$ satisfy A1, A2 and (4.20) for all $\mu \in I_\mu$ where $I_\mu \subseteq \mathbb{R}_+$ is an open interval. If the disease is absent, then new infected individuals cannot be produced so we assume

$$\mathbf{g}_0(\mu, \mathbf{0}_m, \mathbf{x}_1) \equiv \mathbf{0}_m \tag{7.2}$$

for all μ and $\mathbf{x}_1$ and write

$$\mathbf{g}_0(\mu, \mathbf{x}_0, \mathbf{x}_1) = G(\mu, \mathbf{x}_0, \mathbf{x}_1) \mathbf{x}_0$$

where $G = (\gamma_{ij})$ is an $m \times m$ matrix, so that equations (7.1) become

$$\begin{aligned} \mathbf{x}_0(t+1) &= G(\mu, \mathbf{x}_0(t), \mathbf{x}_1(t)) \mathbf{x}_0(t) \\ \mathbf{x}_1(t+1) &= \mathbf{g}_1(\mu, \mathbf{x}_0(t), \mathbf{x}_1(t)). \end{aligned} \tag{7.3}$$

The equilibrium equations associated with (7.3) are

$$\mathbf{x}_0 = G(\mu, \mathbf{x}_0, \mathbf{x}_1) \mathbf{x}_0 \tag{7.4}$$

$$\mathbf{x}_1 = \mathbf{g}_1(\mu, \mathbf{x}_0, \mathbf{x}_1). \tag{7.5}$$

If

$$(\mathbf{x}_0, \mathbf{x}_1)^T = \begin{pmatrix} \mathbf{x}_0 \\ \mathbf{x}_1 \end{pmatrix} \in \mathbb{R}_+^n$$

7.2. ENDEMIC EQUILIBRIA

is a solution of the equilibrium equations (7.4)-(7.5), then we call $\left(\mu, (\mathbf{x}_0, \mathbf{x}_1)^T\right)$ an *equilibrium pair*. We say $\left(\mu, (\mathbf{0}_m, \mathbf{x}_1)^T\right)$ is a *disease-free equilibrium pair* and, if $\mathbf{x}_0 \in \mathring{\mathbb{R}}_+^m$, then $(\mu, (\mathbf{x}_0, \mathbf{x}_1)^T)$ is an *endemic equilibrium pair*. If $(\mathbf{x}_0, \mathbf{x}_1)^T$ is a locally asymptotically stable (or unstable) equilibrium of (7.3), then we say $(\mu, (\mathbf{x}_0, \mathbf{x}_1)^T)$ is a *stable (unstable) equilibrium pair*.

The equilibrium equation associated with the disease-free equation

$$\mathbf{x}_1(t+1) = \mathbf{g}_1(\mu, \mathbf{0}_m, \mathbf{x}_1(t)) \tag{7.6}$$

is

$$\mathbf{x}_1 = \mathbf{g}_1(\mu, \mathbf{0}_m, \mathbf{x}_1) \tag{7.7}$$

for which we make the following assumption.

A4: Suppose $\mathbf{x}_1^*$ satisfies the Equation (7.7) for $\mu \in I_\mu$ where I_μ is an open interval in $\mathbb{R}_+$. Assume (7.2) holds and that the $m \times m$ matrix $G(\mu, \mathbf{x}_0, \mathbf{x}_1) = (\gamma_{ij}(\mu, \mathbf{x}_0, \mathbf{x}_1))$ is nonnegative and irreducible and its entries $\gamma_{ij}(\mu, \mathbf{x}_0, \mathbf{x}_1)$, as well as those of the vector $\mathbf{g}_1(\mu, \mathbf{x}_0, \mathbf{x}_1)$, are twice continuously differentiable functions of $\mu \in I_\mu$ and of $\mathbf{x} = (\mathbf{x}_0, \mathbf{x}_1)^T$ in an open neighborhood of $(\mathbf{0}_m, \mathbf{x}_1^*)^T$ in $\mathbb{R}_+^n$.

A5: There exists a $\mu_0 \in I_\mu$ such that $\rho(G(\mu_0, \mathbf{0}_m, \mathbf{x}_1^*)) = 1$ and

$$\rho(J_{\mathbf{x}_1} \mathbf{g}_1(\mu, \mathbf{0}_m, \mathbf{x}_1^*)) < 1 \tag{7.8}$$

for $\mu \in I_\mu$.

Inequality (7.8) and the Linearization Principle imply that $\mathbf{x}_1^*$ is a locally asymptotically stable equilibrium of the disease-free equation (7.6). We next turn our attention to the stability properties of the disease-free equilibrium pair $(\mu, (\mathbf{0}_m, \mathbf{x}_1^*)^T)$ of epidemic model equation (7.3) as they depend on μ.

Definition 29 *A superscript "0" denotes evaluation at $\mu = \mu_0$ and the disease-free equilibrium $(\mathbf{x}_0, \mathbf{x}_1)^T = (\mathbf{0}_m, \mathbf{x}_1^*)^T$. Thus, $G^0 := G(\mu_0, \mathbf{0}_m, \mathbf{x}_1^*)$, $J_{\mathbf{x}_1}^0 \mathbf{g}_1 := J_{\mathbf{x}_1} \mathbf{g}_1(\mu_0, \mathbf{0}_m, \mathbf{x}_1^*)$, and so on. Let $\mathbf{w}_L^T$ and $\mathbf{w}_R$ be positive left and right eigenvectors of G^0 associated with the eigenvalue 1.*

Two quantities that will be important in our study of endemic equilibria:

$$\kappa := -\mathbf{w}_L^T (\kappa_{ij}) \mathbf{w}_R \quad \text{and} \quad \alpha := \mathbf{w}_L^T (\alpha_{ij}) \mathbf{w}_R \tag{7.9}$$

where (κ_{ij}) and (α_{ij}) are $m \times m$ matrices whose entries are

$$\kappa_{ij} := \left(\nabla_{\mathbf{x}_0}^0 \gamma_{ij} + \nabla_{\mathbf{x}_1}^0 \gamma_{ij} \left(I - J_{\mathbf{x}_1}^0 \mathbf{g}_1 \right)^{-1} J_{\mathbf{x}_0}^0 \mathbf{g}_1 \right) \mathbf{w}_R$$

$$\alpha_{ij} := \partial_\mu^0 \gamma_{ij} + \nabla_{\mathbf{x}_1}^0 \gamma_{ij} \left(I - J_{\mathbf{x}_1}^0 \mathbf{g}_1 \right)^{-1} \partial_\mu^0 \mathbf{g}_1.$$

Here $\nabla_{\mathbf{x}}$ denotes a (*row*) vector gradient of a scalar valued function of $\mathbf{x}$ and ∂_μ denotes partial differentiation with respect to μ. Note that the matrix $I - J_{\mathbf{x}_1}^0 \mathbf{g}_1$ is invertible by assumption A5.

In the next section, we consider the existence of endemic equilibrium pairs (7.3) that arise from a bifurcation at the disease-free equilibrium pair $(\mu_0, \mathbf{0}_m, \mathbf{x}_1^*)$ when it destabilizes.

7.2.1 The Bifurcation of Endemic Equilibria

We first take up the question of the existence of endemic equilibria of Equations (7.1) as they depend on μ. In Section 7.2.2, we consider their stability properties.

Theorem 64 *Assume A4 and A5 hold and that $\alpha \neq 0$.*

(a) For ε in open interval I_ε in R^1 containing $\varepsilon = 0$ there exist endemic equilibrium pairs $\left(\mu(\varepsilon), (\mathbf{x}_0(\varepsilon), \mathbf{x}_1(\varepsilon))^T\right)$ of the Equations (7.3) that have the form

$$\mu(\varepsilon) = \mu_0 + \eta(\varepsilon) \tag{7.10}$$

$$\mathbf{x}_0(\varepsilon) = \varepsilon \mathbf{w}_R + \mathbf{z}_0(\varepsilon) \tag{7.11}$$

$$\mathbf{x}_1(\varepsilon) = \mathbf{x}_1^* + \mathbf{z}_1(\varepsilon) \tag{7.12}$$

where

$$\eta \in C^2\left(I_\varepsilon \to \mathbb{R}^1\right), \quad \mathbf{z}_0 \in C^2\left(I_\varepsilon \to \mathbb{R}^m\right), \quad \mathbf{z}_1 \in C^2\left(I_\varepsilon \to \mathbb{R}^{n-m}\right)$$

with $\mathbf{z}_0(\varepsilon)^T \mathbf{w}_R = 0$ for $\varepsilon \in I_\varepsilon$ and where

$$\eta(0) = 0, \quad \mathbf{z}_0(0) = \mathbf{z}_0'(0) = \mathbf{0}_m, \quad \mathbf{z}_1(0) = \mathbf{0}_{n-m}.$$

(b) Furthermore,

$$\eta'(0) = \frac{\kappa}{\alpha} \tag{7.13}$$

and, as a result, the local bifurcation of positive equilibria described in (a) is μ-forward if α and κ have the same sign and μ-backward if they have opposite signs.

Proof. (a) The proof makes use of Theorem 54 in Chapter 6. Because (7.8) implies the invertibility of $I - J_{\mathbf{x}_1}^0 \mathbf{g}_1$, we can apply the implicit function theorem to solve the second equilibrium equation (7.5) for

$$\mathbf{x}_1 = \xi_1(\mu, \mathbf{x}_0) \text{ where } \xi_1(\mu_0, \mathbf{0}_m) = \mathbf{x}_1^* \tag{7.14}$$

and where $\xi_1(\mu, \mathbf{x}_0)$ is twice continuously differentiable for μ near μ_0 and $\mathbf{x}_0$ near $\mathbf{0}_m$; thus

$$\xi_1(\mu, \mathbf{x}_0) = \mathbf{g}_1(\mu, \mathbf{x}_0, \xi_1(\mu, \mathbf{x}_0)). \tag{7.15}$$

The first equilibrium equation (7.4) then becomes an equation

$$\mathbf{x}_0 = G(\mu, \mathbf{x}_0, \xi_1(\mu, \mathbf{x}_0)) \mathbf{x}_0 \tag{7.16}$$

to solve for $\mathbf{x}_0 = \mathbf{x}_0(\mu)$. To do this we can apply Theorem 54 to Equation (7.16) with the roles of $\mathbf{x}$ and $A(\mu, \mathbf{x})$ played instead by $\mathbf{x}_0$ and

$$G(\mu, \mathbf{x}_0, \xi_1(\mu, \mathbf{x}_0)) = (\gamma_{ij}(\mu, \mathbf{x}_0, \xi_1(\mu, \mathbf{x}_0)))$$

respectively. This will then prove Theorem 64.

To apply Theorem 54 to Equation (7.16) all we need to do is verify that

$$\mathbf{w}_L^T \partial_\mu^0 G \mathbf{w}_R = \mathbf{w}_L^T \left(\partial_\mu^0 \gamma_{ij}\right) \mathbf{w}_R \neq 0.$$

7.2. ENDEMIC EQUILIBRIA

The partial derivative of $\gamma_{ij}(\mu, \mathbf{x}_0, \xi_1(\mu, \mathbf{x}_0))$ with respect to μ is

$$\partial_\mu \gamma_{ij}(\mu, \mathbf{x}_0, \xi_1(\mu, \mathbf{x}_0)) + \nabla_{\mathbf{x}_1} \gamma_{ij}(\mu, \mathbf{x}_0, \xi_1(\mu, \mathbf{x}_0)) \partial_\mu \xi_1(\mu, \mathbf{x}_0).$$

To calculate $\partial_\mu^0 G$ we evaluate this expression at $\mu = \mu_0$ and $\mathbf{x}_0 = \mathbf{0}_m$ (recall $\xi_1(\mu_0, \mathbf{0}_m) = \mathbf{x}_1^*$) to obtain

$$\partial_\mu^0 \gamma_{ij} + \nabla_{\mathbf{x}_1}^0 \gamma_{ij} \partial_\mu \xi_1(\mu_0, \mathbf{0}_m) \tag{7.17}$$

which requires us to calculate the derivative $\partial_\mu \xi_1(\mu_0, \mathbf{0}_m)$. This we do by implicit differentiation of Equation (7.15) with respect to μ

$$\partial_\mu \xi_1(\mu, \mathbf{x}_0) = \partial_\mu \mathbf{g}_1(\mu, \mathbf{x}_0, \xi_1(\mu, \mathbf{x}_0)) + J_{\mathbf{x}_1} \mathbf{g}_1(\mu, \mathbf{x}_0, \xi_1(\mu, \mathbf{x}_0)) \partial_\mu \xi_1(\mu, \mathbf{x}_0)$$

followed by an evaluate at $\mu = \mu_0$ and $\mathbf{x}_0 = \mathbf{0}_m$. This results in the equation

$$\partial_\mu \xi_1(\mu_0, \mathbf{0}_m) = \partial_\mu^0 \mathbf{g}_1 + J_{\mathbf{x}_1}^0 \mathbf{g}_1 \partial_\mu \xi_1(\mu_0, \mathbf{0}_m)$$

which we solve for

$$\partial_\mu \xi_1(\mu_0, \mathbf{0}_m) = \left(I - J_{\mathbf{x}_1}^0 \mathbf{g}_1\right)^{-1} \partial_\mu^0 \mathbf{g}_1.$$

Using this result, we see that (7.17) equals α_{ij} and hence

$$\mathbf{w}_L^T \partial_\mu^0 G \mathbf{w}_R = \alpha \tag{7.18}$$

which is nonzero by assumption.

Theorem 54 now implies that Equation (7.16) has solutions μ and $\mathbf{x}_0$ in the parametric form (7.10)-(7.11) which, together with $\mathbf{x}_1(\varepsilon) = \xi_1(\mu(\varepsilon), \mathbf{x}_0(\varepsilon))$, give endemic equilibrium pairs $\left(\mu(\varepsilon), (\mathbf{x}_0(\varepsilon), \mathbf{x}_1(\varepsilon))^T\right)$ of the equilibrium equations (7.3).

(b) To obtain the formula for $\eta'(0)$ we apply Theorem 55 in Chapter 6 to Equation (7.16) with the roles of $\mathbf{x}$ and $A(\mu, \mathbf{x})$ played, respectively, by $\mathbf{x}_0$ and

$$G(\mu, \mathbf{x}_0, \xi_1(\mu, \mathbf{x}_0)) = (\gamma_{ij}(\mu, \mathbf{x}_0, \xi_1(\mu, \mathbf{x}_0))).$$

This gives

$$\eta'(0) = \frac{-\mathbf{w}_L^T \left(\nabla_{\mathbf{x}_0}^0 \gamma_{ij}(\mu_0, \mathbf{x}_0, \xi_1(\mu_0, \mathbf{x}_0))\big|_{\mathbf{x}_0 = \mathbf{0}_m} \mathbf{w}_R\right) \mathbf{w}_R}{\mathbf{w}_L^T \partial_\mu^0 G \mathbf{w}_R}. \tag{7.19}$$

By (7.18) the denominator equals α and we have the numerator left to calculate. To do this we need to calculate the gradient of $\gamma_{ij}(\mu, \mathbf{x}_0, \xi_1(\mu, \mathbf{x}_0))$ with respect of $\mathbf{x}_0$ and evaluate the answer at $\mu = \mu_0$ and $\mathbf{x}_0 = \mathbf{0}_m$. This gradient

$$\nabla_{\mathbf{x}_0}^0 \gamma_{ij}(\mu_0, \mathbf{x}_0, \xi_1(\mu_0, \mathbf{x}_0))\big|_{\mathbf{x}_0 = \mathbf{0}_m} = \nabla_{\mathbf{x}_0}^0 \gamma_{ij} + \nabla_{\mathbf{x}_1}^0 \gamma_{ij} J_{\mathbf{x}_0} \xi_1(\mu_0, \mathbf{0}_m) \tag{7.20}$$

requires the calculation of $J_{\mathbf{x}_0} \xi_1(\mu_0, \mathbf{0}_m)$, which find by taking the Jacobian of equation (7.15) with respect to $\mathbf{x}_0$:

$$J_{\mathbf{x}_0} \xi_1(\mu, \mathbf{x}_0) = J_{\mathbf{x}_0} \mathbf{g}_1(\mu, \mathbf{x}_0, \xi_1(\mu, \mathbf{x}_0)) + J_{\mathbf{x}_1} \mathbf{g}_1(\mu, \mathbf{x}_0, \xi_1(\mu, \mathbf{x}_0)) J_{\mathbf{x}_0} \xi_1(\mu, \mathbf{x}_0)$$

followed by an evaluate at $\mu = \mu_0$ and $\mathbf{x}_0 = \mathbf{0}_m$. From the result

$$J_{\mathbf{x}_0} \xi_1(\mu_0, \mathbf{0}_m) = J_{\mathbf{x}_0}^0 \mathbf{g}_1 + J_{\mathbf{x}_1}^0 \mathbf{g}_1 J_{\mathbf{x}_0} \xi_1(\mu_0, \mathbf{0}_m)$$

we get
$$J_{\mathbf{x}_0}\xi_1(\mu_0, \mathbf{0}_m) = \left(I - J^0_{\mathbf{x}_1}\mathbf{g}_1\right)^{-1} J^0_{\mathbf{x}_0}\mathbf{g}_1.$$
and from (7.20) we then obtain
$$\nabla^0_{\mathbf{x}_0}\gamma_{ij}(\mu_0, \mathbf{x}_0, \xi_1(\mu_0, \mathbf{x}_0))\big|_{\mathbf{x}_0=\mathbf{0}_m} = \nabla^0_{\mathbf{x}_0}\gamma_{ij} + \nabla^0_{\mathbf{x}_1}\gamma_{ij}\left(I - J^0_{\mathbf{x}_1}\mathbf{g}_1\right)^{-1} J^0_{\mathbf{x}_0}\mathbf{g}_1.$$
Placing this result into the numerator of (7.19) and noting that the denominator is α, we obtain (7.13). ■

Since $\mathbf{w}_R \in \mathring{\mathbb{R}}_+^m$, for ε small the bifurcating equilibrium pairs guaranteed by Theorem 64 are endemic equilibrium pairs for (and only for) $\varepsilon > 0$. These endemic equilibria collapse to the disease-free equilibrium pair $(\mu_0, (\mathbf{0}_m, \mathbf{x}_1^*)^T)$ as $\varepsilon \to 0$ and in this way they bifurcate from the disease-free equilibrium at $\mu = \mu_0$.

Definition 30 *We say the bifurcation of the endemic equilibrium pairs from the disease-free equilibrium at $\mu = \mu_0$ (as assured by Theorem 64) is μ-**forward** (respectively μ-**backward**) if $\mu > \mu_0$ (respectively $\mu < \mu_0$) in a neighborhood of the disease-free equilibrium pair at $\mu = \mu_0$.*

Assuming $\kappa \neq 0$, the direction of bifurcation is determined by the sign of $\eta'(0)$ given by formula (7.13).

Corollary 8 *The bifurcation of endemic equilibria in Theorem 64 is μ-forward if κ and α have the same signs and μ-backward if κ and α have opposite signs.*

7.2.2 Stability of Endemic Equilibria

To study the (local asymptotic) stability of an equilibrium pair $(\mu, (\mathbf{x}_0, \mathbf{x}_1)^T)$ of the Equations (7.3) by means of the linearization principle, we consider the associated Jacobian

$$J_{\mathbf{x}}\mathbf{g}(\mu, \mathbf{x}_0, \mathbf{x}_1) = \begin{pmatrix} J_{\mathbf{x}_0}(G(\mu, \mathbf{x}_0, \mathbf{x}_1)\mathbf{x}_0) & J_{\mathbf{x}_1}(G(\mu, \mathbf{x}_0, \mathbf{x}_1)\mathbf{x}_0) \\ J_{\mathbf{x}_0}\mathbf{g}_1(\mu, \mathbf{x}_0, \mathbf{x}_1) & J_{\mathbf{x}_1}\mathbf{g}_1(\mu, \mathbf{x}_0, \mathbf{x}_1) \end{pmatrix}.$$

When evaluated at a disease-free equilibrium pair $(\mu, (\mathbf{0}_m, \mathbf{x}_1)^T)$, this Jacobian is block diagonal

$$J_{\mathbf{x}}\mathbf{g}(\mu, \mathbf{0}_m, \mathbf{x}_1) = \begin{pmatrix} G(\mu, \mathbf{0}_m, \mathbf{x}_1) & O \\ J_{\mathbf{x}_0}\mathbf{g}_1(\mu, \mathbf{0}_m, \mathbf{x}_1) & J_{\mathbf{x}_1}\mathbf{g}_1(\mu, \mathbf{0}_m, \mathbf{x}_1) \end{pmatrix}$$

and its eigenvalues are those of the two diagonal blocks $G(\mu, \mathbf{0}_m, \mathbf{x}_1)$ and $J_{\mathbf{x}_1}\mathbf{g}_1(\mu, \mathbf{0}_m, \mathbf{x}_1)$.

Definition 31 *We say the bifurcation of the endemic equilibrium pairs from the disease-free equilibrium at $\mu = \mu_0$ (as assured by Theorem 64) is **stable** (respectively **unstable**) if, for those endemic equilibria $(\mu, (\mathbf{x}_0^*, \mathbf{x}_1^*)^T)$ in a neighborhood of the disease-free equilibrium pair at $\mu = \mu_0$, the equilibrium $(\mathbf{x}_0^*, \mathbf{x}_1^*)^T$ of the Equations (7.3) with parameter μ is locally asymptotically stable (respectively unstable).*

Theorem 65 *Assume A4 and A5 hold and that $\alpha \neq 0$.*

(a) *If $\alpha > 0$ then the disease-free equilibrium $(\mathbf{0}_m, \mathbf{x}_1^*)^T$ of (7.3) destabilizes as μ increases through μ_0. If $\alpha < 0$ then the disease-free equilibrium $(\mathbf{0}_m, \mathbf{x}_1^*)^T$ destabilizes as μ decreases through μ_0.*

7.2. ENDEMIC EQUILIBRIA

(b) Assume $G(\mu_0, \mathbf{0}_m, \mathbf{x}_1^*)$ is primitive. If $\kappa > 0$ then for $\varepsilon > 0$ sufficiently small the endemic equilibrium pairs (6.21) in Theorem 64 are locally asymptotically stable. If $\kappa < 0$ then for $\varepsilon > 0$ sufficiently small, the endemic equilibrium pairs (6.21) are unstable.

Proof.

(a) Inequality (7.8) in A5 implies all eigenvalues of the lower diagonal block $J_{\mathbf{x}_1}\mathbf{g}_1(\mu, \mathbf{0}_m, \mathbf{x}_1)$ in the Jacobian $J_{\mathbf{x}}\mathbf{g}(\mu, \mathbf{0}_m, \mathbf{x}_1^*)$ satisfy $|\lambda| < 1$ for small ε. The remaining eigenvalues of $J_{\mathbf{x}}\mathbf{g}(\mu, \mathbf{0}_m, \mathbf{x}_1^*)$ are those of the upper diagonal block $G(\mu, \mathbf{0}_m, \mathbf{x}_1)$. Arguing as in Section 6.1 of Chapter 6 we have

$$\left.\frac{d\rho(G(\mu, \mathbf{0}_m, \mathbf{x}_1^*))}{d\mu}\right|_{\mu=\mu_0} = \frac{\mathbf{w}_L^T \partial_\mu^0 G \mathbf{w}_R}{\mathbf{w}_L^T \mathbf{w}_R} = \frac{\alpha}{\mathbf{w}_L^T \mathbf{w}_R} \qquad (7.21)$$

(see (6.3)). Since $\rho(G^0) = 1$ by assumption A5, it follows that $\rho(G(\mu, \mathbf{0}_m, \mathbf{x}_1^*))$ increases through 1 as μ increases through μ_0 when $\alpha > 0$ or as μ increases through μ_0 when $\alpha < 0$. This proves (a).

(b) Since the upper right block of the Jacobian $J_{\mathbf{x}}\mathbf{g}(\mu(\varepsilon), \mathbf{x}_0(\varepsilon), \mathbf{x}_1(\varepsilon))$ approaches the zero matrix as $\varepsilon \to 0$, the eigenvalues of the Jacobian (and their ε derivatives) approach the eigenvalues (and their ε derivatives) of the diagonal block matrices

$$J_{\mathbf{x}_0}(G(\mu(\varepsilon), \mathbf{x}_0(\varepsilon), \mathbf{x}_1(\varepsilon))\mathbf{x}_0(\varepsilon)) \quad \text{and} \quad J_{\mathbf{x}_1}\mathbf{g}_1(\mu(\varepsilon), \mathbf{x}_0(\varepsilon), \mathbf{x}_1(\varepsilon)).$$

It follows from A5 and the primitivity of $G(\mu_0, \mathbf{0}_m, \mathbf{x}_1^*)$ that $n-1$ of the eigenvalues of these two matrices satisfy $|\lambda| < 1$ except for the eigenvalue $\lambda(\varepsilon)$ of the first matrix for which $\lambda(0) = \rho(G(\mu_0, \mathbf{0}_m, \mathbf{x}_1^*)) = 1$ by A5. The determination of the stability (by linearization) of the bifurcating endemic equilibria reduces to an investigation of the eigenvalue $\lambda(\varepsilon)$ for $\varepsilon \gtrless 0$. Is $\lambda(\varepsilon)$ greater than or less than 1 for $\varepsilon \gtrless 0$? We obtain an answer by calculating $\lambda'(0)$ from a differentiation of

$$\lambda(\varepsilon)\mathbf{w}_R(\varepsilon) - M(\varepsilon)\mathbf{w}_R(\varepsilon) = \mathbf{0}$$

where

$$M(\varepsilon) := J_{\mathbf{x}_0}(G(\mu(\varepsilon), \mathbf{x}_0(\varepsilon), \mathbf{x}_1(\varepsilon))\mathbf{x}_0(\varepsilon))$$

and $\mathbf{w}_R(\varepsilon)$ is an (continuously differentiable) eigenvector chose so that $\mathbf{w}_R(0) = \mathbf{w}_R$ followed by an evaluation at $\varepsilon = 0$. This leads to

$$\lambda(\varepsilon)\mathbf{w}_R'(\varepsilon) + \lambda'(\varepsilon)\mathbf{w}_R(\varepsilon) - M(\varepsilon)\mathbf{w}_R'(\varepsilon) - M'(\varepsilon)\mathbf{w}_R(\varepsilon) = \mathbf{0}$$

and

$$\mathbf{w}_R'(0) - G(\mu_0, \mathbf{0}_m, \mathbf{x}_1^*)\mathbf{w}_R'(0) = -\lambda'(0)\mathbf{w}_R + M'(0)\mathbf{w}_R$$

which implies the right side is orthogonal to $\mathbf{w}_L$:

$$\lambda'(0) = \frac{\mathbf{w}_L^T M'(0)\mathbf{w}_R}{\mathbf{w}_L^T \mathbf{w}_R}.$$

The calculation of $M'(0)$, along the lines of the proof of Theorem 70, leads to $\lambda'(0) = -\kappa$ and it follows that the sign of $\lambda'(0)$ is the opposite of the sign of κ. If $\kappa > 0$, then $\lambda'(0) < 0$ and hence $\lambda(\varepsilon) < 1$ for $\varepsilon > 0$ small. On the other hand, if $\kappa < 0$, then $\lambda'(0) > 0$ and hence $\lambda(\varepsilon) > 1$ for $\varepsilon > 0$ small.

∎

From Theorems 64 and 65, we obtain the following corollary.

Corollary 9 *Suppose $\alpha > 0$ in Theorems 64 and 65 so that the disease-free equilibrium destabilizes as μ increases through μ_0. Then $\kappa > 0$ implies the bifurcation of endemic equilibria at $\mu = \mu_0$ is forward and stable and $\kappa < 0$ implies the bifurcation is backward and unstable.*

On the other hand, if $\alpha < 0$ so that the disease-free equilibrium destabilizes as μ decreases through μ_0, then $\kappa > 0$ implies the bifurcation of endemic equilibria at $\mu = \mu_0$ is backward and stable and $\kappa < 0$ implies the bifurcation is forward and unstable.

Theorems 64 and 65 focus on the destabilization of a disease-free equilibrium and the resulting bifurcation of endemic equilibria with respect to a parameter μ appearing in the model Equations (7.3). If $G(\mu, \mathbf{0}_m, \mathbf{x}_1) = J_{\mathbf{x}_0} \mathbf{g}_0(\mu, \mathbf{0}_m, \mathbf{x}_1^*)$ satisfies the conditions in Section 4.6 sufficient to define the reproduction number $\mathcal{R}_0(\mu)$ at the disease-free equilibrium, namely if $G(\mu, \mathbf{0}_m, \mathbf{x}_1) = F(\mu) + T(\mu)$ where $F(\mu)$ and $T(\mu)$ satisfy A3 for all μ, then the bifurcation scenario described in Theorem 65 can be re-formulated in terms of $\mathcal{R}_0(\mu)$. According to the proof of Theorem 65, the destabilization of the disease-free equilibrium occurs when the spectral radius of $G(\mu, \mathbf{0}_m, \mathbf{x}_1)$, increases through 1, which is equivalent to $\mathcal{R}_0(\mu)$ increasing through 1 (Theorem 18). Referring to (7.21), we see that the condition $\alpha \neq 0$ means that the passage of $\rho(G(\mu, \mathbf{0}_m, \mathbf{x}_1))$ through 1 is transversal. By Theorem 56 this is equivalent to the transversal increase of $\mathcal{R}_0$ through 1. This leads to an alternative formulation of Theorem 65 in terms of the net reproduction number.

Corollary 10 *Assume A4 and A5 hold and that $G(\mu_0, \mathbf{0}_m, \mathbf{x}_1^*)$ is primitive. Assume $G(\mu, \mathbf{0}_m, \mathbf{x}_1) = F(\mu) + T(\mu)$ where $F(\mu)$ and $T(\mu)$ satisfy A3 for all μ. If $\mathcal{R}_0(\mu)$ increases transversely through 1 (i.e., $\partial_\mu \mathcal{R}_0(\mu_0) > 0$), then the disease-free equilibrium destabilizes and bifurcation of endemic equilibria occurs (from the disease-free equilibrium) that is forward and stable if $\kappa > 0$ and backward and unstable if $\kappa < 0$.*

In Corollary 10, we see that the *direction of bifurcation with respect to $\mathcal{R}_0$ determines the stability* of the bifurcating endemic equilibria. Thus, under the circumstances of Corollary 10, a stability analysis is not necessary for the bifurcating endemic analysis near the bifurcation point. Only the direction of bifurcation need be ascertained, which can be done by determining the sign of κ. The power of this result is its generality. The limitation of the result is that it is local. That is to say, Theorem 64 and Corollary 10 guarantee the existence of endemic equilibria only near the bifurcation point and that the stability assertion is local (being determined by the Linearization Principle). It is often the case in specific models of interest in applications that endemic equilibria also exist outside a neighborhood of the bifurcation point. Moreover, even if the branch of endemic equilibria extends outside a neighborhood of the bifurcation point, the stability properties near the bifurcation point might not persist globally. Secondary bifurcations, resulting from destabilization of the endemic equilibria can occur, resulting in non-equilibrium, even chaotic, attractors. (See Example 99.) Furthermore, the global stability of equilibria is also often of interest, which requires further analysis.

Example 98 *Consider the ($n = 2$ dimensional) SI model*

$$\begin{array}{rl} S(t+1) = & b\frac{1}{1+cS(t)}S(t) + \exp\left(-c_\varphi \frac{I(t)}{S(t)+I(t)}\right)\sigma_S S(t) \\ I(t+1) = & \left(1 - \exp\left(-c_\varphi \frac{I(t)}{S(t)+I(t)}\right)\right)\sigma_S S(t) + \sigma_I I(t) \end{array} \quad (7.22)$$

7.2. ENDEMIC EQUILIBRIA

with $0 < \sigma_S < 1$, $0 \leq \sigma_I < 1$, $c_\varphi > 0$, and $c > 0$. This is the SI model studied in Example 77 with a discrete logistic recruitment f, complete immunity ($v = 0$), and exponential escape function $\varphi(z) = \exp(-c_\varphi z)$. The disease-free equation

$$S(t+1) = b\frac{1}{1+cS(t)}S(t) + \sigma_S S(t)$$

has a (unique) positive equilibrium

$$S^* = \frac{1}{c}\left(\frac{b}{1-\sigma_S} - 1\right) \tag{7.23}$$

provided

$$b > 1 - \sigma_S \tag{7.24}$$

which we assume holds. This positive equilibrium is stable by linearization since (7.24) implies

$$0 < \frac{d}{dS}\left(b\frac{1}{1+cS}S + \sigma_S S\right)\bigg|_{S=S^*} = \frac{1}{b}\left(b\sigma_S + (1-\sigma_S)^2\right) < 1.$$

From (4.31) in Example 77 we have

$$\mathcal{R}_0 = c_\varphi \frac{\sigma_S}{1-\sigma_I} \tag{7.25}$$

and from Theorem 44 we know that the disease-free equilibrium $(0, S^*)^T$ of the SI model (7.22) loses stability as $\mathcal{R}_0$ increases through 1.

Corollary 10 implies the bifurcation of endemic equilibria if $\mathcal{R}_0$ is transversely increased through 1. From the formula (7.25) we see that this can be accomplished by manipulating any one of the three parameters appearing in the formula. For example, if we choose $\mu = c_\varphi$ and write

$$\mathcal{R}_0(\mu) = \mu \frac{\sigma_S}{1-\sigma_I}$$

we see that $\mathcal{R}_0(\mu_0) = 1$ implies $\mu_0 = (1-\sigma_I)/\sigma_S$ and $\partial_\mu \mathcal{R}_0(\mu_0) = \sigma_S/(1-\sigma_I) > 0$. To apply Corollary 10 we need to check the required hypotheses A4 and A5 and to determine the sign of κ.

For the SI model (7.22)

$$\mathbf{x}_0 = I, \quad \mathbf{x}_1 = S$$

are both 1 dimensional vectors ($m = 1$ and $n - m = 1$). For notational convenience, in this example we drop the unnecessary parentheses for 1 dimensional vectors and 1×1 matrices. With the terms

$$\mathbf{g}_0(\mu, I, S) = G(\mu, I, S) I$$

$$\mathbf{g}_1(\mu, I, S) = bS\frac{1}{1+cS} + \exp\left(-\mu\frac{I}{S+I}\right)\sigma_S S$$

where

$$G(\mu, I, S) = \begin{cases} \frac{1-\exp\left(-\mu\frac{I}{S+I}\right)}{I}\sigma_S S + \sigma_I & \text{if } I \neq 0 \\ \mu\sigma_S + \sigma_I & \text{if } I = 0 \end{cases}$$

is an analytic function with a removable singularity at $I = 0$, it is straightforward to verify that assumption A4 holds. Furthermore, assumption A5 holds because

$$\rho(G(\mu_0, 0, S^*)) = G(\mu_0, 0, S^*) = \mu_0\sigma_S + \sigma_I = 1$$

and, by (7.24),

$$\rho\left(J_{\mathbf{x}_1}\mathbf{g}_1\left(\mu,0,\mathbf{x}_1^*\right)\right) = \partial_S\left(bS\frac{1}{1+cS} + \exp\left(-\mu\frac{I}{S+I}\right)\sigma_S S\right)\Big|_{I=0,S=S^*}$$

$$= \frac{1}{b}\left(b\sigma_S + (1-\sigma_S)^2\right) < 1.$$

Since $G(\mu_0, 0, S^*) = \mu_0\sigma_S + \sigma_I$, as a 1×1 matrix, is positive, it is primitive and the requirements for Corollary 10 are met once we determine the sign of κ. Since $G(\mu, I, S) = \gamma_{11}(\mu, I, S)$ is a 1×1 positive matrix, the eigenvectors of $G(\mu_0, 0, S^*)$ in the formula (7.9) for κ are $\mathbf{w}_L^T = \mathbf{w}_R = 1$. Since

$$\nabla^0_{\mathbf{x}_1}\gamma_{11} = \partial_S G(\mu_0, 0, S)|_{S=S^*}$$

$$= \lim_{I\to 0}\frac{d}{dS}\left(\left(\frac{1-\exp\left(-\mu\frac{I}{S+I}\right)}{I}\right)\sigma_S S + \sigma_I\right)\Big|_{S=S^*} = 0$$

the formula for κ becomes $\kappa = -\kappa_{11}$ where κ_{11} is simply $\kappa_{11} = \nabla^0_{\mathbf{x}_0}\gamma_{11}\mathbf{w}_R$, i.e.,

$$\kappa_{11} = \partial_I G(\mu_0, I, S^*)|_{I=0}$$

$$= \lim_{I\to 0}\frac{d}{dI}\left(\left(\frac{1-\exp\left(-\mu\frac{I}{S^*+I}\right)}{I}\right)\sigma_S S^* + \sigma_I\right)$$

$$= -\frac{1}{2}\mu\frac{\mu+2}{S^*}\sigma_S.$$

Thus

$$\kappa = \frac{1}{2}\mu\frac{\mu+2}{S^*}\sigma_S > 0$$

and the bifurcation of endemic equilibria is forward and stable.

The sample simulations in Figure 7.1 illustrate this bifurcation. Numerical simulations also suggest that endemic equilibria exist for all $\mathcal{R}_0 > 1$ and are globally asymptotically stable on $\mathring{\mathbb{R}}_+^2$, but we do not study that problem here.

Example 99 *As a variant of the SI model (7.22) in Example 98 consider the SI model*

$$\begin{align} S(t+1) &= bS(t)\exp\left(-cS(t)\right) + \exp\left(-c_\varphi\frac{I(t)}{N(t)}\right)\sigma_S S(t) \\ I(t+1) &= \left(1-\exp\left(-c_\varphi\frac{I(t)}{N(t)}\right)\right)\sigma_S S(t) + \sigma_I I(t) \end{align} \quad (7.26)$$

in which the logistic recruitment function has been replaced by an exponential (Ricker-type) function. The disease-free equation

$$S(t+1) = bS(t)\exp\left(-cS(t)\right) + \sigma_S S(t)$$

has a positive equilibrium

$$S^* = \frac{1}{c}\ln\left(\frac{b}{1-\sigma_S}\right) \text{ for } 1 < \frac{b}{1-\sigma_S}$$

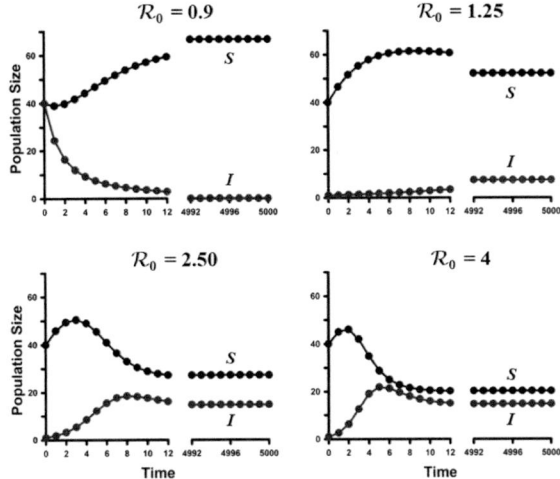

Figure 7.1: Sample solution time series obtained from (7.22) with $b = 3/2$, $c = 3/100$, $\sigma_S = 1/2$ and $\sigma_I = 2/5$. Upper left: $c_\varphi = 1.08$. Upper right: $c_\varphi = 1.5$. Lower left: $c_\varphi = 3$. Lower right: $c_\varphi = 4.8$.

that satisfies the stability constraint (7.8) *provided*

$$1 < \frac{b}{1 - \sigma_S} < \exp\left(\frac{2}{1 - \sigma_S}\right).$$

(See (4.34).)

In this example, numerical simulations show, however, that a stable endemic equilibrium does not exist for all $\mathcal{R}_0 > 1$. The bifurcation diagram in Figure 7.2 shows a sample case in which the endemic equilibrium destabilizes and results in a period-doubling cascade to chaos as $\mathcal{R}_0$ increases. Figure 7.3 shows a sample strange attractor.

It is interesting to note in Example 99 that non-equilibrium disease dynamics (and, apparently, strange attractors) can occur for large reproduction numbers $\mathcal{R}_0$ even though in the absence of the disease the population has (globally) stable equilibrium dynamics. Consequently, in the case of a disease outbreak, the dynamics of the disease-free susceptible fixed-point population do not always predict the disease pattern. This could have important practical implications, such as in designing vaccination protocols for controlling the infection.

7.3 Models with Multiple Infected Classes

In the example models used in previous sections, the population is structured into only two classes: susceptible and infected individuals. More sophisticated models designate additional disease-related classes and the transitions that can occur among them. For example, a modeler might want to distinguish different classes of infected individuals, such as those who are infectious to susceptible

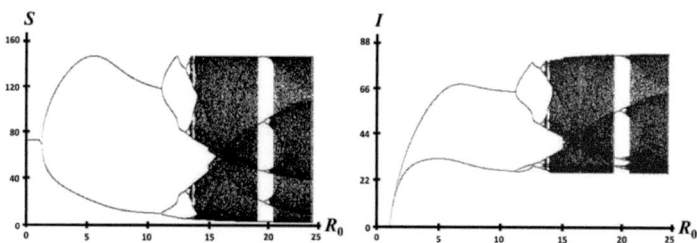

Figure 7.2: The I and S components of the attractors from the SI model (7.26) plotted against R_0 with $\sigma_S = 0.5$, $\sigma_I = 0.35$, $b = 20$, and $c = 0.05$.

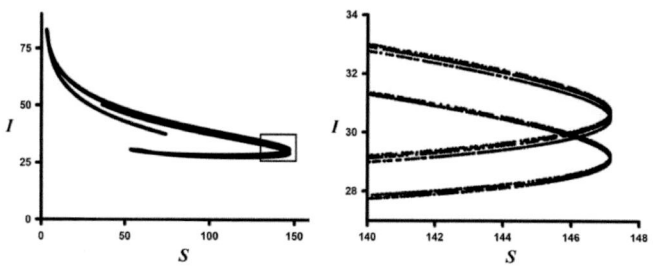

Figure 7.3: A strange attractor from the bifurcation diagram in Figure 7.2 with $\mathcal{R}_0 = 18$. Shown are 50,000 points from the orbit with initial condition $(S(0), I(0))^T = (50, 40)$ after the first 1,000 initial transient points are excluded. Also shown is a blow-up view of the indicated piece of the attractor.

7.3. MODELS WITH MULTIPLE INFECTED CLASSES

individuals, are infected but not infectious, are infectious but asymptomatic, are mildly or severely ill, are quarantined, are medicated, are mobile or stationary, are located in different regions, and so on. Non-infected individuals might also need to be categorized into several classes, such as those who are susceptible and those who are not, those who have recovered from the disease, and so on. Our primary goal is to illustrate basic methodologies for constructing discrete-time epidemic models and do so by presenting some basic types of models that researchers use. Therefore, we will restrict attention to models that have only one or two non-infected classes. However, we do allow for any number m of classes $I_1, \ldots, I_m$ of infected individuals:

$$\mathbf{x}_0 = (I_1, \ldots, I_m)^T = \begin{pmatrix} I_1 \\ \vdots \\ I_m \end{pmatrix}.$$

The list of non-infected classes in the models we study here will consist of either a single class of susceptible individuals $\mathbf{x}_1 = S$ (for notational simplicity we drop the parentheses for one-dimensional vectors) or two classes $\mathbf{x}_1 = (S, R)^T$ where R is the class of individuals recovered and immune from the disease. To build a model by using the methodology and notation in Section 4.7, we need to mathematically specify the dynamics of the infected and non-infected classes $\mathbf{x}_0$ and $\mathbf{x}_1$, i.e., the two expressions $\mathbf{g}_0(\mathbf{x}_0, \mathbf{x}_1)$ and $\mathbf{g}_1(\mathbf{x}_0, \mathbf{x}_1)$ in Equations (4.19). To calculate $\mathcal{R}_0$ we need also to identity the additive components $\mathbf{n}(\mathbf{x}_0, \mathbf{x}_1)$ and $\mathbf{s}(\mathbf{x}_0, \mathbf{x}_1)$ in $\mathbf{g}_0(\mathbf{x}_0, \mathbf{x}_1) = \mathbf{n}(\mathbf{x}_0, \mathbf{x}_1) + \mathbf{s}(\mathbf{x}_0, \mathbf{x}_1)$.

Let τ_{ij} ($0 \leq \tau_{ij} \leq 1$) denote the fraction of j-class infectious individuals who move to the ith infectious class per unit time. In some circumstances, these fractions (or probabilities) could depend on (class-specific) population densities, but here we assume they do not and that they remain constant over time. Then

$$I_i(t+1) = \sum_{j=1}^{m} \tau_{ij} I_j(t) \quad \text{for } i = 1, 2, \ldots, m$$

and the component $\mathbf{s}(\mathbf{x}_0, \mathbf{x}_1)$ in $\mathbf{g}_0(\mathbf{x}_0, \mathbf{x}_1)$ is

$$\mathbf{s}(\mathbf{x}_0, \mathbf{x}_1) = T\mathbf{x}_0$$

where $T = (\tau_{ij})$ is the $m \times m$ matrix of (transition) probabilities. Note that this matrix $T = J_{\mathbf{x}_0} \mathbf{s}(\mathbf{0}_m, \mathbf{x}_1^*)$ is used in defining $\mathcal{R}_0 = \rho(F(I-T)^{-1})$ associated with a disease-free equilibrium $(\mathbf{0}_m, \mathbf{x}_1^*)^T$.

We will not strive for the upmost generality here but only consider the case when an individual passes through the infectious states sequentially (assuming it survives each stage). By this we mean all newly infected individuals are in I_1 and, if they survive, will either remain in this class or advanced to I_2, and so on. (It should be clear how mathematically to modify the presentation here for other circumstances and types of transitions among infected states.) In this case, T has the bi-diagonal form

$$T = \begin{pmatrix} \tau_{11} & 0 & 0 & \cdots & 0 & 0 \\ \tau_{21} & \tau_{22} & 0 & \cdots & 0 & 0 \\ 0 & \tau_{32} & \tau_{33} & \cdots & 0 & 0 \\ \vdots & \vdots & \vdots & & \vdots & \vdots \\ 0 & 0 & 0 & \cdots & \tau_{m-1,m-1} & 0 \\ 0 & 0 & 0 & \cdots & \tau_{m,m-1} & \tau_{mm} \end{pmatrix} \quad (7.27)$$

where, after a unit of time, τ_{ii} is the probability an i-class infected individual remains in the i-class and $\tau_{i+1,i}$ is the probability progresses to the next infected class. Note that this T is the same transition matrix that arises in the standard size-structured (or Usher) structured population model discussed in Chapter 2. As pointed out there, these transition probabilities could be written

$$\tau_{i,i} = \sigma_I (1 - p_i), \quad \tau_{i+1,i} = \sigma_I p_i \text{ for } i = 1, 2, \ldots, m-1, \quad \text{and} \quad \tau_{mm} = s_m$$

where σ_I is the probability of survival (per unit time) and p_i is the probability of advancing to the next infectious class given survival. If $p_i = 1$ then T takes the form from a classic Leslie age structured population model in Chapter 2. We assume

$$0 \le \tau_{ii} < 1, \quad 0 < \tau_{i+1,i} \le 1.$$

The matrix $(I - T)^{-1}$ needed in the definition and calculation of $\mathcal{R}_0$ is [67]

$$(I - T)^{-1} = \begin{pmatrix} \frac{1}{1-\tau_{11}} & 0 & \cdots & 0 \\ \frac{\tau_{21}}{(1-\tau_{11})(1-\tau_{22})} & \frac{1}{1-\tau_{22}} & \cdots & 0 \\ \vdots & \vdots & \cdots & \vdots \\ \prod_{j=1}^{i} \frac{\tau_{j,j-1}}{1-\tau_{jj}} & \prod_{j=2}^{i} \frac{\tau_{j,j-1}}{1-\tau_{jj}} & \cdots & 0 \\ \vdots & \vdots & \cdots & \vdots \\ \prod_{j=1}^{m} \frac{\tau_{j,j-1}}{1-\tau_{jj}} & \prod_{j=2}^{m} \frac{\tau_{j,j-1}}{1-\tau_{jj}} & \cdots & \frac{1}{1-\tau_{mm}} \end{pmatrix} \quad (7.28)$$

where for notational convenience we define

$$\tau_{10} := 1.$$

The entry

$$\prod_{j=1}^{i} \frac{\tau_{j,j-1}}{1 - \tau_{jj}}$$

in the first column of this matrix is the expected time an infected individual (i.e., an individual in class I_i) will spend in the infected class I_i.

To calculate $\mathcal{R}_0$ associated with a disease-free equilibrium $(\mathbf{0}_m, \mathbf{x}_1^*)$ we need the matrix F, which this requires that we model how new infections occur (i.e., we need to define the term $\mathbf{n}(\mathbf{x}_0, \mathbf{x}_1)$). Since, under our assumption, all newly infected individuals enter I_1, we have

$$\mathbf{n}(\mathbf{x}_0, \mathbf{x}_1) = \begin{pmatrix} n_1(\mathbf{x}_0, \mathbf{x}_1) \\ 0 \\ \vdots \\ 0 \end{pmatrix}$$

$$\mathbf{g}_0(\mathbf{x}_0, \mathbf{x}_1) = \mathbf{n}(\mathbf{x}_0, \mathbf{x}_1) + \mathbf{s}(\mathbf{x}_0, \mathbf{x}_1) = \begin{pmatrix} n_1(\mathbf{x}_0, \mathbf{x}_1) \\ 0 \\ \vdots \\ 0 \end{pmatrix} + T\mathbf{x}_0$$

7.3. MODELS WITH MULTIPLE INFECTED CLASSES

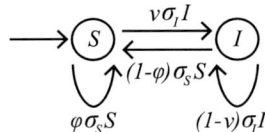

Figure 7.4: The flow diagram for the model (7.31).

and it follows that all rows in the Jacobian matrix $F = J_{x_0}\mathbf{n}(\mathbf{0}_m, \mathbf{x}_1^*)$ are zero rows except for the first row:

$$F = \begin{pmatrix} \frac{\partial n_1(\mathbf{0}_m, \mathbf{x}_1^*)}{\partial I_1} & \frac{\partial n_1(\mathbf{0}_m, \mathbf{x}_1^*)}{\partial I_2} & \cdots & \frac{\partial n_1(\mathbf{0}_m, \mathbf{x}_1^*)}{\partial I_m} \\ 0 & 0 & \cdots & 0 \\ \vdots & \vdots & & \vdots \\ 0 & 0 & \cdots & 0 \\ 0 & 0 & \cdots & 0 \end{pmatrix} \qquad (7.29)$$

It follows that all rows in the next-generation matrix $F(I-T)^{-1}$ are zero rows, except for the first row. Consequently, the only nonzero eigenvalue of $F(I-T)^{-1}$ is the upper left corner entry, which is the inner product of the first row of F with the first column of $(I-T)^{-1}$, i.e.,

$$\mathcal{R}_0 = \sum_{i=1}^m \frac{\partial n_1(\mathbf{0}_m, \mathbf{x}_1^*)}{\partial I_i} \prod_{j=1}^i \frac{\tau_{j,j-1}}{1-\tau_{jj}}. \qquad (7.30)$$

(Recall $\tau_{10} = 1$.)

Note that T and $(I-T)^{-1}$ are both nonnegative and that $\rho(T) = \max\{\tau_{ii}\} < 1$. This means T and F satisfy the requirements in A3 needed for the application of Theorem 44 that relates $\mathcal{R}_0$ to the spectral radius $\rho^* = \rho(F+T)$.

With this setup of infected transitions, in order to complete the specification of an epidemic model one needs to define the non-infected states $\mathbf{x}_1$, prescribe their dynamics $\mathbf{g}_1(\mathbf{x}_0, \mathbf{x}_1)$, and account for their interactions with the disease states $\mathbf{n}(\mathbf{x}_0, \mathbf{x}_1)$. In the following sections, we look at some examples. In Section 7.3.1 we consider a basic SI model in which individuals are classified into a *single* susceptible class; see Figure 7.4. In Section 7.3.2 we show how simple modifications of this model yield classic type models with the additional classifications of recovered and/or exposed classes.

7.3.1   A Basic SI Model

In this section, we consider a basic SI model in which individuals are classified into a *single* susceptible class; see Figure 7.4. Let $\mathbf{x}_1(t) = S(t)$ denote the class of all susceptible individuals at time t. Let σ_S denote the fraction of susceptible individuals that survives a time unit. Let $\theta_i \geq 0$ denote the fraction of the surviving susceptibles $\sigma_S S(t)$ that are infected (per unit time) have been by individuals from class $I_i(t)$. Then the total number of new infections at time $t+1$ is $\sum_{i=1}^m \theta_i \sigma_S S(t)$, where we must assume that $\sum_{i=1}^m \theta_i \leq 1$. The number of surviving susceptibles that do not become infected is

$$\sigma_S S(t) - \Sigma_{i=1}^m \theta_i \sigma_S S(t) = \left(1 - \Sigma_{i=1}^m \theta_i\right) \sigma_S S(t).$$

Assuming $\theta_i = \theta_i(S, I_i)$ and $f = f(S, \mathbf{I})$ are functions of $(S, \mathbf{I})$ where $\mathbf{I} = (I_1, I_2, \ldots, I_m)^T$ we obtain the model equations

$$\begin{aligned} S(t+1) &= f(S(t), \mathbf{I}(t)) + \varphi(S(t), \mathbf{I}(t)) \sigma_S S(t) \\ I_1(t+1) &= (1 - \varphi(S(t), \mathbf{I}(t))) \sigma_S S(t) + \tau_{11} I_1(t) \\ I_i(t+1) &= \tau_{i,i-1} I_{i-1}(t) + \tau_{ii} I_i(t) \quad \text{for } i = 2, \ldots, m \end{aligned} \tag{7.31}$$

where $n = m + 1$ and

$$\varphi(S, \mathbf{I}) := 1 - \sum_{i=1}^m \theta_i(S, I_i) \tag{7.32}$$

is the infection escape function. We assume

$$0 < \sigma_S < 1,\ 0 \le \tau_{ii} < 1,\ 0 < \tau_{i+1,i} \le 1$$

and the infection fractions $\theta_j(S, I_j)$ are twice continuously differentiable and satisfy

$$\begin{array}{l} 0 \le \theta_i(S, I_i),\quad \sum_{i=1}^m \theta_i(S, I_i) \le 1,\ \text{and } \theta_i(S, 0) \equiv 0 \text{ for all } (S, I_i) \in \mathbb{R}_+ \times \mathbb{R}_+ \\ \frac{\partial \theta_i(S,0)}{\partial I_i} \ge 0 \quad \text{for all } S \ge 0 \text{ with at least one derivative positive.} \end{array} \tag{7.33}$$

Example infection fractions are

$$\theta_i = 1 - \exp\left(-c_i \frac{I_i}{N}\right),\quad c_i > 0 \tag{7.34}$$

where

$$N = S + \Sigma_{i=1}^m I_i$$

is total population size.

In the notation of Section 4.6, the dynamics of the infected classes $\mathbf{x}_0 = \mathbf{I}$ are described by $\mathbf{g}_0(\mathbf{x}_0, \mathbf{x}_1) = \mathbf{n}(\mathbf{x}_0, \mathbf{x}_1) + \mathbf{s}(\mathbf{x}_0, \mathbf{x}_1)$ with

$$\mathbf{n}(\mathbf{x}_0, \mathbf{x}_1) = \begin{pmatrix} \sigma_S(1 - \varphi(S, \mathbf{I})) S \\ 0 \\ \vdots \\ 0 \end{pmatrix},\quad \mathbf{s}(\mathbf{x}_0, \mathbf{x}_1) = \begin{pmatrix} \tau_{11} I_1 \\ \tau_{21} I + \tau_{22} I_2 \\ \vdots \\ \tau_{m,m-1} I_{m-1} + \tau_{mm} I_m \end{pmatrix}$$

and those of the non-infected class $\mathbf{x}_1 = S$ by

$$\mathbf{g}_1(\mathbf{x}_0, \mathbf{x}_1) = f(S, \mathbf{I}) + \sigma_S \varphi(S, \mathbf{I}) S.$$

The first (and only nonzero) row of the Jacobian $F = J_{x_0} \mathbf{n}(S, \mathbf{0}_m)$ is

$$\begin{pmatrix} \frac{\partial \theta_1(S^*, 0)}{\partial I_1} \sigma_S S^* & \frac{\partial \theta_2(S^*, 0)}{\partial I_2} \sigma_S S^* & \cdots & \frac{\partial \theta_m(S^*, 0)}{\partial I_m} \sigma_S S^* \end{pmatrix} \tag{7.35}$$

and thus by (7.30)

$$\mathcal{R}_0 = \sigma_S S^* \sum_{i=1}^m \frac{\partial \theta_i(S^*, 0)}{\partial I_i} \prod_{j=1}^i \frac{\tau_{j,j-1}}{1 - \tau_{jj}}. \tag{7.36}$$

7.3. MODELS WITH MULTIPLE INFECTED CLASSES

By (7.33) it follows that $\mathcal{R}_0 > 0$.

The disease-free dynamics are governed by the scalar difference equation

$$S(t+1) = f(S(t), \mathbf{0}_m) + \sigma_S S(t). \tag{7.37}$$

If we assume this equation has a positive equilibrium S^* which is locally stable by linearization, i.e., for which

$$\left| \frac{\partial f(S, \mathbf{0}_m)}{\partial S} + \sigma_S \right|_{S=S^*} < 1 \tag{7.38}$$

holds, then all the requirements A1, A2 and A3 needed to invoke Theorem 44 are satisfied.

Theorem 66 *Assume the infection rates θ_i satisfy (7.33) and that the disease-free model Equation (7.37) has an equilibrium $S^* > 0$ that satisfies the inequality (7.38). The disease-free equilibrium $\mathbf{x}^* = (S^*, \mathbf{0}_m)^T$ of the SI model (7.31) with (7.32) is locally asymptotically stable on $\mathbb{R}_+^{m+1}$ for $\mathcal{R}_0 < 1$ and unstable for $\mathcal{R}_0 > 1$, where $\mathcal{R}_0$ is given by (7.36).*

A common modeling assumption is that the infection rate θ_i depends on the proportion I_i/N of I_i class individuals in the total population. In this case, $\theta_i = \theta_i(I_i/N)$ where we assume $\theta_i(z)$ is twice continuously differentiable and satisfies

$$\begin{array}{c} 0 \leq \theta_i(z) \leq 1 \text{ for } 0 \leq z \leq 1 \\ \theta_i(0) = 0, \, \theta_i'(0) \geq 0, \text{ and } \sum_{i=1}^{m} \theta_i'(0) > 0. \end{array} \tag{7.39}$$

Then formula (7.36) yields

$$\mathcal{R}_0 = \sigma_S \sum_{i=1}^{m} \theta_i'(0) \prod_{j=1}^{i} \frac{\tau_{j,j-1}}{1 - \tau_{jj}}. \tag{7.40}$$

Theorem 66 asserts the local stability of the disease-free equilibrium when $\mathcal{R}_0 < 1$. We can obtain global stability by making use of Theorem 45.

Theorem 67 *Assume $\theta_i = \theta_i(I_i/N)$ the SI model (7.31) with (7.32) where $\theta_i(z)$ satisfies (7.39) with $\theta_i'(0) > 0$ and $\theta_i''(z) \leq 0$ for $z \geq 0$ and all i. If the $S^* > 0$ is a locally asymptotically stable equilibrium of the disease-free model equation (7.37) that satisfies (7.38), then the disease-free equilibrium $\mathbf{x}^* = (S^*, 0, \ldots, 0)^T$ of the SI model (7.31) with (7.32) is globally asymptotically stable on $\mathbb{R}_+^{m+1}$ when $\mathcal{R}_0 < 1$, where $\mathcal{R}_0$ is given by (7.36).*

Proof. To apply Theorem 45 we need only to show that A6 holds. The first requirement A6(a) is that $(I-T)^{-1}$ is a nonnegative matrix, which we see is true by the formula (7.28). The first row (7.35) in F consists entirely of positive entries since $\theta_i'(0) > 0$ for all i. As a result, $(I-T)^{-1}F$ is a positive matrix and therefore irreducible. Thus, A6(a) holds and we turn our attention to A6(b), which requires that the vector $\mathbf{f}(\mathbf{x}_0, \mathbf{x}_1) = (F+T)\mathbf{x}_0 - \mathbf{g}_0(\mathbf{x}_0, \mathbf{x}_1)$ be nonnegative. Now

$$F + T = \begin{pmatrix} \tau_{11} + \sigma_S \theta_1'(0) & \sigma_S \theta_2'(0) & \cdots & \sigma_S \theta_m'(0) \\ \tau_{21} & \tau_{22} & \cdots & 0 \\ 0 & \tau_{32} & \cdots & 0 \\ \vdots & \vdots & & \vdots \\ 0 & 0 & \cdots & \tau_{mm} \end{pmatrix}$$

implies
$$(F+T)\mathbf{x}_0 = \begin{pmatrix} \tau_{11}I_1 + \sigma_S \sum_{i=1}^m \theta_i'(0) I_i \\ \tau_{21}I_1 + \tau_{22}I_2 \\ \vdots \\ \tau_{m,m-1}I_{m-1} + \tau_{mm}I_m \end{pmatrix}$$

which together with
$$\mathbf{g}_0(\mathbf{x}_0,\mathbf{x}_1) = \begin{pmatrix} \tau_{11}I_1 + \sigma_S \sum_{i=1}^m \theta_i\left(\frac{I_i}{N}\right) S \\ \tau_{21}I_1 + \tau_{22}I_2 \\ \vdots \\ \tau_{m,m-1}I_{m-1} + \tau_{mm}I_m \end{pmatrix}$$

gives
$$\mathbf{f}(\mathbf{x}_0,\mathbf{x}_1) = \sigma_S \begin{pmatrix} \sum_{i=1}^m \left(\theta_i'(0)I_i - \theta_i\left(\frac{I_i}{N}\right) S\right) \\ 0 \\ \vdots \\ 0 \end{pmatrix}.$$

The nonnegativity nonnegative requirement for $\mathbf{f}(\mathbf{x}_0,\mathbf{x}_1)$ reduces to the nonnegativity of its first component. The Mean Value Theorem implies
$$\theta_i\left(\frac{I_i}{N}\right) = \theta_i'(\zeta)\frac{I_i}{N} \quad \text{for} \quad 0 \leq \zeta \leq \frac{I_j}{N}$$

(recall $\theta_i(0) = 0$) and hence
$$\theta_i'(0)I_i - \theta_i\left(\frac{I_i}{N}\right) S = I_i\left(\theta_i'(0) - \theta_i'(\zeta)\frac{S}{N}\right) \geq I_i\left(\theta_i'(0) - \theta_i'(\zeta)\right) \geq 0.$$

A6 holds and an application of Theorem 45 finishes the proof. ∎

Under the assumptions in Theorem 67 we can apply Corollary 10 to conclude that the destabilization of the disease-free equilibrium, as any model parameter is manipulated so that the reproduction number
$$\mathcal{R}_0 = \sigma_S \sum_{i=1}^m \theta_i'(0) \prod_{j=1}^i \frac{\tau_{j,j-1}}{1 - \tau_{jj}}$$

increases through 1, will result in bifurcation of endemic equilibria at $\mathcal{R}_0 = 1$. The direction, and hence the stability, of the bifurcation is determined by the sign of the quantity κ defined by (7.9). Writing
$$\mathbf{g}_0(\mathbf{x}_0,\mathbf{x}_1) = \begin{pmatrix} \tau_{11}I_1 + \sigma_S \sum_{i=1}^m \theta_i\left(\frac{I_i}{N}\right) S \\ \tau_{21}I_1 + \tau_{22}I_2 \\ \vdots \\ \tau_{m,m-1}I_{m-1} + \tau_{mm}I_m \end{pmatrix}$$

7.3. MODELS WITH MULTIPLE INFECTED CLASSES

as $\mathbf{g}_0(\mathbf{x}_0, \mathbf{x}_1) = G(\mathbf{x}_0, \mathbf{x}_1)\mathbf{x}_0$ with $\mathbf{x}_0 = \begin{pmatrix} I_1 & I_2 & \cdots & I_m \end{pmatrix}^T$ and $\mathbf{x}_1 = (S)$ and with

$$G(\mathbf{x}_0, \mathbf{x}_1) = (\gamma_{ij}(\mathbf{x}_0, \mathbf{x}_1)) = \begin{pmatrix} \tau_{11} + \sigma_S S \frac{\theta_1\left(\frac{I_1}{N}\right)}{I_1} & \sigma_S S \frac{\theta_2\left(\frac{I_2}{N}\right)}{I_2} & \cdots & \sigma_S S \frac{\theta_m\left(\frac{I_m}{N}\right)}{I_m} \\ \tau_{21} & \tau_{22} & \cdots & 0 \\ 0 & \tau_{32} & \cdots & 0 \\ \vdots & \vdots & & \vdots \\ 0 & 0 & \cdots & \tau_{mm} \end{pmatrix}$$

we use the matrix entries γ_{ij} in G to calculate κ from the formula (7.9). Note that

$$\partial_S \left(\sigma_S S \frac{\theta_i\left(\frac{I_i}{N}\right)}{I_i} \right) \Bigg|_{I_i = 0} = 0$$

for all i and therefore $\nabla^0_{\mathbf{x}_1} \gamma_{ij} = \mathbf{0}^T$ and the formula for κ simplifies to

$$\kappa = -\mathbf{w}_L^T \left(\nabla^0_{\mathbf{x}_0} \gamma_{ij} \mathbf{w}_R \right) \mathbf{w}_R. \tag{7.41}$$

The entries in rows 2 through m in the matrix $G(\mathbf{x}_0, \mathbf{x}_1)$ are constants and therefore

$$\nabla^0_{\mathbf{x}_0} \gamma_{ij} = \mathbf{0}_m^T \text{ for all } j \text{ and all } i = 2, 3, \cdots, m.$$

This means only the first row in the matrix $\left(\nabla^0_{\mathbf{x}_0} \gamma_{ij} \mathbf{w}_R \right)$ is a nonzero row:

$$\left(\nabla^0_{\mathbf{x}_0} \gamma_{ij} \mathbf{w}_R \right) = \begin{pmatrix} \nabla^0_{\mathbf{x}_0} \gamma_{11} \mathbf{w}_R & \nabla^0_{\mathbf{x}_0} \gamma_{12} \mathbf{w}_R & \cdots & \nabla^0_{\mathbf{x}_0} \gamma_{1m} \mathbf{w}_R \\ 0 & 0 & \cdots & 0 \\ \vdots & \vdots & & \vdots \\ 0 & 0 & \cdots & 0 \end{pmatrix}.$$

Moreover

$$\partial^0_{I_i}(\gamma_{11}) - \partial_{I_i}\left(\sigma_S S \frac{\theta_j\left(\frac{I_j}{N}\right)}{I_j} \right) \Bigg|_{(S, \mathbf{I}) = (S^*, \mathbf{0}_m)} = 0 \text{ for } i \neq j$$

which implies

$$\partial^0_{I_i} \gamma_{1j} = 0 \text{ for } i \neq j.$$

This means that the gradient

$$\nabla^0_{\mathbf{x}_0} \gamma_{1j} = \begin{pmatrix} \partial^0_{I_1} \gamma_{1j} & \partial^0_{I_2} \gamma_{1j} & \cdots & \partial^0_{I_m} \gamma_{1j} \end{pmatrix}$$

consists of all zeros except at the jth location. Therefore

$$\left(\nabla^0_{\mathbf{x}_0} \gamma_{ij} \mathbf{w}_R \right) = \begin{pmatrix} \partial^0_{I_1} \gamma_{11} w_{R_1} & \partial^0_{I_2} \gamma_{12} w_{R_2} & \cdots & \partial^0_{I_m} \gamma_{1m} w_{R_m} \\ 0 & 0 & \cdots & 0 \\ \vdots & \vdots & & \vdots \\ 0 & 0 & \cdots & 0 \end{pmatrix}$$

where w_{R_i} are the positive entries of the positive right eigenvector $\mathbf{w}_R$ of $G\left(\mathbf{0}_m, \mathbf{x}_1^*\right)$. From formula (7.41) we get

$$\kappa = -\begin{pmatrix} w_{L_1} & w_{L_2} & \cdots & w_{L_m} \end{pmatrix} \begin{pmatrix} \partial_{I_1}^0 \gamma_{11} w_{R_1} & \partial_{I_2}^0 \gamma_{12} w_{R_2} & \cdots & \partial_{I_m}^0 \gamma_{1m} w_{R_m} \\ 0 & 0 & \cdots & 0 \\ \vdots & \vdots & & \vdots \\ 0 & 0 & \cdots & 0 \end{pmatrix} \begin{pmatrix} w_{R_1} \\ w_{R_2} \\ \vdots \\ w_{R_m} \end{pmatrix}$$

$$= -w_{L_1} \sum_{i=1}^m \partial_{I_i}^0 \gamma_{1i} w_{R_i}^2.$$

Finally we need to calculate

$$\partial_{I_i}^0 \gamma_{1i} = \partial_{I_i} \left(\frac{\theta_i\left(\frac{I_i}{N}\right)}{I_i} \sigma_S S \right) \bigg|_{(S,\mathbf{I})=(S^*, \mathbf{0}_m)}.$$

Given the removal singularity at $I_i = 0$, we set $S = S^*$ and $I_j = 0$ for all $j \neq i$ in this expression and obtain

$$\partial_{I_i}^0 \gamma_{1i} = \sigma_S S^* \lim_{I_i \to 0} \partial_{I_i} \left(\frac{\theta_i\left(\frac{I_i}{S^*+I_i}\right)}{I_i} \right).$$

Using $\theta_i(0) = 0$ and a Taylor series expansion of the parenthetical expression in I_i centered on $I_i = 0$ to calculate the indicated limit, we find that

$$\partial_{I_i}^0 \gamma_{1i} = \left(-\theta_i'(0) + \frac{1}{2} \theta_i''(0) \right)$$

and finally

$$\kappa = w_{L_1} \frac{\sigma_S}{S^*} \sum_{i=1}^m \left(\theta_i'(0) - \frac{1}{2} \theta_i''(0) \right) w_{R_i}^2. \tag{7.42}$$

Theorem 68 *Assume $\theta_i = \theta_i(I_i/N)$ the SI model (7.31) with (7.32) where $\theta_i(z)$ satisfies (7.39). If the $S^* > 0$ is a locally asymptotically stable equilibrium of the disease-free model equation (7.37) that satisfies (7.38), then the disease-free equilibrium $\mathbf{x}^* = (S^*, 0, \ldots, 0)^T$ destabilizes and a bifurcation of endemic equilibria occurs as $\mathcal{R}_0$, given by (7.40), transversely increases through 1. The bifurcation is forward and stable if $\kappa > 0$ and backward and unstable if $\kappa < 0$ where κ is given by (7.42),*

For example if $\theta_i''(0) \leq 0$ (as in Theorem 67 concerning the global asymptotic stability of the disease-free equilibrium) then $\kappa > 0$ and the bifurcation will be forward and stable. A specific case is when the infection rates $\theta_i(z)$ given are by the exponential expressions (7.34), for which

$$\kappa = w_{L1} \frac{\sigma_S}{S^*} \sum_{i=1}^m \left(c_i + \frac{1}{2} c_i^2 \right) w_{Ri}^2 > 0$$

and

$$\mathcal{R}_0 = \sigma_S \sum_{i=1}^m c_i \prod_{j=1}^i \frac{\tau_{j,j-1}}{1 - \tau_{jj}}. \tag{7.43}$$

Both SI models in Examples 98 and 99 serve as examples of forward and stable bifurcation of endemic equilibria at $\mathcal{R}_0 = 1$.

7.3. MODELS WITH MULTIPLE INFECTED CLASSES

Remark 20 *In some epidemic models, the infection rates and the escape function φ are assumed dependent on the number of infected individuals rather that their proportion in the population, as in Theorem 67 [37]. In this case $\theta_i = \theta_i(I)$. Under the conditions (7.39) on $\theta_i(I)$, the local stability Theorem 66 applies to such models with*

$$\mathcal{R}_0 = \sigma_S S^* \sum_{i=1}^{m} \theta_i'(0) \prod_{j=1}^{i} \frac{\tau_{j,j-1}}{1 - \tau_{jj}}.$$

However, the global stability Theorem 45 does not apply to such models.

7.3.2 SIR and SEIR Models

An SIR Model. In the basic SI model (7.31) with (7.32) in Section 7.3.1, individuals that survive the infection (namely, those that leave the class I_m, which are $(1 - \tau_{mm})I_m$ in number) are removed from the population. For a population in which this is not the case, a modeler must decide what role recovered individuals play in the model. There are many possibilities, of course. For example, these individuals could be immune to further infection or they could be susceptible again. In the latter case, they return to the S class, a case we consider below. In the former case, we define an additional state variable $R(t)$ for recovered individuals and supplement the basic SI model with an equation for R and we re-define total population size to be

$$N = S + \sum_{j=1}^{m} I_j + R.$$

We assume no one recovers without passing through all infected states. Then $R(t+1)$ equals the new individuals that arrive from $I_m(t)$ plus those recovered individuals at time t that survive. The former is $(1 - \tau_{mm})I_m(t)$ and the latter we assume is $\sigma_R R(t)$ where $\sigma_R < 1$ is the probability a recovered individual survives a unit of time. We adjoin the resulting equation for $R(t+1)$ to the equations in the basic SI model (7.31) with (7.32) to form an SIR model

$$\begin{aligned} S(t+1) &= f(S(t), \mathbf{I}(t), R(t)) + \varphi(S(t), \mathbf{I}(t), R(t)) \sigma_S S(t) \\ I_1(t+1) &= (1 - \varphi(S(t), \mathbf{I}(t), R(t))) \sigma_S S(t) + \tau_{11} I_1(t) \\ I_i(t+1) &= \tau_{i,i-1} I_{i-1}(t) + \tau_{ii} I_i(t) \quad \text{for } i = 2, \ldots, m \\ R(t+1) &= \sigma_R R(t) + (1 - \tau_{mm}) I_m(t) \end{aligned} \quad (7.44)$$

with

$$\varphi(S, \mathbf{I}, R) = 1 - \sum_{i=1}^{m} \theta_i(S, I_i, R). \quad (7.45)$$

Notice that R is allowed to play a role in the susceptible recruitment function f. The infected class vector $\mathbf{x}_0$ and the transition matrix T remain the same as in basic SI model in Section 7.3.1. The disease-free dynamics are governed by the two difference equations

$$\begin{aligned} S(t+1) &= f(S(t), \mathbf{0}_m, R(t)) + \sigma_S S(t) \\ R(t+1) &= \sigma_R R(t). \end{aligned} \quad (7.46)$$

(Note that the uncoupled equation for $R(t)$ implies $\lim_{t \to \infty} R(t) = 0$.) If we assume the scalar difference equation

$$S(t+1) = f(S(t), \mathbf{0}_m, 0) + \sigma_S S(t)$$

for susceptibles in the absence of recovered individuals has a positive equilibrium $S^* > 0$ that satisfies

$$\left| \frac{\partial f(S, \mathbf{0}_m, 0)}{\partial S} + \sigma_S \right|_{S=S^*} < 1$$

then a straightforward application of the Linearization Principle shows that the equilibrium $\mathbf{x}_1^* = (S^*, \mathbf{0})^T$ is locally asymptotically stable if $\sigma_R < 1$ of the disease-free equations (7.46). The calculation of $\mathcal{R}_0$ is identical to that for the basic SI, yielding the same formula (7.36). The result is that *Theorems 66 and 67 hold for this SIR model* with $\mathbf{x}^* = (S^*, \mathbf{0}_m)^T \in \mathbb{R}_+^{m+1}$ replaced by $\mathbf{x}^* = (S^*, \mathbf{0}_m, 0)^T \in \mathbb{R}_+^{m+2}$. Here are two examples.

Example 100 *Adding the equation for $R(t)$ to the SI model equations in Example 98, we obtain the SIR model*

$$\begin{aligned} S(t+1) &= bS(t)\frac{1}{1+cS(t)} + \exp\left(-c_\varphi \frac{I(t)}{N(t)}\right)\sigma_S S(t) \\ I(t+1) &= \left(1 - \exp\left(-c_\varphi \frac{I(t)}{N(t)}\right)\right)\sigma_S S(t) + \tau_{11} I(t) \\ R(t+1) &= \sigma_R R(t) + (1 - \tau_{11}) I(t) \end{aligned}$$

where

$$N(t) = S(t) + I(t) + R(t).$$

From our observations above, the analysis in Example 98, and Theorems 66 and 67 we have that the disease-free equilibrium

$$S^* = \frac{1}{c}\left(\frac{b}{1-\sigma_S} - 1\right) \text{ for } 1 - \sigma_S < b$$

$$\mathbf{x}^* = (S^*, 0, 0)^T = \begin{pmatrix} \frac{1}{c}\left(\frac{b}{1-\sigma_S} - 1\right) \\ 0 \\ 0 \end{pmatrix} \text{ for } 1 - \sigma_S < b$$

is globally asymptotically stable if

$$\mathcal{R}_0 = c_\varphi \frac{\sigma_S}{1 - \tau_{11}} < 1$$

and unstable if $\mathcal{R}_0 > 1$. (See (4.33), (4.34) and Problem 1 in Section 4.6.) Furthermore, from formula (7.42) we calculate

$$\kappa = \frac{\sigma_S}{S^*}\left(c_\varphi + \frac{1}{2}c_\varphi^2\right) > 0$$

and hence a forward and stable bifurcation of endemic equilibria occurs at $\mathcal{R}_0 = 1$, which is to say that there exist locally asymptotically stable endemic equilibria at least for $\mathcal{R}_0 > 1$ close to 1.

The same conclusions hold for the SIR model

$$\begin{aligned} S(t+1) &= bS(t)\exp(-cS(t)) + \exp\left(-c_\varphi \frac{I(t)}{N(t)}\right)\sigma_S S(t) \\ I(t+1) &= \left(1 - \exp\left(-c_\varphi \frac{I(t)}{N(t)}\right)\right)\sigma_S S(t) + \tau_{11} I(t) \\ R(t+1) &= \sigma_R R(t) + (1 - \tau_{11}) I_1(t) \end{aligned}$$

7.3. MODELS WITH MULTIPLE INFECTED CLASSES

obtained from the SI model in Example 99 where the disease-free equilibrium is

$$\mathbf{x}^* = (S^*, 0, 0)^T = \begin{pmatrix} \frac{1}{c} \ln\left(\frac{b}{1-\sigma_S}\right) \\ 0 \\ 0 \end{pmatrix} \quad \text{for } 1 - \sigma_S < b$$

provided

$$1 < \frac{b}{1-\sigma_S} < \exp\left(\frac{2}{1-\sigma_S}\right).$$

(See (4.33), (4.34) and Problem 1 in Section 4.6.)

An SEIR Model. In the SI and SIR models considered so far, we assumed no distinguishing features of the infected classes I_i other than that they are visited sequentially. Modelers often consider any number of different categories of infected states. One such distinction involves a latency period, in which the first class I_1 consists of newly infected, but not yet infectious, individuals. In this case, the first class is often denoting by E (to stand for exposed, but not infectious individuals), so that $I_1(t)$ could instead be denoted by $E(t)$ in the model equations. If individuals in the $I_1 = E$ class are not infectious, then $\theta_1(S, \mathbf{I}) = 0$ in the basic SI model (7.31) with (7.32). Theorem 66 concerning the stability of a disease-free equilibrium applies with reproduction number

$$\mathcal{R}_0 = \sigma_S S^* \sum_{i=2}^{m} \frac{\partial \theta_i(S^*, 0)}{\partial I_i} \prod_{j=1}^{i} \frac{\tau_{j,j-1}}{1 - \tau_{jj}}. \tag{7.47}$$

However, the global stability Theorem 67 for the disease-free equilibrium does not apply to this model. The reason is that $\partial \theta_1(S^*, 0)/\partial I_1 = 0$ implies the first column of the matrix $(I - T)^{-1} F$ consists entirely of zeros and hence this matrix is not irreducible, a requirement of assumption A6(a).

Similar remarks hold for an SIR model (7.44) with (7.45) that includes an exposed class $I_1 = E$, which we refer to as an SEIR model. Theorem 66 applies with $\mathbf{x} = (S, E, I_2, \ldots, I_m)^T \in \mathbb{R}_+^{m+1}$ replaced by $\mathbf{x}^* = (S, E, I_2, \ldots, I_m, R)^T \in \mathbb{R}_+^{m+2}$ and $\mathcal{R}_0$ given by (7.47). (Again, Theorem 67 does not apply.) The disease-free equations are now

$$\begin{aligned} S(t+1) &= f(S(t), 0, \ldots, 0, R(t)) + \sigma_S S(t) \\ R(t+1) &= \sigma_R R(t). \end{aligned} \tag{7.48}$$

An equilibrium $\mathbf{x}^* = (S^*, R^*)$ of these disease-free equations obviously must have $R^* = 0$ and an $S^* > 0$ that satisfies the algebraic equation

$$S = f(S(t), 0, \ldots, 0, 0) + \sigma_S S.$$

An equilibrium $\mathbf{x}^* = (S^*, 0)$ of the disease-free equations is locally asymptotically stable if

$$\left| \frac{\partial f(S, 0, \ldots, 0, R)}{\partial S} + \sigma_S \right|_{(S,R)=(S^*,0)} < 1 \tag{7.49}$$

(see Problem 2).

Example 101 *Consider the SEIR model obtained from the SIR model (7.44) with (7.45) and $m = 2$ infected classes in which class $I_1 = E$ is not infectious (i.e., $\theta_1 = 0$). Suppose the infection rate of class I_2 is $1 - \exp(-c_\varphi I/N)$ or equivalently that the escape function φ is $\exp(-c_\varphi I/N)$. This yields, by relabeling I_2 as I, the SEIR model*

$$\begin{aligned} S(t+1) &= be^{-cS(t)}S(t) + \sigma_S \exp\left(-c_\varphi \tfrac{I(t)}{N(t)}\right)S(t) \\ E(t+1) &= \sigma_S\left(1 - \exp\left(-c_\varphi \tfrac{I(t)}{N(t)}\right)\right)S(t) + \tau_{11}E(t) \\ I(t+1) &= \tau_{21}E(t) + \tau_{22}I(t) \\ R(t+1) &= \sigma_R R(t) + (1 - \tau_{22})I(t) \end{aligned} \quad (7.50)$$

with total population size $N = S + E + I + R$. The disease-free equations are

$$\begin{aligned} S(t+1) &= be^{-cS(t)}N(t) + \sigma_S S(t) \\ R(t+1) &= \sigma_R R(t). \end{aligned}$$

Formula (7.47) yields

$$\mathcal{R}_0 = \sigma_S c_\varphi \frac{1}{1 - \tau_{11}} \frac{\tau_{21}}{1 - \tau_{22}}.$$

The equilibrium

$$\begin{pmatrix} S^* \\ R^* \end{pmatrix} = \begin{pmatrix} \frac{1}{c}\ln\left(\frac{b}{1-\sigma_S}\right) \\ 0 \end{pmatrix}$$

of these equations satisfies the stability condition (7.49), and this equilibrium is therefore locally asymptotically stable, provided

$$1 < \frac{b}{1 - \sigma_S} < \exp\left(\frac{2}{1 - \sigma_S}\right).$$

(See (4.33), (4.34), and Problem 1 in Section 4.6.)

Theorem 66 implies that the disease-free equilibrium

$$\mathbf{x}^* = (S^*, 0, 0, 0)^T = \begin{pmatrix} \frac{1}{c}\ln\left(\frac{b}{1-\sigma_S}\right) \\ 0 \\ 0 \\ 0 \end{pmatrix}$$

of the SEIR model is locally asymptotically stable if $\mathcal{R}_0 < 1$ and unstable if $\mathcal{R}_0 > 1$. Furthermore, from formula (7.42) we calculate

$$\kappa = \frac{\sigma_S}{S^*}\left(c_\varphi + \frac{1}{2}c_\varphi^2\right) > 0$$

and hence a forward and stable bifurcation of endemic equilibria occurs at $\mathcal{R}_0 = 1$, which is to say that there exist locally asymptotically stable endemic equilibria at least for $\mathcal{R}_0 > 1$ close to 1.

Exercises 7.3

1. The basic SI model (7.31)–(7.32) assumes that the $(1 - \tau_{mm}) I_m(t)$ individuals who leave the last infected class I_m are removed from the population. Suppose instead these individuals remain in the population and become again susceptible creating an SIS model. Calculate $\mathcal{R}_0$ and apply Theorems 66 and 67.

2. Use the Linearization Principle to show that the inequality (7.49) implies the local asymptotic stability of an equilibrium $(S^*, 0)$ of the disease-free equations (7.48) for the SEIR model.

3. Using the formula (7.43) for the SI model (7.31)–(7.32)–(7.34), calculate the elasticities (sensitivity indices) of the reproduction number $\mathcal{R}_0$ with respect to the infectiousness coefficients c_i and the infectious state transitions $\tau_{i,i-1}$ and τ_{ii}.

7.4 Models with Disease-Free Population Cycles

In Section 4.7, the next-generation matrix and its spectral radius $\mathcal{R}_0$ are defined with respect to an equilibrium $\mathbf{x}_1^*$ of the population in the absence of the disease. To do this, we assume the population has a static (non-cyclic) steady state in the absence of the disease [11], [251]. Under the additional assumption that this equilibrium is locally asymptotically stable by linearization (i.e., that (4.25) holds), Theorem 44 justifies that the inherent reproduction number $\mathcal{R}_0$ determines the stability properties of the disease-free equilibrium $(\mathbf{0}, \mathbf{x}_1^*)^T$ in the presence of the disease. Many populations, however, exhibit sustained fluctuations rather than a steady equilibrium. In this section, we consider epidemic models in which the disease-free population has a stable k-cycle rather than a stable equilibrium.

Returning to the general modeling methodology and notation presented in Section 4.6, we consider a system of the form

$$\begin{aligned} \mathbf{x}_0(t+1) &= \mathbf{g}_0(\mathbf{x}_0(t), \mathbf{x}_1(t)) \\ \mathbf{x}_1(t+1) &= \mathbf{g}_1(\mathbf{x}_0(t), \mathbf{x}_1(t)) \end{aligned} \quad (7.51)$$

under assumptions A1 and (4.22). In place of assumption A2, we assume the following

A2_k : The Equation (4.24) has a k-cycle $\{\bar{\mathbf{x}}_{11}, \bar{\mathbf{x}}_{12}, \ldots, \bar{\mathbf{x}}_{1k}\}$ for which

$$\rho\left(\Pi_{j=1}^k J_{\mathbf{x}_1} \mathbf{g}_1(\mathbf{0}, \bar{\mathbf{x}}_{1j})\right) < 1.$$

The k-cycle $\{\bar{\mathbf{x}}_{11}, \bar{\mathbf{x}}_{12}, \ldots, \bar{\mathbf{x}}_{1k}\}$ of Equation (4.24), which is locally asymptotically stable under assumption A2_k, corresponds to a k-cycle

$$\{\bar{\mathbf{x}}_{11}, \bar{\mathbf{x}}_{12}, \ldots, \bar{\mathbf{x}}_{1k}\} = \left\{(\mathbf{0}, \bar{\mathbf{x}}_{11})^T, (\mathbf{0}, \bar{\mathbf{x}}_{12})^T, \ldots, (\mathbf{0}, \bar{\mathbf{x}}_{1k})^T\right\}$$

of Equations (7.51). To determine the stability properties of this disease-free k-cycle we can apply the Linearization Principle to the composite of system (7.51) (see Chapter 3). The Jacobian of (7.51), when evaluated at points on the k-cycle, is

$$J_{\mathbf{x}} \mathbf{g}(\mathbf{0}, \bar{\mathbf{x}}_{1j}) = \begin{pmatrix} J_{\mathbf{x}_0} \mathbf{g}_0(\mathbf{0}, \bar{\mathbf{x}}_{1j}) & O \\ J_{\mathbf{x}_0} \mathbf{g}_1(\mathbf{0}, \bar{\mathbf{x}}_{1j}) & J_{\mathbf{x}_1} \mathbf{g}_1(\mathbf{0}, \bar{\mathbf{x}}_{1j}) \end{pmatrix}$$

(see (4.23)) and the Jacobian of the composite at the k-cycle is the product

$$\prod_{j=k}^{1} J_{\mathbf{x}}\mathbf{g}(\mathbf{0}, \bar{\mathbf{x}}_{1j}) = J_{\mathbf{x}}\mathbf{g}(\mathbf{0}, \bar{\mathbf{x}}_{1k}) \cdots J_{\mathbf{x}}\mathbf{g}(\mathbf{0}, \bar{\mathbf{x}}_{12}) J_{\mathbf{x}}\mathbf{g}(\mathbf{0}, \bar{\mathbf{x}}_{11})$$

$$= \begin{pmatrix} \Pi_{j=k}^{1} J_{\mathbf{x}_0} g_0(\mathbf{0}, \bar{\mathbf{x}}_{1j}) & O \\ * & \Pi_{j=k}^{1} J_{\mathbf{x}_1} g_1(\mathbf{0}, \bar{\mathbf{x}}_{1j}) \end{pmatrix}$$

(the asterisk denotes an unneeded entry). The eigenvalues of this block diagonal matrix are the eigenvalues of the diagonal blocks. Defining

$$\rho_k^* := \rho\left(\Pi_{j=k}^{1} J_{\mathbf{x}_0} g_0(\mathbf{0}, \bar{\mathbf{x}}_{1j})\right)$$

we obtain the following generalization of Theorem 43 from Linearization Principle (Section 3.8 in Chapter 3) and assumption A2$_k$.

Theorem 69 *Assume A1 and A2$_k$ hold. Then the disease-free k-cycle*

$$\left\{(\mathbf{0}, \bar{\mathbf{x}}_{11})^T, (\mathbf{0}, \bar{\mathbf{x}}_{12})^T, \ldots, (\mathbf{0}, \bar{\mathbf{x}}_{1k})^T\right\}$$

of Equations (7.51) is locally asymptotically stable if $\rho_k^* < 1$ and unstable if $\rho_k^* > 1$.

Example 102 *Consider the SI model in Examples 77 and 99*

$$\begin{aligned} S(t+1) &= be^{-cS(t)}S(t) + \varphi\left(\tfrac{I(t)}{N(t)}\right)\sigma_S S(t) \\ I(t+1) &= \left(1 - \varphi\left(\tfrac{I(t)}{N(t)}\right)\right)\sigma_S S(t) + \sigma_I I(t) \end{aligned} \quad (7.52)$$

with $0 < \sigma_S < 1$, $0 \leq \sigma_I < 1$ and positive b, c and an escape function that satisfies conditions (4.28), namely

$$0 \leq \varphi(z) \leq 1 \text{ for all } 0 \leq z \leq 1 \text{ and } \varphi(0) = 1, \; \varphi'(0) < 0.$$

The equation for the disease-free population dynamics is

$$S(t+1) = be^{-cS(t)}S(t) + \sigma_S S(t). \quad (7.53)$$

This equation has one positive equilibrium

$$S^* = \frac{1}{c}\ln\left(\frac{b}{1-\sigma_S}\right) \text{ when } 1 < \frac{b}{1-\sigma_S}.$$

which is

locally asymptotically stable when $1 < \frac{b}{1-\sigma_S} < \exp\left(\frac{2}{1-\sigma_S}\right)$
unstable when $\exp\left(\frac{2}{1-\sigma_S}\right) < \frac{b}{1-\sigma_S}$.

(See Problem 1 in Section 4.6.) When S^* is locally asymptotically stable, we know from Example 77 that the disease-free equilibrium $(S^*, 0)^T$ of the SI model (7.52) is locally asymptotically stable when $\mathcal{R}_0 < 1$ and unstable when $\mathcal{R}_0 > 1$ where

$$\mathcal{R}_0 = -\sigma_S \varphi'(0) \frac{1}{1-\sigma_I}. \quad (7.54)$$

7.4. MODELS WITH DISEASE-FREE POPULATION CYCLES

On the other hand, when S^* is unstable the disease-free equation (7.53) exhibits a period-doubling cascade to chaos as the quantity $b/(1-\sigma_S)$ increases without bound. See Figure 7.5 for an example bifurcation diagram.

Consider the case when the disease-free equilibrium $(S^*, 0)^T$ is unstable and the disease-free equation (7.53) has a k-cycle $\{S_1^*, S_2^*, \ldots, S_k^*\}$ that is stable by linearization, i.e., for which

$$\left| \prod_{j=k}^{1} \left(b\left(1 - cS_j^*\right) e^{-cS_j^*} + \sigma_S \right) \right| < 1.$$

We can apply Theorem 69 to determine the stability properties of the disease-free k-cycle

$$\left\{ (S_1^*, 0)^T, (S_2^*, 0)^T, \ldots, (S_k^*, 0)^T \right\} \tag{7.55}$$

for the SI model (7.52). For that model we have $\mathbf{x}_0 = I$, $\mathbf{x}_1 = S$ are one-dimensional column vectors and

$$\mathbf{g}_0(\mathbf{x}_0, \mathbf{x}_1) = \left(1 - \varphi\left(\frac{I}{N}\right)\right) \sigma_S S + \sigma_I I.$$

Thus,

$$J_{\mathbf{x}_0} \mathbf{g}_0(\mathbf{x}_0, \mathbf{x}_1) = -\varphi'\left(\frac{I}{N}\right) \frac{S}{(S+I)^2} \sigma_S S + \sigma_I$$

from which it follows that $J_{\mathbf{x}_0} \mathbf{g}_0(0, \mathbf{x}_1) = -\varphi'(0)\sigma_S + \sigma_I$ for all $\mathbf{x}_1 = S$ and, in particular, for all points $\mathbf{x}_1 = S_j^*$ on the k-cycle. As a result $\rho_k^* = (-\varphi'(0)\sigma_S + \sigma_I)^k$ in Theorem 69 and we conclude that the disease-free k-cycle (7.55) of the SI model (7.52) is

$$\begin{aligned} &\text{is locally asymptotically stable if } -\varphi'(0)\sigma_S + \sigma_I < 1 \\ &\text{is unstable if } -\varphi'(0)\sigma_S + \sigma_I > 1. \end{aligned} \tag{7.56}$$

It is a feature of this model that the stability constraints in (7.56) are the same for all periods k (including the equilibrium case when $k = 1$) and could be re-written as $\mathcal{R}_0 < 1$ and $\mathcal{R}_0 > 1$ respectively where, as we saw in Section 7.3.1,

$$\mathcal{R}_0 = -\varphi'(0) \frac{\sigma_S}{1 - \sigma_I}$$

is the reproduction number associated with the disease-free equilibrium.

The conclusions reached in Example 102 are illustrated in Figure 7.6 by the sample solution time series of the SI model (7.52) with escape function

$$\varphi\left(\frac{I}{N}\right) = \exp\left(-c_\varphi \frac{I}{N}\right) \tag{7.57}$$

and the parameter values are given in the caption. The disease-free equation (7.53) has a stable 2-cycle (see Figure 7.5) and the left plot in Figure 7.6 illustrates the stability of the associated disease-free 2-cycle of the SI model (7.52) in a case when $\rho_2^* = (c_\varphi \sigma_S + \sigma_I)^2 < 1$ (equivalently $\mathcal{R}_0 = c_\varphi \sigma_S / (1 - \sigma_I) < 1$). The right plot in Figure 7.6 illustrates the instability of the disease-free 2-cycle when $\rho_2^* > 1$ (equivalently $\mathcal{R}_0 > 1$), in which case the sample solution shown approaches an endemic

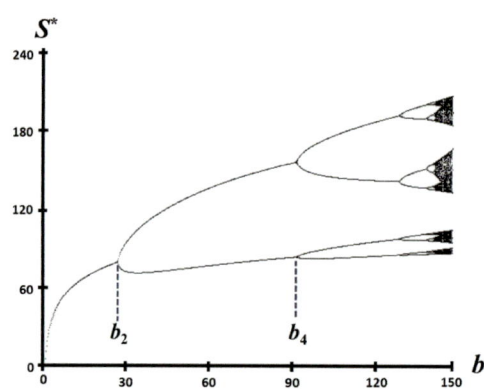

Figure 7.5: Period-doubling bifurcations in model (7.53) with escape function (7.57) and parameter values $c = 0.05$ and $\sigma_S = 0.5$. For $0.5 < b < b_2$ there is a stable equilibrium and for $b_2 < b < b_4$ there is a stable 2-cycle where $b_2 \approx 27.3$ and $b_4 \approx 92.2$.

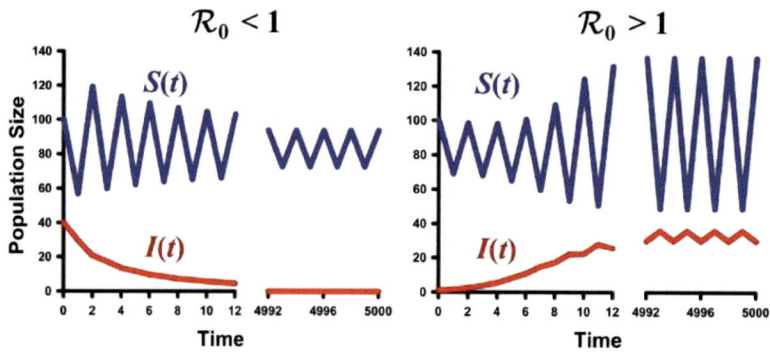

Figure 7.6: Sample solution times series from the SI model (7.53) with escape function (7.57) and parameter values $b = 30$, $c = 5/100$ and $c_S = 1/2$, $c_I = 2/5$. Left plot: $c_\varphi = 11/10$ and $\rho_2^* = 361/400 \approx 0.90$ ($\mathcal{R}_0 = 11/12 \approx 0.92$). Right plot: $c_\varphi = 12/5 = 2.4$ and $\rho_2^* = 64/25 \approx 2.56$ ($\mathcal{R}_0 = 2$).

7.4. MODELS WITH DISEASE-FREE POPULATION CYCLES

2-cycle. (Endemic 2-cycles bifurcate from the disease-free 2-cycle as $\mathcal{R}_0$ increases through 1 in direct analogy with the bifurcation of endemic equilibria when a disease-free equilibrium destabilizes as seen in Section 7.2.)

In Example 102, the stability determining quantity ρ_k^* in Theorem 69 happened to be independent of the period k of the disease-free cycle. This (perhaps unexpected) result is due to the dependence of φ on the ratio I/N in that example, which resulted in the same Jacobian $J_{\mathbf{x}_0}\mathbf{g}_0\left(\mathbf{0},\bar{\mathbf{x}}_{1j}\right)$ at each point $\bar{\mathbf{x}}_{1j}$ of the k-cycle. This will not necessarily be the case for other types of dependencies of the escape function φ on I and S.

Example 103 *If in Example 102 we replace the escape function $\varphi(I/N)$ in the SI model (7.52) by $\varphi(I)$ we get the SI model*

$$\begin{array}{rl} S(t+1) = & be^{-cS(t)}S(t) + \varphi(I(t))\sigma_S S(t) \\ I(t+1) = & (1-\varphi(I(t)))\sigma_S S(t) + \sigma_I I(t) \end{array}.$$

Everything in Example 102 remains unchanged except now $\mathbf{g}_0(\mathbf{x}_0,\mathbf{x}_1) = \hat{\varphi}(I)\sigma_S S + \sigma_I I$ and $J_{\mathbf{x}_0}\mathbf{g}_0(\mathbf{x}_0,\mathbf{x}_1) = -\varphi'(I)\sigma_S S + \sigma_I$ and

$$\rho_k^* = \prod_{j=k}^{1}\left(-\varphi'(0)\sigma_S S_j^* + \sigma_I\right)$$

where $\{S_1^*,\ldots,S_k^*\}$ is a k-cycle of the equation

$$S(t+1) = be^{-cS(t)}S(t) + \sigma_S S(t). \tag{7.58}$$

Note that in this example ρ_k^* depends on the cycle points S_j^*.

As a specific example take exponential escape function $\varphi(I) = \exp(-c_\varphi I)$ and suppose there exists a (locally asymptotically stable) 2-cycle of the disease-free equation (7.58). Since the 2-cycle $\{S_1^*,S_2^*\}$ arises by bifurcation from the equilibrium

$$S^* = \frac{1}{c}\ln\left(\frac{b}{1-\sigma_S}\right).$$

it follows that both cycle points S_1^* and S_2^* are close to S^* for

$$\frac{b}{1-\sigma_S} \gtrapprox \exp\left(\frac{2}{1-\sigma_S}\right)$$

and

$$\rho_2 \approx (c_\varphi \sigma_S S^* + \sigma_I)^2 = \left(c_\varphi \sigma_S \frac{1}{c}\frac{2}{1-\sigma_S} + \sigma_I\right)^2.$$

Under these circumstances the disease-free 2-cycle $\{(S_1^*,0)^T,(S_2^*,0)^T\} \approx \{(S^*,0),(S^*,0)\}$ of the SI model is locally asymptotically stable if

$$c_\varphi \sigma_S \frac{1}{c}\frac{2}{1-\sigma_S} + \sigma_I < 1$$

and unstable if the inequality is reversed.

The destabilization of the disease-free cycle as ρ_k^* increases through 1 in Theorem 69 suggests, in analog with Theorem 64 when $k = 1$, that a bifurcation of endemic cycles will occur. One way to study this bifurcation is to apply Theorems 64 and 65 in Section 7.2 to the k-composite of Equations (7.51) whose fixed points (equilibria) correspond to k-cycles of (7.51). Toward this end we assume, as in Section 7.2, that a model parameter μ has been designated as a bifurcation parameter and (under assumption (7.2)) consider Equations (7.3) with their disease-free equation (7.6), i.e.,

$$\begin{aligned} \mathbf{x}_0(t+1) &= G(\mu, \mathbf{x}_0(t), \mathbf{x}_1(t))\mathbf{x}_0(t) \\ \mathbf{x}_1(t+1) &= \mathbf{g}_1(\mu, \mathbf{x}_0(t), \mathbf{x}_1(t)) \end{aligned} \tag{7.59}$$

and

$$\mathbf{x}_1(t+1) = \mathbf{g}_1(\mu, \mathbf{0}, \mathbf{x}_1(t)). \tag{7.60}$$

The k-composite equations associated with these equations have the forms

$$\begin{aligned} \mathbf{y}_0(t+1) &= G^k(\mu, \mathbf{y}_0(t), \mathbf{y}_1(t))\mathbf{y}_0(t) \\ \mathbf{y}_1(t+1) &= \mathbf{g}_1^k(\mu, \mathbf{y}_0(t), \mathbf{y}_1(t)) \end{aligned} \tag{7.61}$$

and

$$\mathbf{y}_1(t+1) = \mathbf{g}_1^k(\mu, \mathbf{0}, \mathbf{y}_1(t)) \tag{7.62}$$

where $(\mathbf{y}_0(t), \mathbf{y}_1(t))^T$ equals $(\mathbf{x}_0(tk), \mathbf{x}_1(tk))^T$. By Theorem 24 in Chapter 3, the existence and stability of k-cycles of Equation (7.59) correspond to the existence and stability of equilibria of Equation (7.61) to which we can apply Theorems 64 and 65. To do this, we assume the following extensions of A4 and A5 in Section 7.2.

A4$_k$: Suppose $\mathbf{y}_1^*$ is an equilibrium of Equation (7.62) for all $\mu \in I_\mu$ where I_μ is an open interval in $\mathbb{R}_+$. Assume (7.2) holds and that the $m \times m$ matrix G^k is nonnegative and irreducible and its entries γ_{ij}^k, as well as those of the vector $\mathbf{g}_1^k$, are twice continuously differentiable functions of $\mu \in I_\mu$ and of $\mathbf{y} = (\mathbf{y}_0, \mathbf{y}_1)^T$ in an open neighborhood of $(\mathbf{0}, \mathbf{y}_1^*)^T$ in $\mathbb{R}_+^n$. Assume

$$\rho\left(J_{\mathbf{x}_1}\mathbf{g}_1^k(\mu, \mathbf{0}, \mathbf{y}_1^*)\right) < 1. \tag{7.63}$$

$\mathbf{y}_1^*$ in A4$_k$ corresponds to a k-cycle of the disease-free equation (7.60) and inequality (7.63) guarantees it is locally asymptotically stable.

A5$_k$: There exists a $\mu_0 \in I_\mu$ such that $\rho\left(G^k(\mu_0, \mathbf{0}, \mathbf{y}_1^*)\right) = 1$.

Applying the notation used in Section 7.2 to the composite equations, we define the diagnostic quantities

$$\kappa_k := -\mathbf{w}_L^T(\kappa_{ij}^k)\mathbf{w}_R \quad \text{and} \quad \alpha_k := \mathbf{w}_L^T(\alpha_{ij}^k)\mathbf{w}_R \tag{7.64}$$

where (κ_{ij}^k) and (α_{ij}^k) are $m \times m$ matrices whose entries are

$$\kappa_{ij}^k := \left(\nabla_{\mathbf{y}_0}^0 \gamma_{ij}^k + \nabla_{\mathbf{y}_1}^0 \gamma_{ij}^k \left(I - J_{\mathbf{y}_1}^{k0}\mathbf{g}_1^k\right)^{-1} J_{\mathbf{y}_0}^{k0}\mathbf{g}_1^k\right)\mathbf{w}_R \tag{7.65}$$

$$\alpha_{ij}^k := \partial_\mu^0 \gamma_{ij}^k + \nabla_{\mathbf{y}_1}^0 \gamma_{ij}^k \left(I - J_{\mathbf{y}_1}^{k0}\mathbf{g}_1^k\right)^{-1} \partial_\mu^0 \mathbf{g}_1^k. \tag{7.66}$$

Here J_1^{k0} is the Jacobian of $\mathbf{g}_1^k$ evaluated at the bifurcation point $\mu = \mu_0$, $\mathbf{y}_0 = \mathbf{0}$, and $\mathbf{y}_1 = \mathbf{y}_1^*$ and $\mathbf{w}_L^T$ and $\mathbf{w}_R$ are left and right eigenvectors of $G^k(\mu_0, \mathbf{0}, \mathbf{y}_1^*)$.

7.4. MODELS WITH DISEASE-FREE POPULATION CYCLES

Applying Theorem 64 we have that $\alpha_k \neq 0$ implies the bifurcation of equilibria of the composite equations (7.61) which are parameterized by small ε of the form

$$\mu(\varepsilon) = \mu_0 + \eta(\varepsilon) \tag{7.67}$$

$$\mathbf{y}_0(\varepsilon) = \varepsilon \mathbf{w}_R + \mathbf{z}_0(\varepsilon) \tag{7.68}$$

$$\mathbf{y}_1(\varepsilon) = \mathbf{y}_1^* + \mathbf{z}_1(\varepsilon) \tag{7.69}$$

where $\eta(0) = 0$, $\mathbf{z}_0(0) = d\mathbf{z}_0(0)/d\varepsilon = \mathbf{0}$ and $\mathbf{z}_1(0) = \mathbf{0}$. Furthermore,

$$\eta'(0) = \frac{\kappa_k}{\alpha_k}. \tag{7.70}$$

By Theorem 65 and Corollary 9, if $\kappa_k \neq 0$ then the signs of κ_k and α_k determine both the direction of bifurcation and the local stability of the bifurcating endemic equilibria of the composite.

Theorem 70 *In addition to $A4_k$ and $A5_k$ assume $\alpha_k > 0$. The disease-free k-cycle destabilizes as μ increases through μ_0 and results in the bifurcation of endemic k-cycles which exist and are locally asymptotically stable for $\mu \gtrsim \mu_0$ when $\kappa_k > 0$ or exist and are unstable for $\mu \lesssim \mu_0$ when $\kappa_k < 0$.*

If $\alpha_k < 0$ then the inequalities involving μ in Theorem 70 are reversed.

Example 104 *The disease-free equation associated with the SI model in Example 102 with an exponential escape function*

$$\begin{aligned} S(t+1) &= be^{-cS(t)}S(t) + \exp\left(-c_\varphi \frac{I(t)}{N(t)}\right)\sigma_S S(t) \\ I(t+1) &= \left(1 - \exp\left(-c_\varphi \frac{I(t)}{N(t)}\right)\right)\sigma_S S(t) + \sigma_I I(t) \end{aligned}$$

is

$$S(t+1) = be^{-cS(t)}S(t) + \sigma_S S(t)$$

which we saw in Example 103 has a stable 2-cycle $\{(S_1^*, 0)^T, (S_2^*, 0)^T\} \approx \{(S^*, 0), (S^*, 0)\}$ for

$$\frac{b}{1-\sigma_S} \gtrsim \exp\left(\frac{2}{1-\sigma_S}\right) \tag{7.71}$$

where

$$S^* = \frac{1}{c}\ln\left(\frac{b}{1-\sigma_S}\right) \quad \text{when } 1 < \frac{b}{1-\sigma_S}$$

is its equilibrium. If we use $\mu = c_\varphi$ as a bifurcation parameter, then this equilibrium destabilizes at

$$\mu_0 = c\frac{(1-\sigma_I)(1-\sigma_S)}{2\sigma_S}. \tag{7.72}$$

To apply Theorem 70 we identify

$$\mathbf{g}_1(\mu, I, S) = be^{-cS}S + \exp\left(-c_\varphi \frac{I}{N}\right)\sigma_S S$$

$$\mathbf{g}_0(\mu, I, S) = G(\mu, I, S)I \quad \text{where}$$

$$G(\mu, I, S) = \gamma_{11}(\mu, I, S) = \begin{cases} \frac{1-\exp(-\mu\frac{I}{N})}{I}\sigma_S S + \sigma_I & \text{for } I \neq 0 \\ \frac{1}{N}\mu\sigma_S S & \text{for } I = 0 \end{cases}.$$

$A4_2$ and $A5_2$ are satisfied under condition (7.71) and with μ_0 defined by (7.72). Calculations show (an algebraic computer program is helpful) $\gamma_{11}^2(\mu, 0, S) = (\sigma_2 + \mu\sigma_1)^2$ and hence $\partial_S^0 \gamma_{11}^2 = 0$. In this case where the $\mathbf{g}_i$ is one dimensional, we have $\mathbf{w}_R = \mathbf{w}_L = 1$ and

$$\alpha_2 = \partial_\mu^0 \gamma_{11}^2 = 2\sigma_1(\sigma_2 + \mu_0\sigma_1) > 0,$$

as required in Theorem 70. The calculation of $\kappa_2 = -\partial_I^0 \gamma_{11}^2$ is tedious, but straightforward, with the result

$$\kappa_2 = c(1-\sigma_I)(1-\sigma_S)[c(1-\sigma_S)(1-\sigma_I) + 2\sigma_2]$$
$$\times \frac{[c(1-\sigma_S)(1-\sigma_2) + 2 + 2\sigma_2][c(1-\sigma_2)(1-\sigma_S) + 4\sigma_S]}{32\sigma_I S^*}.$$

Since $\kappa_2 > 0$ a forward and stable bifurcation of endemic 2-cycles bifurcates from the disease-free 2-cycle for $\mu \gtrsim \mu_0$.

Exercises 7.4

1. Use the Linearization Principle to establish the stability results in Example 102 for the two equilibria S^* of the disease-free equation (7.53).

2. (a) Prove that solutions of the SI model equations (7.52) with positive initial conditions remain positive and bounded for all $t > 0$.

   (b) When $\mathcal{R}_0 > 1$ in model (7.52), use simulations to investigate the possible endemic states (equilibria, cycles, etc.).

3. Consider the following SI model with Hassell Recruitment function

$$\begin{aligned} S(t+1) &= \frac{bS(t)}{(1+S(t))^\alpha} + (1-d)S(t)e^{-\beta\frac{I(t)}{N(t)}} \\ I(t+1) &= (1-d)S(t)\left(1 - e^{-\beta\frac{I(t)}{N(t)}}\right) + (1-d)I(t), \end{aligned} \quad (7.73)$$

where $\alpha, \beta, b > 0$ and d is in the open interval $(0, 1)$.

   (a) Prove that solutions with positive initial conditions remain positive and bounded for all $t > 0$.

   (b) Use a bifurcation diagram to discuss the bifurcations that occur in the disease-free equation of model (7.73).

   (c) Calculate all the equilibrium points of the disease-free equation of model (7.73) and determine their stability properties using the Linearization Principle.

   (d) Assume the disease-free equation has an equilibrium that is locally asymptotically stable by linearization. What is $\mathcal{R}_0$ for the diseases-free equilibrium of (7.73)?

   (e) Assume the disease-free equation has a k-cycle that is locally asymptotically stable by linearization. What is ρ_k^* and how does it relate to $\mathcal{R}_0$?

7.5. TWO CASE STUDIES

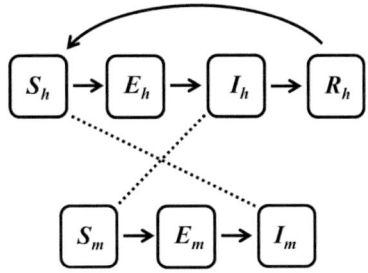

Figure 7.7: The dashed lines shown represent the interactions between infectious and susceptible individuals. Not shown are the in-flows into the susceptible classes due to births.

4. Repeat Problem 3 for the SI model (with the so-called Bob-White Quail recruitment function)

$$\begin{align}
S(t+1) &= S(t)\left(b + \frac{\sigma}{1+(S(t))^\alpha}\right) + (1-d)S(t)e^{-\beta\frac{I(t)}{N(t)}} \\
I(t+1) &= (1-d)S(t)\left(1 - e^{-\beta\frac{I(t)}{N(t)}}\right) + (1-d)I(t)
\end{align} \tag{7.74}$$

where $\alpha, \beta, b, \sigma > 0$ and $0 < d < 1$.

5. Repeat Problem 3 for the SI model

$$\begin{align}
S(t+1) &= S(t)g(S(t)) + (1-d)S(t)e^{-\beta\frac{I(t)}{N(t)}} \\
I(t+1) &= (1-d)S(t)\left(1 - e^{-\beta\frac{I(t)}{N(t)}}\right) + (1-d)I(t)
\end{align} \tag{7.75}$$

where the growth function $g \in C^1(\mathbb{R}_+ \to \mathbb{R}_+)$ is a decreasing function satisfying $g(0) > 1$, $\lim_{S\to\infty} g(S) < 1$ Assume $\beta > 0$ and $0 < d < 1$.

7.5 Two Case Studies

We conclude this chapter with applications of the modeling methodology and analysis presented in this chapter made to two specific diseases, namely malaria, and SARS. The models, their derivations, and analyses closely follow that in [49] and [50] for malaria and in [230] and [52] for SARS.

7.5.1 A Malaria Model

The malaria model we consider, which is structured similarly to the differential equation model studied in [49] and [50], is described by the compartmental diagram shown in Figure 7.7. Malaria is caused by the *Plasmodium* parasite that is transmitted to humans by a female mosquito (sp. *Anopheles*) bites. The model structures both the human and mosquito populations into disease classes as is done in Section 7.3. Humans are classified as susceptible S_h, exposed (but not infectious) E_h, infectious I_h and recovered (immune) R_h while female mosquitoes are classified as susceptible S_m, exposed E_m and infectious (where they remain for life) I_m.

Transmission of the parasite between humans and mosquitoes occurs as follows. With the bite of a susceptible human by an infectious mosquito, there is a certain probability that the parasite will be transmitted and the human victim will move to the exposed class. After a latency period during which the parasite passes through some life cycle stages in the liver, it enters the bloodstream which usually signals the onset of malaria and the victim moves to the infectious class. After some period of time, an infectious human recovers and moves to the recovered class where they have immunity to infection which, however, is lost over time and the victim returns to the susceptible class. With the bite of an infectious human by a susceptible mosquito,[1] there is a certain probability the parasite will be transmitted and the mosquito will move to the exposed class. After a latency period during which the parasite passes through some life cycle stages, it enters the salivary glands of the mosquito and the mosquito moves to the infectious class, where it remains for life. We assume that both populations grow logistically in the absence of the parasite and infection.

Let

$$N_h := S_h + E_h + I_h + R_h$$
$$N_m := S_m + E_m + I_m$$

denote the total human and mosquito population sizes, respectively. Following the notation in Section 4.6, we list the infected classes first, in the following order (exposed classes, infectious classes, susceptible classes, and the recovered human class)

$$\mathbf{x} = \begin{pmatrix} E_h & E_m & I_h & I_m & S_h & S_m & R_h \end{pmatrix}^T$$

with $\mathbf{x} = (\mathbf{x}_0, \mathbf{x}_1)^T$ where

$$\mathbf{x}_0 = \begin{pmatrix} E_h & E_m & I_h & I_m \end{pmatrix}^T, \quad \mathbf{x}_1 = \begin{pmatrix} S_h & S_m & R_h \end{pmatrix}^T.$$

The infection escape functions are given by

$$\varphi_h(\mathbf{x}) := \exp\left(-\beta_h(N_h, N_m)\beta_{hm}\frac{I_m}{N_m}\right)$$
$$\varphi_m(\mathbf{x}) := \exp\left(-\beta_m(N_h, N_m)\beta_{mh}\frac{I_h}{N_h}\right)$$

where, following [49],

$$\beta_h(N_h, N_m) := \frac{\sigma_m N_m \sigma_h N_h}{\sigma_m N_m + \sigma_h N_h}\frac{1}{N_h}$$
$$\beta_m(N_h, N_m) := \frac{\sigma_m N_m \sigma_h N_h}{\sigma_m N_m + \sigma_h N_h}\frac{1}{N_m}$$

are, respectively, the number of bites per human and per mosquito per unit time (which is one day). Note that, unlike models in previous sections, the infectiousness coefficients (the coefficients of the fractions I_h/N_h and I_m/N_m) are not constant through time because they depend on the total populations of humans and mosquitoes.

---

[1] Mosquitos can become infected by also biting a recovered human, but this transmission occurs at a much lower rate and we ignore it in this model.

7.5. TWO CASE STUDIES

The parameters in the model equations

$$\begin{aligned}
E_h(t+1) &= (1-\varphi_h(\mathbf{x}(t)))\sigma_{Sh}S_h(t) + (1-v_{Eh})\sigma_{Eh}E_h(t) \\
E_m(t+1) &= (1-\varphi_m(\mathbf{x}(t)))\sigma_{Sm}S_m(t) + (1-v_{Em})\sigma_{Em}E_m(t) \\
I_h(t+1) &= v_{Eh}\sigma_{Eh}E_h(t) + (1-v_{Ih})\sigma_{Ih}I_h(t) \\
I_m(t+1) &= v_{Em}\sigma_{Em}E_m(t) + \sigma_{Im}I_m(t) \\
S_h(t+1) &= b_h\frac{1}{1+c_h N_h(t)}N_h(t) + v_{Rh}\sigma_{Rh}R_h(t) + \varphi_h(\mathbf{x}(t))\sigma_{Sh}S_h(t) \\
S_m(t+1) &= b_m\frac{1}{1+c_m N_m(t)}N_m(t) + \varphi_m(\mathbf{x}(t))\sigma_{Sm}S_m(t) \\
R_h(t+1) &= v_{Ih}\sigma_{Ih}I_h(t) + (1-v_{Rh})\sigma_{Rh}R_h(t)
\end{aligned} \qquad (7.76)$$

are described in Table 7.1.

Survival rates	Transition and infection probabilities
σ_{Sh} susceptible human	v_{Eh} exposed human becomes infectious
σ_{Sm} susceptible mosquito	v_{Em} exposed mosquito becomes infectious
σ_{Eh} exposed human	v_{Ih} infectious human recovers
σ_{Em} exposed mosquito	v_{Rh} recovered human becomes susceptible
σ_{Ih} infectious human	β_{mh} mosquito becomes infected by biting a human
σ_{Im} infectious mosquito	β_{hm} human becomes infected by a mosquito bite
σ_{Rh} recovered human	
Logistic growth coefficients	Biting rates
b_h, b_m birth rates	σ_h average bites human can receive per day
c_h, c_m density coefficients	σ_m average bites a mosquito can give per day

Table 7.1: Parameters for the Malaria Model (7.76)

We begin with some properties of the model equations when the human and mosquito populations have no interaction. The reader is asked to verify these properties in Exercise 1 and 2. Equation

$$S_m(t+1) = b_m \frac{1}{1+c_m S_m(t)} S_m(t) + \sigma_{Sm} S_m(t) \qquad (7.77)$$

for the mosquito population dynamics in the absence of contacts with human has a unique equilibrium

$$S_m^* = \frac{1}{c_m}\left(\frac{b_m}{1-\sigma_{Sm}} - 1\right) \qquad (7.78)$$

which is nonnegative and globally asymptotically stable for $S_m(0) > 0$ provided

$$\frac{b_m}{1-\sigma_{Sm}} > 1. \qquad (7.79)$$

The quotient in (7.79) is the mosquito reproduction number in absence of contacts with mosquitoes. The (uncoupled) equations

$$\begin{aligned}
S_h(t+1) &= b_h \frac{1}{1+c_h S_h(t)} S_h(t) + \sigma_{Sh} S_h(t) \\
R_h(t+1) &= (1-v_{Rh})\sigma_{Rh} R_h(t)
\end{aligned} \qquad (7.80)$$

describe human population dynamics in the absence of contacts with mosquitoes (and hence of malaria). They have a unique equilibrium

$$\begin{pmatrix} S_h^* \\ R_h^* \end{pmatrix} = \begin{pmatrix} \frac{1}{c_h}\left(\frac{b_h}{1-\sigma_{Sh}}-1\right) \\ 0 \end{pmatrix} \tag{7.81}$$

which is nonnegative provided

$$\frac{b_h}{1-\sigma_{Sh}} > 1. \tag{7.82}$$

This inequality also implies that the nonnegative equilibrium is globally asymptotically stable for initial conditions $S_0(0) > 0$ and $R_h(0) \geq 0$. The quotient in (7.82) is the human reproduction number in the absence of contacts with mosquitoes.

Thus, under inequalities (7.82) and (7.79), each population has a globally stable equilibrium in the absence of contact between them.

It follows that the disease-free equations

$$\begin{aligned} S_h(t+1) &= b_h \frac{1}{1+c_h S_h(t)} S_h(t) + \sigma_{Sh} S_h(t) \\ S_m(t+1) &= b_m \frac{1}{1+c_m S_m(t)} S_m(t) + \sigma_{Sm} S_m(t) \\ R_h(t+1) &= (1-v_{Rh})\sigma_{Rh} R_h(t) \end{aligned}$$

for the malaria model (7.76) have a unique, disease-free equilibrium $\mathbf{x} = (\mathbf{0}, \mathbf{x}_1^*)^T$ with $\mathbf{x}_1^* = (S_h^*, R_h^*, S_m^*)^T$ provided (7.82) and (7.79) hold, which we assume from now on. We can study the stability of this disease-free equilibrium, and the existence and stability of endemic equilibrium, by making use of the theorems in Sections 4.7 and 7.2, once we have calculated the reproduction number $\mathcal{R}_0$ associated with this disease-free equilibrium.

From the model equations, we first identify

$$\mathbf{g}_0 = \begin{pmatrix} (1-\varphi_h(\mathbf{x}))\sigma_{Sh}S_h + (1-v_{Eh})\sigma_{Eh}E_h \\ (1-\varphi_m(\mathbf{x}))\sigma_{Sm}S_m + (1-v_{Em})\sigma_{Em}E_m \\ v_{Eh}\sigma_{Eh}E_h + (1-v_{Ih})\sigma_{Ih}I_h \\ v_{Em}\sigma_{Em}E_m + \sigma_{Im}I_m \end{pmatrix}$$

$$\mathbf{g}_1 = \begin{pmatrix} b_h \frac{1}{1+c_h N_h} N_h + v_{Rh}\sigma_{Rh}R_h + \varphi_h(\mathbf{x})\sigma_{Sh}S_h \\ b_m \frac{1}{1+c_m N_m} N_m + \varphi_m(\mathbf{x})\sigma_{Sm}S_m \\ v_{Ih}\sigma_{Ih}I_h + (1-v_{Rh})\sigma_{Rh}R_h \end{pmatrix}.$$

Note that $\mathbf{g}_0$ identically vanishes when $\mathbf{x}_0 = \mathbf{0}$, and that $\mathbf{g}_0 = \mathbf{n}_0 + \mathbf{s}_0$

$$\mathbf{n}_0 = \begin{pmatrix} (1-\varphi_h(\mathbf{x}))\sigma_{Sh}S_h \\ (1-\varphi_m(\mathbf{x}))\sigma_{Sm}S_m \\ 0 \\ 0 \end{pmatrix}, \quad \mathbf{s}_0 = \begin{pmatrix} (1-v_{Eh})\sigma_{Eh}E_h \\ (1-v_{Em})\sigma_{Em}E_m \\ v_{Eh}\sigma_{Eh}E_h + (1-v_{Ih})\sigma_{Ih}I_h \\ v_{Em}\sigma_{Em}E_m + \sigma_{Im}I_m \end{pmatrix}$$

where $\mathbf{n}_0$ gives the new infections (entries into the exposed classes) and $\mathbf{s}_0$ describes the transitions between the infected classes. Recall from Section 4.6 that $\mathcal{R}_0$ is the spectral radius of the next-

7.5. TWO CASE STUDIES

generation matrix $F(I-T)^{-1}$ where $F = J_{x_0}\mathbf{n}(\mathbf{0}, \mathbf{x}_1^*)$ and $T = J_{x_0}\mathbf{s}(\mathbf{0}, \mathbf{x}_1^*)$. Calculations show that

$$F = \begin{pmatrix} 0 & 0 & 0 & \beta_h(S_h^*, S_m^*)\beta_{hm}\frac{1}{S_m^*}\sigma_{Sh}S_h^* \\ 0 & 0 & \beta_m(S_h^*, S_m^*)\beta_{mh}\frac{1}{S_h^*}\sigma_{Sm}S_m^* & 0 \\ 0 & 0 & 0 & 0 \\ 0 & 0 & 0 & 0 \end{pmatrix}$$

$$T = \begin{pmatrix} (1-v_{Eh})\sigma_{Eh} & 0 & 0 & 0 \\ 0 & (1-v_{Em})\sigma_{Em} & 0 & 0 \\ v_{Eh}\sigma_{Eh} & 0 & (1-v_{Ih})\sigma_{Ih} & 0 \\ 0 & v_{Em}\sigma_{Em} & 0 & \sigma_{Im} \end{pmatrix}$$

and that the next-generation matrix has the block diagonal form

$$F(I-T)^{-1} = \begin{pmatrix} M_1 & M_2 \\ \mathbf{0}_{2\times 2} & \mathbf{0}_{2\times 2} \end{pmatrix}$$

where

$$M_1 = \begin{pmatrix} 0 & \frac{1}{S_m^*}\frac{\beta_h(S_h^*, S_m^*)S_h^*\beta_{hm}\sigma_{Sh}v_{Em}\sigma_{Em}}{(1-(1-v_{Em})\sigma_{Em})(1-\sigma_{Im})} \\ \frac{1}{S_h^*}\frac{\beta_m(S_h^*, S_m^*)S_m^*\beta_{mh}v_{\sigma_{Sm}Eh}\sigma_{Eh}}{(1-(1-v_{Eh})\sigma_{Eh})(1-(1-v_{Ih})\sigma_{Ih})} & 0 \end{pmatrix}.$$

The eigenvalues of $F(I-T)^{-1}$ are the double eigenvalue 0 and the eigenvalues of M_1, which are the square roots of the product of its anti-diagonal entries. With some re-grouping of terms, this gives

$$\mathcal{R}_0 = \sqrt{\mathcal{R}_{0h}\mathcal{R}_{0m}} \tag{7.83}$$

where

$$\begin{aligned} \mathcal{R}_{0h} &:= \beta_h(S_h^*, S_m^*)\beta_{mh}\left[\sigma_{Sh}\frac{v_{Eh}\sigma_{Eh}}{(1-(1-v_{Eh})\sigma_{Eh})(1-(1-v_{Ih})\sigma_{Ih})}\right] \\ \mathcal{R}_{0m} &:= \beta_m(S_h^*, S_m^*)\beta_{hm}\left[\sigma_{Sm}\frac{v_{Em}\sigma_{Em}}{(1-(1-v_{Em})\sigma_{Em})(1-\sigma_{Im})}\right] \end{aligned}. \tag{7.84}$$

Note that

$$\begin{aligned} \mathcal{R}_{0h} &= \text{[bites per human per unit time]} \\ &\quad \times \text{[probability of transmission from human to mosquito]} \\ &\quad \times \text{[expected time a human is infectious]} \\ &= \text{[number of mosquitoes infected by a human per unit time]} \\ &\quad \times \text{[expected time a human is infectious]} \\ &= \text{expected number of mosquitoes infected by a human per lifetime} \end{aligned}$$

and, similarly, $\mathcal{R}_{0m}$ is number of humans infected by a mosquito per life time. The reproduction number (7.83) associated with the disease-free equilibrium is the geometric mean of these two expectations.

Theorem 44 implies that the disease-free equilibrium is locally asymptotically stable if $\mathcal{R}_0 < 1$ and is unstable if $\mathcal{R}_0 > 1$. Corollary 10 in Section 7.2 implies the bifurcation of endemic equilibria (an outbreak of malaria) at $\mathcal{R}_0 = 1$. Applying this corollary rigorously requires a calculation of κ. However, a general formula for κ is not feasible for this model because of its high dimension. Numerical simulations for specific parameter values can verify a predicted outbreak. An example

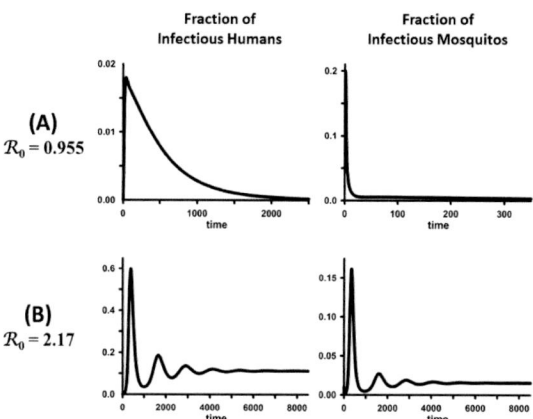

Figure 7.8: Plotted are $I_h(t)/N_h(t)$ and $I_m(t)/N_m(t)$ for a sample solution of the malaria model (7.76) with the unit of time equal to one day. Parameters (adapted from those in [50]) are: $v_{Rh} = 0.005$, $v_{Eh} = 0.1$, $v_{Ih} = 0.0015$, $v_{Em} = 0.1$, $\sigma_{Sh} = 0.999$, $\sigma_{Rh} = 0.999$, $\sigma_{Eh} = 0.999$, $\sigma_{Ih} = 0.98$, $\sigma_{Sm} = 0.95$, $\sigma_{Em} = 0.95$, $\sigma_{I_m} = 0.95$, $b_h = 0.005$, $b_m = 10$, $c_h = 0.0001$, $c_m = 0.001$, $\beta_{hm} = 0.2$, $\beta_{mh} = 0.4$, $\sigma_m = 0.5$. **(A)** $\sigma_h = 0.36$ with all initial conditions equal to 0 except $S_h(0) = 40,000$, $S_m(0) = 9,900$, and $I_m(0) = 100$. **(B)** $\sigma_h = 1$ with all initial conditions equal to 0 except $S_h(0) = 40,000$, $S_m(0) = 8,000$, and $I_m(0) = 2,000$.

appears in Figure 7.8 for parameter estimates commensurate with those given in [50]. In Figure 7.8, we see the stability of the disease-free equilibrium, and the resulting failure of the disease to establish itself in the population, for a low biting rate of mosquitoes on humans σ_h, which results in an $\mathcal{R}_0 < 1$. For a higher biting rate, $\mathcal{R}_0 > 1$ and there is a stable endemic equilibrium. In that case, the fraction of infected individuals does not tend to zero in either population.

Efforts to prevent or mitigate epidemics often focus on how to decrease $\mathcal{R}_0$ and, if possible, decrease it below 1 so that the disease-free equilibrium is stable. The proportional effect on $\mathcal{R}_0$ that a proportional change in a model parameter μ has is measured by the elasticity

$$\frac{\partial \mathcal{R}_0}{\partial \mu} \frac{\mu}{\mathcal{R}_0}$$

defined in Chapter 2 (which, in epidemic modeling, is often called the *sensitivity index*). From (7.83), we find that

$$\frac{\partial \mathcal{R}_0}{\partial \mu} \frac{\mu}{\mathcal{R}_0} = \frac{1}{2}\left(\frac{\partial \mathcal{R}_{0h}}{\partial \mu} \frac{\mu}{\mathcal{R}_{0h}} + \frac{\partial \mathcal{R}_{0m}}{\partial \mu} \frac{\mu}{\mathcal{R}_{0m}}\right) \tag{7.85}$$

(Problem 3). Thus, while $\mathcal{R}_0$ is the geometric mean of $\mathcal{R}_{0h}$ and $\mathcal{R}_{0m}$, the elasticity of $\mathcal{R}_0$ is the arithmetic mean of the elasticities of $\mathcal{R}_{0h}$ and $\mathcal{R}_{0m}$. Since we have an explicit formulas (7.84) for $\mathcal{R}_{0h}$ and $\mathcal{R}_{0m}$ in terms of all the model parameters, we can, in principle, calculate their elasticities, and hence that of $\mathcal{R}_0$, with respect to any parameter and compare them to see which has the greater proportional effect. For some parameters, the formulas are quite complicated, however, so here we will look at just a couple of elasticities by way of illustration.

From Problem 4 we find that

$$\frac{\partial \mathcal{R}_0}{\partial \beta_{mh}} \frac{\beta_{mh}}{\mathcal{R}_0} = \frac{\partial \mathcal{R}_0}{\partial \beta_{hm}} \frac{\beta_{hm}}{\mathcal{R}_0} = \frac{1}{2} \qquad (7.86)$$

for all parameters. Then a 10% increase (or decrease), for example, in either disease transmission probability β_{mh} or β_{hm} will increase (respectively decrease) $\mathcal{R}_0$ by about 5%.

By controlling mosquito bites (clothing, repellent, bed nets, etc.), one hopes that the expected number σ_h of bites per human per day will decrease. A calculation shows (Problem 4) that the elasticity of $\mathcal{R}_0$ with respect to σ_h is

$$0 < \frac{\partial \mathcal{R}_0}{\sigma_h} \frac{\sigma_h}{\mathcal{R}_0} = \frac{\sigma_m S_m^*}{\sigma_m S_m^* + \sigma_h S_h^*} = \frac{1}{1 + \frac{\sigma_h S_h^*}{\sigma_m S_m^*}} < 1. \qquad (7.87)$$

Such preventive measures would have their greatest effect when this elasticity is closest to 1. This occurs when the ratio $\sigma_h S_h^*/\sigma_m S_m^*$ is small, i.e., when the total number of bites humans are expected to receive per day, $\sigma_h S_h^*$, is small when compared to the total number of bites $\sigma_m S_m^*$ the mosquito population could give per day. As a specific example, for the parameter values used in Figure 7.8(B) (adapted from those in [50]) we find that

$$\frac{\partial \mathcal{R}_0}{\sigma_h} \frac{\sigma_h}{\mathcal{R}_0} \approx 0.713.$$

Thus, we can expect that a reduction of σ_h by one half (by 50%), from 1 to 0.5, will approximately reduce $\mathcal{R}_0$ by the factor of $0.713 \times 0.5 = 0.3565$ (or about 35.65%), that is to say from $\mathcal{R}_0 = 2.17 > 1$ to about $\mathcal{R}_0 = 0.3565 \times 2.17 = 0.77361 < 1$.

7.5.2 Isolation and Quarantine: SARS Model

Global emerging and re-emerging infectious disease (ERID) events are dominated by zoonotic events. ERID events have been rising significantly over time [178]. The increasing trends of ERID are a significant burden on global economies and public health [178]. Recent novel coronavirus (COVID-19) pandemic and the last severe acute respiratory syndrome (SARS) outbreak have drawn attention to the disease control and prevention strategies of *isolation* (that is, separation of sick individuals with an infectious disease from individuals who are not sick) and *quarantine* (that is, separation and restriction of the movement of individuals who were exposed to an infectious disease to see if they eventually become sick with the disease).

The fundamental dilemma associated with the implementation of isolation and quarantine (I&Q) as disease control strategies is how to predict the population level efficacy of an individual quarantine. That is, which and how many individuals need to be quarantined to achieve effective control of the disease at the population level? I&Q mitigation strategies have proven effective in the control of some diseases, such as SARS. However, they are not appropriate as effective control strategies for all infectious diseases. For example, diseases like varicella, for which the cost of I&Q may be high and the return minimal, may require a different disease control strategy. In some cases, the use of I&Q as control strategy may be not only costly but harmful.

In [52], a continuous-time model was developed and used to predict when I&Q strategies could stop the spread of SARS in Greater Toronto, Ontario region of Canada. A discrete-time version of the SARS model was constructed in [230]. The discrete-time SARS model has the following 6 epidemiological classes at each time $t \in Z_+$:

- Susceptible ($S(t)$), population density of susceptible individuals capable of catching SARS;
- Exposed ($E(t)$), population density of individuals infected with SARS but in a latent state;
- Infected ($I(t)$), population density of infectious individuals who capable of infecting others with SARS and showing symptoms of SARS;
- Quarantined ($Q(t)$), population density of individuals who are quarantined;
- Isolated ($J(t)$), population density of individuals who are isolated;
- Recovered ($R(t)$), population density of individuals who have recovered from SARS.

The total population at each time $t \in Z_+$ is
$$N(t) = S(t) + E(t) + I(t) + Q(t) + J(t) + R(t).$$
For SARS, the exposed individuals are typically mildly infectious with a reduced coefficient of latent infectivity denoted by $q \in (0,1)$. The model parameters are given in the accompanying Table 7.2 and the flowchart for our model is shown in Figure 7.9.

Parameter	Description
Λ	Recruitment
q	Reduction in E infectivity
$1-\sigma$	Probability of I
$1-r_1$	Probability of transition from J to R
$1-r_2$	Probability of transition from I to R
$1-\eta_1$	Probability of transition from Q to S
$1-\eta_2$	Probability of transition from Q to J
γ	Probability of survival
$1-\rho$	Fraction of quarantined S
α_i	Convex combination coefficients

Table 7.2: Table of Model Parameters

The difference equations for this discrete-time SARS model are

$$\begin{aligned}
S(t+1) &= \Lambda + S(t)\left(\gamma\alpha_1\phi\left(\tfrac{I(t)+qE(t)}{S(t)+E(t)+I(t)+R(t)}\right) + \gamma\rho(1-\alpha_1)\right) + \gamma\alpha_4(1-\eta_1)Q(t) \\
E(t+1) &= \gamma\alpha_1 S(t)\left(1 - \phi\left(\tfrac{I(t)+qE(t)}{S(t)+E(t)+I(t)+R(t)}\right)\right) + \gamma E(t)(\alpha_2\sigma + (1-\alpha_2)\rho) \\
I(t+1) &= \alpha_2(1-\sigma)\gamma E(t) + \gamma I(t)(\alpha_3\sigma + (1-\alpha_3)r_2) \\
Q(t+1) &= \gamma(1-\rho)((1-\alpha_1)S(t) + (1-\alpha_2)E(t)) + \gamma Q(t)(\alpha_4\eta_1 + (1-\alpha_4)\eta_2) \\
J(t+1) &= \alpha_3\gamma(1-\sigma)I(t) + (1-\alpha_4)(1-\eta_2)\gamma Q(t) + \gamma r_1 J(t) \\
R(t+1) &= \gamma(1-r_1)J(t) + \gamma(1-\alpha_3)(1-r_2)I(t) + \gamma R(t),
\end{aligned} \tag{7.88}$$

with which we have the initial condition
$$(S(0), E(0), I(0), Q(0), J(0), R(0))^T \in \mathbb{R}_+^6 \setminus \{\mathbf{0}\}.$$

7.5. TWO CASE STUDIES

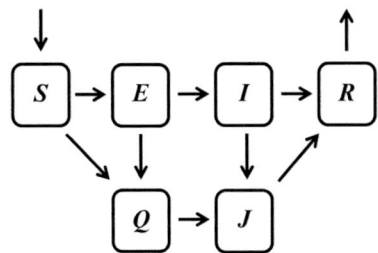

Figure 7.9: Flowchart for Discrete-Time SARS model.

Since each term on the right side of the model equations (7.88) is nonnegative,
$$(S(t), E(t), I(t), Q(t), J(t), R(t))^T \in \mathbb{R}_+^6$$
for all $t \in Z_+$.

The total population in is model is not constant, but it is bounded. From the inequality
$$\begin{aligned} N(t+1) &= S(t+1) + E(t+1) + I(t+1) + Q(t+1) + J(t+1) + R(t+1) \\ &\leq \Lambda + \gamma\left(S(t) + E(t) + I(t) + Q(t) + J(t) + R(t)\right) \\ &\leq \Lambda + \gamma N(t) \end{aligned}$$
we see, by mathematical induction, that
$$0 \leq N(t) \leq x(t),$$
for $t \in Z_+$ where $x(t)$ solves the (linear) difference equation
$$x(t+1) = \Lambda + \gamma x(t)$$
and $x(0) = N(0)$. Since $0 < \gamma < 1$ it follows that
$$\lim_{t \to \infty} x(t) = \frac{\Lambda}{1-\gamma}.$$
Thus, $x(t)$ and hence $N(t)$, are bounded. Specifically
$$\limsup_{t \to \infty} N(t) \leq \lim_{t \to \infty} x(t) = \frac{\Lambda}{1-\gamma},$$
and there is no unbounded population growth.

The linear equation
$$S(t+1) = \Lambda + (\gamma \alpha_1 + \gamma \rho (1-\alpha_1)) S(t)$$
describes the dynamics of the susceptibles in the disease-free case. Since
$$0 < \gamma \alpha_1 + \gamma \rho (1-\alpha_1) < 1,$$

in the absence of SARS, the disease-free equilibrium of the susceptible population

$$S^* = \frac{\Lambda}{1 - (\gamma\alpha_1 + \gamma\rho(1 - \alpha_1))}.$$

is globally asymptotically stable. Next, we apply the next-generation matrix method to compute the basic reproduction number, $\mathcal{R}_0$.

We write the state variable for the model as $\mathbf{x} = (\mathbf{x}_0, \mathbf{x}_1)$ with

$$\mathbf{x}_0 = (E, I, Q, J)^T, \quad \mathbf{x}_1 = (S, R)^T$$

and from Equations (7.88) identify

$$\mathbf{n}(\mathbf{x}_0, \mathbf{x}_1) = \begin{pmatrix} \gamma\alpha_1 S \left(1 - \phi\left(\frac{I+qE}{S+E+I+R}\right)\right) \\ 0 \\ 0 \\ 0 \end{pmatrix}$$

$$\mathbf{s}(\mathbf{x}_0, \mathbf{x}_1) = \begin{pmatrix} \gamma E(\alpha_2\sigma + (1 - \alpha_2)\rho) \\ \alpha_2(1 - \sigma)\gamma E + \gamma I(\alpha_3\sigma + (1 - \alpha_3)r_2) \\ \gamma(1 - \rho)((1 - \alpha_1)S + (1 - \alpha_2)E) + \gamma Q(\alpha_4\eta_1 + (1 - \alpha_4)\eta_2) \\ \alpha_3\gamma(1 - \sigma)I + (1 - \alpha_4)(1 - \eta_2)\gamma Q + \gamma r_1 J \end{pmatrix}$$

for $\mathbf{g}_0(\mathbf{x}_0, \mathbf{x}_1) = \mathbf{n}(\mathbf{x}_0, \mathbf{x}_1) + \mathbf{s}(\mathbf{x}_0, \mathbf{x}_1)$. Thus, in the Jacobian

$$J_{\mathbf{x}_0}\mathbf{g}_0(\mathbf{0}, \mathbf{x}_1^*) = \begin{pmatrix} \gamma(-\alpha_1 q\phi'(0) + \alpha_2\sigma + (1 - \alpha_2)\rho) & -\gamma\alpha_1\phi'(0) & 0 & 0 \\ \gamma\alpha_2(1 - \sigma) & \gamma(\alpha_3\sigma + (1 - \alpha_3)r_2) & 0 & 0 \\ \gamma(1 - \alpha_2)(1 - \rho) & 0 & \gamma(\alpha_4\eta_1 + (1 - \alpha_4)\eta_2) & 0 \\ 0 & \gamma\alpha_3(1 - \sigma) & \gamma(1 - \alpha_4)(1 - \eta_2) & \gamma r_1 \end{pmatrix}$$

which we write as $J_{\mathbf{x}_0}\mathbf{g}_0(\mathbf{0}, \mathbf{x}_1^*) = F + T$ where

$$F = \begin{pmatrix} -\gamma\alpha_1 q\phi'(0) & -\gamma\alpha_1\phi'(0) & 0 & 0 \\ 0 & 0 & 0 & 0 \\ 0 & 0 & 0 & 0 \\ 0 & 0 & 0 & 0 \end{pmatrix}$$

$$T = \begin{pmatrix} \gamma(\alpha_2\sigma + (1 - \alpha_2)\rho) & 0 & 0 & 0 \\ \gamma\alpha_2(1 - \sigma) & \gamma(\alpha_3\sigma + (1 - \alpha_3)r_2) & 0 & 0 \\ \gamma(1 - \alpha_2)(1 - \rho) & 0 & \gamma(\alpha_4\eta_1 + (1 - \alpha_4)\eta_2) & 0 \\ 0 & \gamma\alpha_3(1 - \sigma) & \gamma(1 - \alpha_4)(1 - \eta_2) & \gamma r_1 \end{pmatrix}.$$

and hence

$$I - T = \begin{pmatrix} 1 - \gamma(\alpha_2\sigma + (1 - \alpha_2)\rho) & 0 & 0 & 0 \\ -\gamma\alpha_2(1 - \sigma) & 1 - \gamma(\alpha_3\sigma + (1 - \alpha_3)r_2) & 0 & 0 \\ -\gamma(1 - \alpha_2)(1 - \rho) & 0 & 1 - \gamma(\alpha_4\eta_1 + (1 - \alpha_4)\eta_2) & 0 \\ 0 & -\gamma\alpha_3(1 - \sigma) & -\gamma(1 - \alpha_4)(1 - \eta_2) & 1 - \gamma r_1 \end{pmatrix}.$$

Due to the structure of the matrix F, only the first 2×2 block of $(I - T)^{-1}$ is important in the next-generation matrix $F(I - T)^{-1}$ for the computation of $\mathcal{R}_0$. Since $I - T$ has a block-triangular

7.5. TWO CASE STUDIES

form, the first 2×2 block of $(I-T)^{-1}$ is obtained from inverting the first 2×2 block of $(I-T)$. Hence,

$$(I-T)^{-1} = \begin{pmatrix} \frac{1}{1-\gamma(\alpha_2\sigma+(1-\alpha_2)\rho)} & 0 & 0 & 0 \\ \frac{\gamma\alpha_2(1-\sigma)}{\Delta} & \frac{1}{1-\gamma(\alpha_3\sigma+(1-\alpha_3)r_2)} & 0 & 0 \\ * & * & * & * \\ * & * & * & * \end{pmatrix}$$

where

$$\Delta := (1-\gamma(\alpha_2\sigma+(1-\alpha_2)\rho))(1-\gamma(\alpha_3\sigma+(1-\alpha_3)r_2)).$$

Consequently, the matrix $F(I-T)^{-1}$ has the relatively simple form,

$$F(I-T)^{-1} = \begin{pmatrix} \frac{f_{11}}{1-\gamma(\alpha_2\sigma+(1-\alpha_2)\rho)} + \frac{f_{12}\gamma\alpha_2(1-\sigma)}{\Delta} & \frac{f_{12}}{1-\gamma(\alpha_3\sigma+(1-\alpha_3)r_2)} & 0 & 0 \\ 0 & 0 & 0 & 0 \\ 0 & 0 & 0 & 0 \\ 0 & 0 & 0 & 0 \end{pmatrix},$$

and whose eigenvalues appear along the diagonal. Thus

$$\mathcal{R}_0 = \rho\left(F(I-T)^{-1}\right) = \frac{-\gamma\alpha_1 q\phi'(0)}{1-\gamma(\alpha_2\sigma+(1-\alpha_2)\rho)} + \frac{-\gamma\alpha_1\phi'(0)\gamma\alpha_2(1-\sigma)}{\Delta}.$$

It follows that the disease-free equilibrium

$$(S, E, I, Q, J, R)^T = \left(\frac{\Lambda}{1-(\gamma\alpha_1+\gamma\rho(1-\alpha_1))}, 0, 0, 0, 0, 0\right)^T$$

is locally asymptotically stable, and there is no SARS outbreak, if $\mathcal{R}_0 < 1$. However, the disease-free equilibrium is unstable and there is a SARS outbreak if $\mathcal{R}_0 > 1$.

Note that the first term appearing in $\mathcal{R}_0$, namely,

$$\frac{-\gamma\alpha_1 q\phi'(0)}{1-\gamma(\alpha_2\sigma+(1-\alpha_2)\rho)},$$

equals the average number of secondary SARS infections produced by an exposed individual, while the second term

$$\frac{-\gamma\alpha_1\phi'(0)\gamma\alpha_2(1-\sigma)}{\Delta}$$

equals the average number of secondary SARS infections produced by an infected individual.

Exercises 7.5

1. Prove that inequality (7.82) implies the equilibrium (7.81) is globally asymptotically stable for initial conditions $S_h(0) > 0$ and $R_h(0) \geq 0$. Show that the quotient in (7.82) is the human reproduction number in the absence of contacts with mosquitoes.

2. Prove that inequality (7.79) implies the equilibrium (7.78) is globally asymptotically stable for initial conditions $S_m(0) > 0$. Show that the quotient in (7.79) is the mosquito reproduction number in the absence of contacts with humans.

3. Prove (7.85).

4. Prove (7.86) and (7.87).

Chapter 8

Evolutionary Models

8.1 Introduction

The models studied in the previous chapters, the $k \times k$ matrix model
$$\mathbf{x}(t+1) = A\left(\mathbf{x}(t)\right)\mathbf{x}(t) \tag{8.1}$$
for structured populations in Chapter 6 and its special $k = 1$ case
$$x(t+1) = r\left(x(t)\right)x(t) \tag{8.2}$$
considered in Chapter 1, are time-autonomous models. In these models coefficients representing birth rates, survival probabilities, competitive and predation interactions, resource consumption rates, and so on are time t-dependent (if they change overtime at all) only through dependence on population density. This feature provides a great deal of mathematical tractability, but it is a simplifying assumption that often does not occur for biological populations. There are many reasons why model coefficients might change over time. For example, a (per capita) birth rate or survival rate might change due to climate change, or periodically oscillate due to seasonal changes, irregularly fluctuating due to environmental stochasticity, and so on. In this chapter, we consider changes in model parameters that result from evolution by natural selection. In Section 8.1.1, we discuss one methodology available for modeling the evolution of model coefficients and provide an introductory example in the following Section 8.1.2. In Section 8.2, we discuss the fundamental bifurcation theorem for a general Darwinian model for structured populations (a generalization of the results for non-evolutionary models in Chapter 6, Section 6.2). In Section 8.3, we give several examples to illustrate the methodology and the application of the theorems in Section 8.2.

8.1.1 Darwinian Dynamic Models

The methodology we use assumes changes in the model equation, coefficients are due to a dependency on a (continuous) phenotypic trait v distributed among the individuals in a population (or, more generally, on a vector of traits $\mathbf{v} = (v_1, ..., v_m)^T$) that is subject to the three Darwinian principles of variability, heritability, and differential fitness [99], [216], [285]. The modeling approach we take, called Darwinian dynamics or evolutionary game theory, closely follows that derived in [298]. In this approach to evolutionary dynamics, each trait v_i is assumed to have a heritable component that is normally distributed in the population with a mean $u_i(t)$ that changes over time due to evolutionary principles. The entries a_{ij} in the projection matrix A in (8.1) (which is simply r in (8.2) are per capita (i.e., individual) rates per unit time that we assume can depend on the individual's trait vector $\mathbf{v}$. Those coefficients that are affected by interactions among individuals will also depend on the traits of other individuals. In Darwinian dynamics, this is modeled by assuming that a_{ij} can also depend on the population mean trait vector $\mathbf{u} = (u_1, ..., u_m)^T$. Thus, we write $a_{ij}(\mathbf{x}, \mathbf{u}, \mathbf{v})$. The method of Darwinian dynamics postulates a coupled eco-evolutionary model with (so-called first order) trait and population dynamics given by the equations

$$
\begin{aligned}
&(a) \quad \mathbf{x}(t+1) = A(\mathbf{x}(t), \mathbf{u}(t), \mathbf{v})|_{\mathbf{v}=\mathbf{u}(t)} \mathbf{x}(t) \\
&(b) \quad \mathbf{u}(t+1) = \mathbf{u}(t) + M \nabla_\mathbf{v} \ln r(\mathbf{x}(t), \mathbf{u}(t), \mathbf{v})|_{\mathbf{v}=\mathbf{u}(t)}
\end{aligned}
\quad (8.3)
$$

where

$$\nabla_\mathbf{v} = (\partial_{v_1}, \partial_{v_2}, ..., \partial_{v_m})^T$$

is the gradient with respect to $\mathbf{v}$,

$$r(\mathbf{x}, \mathbf{u}, \mathbf{v}) := \rho(A(\mathbf{x}, \mathbf{u}, \mathbf{v}))$$

is (under the assumption that A is nonnegative and irreducible [30]) the spectral radius of $A(\mathbf{x}, \mathbf{u}, \mathbf{v})$, and

$$
M = \begin{pmatrix}
\sigma_1^2 & \sigma_{12} & \cdots & \sigma_{1m} \\
\sigma_{12} & \sigma_2^2 & \cdots & \sigma_{2m} \\
\vdots & \vdots & & \vdots \\
\sigma_{1m} & \sigma_{2m} & \cdots & \sigma_m^2
\end{pmatrix}
\quad (8.4)
$$

is an $m \times m$ variance–covariance matrix. The diagonal terms σ_i^2 in M are the variances of the trait v_i in the population and the off-diagonal terms σ_{ij} are their covariances, which in the first-order model are assumed constant.

The equation (8.3b) for the mean trait $\mathbf{u}(t)$ is often called Fisher's or Lande's equation (or the canonical equation of evolution). It states, roughly, that the change in the mean trait is proportional to the fitness gradient (with respect to the individual trait $\mathbf{v}$, not the population mean trait $\mathbf{u}$). Here fitness (of an individual with trait $\mathbf{v}$ within a population with density $\mathbf{x}$ and mean trait $\mathbf{u}$) is taken to be (the exponential population growth rate) $\ln r(\mathbf{x}, \mathbf{u}, \mathbf{v})$. For a justification of this equation and for this choice for fitness see [24], [204], [205], [237], [298].

The model tracks the mean trait within a population when its dynamics are coupled with the population dynamics. This is referred to as trait (or strategy)-driven evolution [298]. Another evolutionary process involves the competition with a similar species (a mutant species) with a different mean trait and whether this mutant can invade and even displace the resident species. This involves the notion of ESS (evolutionarily stable strategy), which we take up in Section 8.1.3.

Recall that $\mathbb{R}_+^k$ and $\mathring{\mathbb{R}}_+^k$ denote the nonnegative and positive cones, respectively, in $\mathbb{R}^k$. In general it is assumed that the scalar-valued model components in (8.3) are continuously differentiable functions $\Omega \times \Upsilon \times \Upsilon \to \mathbb{R}_+$, where Ω is an open set containing $\mathbb{R}_+^k$ and Υ is an open set in $\mathbb{R}^m$. Because the equation for $\mathbf{x}(t)$ in (8.3) is of Kolmogorov-type, $\mathbf{x}(0) \in \mathbb{R}_+^k$ implies $\mathbf{x}(t) \in \mathbb{R}_+^k$ for all $t = 0, 1, 2, \cdots$.

The Equations (8.3) couple the population (ecological) dynamics of $\mathbf{x}$ with the (within-population) evolutionary dynamics of the mean trait $\mathbf{u}$. We will be mostly concerned with the equilibria of Darwinian models (8.3) and their stability properties. We refer to an equilibrium $(\mathbf{x}^*, \mathbf{u}^*)^T$ as an *extinction equilibrium* if $\mathbf{x}^* = \mathbf{0}_k$ and a *positive equilibrium* if $\mathbf{x}^* \in \mathring{\mathbb{R}}_+^k$. Most often

8.1. INTRODUCTION

in applications the projection matrix $A\left(\mathbf{x},\mathbf{u},\mathbf{v}\right)$ is nonnegative and irreducible so that, if $\left(\mathbf{x}^{*},\mathbf{u}^{*}\right)^{T}$ is an equilibrium with nonnegative, nonzero component, i.e., if

$$A\left(\mathbf{x}^{*},\mathbf{u}^{*},\mathbf{u}^{*}\right)\mathbf{x}^{*}=\mathbf{x}^{*}\in\mathbb{R}_{+}^{k}\setminus\{\mathbf{0}_{k}\},$$

then, by known results from Perron–Frobenius theory, 1 is the spectral radius of $A\left(\mathbf{x}^{*},\mathbf{u}^{*},\mathbf{u}^{*}\right)$ and $\mathbf{x}^{*}\in\mathring{\mathbb{R}}_{+}^{k}$. Thus, in this case, the only nonnegative, non-extinction equilibria are positive equilibria.

For an unstructured population $k=1$, the Darwinian model (8.3) takes the form (with the subscript $i=1$ dropped)

$$\begin{array}{l} x\left(t+1\right)=\left.r\left(x\left(t\right),u\left(t\right),v\right)\right|_{v=u(t)}x\left(t\right) \\ \mathbf{u}\left(t+1\right)=\mathbf{u}(t)+M\nabla_{\mathbf{v}}\ln r\left(\mathbf{x}(t),\mathbf{u}(t),\mathbf{v}\right)|_{\mathbf{v}=\mathbf{u}(t)} \end{array} \quad (8.5)$$

for multiple traits and for a single trait

$$\begin{array}{l} x\left(t+1\right)=\left.r\left(x\left(t\right),u\left(t\right),v\right)\right|_{v=u(t)}x\left(t\right) \\ u\left(t+1\right)=u\left(t\right)+\sigma^{2}\left.\partial_{v}\ln r\left(x\left(t\right),u\left(t\right),v\right)\right|_{v=u(t)} \end{array}, \quad (8.6)$$

where the coefficient $\sigma^{2}\geq 0$ is often called the speed of evolution. (As in Chapter 6, the symbol ∂_{v} denotes partial differentiation with respect to v.) The trait equation is sometimes called Fisher's fundamental theorem of natural selection [138], Lande's equation [204], the canonical equation of adaptive dynamics [111], or the breeder's equation [229].

8.1.2 An Example: A Darwinian Logistic (Beverton–Holt) Model

Life history strategies are usually driven by trade-offs [263], [286]. One important trade-off concerns the allocation of resources to reproductive effort versus post-reproductive survival in that higher allocation to reproduction leads to lower post-reproduction survival and the opportunity for further reproduction. Suppose f is the fraction of an available resource a consumed by an individual that is allocated to reproduction. Then the density-free birth rate is bfa, where b is the number of births per unit time per unit resource. Assume the post-reproduction survival rate is proportional to the fraction $1-f$ of resources not allocated to reproduction. We assume a is constant in time and (without loss in mathematical generality) choose resource units so that $a=1$. From our basic modeling methodology presented in Chapter 1, for a closed, unstructured population $x(t)$, the census at time $t+1$ consists of newborns plus survivors:

$$x(t+1)=bfs_{n}\beta\left(x\left(t\right)\right)x(t)+s_{a}\left(1-f\right)\sigma\left(x(t)\right)x(t),$$

where s_{n} and s_{a} are the density-free survival rates of newborns and adults, respectively, and $\beta\left(x\right)$ and $\sigma\left(x\right)$ are the density effects on newborn and adult survival. If, following Leslie [210], we assume both newborns and adults are subject to the same effects of density on their survival and, specifically, that these density effects are described by a factor (see Example 89 in Chapter 6)

$$\beta\left(x\right)=\sigma\left(s\right)=\frac{1}{1+cx},\ c>0,$$

then the result is a discrete logistic (Beverton–Holt) model population model (Example 2) in Chapter 1)

$$x(t+1)=\left[bfs_{n}+s_{a}\left(1-f\right)\right]\frac{1}{1+cx\left(t\right)}x\left(t\right) \quad (8.7)$$

$$b, c > 0, \quad 0 \leq f, s_a, s_n < 1.$$

In this example, we assume f and c are subject to evolutionary change.

Specifically, we assume the fraction of resources an individual is able to allocate toward reproduction is dependent on a trait v and write $f = f(v)$. We assume the fraction is maximized at some value of the trait, which we choose as the reference point used for the scale on which v is measured. Without loss in generality, we assume is f is maximized at $v = 0$. We also assume that the trait determines the competitive prowess of the individual and we write $c = c(v - u)$, which implies the competitive intensity experienced by the individual is determined by how much its trait differs from a typical individual's trait in the population. We assume competition intensity is greatest when the individual has the mean trait, i.e., that $c(z)$ has a maximum at $z = 0$. For this example, we take

$$\begin{aligned} f(v) &= \exp\left(-w_0 v^2\right), \quad w_0 > 0 \\ c(v - u) &= c_0 \exp\left(-w_1 (v - u)^2\right), \quad c_0, w_1 > 0. \end{aligned} \tag{8.8}$$

Then

$$r(x, u, v) = [b s_n f(v) + s_a (1 - f(v))] \frac{1}{1 + c(v - u)x}$$

and the result is a Darwinian dynamic version of the discrete logistic (Beverton–Holt) equation:

$$x(t+1) = \left[(b s_n - s_a) \exp\left(-w_0 u^2(t)\right) + s_a\right] \frac{1}{1 + c_0 x(t)} x(t), \tag{8.9}$$

$$u(t+1) = g\left(u^2(t)\right) u(t), \tag{8.10}$$

where

$$g(z) := 1 - 2\sigma^2 w_0 \frac{(b s_n - s_a) \exp(-w_0 z)}{(b s_n - s_a) \exp(-w_0 z) + s_a}.$$

Note that $b s_n \leq s_a$ and Equation (8.9) imply $0 \leq x(t+1) \leq s_a x(t)$ and hence $\lim_{t \to \infty} x(t) \to 0$ for all initial conditions $x(0) \geq 0$. In this case, the population goes extinct regardless of the trait dynamics. Therefore, we assume from now on that

$$b s_n > s_a. \tag{8.11}$$

The equilibrium equations associated with Equations (8.9)–(8.10) are

$$x = \left[(b s_n - s_a) \exp\left(-w_0 u^2\right) + s_a\right] \frac{1}{1 + c_0 x} x \tag{8.12}$$

$$u = g\left(u^2\right) u. \tag{8.13}$$

Since assumption (8.11) implies $g(z) < 1$ for all z, we see from Equation (8.13) $u = 0$. Solving Equation (8.12) for $x \geq 0$ when $u = 0$, we obtain the extinction equilibrium $(x, u)^T = (0, 0)^T$ and the equilibrium

$$\begin{pmatrix} x \\ u \end{pmatrix} = \begin{pmatrix} \frac{b s_n - 1}{c_0} \\ 0 \end{pmatrix}. \tag{8.14}$$

We refer to (8.14) as a *positive equilibrium* when $b s_n > 1$ since then the population component x is positive. Note that this equilibrium collapses to the extinction equilibrium at $b s_n = 1$ and hence this pair of equilibria form a transcritical bifurcation of two equilibrium branches.

8.1. INTRODUCTION

To ascertain the stability properties of these equilibria by means of the linearization principle, we evaluate the Jacobian associated with (8.9)–(8.10)

$$\begin{pmatrix} [(bs_n - s_a)\exp(-w_0 u^2) + s_a]\frac{1}{(1+c_0 x)^2} & * \\ 0 & g(u^2) + 2g'(u^2)u^2 \end{pmatrix}$$

(where the asterisk is an unneeded term) at each equilibrium and calculate its eigenvalues λ_1 and λ_2 (which appear along the diagonal). For the extinction equilibrium, we obtain eigenvalues

$$\lambda_1 = bs_n, \quad \lambda_2 = g(0),$$

and for the positive equilibrium (8.14), we obtain

$$\lambda_1 = \frac{1}{bs_n}, \quad \lambda_2 = g(0),$$

where

$$g(0) = 1 - 2\sigma^2 w_0 \frac{bs_n - s_a}{bs_n} < 1.$$

Define

$$\sigma_0^2 := \frac{1}{w_0}\frac{bs_n}{bs_n - s_a}.$$

Theorem 71 *In addition to* (8.11) *suppose*

$$\sigma^2 < \sigma_0^2. \tag{8.15}$$

The extinction equilibrium of (8.9)–(8.10) *is globally asymptotically stable on* $\mathbb{R}_+ \times \mathbb{R}$ *if* $bs_n < 1$. *When* $bs_n > 1$ *the extinction equilibrium is unstable and the positive equilibrium* (8.14) *is globally asymptotically stable on* $\mathring{\mathbb{R}}_+ \times \mathbb{R}$.

Proof. The inequalities (8.11) and (8.15) imply $-1 < g(0) < 1$ and hence $|\lambda_2| < 1$ for both equilibria. It follows from the linearization principle that the extinction equilibrium is locally asymptotically stable if $bs_n < 1$ and unstable $bs_n > 1$ and that the positive equilibrium (8.14) is locally asymptotically stable when it exists, i.e., when $bs_n > 1$.

By squaring both sides of the trait Equation (8.10) and letting $y(t) := u^2(t)$ we obtain

$$y(t+1) = g^2(y(t))y(t). \tag{8.16}$$

A straightforward calculation shows $g'(z) > 0$ for $z \geq 0$ and since $-1 < g(0)$ (by (8.15)) and $g(z) < 1$ for all $z \geq 0$, it follows that $-1 < g(z) < 1$, Thus $g^2(z) < 1$ for all $z \geq 0$ and Equation (8.16) implies, for all $y(0) \geq 0$, that $0 \leq y(t+1) \leq y(t)$. Since $y(t)$ is a nonnegative decreasing sequence, it approaches a limit $y^* \geq 0$. Taking limits of both sides of (8.16) as $t \to \infty$, we get $y^* = g^2(y^*)y^*$ which implies $y^* = 0$. It follows that $\lim_{t\to\infty} u(t) = \lim_{t\to\infty} \sqrt{y(t)} = 0$ for any solution $(x(t), u(t))^T$ of (8.9)–(8.10). The $x(t)$ component satisfies Equation (8.9), which is an asymptotically autonomous difference equation whose limiting equation is the discrete logistic (Beverton–Holt) equation:

$$x(t+1) = bs_n \frac{1}{1 + c_0 x(t)} x(t)$$

and whose global dynamics we know from Chapter 1. Namely, $x = 0$ is globally asymptotically stable on $\mathbb{R}_+$ if $bs_n < 1$ and $x = (bs_n - 1)/c_0$ is globally asymptotically stable on $\mathring{\mathbb{R}}_+$ if $bs_n > 1$. It follows from theorems on asymptotically autonomous equations that the solution of the population equation (8.9) satisfies the same alternatives [65], [98], [241], (see Theorem (46) in Section 8.4). ∎

Theorem 71 shows that when evolution is not too fast, i.e., $\sigma^2 < \sigma_0^2$, the model equations (8.9)–(8.10) possess properties similar to those of the fundamental bifurcation theorem for non-evolutionary models in Chapter 6, namely that stable positive equilibria are created when the extinction equilibrium destabilizes by a forward bifurcation. It is a simple example of the general theorem for Darwinian models in Section 8.2.

If evolution is too fast, i.e., $\sigma^2 > \sigma_0^2$, then $\lambda_2 < -1$ for both equilibria and consequently both are unstable. We will restrict attention in this chapter to slow evolution. (See problem 1 in Exercise 8.3.)

8.1.3 Evolutionary Stable Strategies

The Darwinian equations (8.3) model the evolutionary adaptation of a trait (synonymously a strategy) within a population (species), as it is coupled with its population dynamics. If $(\mathbf{x}^*, \mathbf{u}^*)^T$ is a (locally asymptotically) stable positive equilibrium, then in this framework within a population (intraspecific) evolutionary adaptation of traits, one might say that evolution has favored or selected for the mean trait $\mathbf{u}^*$. However, there is a broader notion of evolutionary stability that is based on the population being immune to invasion by other similar species (or mutants), in which case $\mathbf{u}^*$ is said to be an *evolutionary stable strategy* (ESS). This notion plays a central role in theories of speciation, a complicated topic that involves many nuances (sympatric and disruptive selection, allopatric and geographical barriers, etc.) into which we will not delve in this book, except to point out the concept called the *adaptive* (or *fitness*) *landscape* plays a fundamental role in these theories. If $(\mathbf{x}(t), \mathbf{u}(t))$ is a solution of the Darwinian equations (8.3), then the associated *adaptive landscape* at time t is the function $\ln r(\mathbf{x}(t), \mathbf{u}(t), \mathbf{v})$ of $\mathbf{v}$. If $(\mathbf{x}^*, \mathbf{u}^*)^T$ is a (locally asymptotically) stable positive equilibrium, then the associated *adaptive landscape at equilibrium* is the function

$$L(\mathbf{v}) := \ln r(\mathbf{x}^*, \mathbf{u}^*, \mathbf{v}).$$

Since the trait equilibrium equation implies $\mathbf{v} = \mathbf{u}^*$ is a critical point of $L(\mathbf{v})$, the equilibrium component $\mathbf{u}^*$ could be located at one of four possibilities: a global maximum, a local maximum, a minimum, or a saddle of the landscape. The *ESS Maximum Principle* asserts that $\mathbf{u}^*$ *is an ESS if and only if* $(\mathbf{x}^*, \mathbf{u}^*)^T$ *is (locally asymptotically) stable and* u^* *is located at a global maximum on the landscape* $L(\mathbf{v})$ *at equilibrium* [39], [298].

In as much as the trait equation in (8.3) implies $\mathbf{u}(t)$ moves in an uphill direction on the landscape $\ln r(\mathbf{x}(t), \mathbf{u}(t), \mathbf{v})$ at each point in time t, it might seem that $\mathbf{u}(t)$ would approach $\mathbf{u}^*$ located at a maximum. However, since the landscape changes with time t, this is not necessarily true, as we will see in Example 105. If $\mathbf{u}^*$ is not located at a global maximum, then it is not an ESS and is not immune to invasion by mutants and $\mathbf{u}^*$, even though it is the trait component of a stable equilibrium of the Darwinian equation (8.3). If $\mathbf{u}^*$ is at a minimum or a saddle, then it is typically located between two maxima and disturbances to nearby, mean traits can evolve into equilibria with different trait components located at these maxima (an example of sympatric speciation). If $\mathbf{u}^*$ is at a local maximum, then a mutant with sufficiently distinct traits near a global maximum can invade (by the creation of a separating barrier for a long period of time followed by a reuniting of the populations when the barrier is removed, called allopatric speciation).

The broad field of adaptive dynamics studies these speciation possibilities, usually by making use of the adaptive landscape as a fundamental tool and often using a multi-time scale analysis by separating long evolution time scales from short population dynamic time scales (a method that does not seem to have been rigorously studied mathematically for discrete-time models) [107], [298]. For a discussion of the complexities involved in the study of evolution and the role of adaptive dynamics as a means of studying them, see [3], [39] and the references therein.

Example 105 *An individual is semelparous if it has one reproductive episode after which it dies. In a 1954 paper, L. C. Cole argued that a semelparous life history strategy should be favored by evolution [53]. This assertion became known as Cole's Paradox since interoparity (a life history strategy that includes more than one reproductive episode) is so common in natural populations. By the results in Example 8.1.2, the Darwinian dynamic model (8.9)–(8.10) seemingly supports this conclusion since the only positive equilibrium (8.14) has trait component $u^* = 0$, which implies $f(u^*) = 0$ and a semelparous life history for the mean population trait. However, is $u^* = 0$ an ESS for the positive equilibrium (8.14)? If not, then the semelparous equilibrium can be invaded by mutants with nonzero (iteroparous) mean traits. When $bs_n > 1$ does the adaptive landscape at the positive equilibrium*

$$L(v) = \ln r\left(\frac{bs_n - 1}{c_0}, 0, v\right)$$
$$= \ln\left(\frac{s_a + (bs_n - s_a)\exp(-w_0 v^2)}{1 + (bs_n - 1)\exp(-w_1 v^2)}\right)$$

have a global maximum at $v = 0$? Calculations show

$$L(0) = 0, \quad L'(0) = 0, \quad L''(0) = 2w_0 \frac{bs_n - 1}{bs_n}\left(\frac{w_1}{w_0} - \frac{bs_n - s_a}{bs_n - 1}\right)$$

and that $v = 0$ is a minimum if

$$\frac{w_1}{w_0} > \frac{bs_n - s_a}{bs_n - 1}. \tag{8.17}$$

In this case, the equilibrium (8.14) is not an ESS, even though the equilibrium is globally stable (Figures 8.1 and 8.2).

Note that inequality (8.17) does not hold near for $bs_n > 1$ near 1 where the landscape is approximately $\ln(s_a + (1 - s_a)\exp(-w_0 v^2))$, which has a global maximum at $v = 0$. Thus, the trait component of the positive equilibrium is an ESS for $bs_n > 1$ near 1, but as bs_n increases it can become a non-ESS trait, depending on the ratio w_1/w_0, specifically, if $w_1/w_0 > 1$. In this case, the model (8.9)–(8.10) supports Cole's assertion for $bs_n > 1$ near 1 but does not support it for larger bs_n when $w_1/w_0 > 1$ (i.e., when the distribution of $f(v)$ is broader than that of $c(z)$). This scenario is illustrated by numerical examples in Figure 8.1. Figure 8.2 shows the evolving landscape and how the equilibrium trait component can arrive at a minimum on the adaptive landscape.

8.2 The Fundamental Bifurcation Theorem for Darwinian Models

The tractability of the Darwinian model in Example 105, made possible primarily because of the availability of formulas for its equilibria and the uncoupling of the trait equation allowed

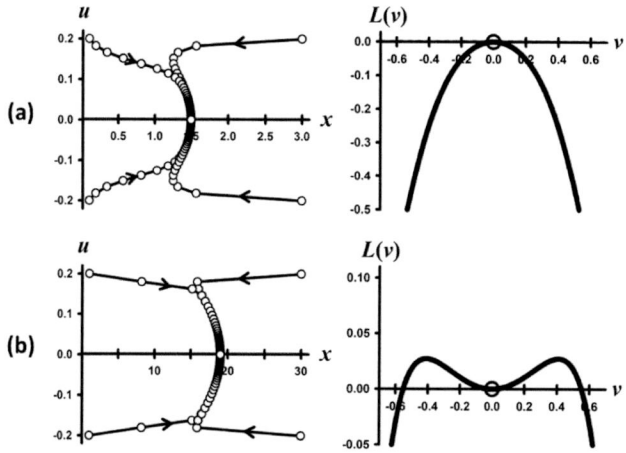

Figure 8.1: Phase plane orbits and adaptive landscapes are shown for two examples of (8.9) and (8.10) with parameter values $w_0 = 5$, $w_1 = 5.5$, $c_0 = 1$, $s_a = 0.25$, $s_n = 0.5$, and $\sigma^2 = 0.01$. (a) $b = 5$: four typical orbits in the x, u-phase plane approach the equilibrium $(x^*, u^*)^T = (1.5, 0)^T$. The adaptive landscape $L(v)$ has a global maximum at $v = 0$ and the trait component of the positive equilibrium is an ESS. In this case, inequality (8.17) fails. (b) $b = 40$: four typical orbits in the x, u-phase plane approach the positive equilibrium $(x^*, u^*)^T = (19.0, 0)^T$. The adaptive landscape $L(v)$ has a local minimum at $v = 0$ and the trait component of the positive equilibrium is not an ESS. In this case inequality (8.17) holds. Figure reproduced from [81] with permission.

8.2. THE FUNDAMENTAL BIFURCATION THEOREM FOR DARWINIAN MODELS

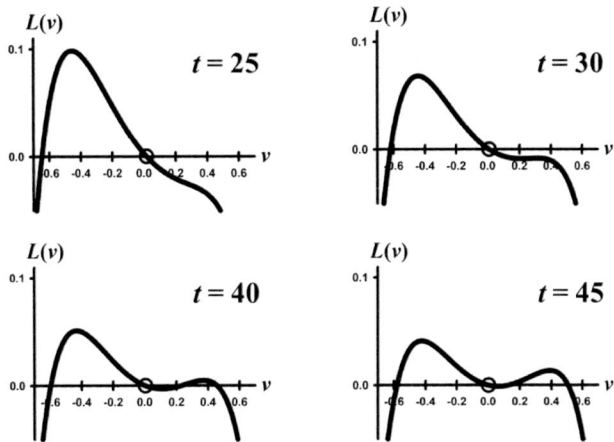

Figure 8.2: Selected time snapshots of the adaptive landscape are shown for the orbit $(x(t), u(t))^T$ with initial condition $(x(0), u(0))^T = (30, 0.2)^T$ in the simulation shown in Figure 8.1(b). The trait component $u(t)$ moves to the left, uphill toward the peak on the left (not easily perceptible on the scale shown), but nonetheless asymptotically arrives at a local minimum on the equilibrium landscape (shown in Figure 8.1(b)). Figure reproduced from [81] with permission.

for a complete global analysis of the dynamics. These features are not typical in Darwinian models, however, and such a complete analysis is generally quite difficult. One feature exhibited in Example 105 that typically occurs in general models (8.3) is the transcritical bifurcation of positive equilibria that occurs at the destabilization of the extinction equilibrium when a model parameter is manipulated (in Example 105 when, for example, b is increased through $1/s_n$). This basic phenomenon is studied in this section where extensions of results for nonlinear matrix models given in Chapter 6, Section 6.2 are made to Darwinian matrix models (8.3).

In this section, we return to the general Darwinian model (8.3) and consider some basic equilibrium properties in the spirit of that done in Chapter 6 for non-evolutionary models. We assume

A1: $A(\mathbf{x}, \mathbf{u}, \mathbf{v}) = (a_{ij}(\mathbf{x}, \mathbf{u}, \mathbf{v}))$ is primitive for $(\mathbf{x}, \mathbf{u}, \mathbf{v}) \in \Omega \times \Upsilon \times \Upsilon$ with entries $a_{ij}(\mathbf{x}, \mathbf{u}, \mathbf{v}) \in C^2(\Omega \times \Upsilon \times \Upsilon \to \mathbb{R}_+)$ and $a_{ij}(\mathbf{0}_k, \mathbf{u}, \mathbf{v})$ is independent of $\mathbf{u}$, i.e., $\nabla_{\mathbf{u}} a_{ij}(\mathbf{0}_k, \mathbf{u}, \mathbf{v}) \equiv 0$ for all $(\mathbf{u}, \mathbf{v}) \in \Upsilon \times \Upsilon$. The variance–covariance matrix M is invertible and diagonally dominant (i.e., $\sigma_i^2 := \sigma_{ii} \geq \sum_{i \neq j} |\sigma_{ij}|$),

where Ω is an open set containing $\mathbb{R}_+^k$ and Υ is an open set in $\mathbb{R}^m$. The diagonally dominant assumption on M places little restriction on the applicability of the model since interest is generally in traits that have little covariance; indeed often it is assumed $\sigma_{ij} = 0$. Entries in A are per-capita vital rates of an individual with trait $\mathbf{v}$ in a population of density $\mathbf{x}$ with mean trait $\mathbf{u}$.

Remark 21 *The restriction on $a_{ij}(\mathbf{0}_k, \mathbf{u}, \mathbf{v})$ in A1 expresses the fact that, in the absence of any density effects, these rates would naturally not depend on the population mean, but only on the trait*

v of the individual. This assumption implies $\partial_{u_i} r(\mathbf{0}_k, \mathbf{u}, \mathbf{v}) \equiv 0$ for all $(\mathbf{u}, \mathbf{v}) \in \Upsilon \times \Upsilon$ and any component u_i of $\mathbf{u}$ where

$$r(\mathbf{x}, \mathbf{u}, \mathbf{v}) := \rho(A(\mathbf{x}, \mathbf{u}, \mathbf{v})) > 0$$

is the dominant eigenvalue of $A(\mathbf{x}, \mathbf{u}, \mathbf{v})$.

The equilibrium equations associated with (8.3) are

$$\begin{aligned}(a) \quad & \mathbf{x} = A(\mathbf{x}, \mathbf{u}, \mathbf{v})|_{\mathbf{v}=\mathbf{u}} \mathbf{x} \\ (b) \quad & \mathbf{0}_m = \nabla_\mathbf{v} \ln r(\mathbf{x}(t), \mathbf{u}(t), \mathbf{v})|_{\mathbf{v}=\mathbf{u}(t)}\end{aligned}$$

or, making use of $\nabla_\mathbf{v} \ln r = r^{-1} \nabla_\mathbf{v} r$,

$$\begin{aligned}(a) \quad & \mathbf{x} = A(\mathbf{x}, \mathbf{u}, \mathbf{v})|_{\mathbf{v}=\mathbf{u}} \mathbf{x} \\ (b) \quad & \mathbf{0}_m = \nabla_\mathbf{v} r(\mathbf{x}(t), \mathbf{u}(t), \mathbf{v})|_{\mathbf{v}=\mathbf{u}(t)}.\end{aligned}$$

Definition 32 $\mathbf{u}^* \in \Upsilon$ *is a critical trait if* $\nabla_\mathbf{v} r(\mathbf{0}_k, \mathbf{u}^*, \mathbf{v})|_{\mathbf{v}=\mathbf{u}^*} = 0$.

The following lemma is trivial.

Lemma 13 $(\mathbf{x}, \mathbf{u})^T = (\mathbf{0}_k, \mathbf{u}^*)$ *is an equilibrium of Equation (8.3) if and only if* $\mathbf{u}^*$ *is a critical trait.*

We assume

A2: there exists a critical trait $\mathbf{u}^* \in \Upsilon$.

The Jacobian associated with Equation (8.3), when evaluated at the extinction equilibrium $(\mathbf{x}, \mathbf{u})^T = (\mathbf{0}_k, \mathbf{u}^*)$, has the block triangular form

$$\begin{pmatrix} A(\mathbf{0}_k, \mathbf{u}^*, \mathbf{v})|_{\mathbf{v}=\mathbf{u}^*} & \mathbf{0}_{m\times k} \\ * & I_{m\times m} + VH(\mathbf{0}_k, \mathbf{u}^*, \mathbf{v})|_{\mathbf{v}=\mathbf{u}^*} \end{pmatrix},$$

where "*" denotes an unneeded block matrix, $\mathbf{0}_{m\times k}$ is the $m \times k$ matrix of zeros, and

$$H(\mathbf{0}_k, \mathbf{u}, \mathbf{v})|_{\mathbf{v}=\mathbf{u}} = \left(\partial_{v_i v_j} r(\mathbf{0}_k, \mathbf{u}, \mathbf{v})|_{\mathbf{v}=\mathbf{u}}\right)$$

(here use is made of Remark 21). Clearly, the eigenvalues of the Jacobian are the eigenvalues of the two diagonal blocks. If all of these eigenvalues λ satisfy $|\lambda| < 1$, then the extinction equilibrium is (locally asymptotically) stable by the Linearization Principle. If at least one eigenvalue satisfies $|\lambda| > 1$, then the extinction equilibrium is unstable. Thus, if $r^* > 1$ where

$$r^* := r(\mathbf{0}_k, \mathbf{u}^*, \mathbf{u}^*),$$

then the extinction equilibrium is unstable. If, on the other hand, $r^* < 1$, then we must consider the eigenvalues of the lower right-hand block in the Jacobian.

Theorem 72 *Assume A1 and A2 hold. Define*

$$H^* := H(\mathbf{0}_k, \mathbf{u}^*, \mathbf{v})|_{\mathbf{v}=\mathbf{u}^*}.$$

(a) If $\rho[I_{m\times m} + VH^*] < 1$, then the extinction equilibrium is (locally asymptotically) stable if $r^* < 1$ and unstable if $r^* > 1$.

(b) If $\rho[I_{m\times m} + VH^*] > 1$, then the extinction equilibrium is unstable for all $r^* > 0$.

8.2. THE FUNDAMENTAL BIFURCATION THEOREM FOR DARWINIAN MODELS

It is often assumed that the covariances σ_{ij} in the variance–covariance matrix V among the traits are equal to 0 or in any case is small. It is proved in [296] (also see [294]) that, if H^* is negative definite and V is diagonally dominant (i.e., $\sigma_i^2 \geq \Sigma |\sigma_{ij}|$), then $\rho[I + VH^*] < 1$ when the variances σ_i^2 are sufficiently small (i.e., evolution is not too fast).

Corollary 11 *Assume A1 and A2 hold. If H^* is negative definite and the variances σ_i^2 in V are small, then the extinction equilibrium $(\mathbf{x}, \mathbf{u})^T = (\mathbf{0}_k, \mathbf{u}^*)^T$ is unstable for $r^* < 1$ and (locally asymptotically) stable if $r^* > 1$.*

Viewing r^* as a bifurcation parameter, we see that when H^* is negative definite the extinction equilibrium destabilizes as r^* increases through 1. Based on the fundamental bifurcation theorems in Chapter 6, Section 6.2, for non-evolutionary matrix models, we anticipate that a transcritical bifurcation will occur at the extinction equilibrium when $r^* = 1$ and that the result will be the creation of a continuum of positive equilibria. If $(\mathbf{x}, \mathbf{u})^T$ is a positive equilibrium associated with a value of r^*, we refer to $\left(r^*, (\mathbf{x}, \mathbf{u})^T\right)$ as a *positive equilibrium pair*. An equilibrium pair $\left(r^*, (\mathbf{0}_k, \mathbf{u}^*)^T\right)$ is called an *extinction equilibrium pair*. By a *stable equilibrium pair* $\left(r^*, (\mathbf{x}, \mathbf{u})^T\right)$, we mean one in which $(\mathbf{x}, \mathbf{u})^T$ is (locally asymptotically) stable equilibrium of (8.3). A crucial diagnostic quantity is needed to describe the occurrence and nature of this bifurcation is the quantity

$$\kappa := -\mathbf{w}_L^T \left[\nabla_\mathbf{x}^0 a_{ij} \mathbf{w}_R\right] \mathbf{w}_R, \tag{8.18}$$

where $\mathbf{w}_L^T$ and $\mathbf{w}_R^T$ are left and right, positive eigenvectors of $A(\mathbf{0}_k, \mathbf{u}^*, \mathbf{v})|_{\mathbf{v}=\mathbf{u}^*}$ associated with $r^* = 1$ and the superscript "0" denotes evaluation at the bifurcation point $\left(r^*, (\mathbf{x}, \mathbf{u})^T\right) = \left(1, (\mathbf{0}_k, \mathbf{u}^*)^T\right)$. (See (6.22) in Section 6.2 of Chapter 6) The following theorem is proved in [93] (see Corollary 3).

Theorem 73 *Assume A1 and A2 hold. Assume further that H^* is nonsingular and $\kappa \neq 0$. Then a continuum of positive equilibrium pairs $\left(r^*, (\mathbf{x}, \mathbf{u})^T\right)$ bifurcates from the extinction equilibrium $(\mathbf{0}_k, \mathbf{u}^*)^T$ at $r^* = 1$. If the variances σ_i^2 in V are small, then the following alternatives hold.*

(a) If H^* is negative definite, then $\kappa > 0$ implies the bifurcation is forward and stable and $\kappa < 0$ implies the bifurcation is backward and unstable.

(b) If H^* is semi-definite or indefinite, then the bifurcation is unstable regardless of its direction.

In Theorem 73 by a forward (or backward) bifurcation, it is meant that, in a neighborhood of the bifurcation point, $r^* > 1$ (respectively, $r^* < 1$) for positive equilibrium pairs on the bifurcating continuum. By a stable (unstable) bifurcation, it is meant that the bifurcating positive equilibria are stable (respectively, unstable). The the direction of bifurcation and hence, by Theorem 73, the stability properties of the bifurcating positive equilibria are determined by the sign of κ. The sign of κ is determined by the partial derivatives, with respect to the components of $\mathbf{x}$, of the entries a_{ij} of the projection matrix evaluated at the bifurcation point. It follows that if, in a specific model, all low-level density effects that are present in the model are negative or deleterious (by which it means that an increase in a low-level density x_n causes a decrease the projection matrix entry a_{ij}, i.e., $\partial_{x_n} a_{ij} < 0$), then $\kappa > 0$ and a *forward-stable bifurcation* will occur. Similarly if, in a specific model, all low-level density effects are present in the model are positive (by which is meant that an increase in a low-level density x_n causes an increase the projection matrix entry a_{ij}, i.e., $\partial_{x_n} a_{ij} > 0$), then $\kappa < 0$ and a *backward-unstable bifurcation* will occur. In these cases, which can usually be determined by simple observation of the model equations, there is no need to calculate κ. However, if there is a mix of negative and positive density effects, then κ needs to be calculated in order to determine its sign.

Remark 22 *The bifurcation point is characterized by the equations:*

$$r\left(\mathbf{0}_k, \mathbf{u}^*, \mathbf{v}\right)\big|_{\mathbf{v}=\mathbf{u}^*} = 1 \tag{8.19}$$
$$\nabla_{\mathbf{v}} r\left(\mathbf{0}_k, \mathbf{u}^*, \mathbf{v}\right)\big|_{\mathbf{v}=\mathbf{u}^*} = 0 \ .$$

If the projection matrix $A(\mathbf{x}, \mathbf{u}, \mathbf{v})$ has, for all $(\mathbf{u}, \mathbf{v}) \in \Upsilon \times \Upsilon$ the additive decomposition

$$A(\mathbf{x}, \mathbf{u}, \mathbf{v}) = F(\mathbf{x}, \mathbf{u}, \mathbf{v}) + T(\mathbf{x}, \mathbf{u}, \mathbf{v})$$

that satisfies the requirements H2 in Chapter 6, Section 6.2 for the definition of the reproduction number

$$\mathcal{R}_0 := \rho\left(F\left(\mathbf{0}_k, \mathbf{u}, \mathbf{u}\right)\left(I - TF\left(\mathbf{0}_k, \mathbf{u}, \mathbf{u}\right)\right)^{-1}\right),$$

then we can invoke Theorem 56 [73] to see that Equations (8.19) locating the bifurcation point can be replaced by the equations:

$$\mathcal{R}_0\left(\mathbf{0}_k, \mathbf{u}^*, \mathbf{v}\right)\big|_{\mathbf{v}=\mathbf{u}^*} = 1 \tag{8.20}$$
$$\nabla_{\mathbf{v}} \mathcal{R}_0\left(\mathbf{0}_k, \mathbf{u}^*, \mathbf{v}\right)\big|_{\mathbf{v}=\mathbf{u}^*} = 0$$

and that r^* and the equilibrium pairs $\left(r^*, (\mathbf{x}, \mathbf{u})^T\right)$ in Theorem 73 and Corollary 11 can be replaced by $\mathcal{R}_0^* := \mathcal{R}_0\left(\mathbf{0}_k, \mathbf{u}^*, \mathbf{v}\right)\big|_{\mathbf{v}=\mathbf{u}^*}$ and equilibrium pairs $\left(\mathcal{R}_0^*, (\mathbf{x}, \mathbf{u})^T\right)$. In this way, $\mathcal{R}_0^*$ can be used in place of r^* as a bifurcation parameter. This is useful in the analysis of many models since analytic formulas are often available for $\mathcal{R}_0$, but are rarely available for r.

Corollary 11 and Theorem 73 apply to the Darwinian logistic (Beverton–Holt) model (8.9)-(8.10) in Section 8.1.2 with $k = m = 1$. In that example

$$r(x, u, v) = [bs_n f(v) + s_a (1 - f(v))] \frac{1}{1 + c(v - u)x},$$

with $f(v) = \exp\left(-w_0 v^2\right)$ and $bs_n - s_a > 0$. From

$$\nabla_{\mathbf{v}} r\left(\mathbf{0}_k, \mathbf{u}, \mathbf{v}\right)\big|_{\mathbf{v}=\mathbf{u}} = (bs_n - s_a) f'(u) = -2(bs_n - s_a) w_0 u \exp\left(-w_0 u^2\right),$$

from which we see that $u^* = 0$ is a critical trait. M and H^* are 1×1 matrices with entries $\sigma^2 > 0$ and $-2w_0(bs_n - s_a) < 0$, respectively. Theorem 72 implies the extinction equilibrium $(x, u)^T = (0, 0)^T$ destabilizes as $r^* = bs_n$ increases through 1. Theorem 73 applies since $\kappa > 0$, as can be seen from the fact that $r(x, u, v)$ is a strictly decreasing function of x (or from a straightforward calculation of $\kappa = c_0 > 0$). Thus, a forward-stable bifurcation occurs. This conclusion is consistent with the analysis of this model carried out in Section 8.1.2, which was based on the availability of formulas for equilibria and eigenvalues. Unlike this example, formulas for equilibria are not, however, available or tractable in most models used in the study of population dynamics.

Corollary 11 and Theorem 73 constitute a basic starting point of any model's analysis, especially since they deal with the fundamental biological question of extinction versus survival. The strength of these theorems lies in their generality (and relative ease of application); their weakness is that they deal only with equilibria near the bifurcation point. Studying the dynamics outside a neighborhood of the bifurcation point requires more analysis and analytic techniques, and the resulting dynamic properties are in general heavily dependent on the particular properties of specific models. While there are general theorems that guarantee the global existence of the bifurcating branch of positive equilibria [67], [238], we would expect that, as with non-evolutionary models, secondary bifurcations, and even routes to chaos will commonly occur outside a neighborhood of the bifurcation point.

8.3 Examples

In this section, we give several examples of Darwinian models for structured populations and/or multiple traits and the use of the general theorems and results in Section 8.2.

8.3.1 A Darwinian Juvenile–Adult Model

The juvenile–adult model studied in Example 93 in Section 6.3 of Chapter 6 has a $k = 2$ dimensional state variable

$$\mathbf{x} = \begin{pmatrix} x_1 \\ x_2 \end{pmatrix}$$

of juvenile and adult individuals with the projection matrix

$$A(\mathbf{x}) = \begin{pmatrix} 0 & b\varphi(x_2) \\ s_1 \alpha(x_2) & s_2 \end{pmatrix},$$

$$0 < s_1, s_2 < 1, \ b > 0$$

and $\varphi, \alpha \in C^2(\Omega \to \mathbb{R}_+)$, where Ω is an open sent containing $\mathbb{R}_+^2$ and where $0 \leq s_1\varphi(x_2) \leq 1$ for $x_2 \geq 0$. We also assume

$$\varphi(0) = \alpha(0) = 1$$

so that b and s_1 retain the biological interpretations as the inherent (i.e., density-free) adult birth rate and juvenile survival rate.

Assume b and s_1 possess a trade-off as a function of a trait v subject to natural selection, namely b and s_1 possess opposite monotonicities as functions of v. When b increases, then s_1 decreases which means a high birth rate is correlated with a low newborn viability and vice versa. As an explicit example, we take

$$b(v) = b\frac{v}{1+v}$$

$$s_1(v) = s_1\frac{1}{1+dv}, \quad d > 0$$

for $v \in \Upsilon = \mathring{\mathbb{R}}_+$. These led to the Darwinian equations

$$\begin{pmatrix} x_1(t+1) \\ x_2(t+1) \end{pmatrix} = \begin{pmatrix} 0 & b\frac{v}{1+v}\varphi(x_2(t)) \\ s_1\frac{1}{1+dv}\alpha(x_2(t)) & s_2 \end{pmatrix}\bigg|_{v=u(t)} \begin{pmatrix} x_1(t) \\ x_2(t) \end{pmatrix}, \quad (8.21)$$

$$u(t+1) = u(t) + \sigma^2 \left.\partial_v \ln r(x_1(t), x_2(t), u(t), v)\right|_{v=u(t)}, \quad (8.22)$$

where $r(\mathbf{x}, \mathbf{u}, \mathbf{v})$ is the dominant eigenvalues of the primitive, population projection matrix

$$A(\mathbf{x}, \mathbf{u}, \mathbf{v}) = \begin{pmatrix} 0 & b\frac{v}{1+v}\varphi(x_2) \\ s_1\frac{1}{1+dv}\alpha(x_2) & s_2 \end{pmatrix}.$$

(Although r and A in do not in this example depend on x_1 or u, we retain them in the argument lists in order to be consistent with the notation in Section 8.2.)

A1 is satisfied with $k=2$, $m=1$ since the 1×1 variance–covariance matrix M has entry $\sigma^2 > 0$. To address assumption A2, we make use of Remark 22 and consider the Equations (8.20) with

$$\mathcal{R}_0(\mathbf{x}, \mathbf{u}, \mathbf{v}) = b\frac{s_1}{1-s_2}\frac{v}{1+v}\frac{1}{1+dv}\varphi(x_2)\alpha(x_2)$$

(see (2.16) in Chapter 2). From

$$\partial_v \mathcal{R}_0(\mathbf{x}, \mathbf{u}, \mathbf{v}) = b\frac{s_1}{1-s_2}\frac{1-dv^2}{(1+dv)^2(1+v)^2}\varphi(x_2)\alpha(x_2),$$

we see that A2 is satisfied with the (unique) critical trait

$$u^* = \frac{1}{\sqrt{d}}.$$

At the bifurcation point

$$\mathcal{R}_0(\mathbf{0}_2, \mathbf{u}^*, \mathbf{v})|_{\mathbf{v}=\mathbf{u}^*} = b\frac{s_1}{1-s_2}\frac{1}{\left(\sqrt{d}+1\right)^2} = 1,$$

we have from Theorem 56 [73] that the single entry $\partial_{vv}\ln r(\mathbf{0}_2, \mathbf{u}^*, \mathbf{v})|_{\mathbf{v}=\mathbf{u}^*}$ in the 1×1 Hessian H^* has the same sign as $\partial_{vv}r(\mathbf{0}_2, \mathbf{u}^*, \mathbf{v})|_{\mathbf{v}=\mathbf{u}^*}$ which has the same sign as

$$\partial_{vv}\mathcal{R}_0(\mathbf{0}_2, \mathbf{u}^*, \mathbf{v})|_{\mathbf{v}=\mathbf{u}^*} = -2b\frac{s_1}{1-s_2}\frac{\left(\sqrt{d}\right)^3}{\left(\sqrt{d}+1\right)^4} < 0.$$

Thus, H^* is negative definite and it follows from Corollary 11 that, for $\sigma^2 > 0$ small, the extinction equilibrium $(\mathbf{x}, \mathbf{u})^T = (\mathbf{0}_2, 0)^T$ loses stability as

$$\mathcal{R}_0^* = b\frac{s_1}{1-s_2}\frac{1}{\left(\sqrt{d}+1\right)^2} \tag{8.23}$$

increases through 1 and that a bifurcation of positive equilibria from the extinction equilibrium occurs whose direction and stability depend on the the direction of bifurcation, which is determined by the sign of κ given by formula (8.18). When $\mathcal{R}_0^* = 1$, we find from

$$A(\mathbf{0}_k, \mathbf{u}^*, \mathbf{v})|_{\mathbf{v}=\mathbf{u}^*} = \begin{pmatrix} 0 & \frac{1-s_2}{s_1}\left(\sqrt{d}+1\right) \\ \frac{s_1}{\sqrt{d}+1} & s_2 \end{pmatrix},$$

8.3. EXAMPLES

which has positive eigenvectors

$$\mathbf{w}_R = \begin{pmatrix} (1-s_2)\left(\sqrt{d}+1\right) \\ s_1 \end{pmatrix}, \qquad \mathbf{w}_L = \begin{pmatrix} s_1 \\ \sqrt{d}+1 \end{pmatrix} \tag{8.24}$$

associated with eigenvalue 1. Calculations show that

$$\nabla_{\mathbf{x}}^0 a_{11} = \nabla_{\mathbf{x}}^0 a_{22} = \begin{pmatrix} 0 & 0 \end{pmatrix},$$

$$\nabla_{\mathbf{x}}^0 a_{12} = \begin{pmatrix} 0 & \frac{1-s_2}{s_1}\left(\sqrt{d}+1\right)\varphi'(0) \end{pmatrix}, \quad \nabla_{\mathbf{x}}^0 a_{21} = \begin{pmatrix} 0 & s_1 \frac{1}{\sqrt{d}+1}\alpha'(0) \end{pmatrix},$$

and thus

$$\kappa = -\mathbf{w}_L^T \left[\nabla_{\mathbf{x}}^0 p_{ij} \mathbf{w}_R \right] \mathbf{w}_R = -s_1^2 (1-s_2) \left(\sqrt{d}+1\right) (\varphi'(0) + \alpha'(0)).$$

From Theorem 73(b), we have the following results for the Darwinian model (8.21)–(8.22):

for σ^2 small, the extinction equilibrium $(\mathbf{x}, \mathbf{u})^T = \left(\mathbf{0}_2, 1/\sqrt{d}\right)^T$ loses stability as $\mathcal{R}_0^*$ (given by (8.23)) increases through 1 with the result that a bifurcation of positive equilibria $(\mathbf{x}, \mathbf{u})^T \in \mathring{\mathbb{R}}_+^2 \times \mathring{\mathbb{R}}_+$ occurs that is forward-stable if $\varphi'(0) + \alpha'(0) < 0$ and backward-unstable if $\varphi'(0) + \alpha'(0) > 0$.

The adaptive landscape at any positive equilibrium $(\mathbf{x}^*, u^*)^T$ is $L(v) = \ln r(\mathbf{x}^*, u^*, v)$ where

$$r(\mathbf{x}^*, u^*, v) = \frac{1}{2} s_2 + \frac{1}{2}\sqrt{s_2^2 + 4b \frac{v}{1+v} \frac{s_1}{1+dv} \alpha(x_2)\varphi(x_2)},$$

from which we calculate

$$\partial_v r(\mathbf{x}^*, u^*, v) = b \frac{s_1}{s_2} \frac{1-dv^2}{(dv+1)^2 (v+1)^2} \alpha(x_2)\varphi(x_2).$$

Since $r(\mathbf{x}^*, u^*, v)$ increases for $0 < v < u^*$ and decreases for $v > u^*$, it follows for $v > 0$ that $r(\mathbf{x}^*, u^*, v)$, and hence $L(v)$, has a global maximum at $v = u^*$. This implies that

the mean trait $u^* = 1/\sqrt{d}$ associated with any (locally asymptotically) stable positive equilibrium of the model (8.21)–(8.22) is an ESS.

This includes, of course, the bifurcating, stable positive equilibria guaranteed by Theorem 73(b) when $\varphi'(0) + \alpha'(0) < 0$. Note that in this case the stable bifurcating positive equilibria near bifurcation has a trait value near u^*, which is inversely proportional to $\sqrt{d}$. This means that, for large values of d, the ESS trait yields low post-reproductive survival, i.e., a semelparous-like life history, whereas a small value of d the ESS trait has maximal survival rate s_1, i.e., a iteroparous-like life history.

The inequality $\varphi'(0) + \alpha'(0) < 0$ that implies a forward-stable bifurcation means that the low adult density effects on the birth rate and newborn survival rate overall are together negative (deleterious), even if one of the two sensitivities $\varphi'(0)$ or $\alpha'(0)$ is positive (representing Allee factors). If a positive effect dominates in the sense that $\varphi'(0) + \alpha'(0) > 0$ and a backward-unstable bifurcation of positive equilibria occurs, then, as seen in Chapter 5, Section 5.5, we would reasonably expect the occurrence of a strong Allee effect (i.e., a multiple attractor scenario in which a survival attractor exists simultaneously with a stable extinction equilibrium). In either case, as pointed out in Section 8.2, the existence and stability properties of positive equilibria that are guaranteed by the Theorem 73 are restricted to a neighborhood of the bifurcation point (i.e., near the extinction equilibrium $(\mathbf{x}, u)^T = (\mathbf{0}_2, u^*)^T$ for $\mathcal{R}_0^*$ near 1) and the dynamics outside of this neighborhood is highly dependent on the particular nonlinearities in the model and they require further analysis.

As an example, consider (8.21)–(8.22) with a discrete logistic (Beverton–Holt) negative density effect on adult fertility

$$\varphi(x_2) = \frac{1}{1 + cx_2}$$

(due, for example, to adult competition among themselves for resources) and a low positive density (an Allee factor) effect on juvenile survival (due, for example, to adult protection of juveniles). To describe this positive effect on juvenile survival, we need to select an expression for $\alpha(x_2)$ that is satisfies $\alpha'(0) > 0$ and the required constraints $\alpha(0) = 1$ and $0 \leq s_1\alpha(x_2) \leq 1$ for $x_2 \geq 0$. For the example here, we use the rational function

$$\alpha(x_2) = \frac{1 + ax_2}{1 + as_1x_2}, \quad a > 0.$$

We refer to a as the Allee coefficient because the rate that juvenile survival increases with small x_2, namely $\alpha'(0) = a(1 - s_1)$, is proportional to a. A calculation shows

$$\varphi'(0) + \alpha'(0) = -c + a(1 - s_1).$$

Thus, a forward and stable bifurcation occurs as $\mathcal{R}_0^*$ increases through 1 if the Allee coefficient is not too large, namely $a < a_0$, where

$$a_0 := \frac{c}{1 - s_1}.$$

On the other hand, a backward and unstable bifurcation occurs if $a > a_0$, in which case we expect the occurrence of a strong Allee effect. The example shown in Figure 8.3 corroborates this expectation for the parameter values given in the figure caption. In this example, $\mathcal{R}_0 = 0.716 < 1$ and the extinction equilibrium $(x_1^*, x_2^*, u^*)^T = (0, 0, 2.236)^T$ is locally asymptotically stable, which is illustrated by the solution with initial condition $(x_1, x_2, u)^T = (0, 3, 1)^T$ shown in Figure 8.3(a). However, since $\varphi'(0) + \alpha'(0) = 0.3 > 0$, a backward and unstable bifurcation occurs in this example. The sample solution shown in Figure 8.3(b) with initial condition $(x_1, x_2, u)^T = (0, 10, 1)^T$ tends to a *positive* equilibrium $(x_1^*, x_2^*, u^*)^T = (6.000, 13.74, 2.236)^T$ (rounded to four significant digits). For the parameter values in this example, the 3×3 Jacobian of the system (8.21)–(8.22) evaluated at this positive equilibrium has eigenvalues

$$\lambda_1 = 0.9858, \ \lambda_2 = 0.9477, \ \lambda_3 = -8.888 \times 10^{-2}$$

(to four significant digits) and hence is locally asymptotically stable by the Linearization Principle. It, therefore, follows, as shown above, that the trait component $u^* = 2.236$ is an ESS. See Figure 8.3(b) for a plot of the adaptive landscape with u^* located at the global maximum.

8.3. EXAMPLES

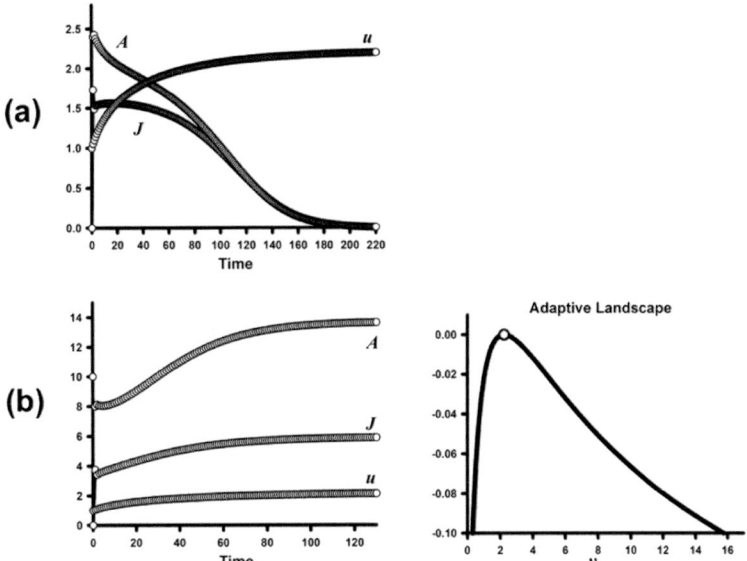

Figure 8.3: The parameter values $b = 1.5$, $s_1 = 0.2$, $s_2 = 0.8$, $c = 0.1$, $a = 0.5$, $d = 0.2$ and $\sigma^2 = 1$ in the model (8.21)–(8.22) yield $\mathcal{R}_0 = 0.716 < 1$, which implies the extinction equilibrium $(0,0,0)^T$ is locally asymptotically stable. This is illustrated by the sample solution with initial condition $(x_1, x_2, u)^T = (0, 3, 1)^T$ shown in (a). On the other hand, the solution with initial condition $(x_1, x_2, u)^T = (0, 10, 1)^T$ tends to a positive equilibrium $(x_1^*, x_2^*, u^*)^T = (6.00, 13.744, 2.236)^T$, as shown in (b). Also shown is the adaptive landscape at equilibrium with the equilibrium trait component (the open circle) located at the global maximum.

8.3.2 A Darwinian LPA Model

A Darwinian version of the LPA model studied in Chapter 6, Section 6.6, was used in [255] to study the evolution of genetic polymorphism in the flour beetle *Tribolium castaneum*. The beetles in the experiments used in the laboratory study was divided into three genotypes based on two possible alleles, *cos* and +, at a single locus: *cos/cos*, *cos/+*, and ++. The *cos* allele confers sensitivity to corn oil that affects fertility and survival rates of individuals in a population grown in an environment with corn oil present (which is known for the three genotypes). In the model used in [255], the trait v is taken to be the frequency of the + allele in the population and in the projection matrix (6.53) of the LPA model, only the density-free coefficients b, μ_l, and μ_a are functions of v (while the coefficients $c_{el}, c_{ea}, c_{pa} > 0$ are constants independent of v). These assumptions yield the projection matrix

$$A(\mathbf{x}, \mathbf{u}, \mathbf{v}) = \begin{pmatrix} 0 & 0 & b(v)\exp(-c_{el}x_1 - c_{ea}x_3) \\ 1 - \mu_l(v) & 0 & 0 \\ 0 & \exp(-c_{pa}x_3) & 1 - \mu_a(v) \end{pmatrix}. \tag{8.25}$$

Specifically, from known mean values of b, μ_l, and μ_a for the three genotypes with $v = 0, 0.5,$ and 1.0, the quadratic fits

$$\begin{aligned} b(v) &= \beta\left(-\tfrac{18}{11}v^2 + \tfrac{21}{11}v + 1\right) \\ \mu_l(v) &= 0.10v^2 - 0.13v + 0.51 \\ \mu_a(v) &= 0.10v^2 - 0.13v + 0.11 \end{aligned} \tag{8.26}$$

were utilized in [255]. Here $\beta > 0$ is the larval recruitment rate per ++ type adult. This produces a $k = 3$-dimensional structured population model (8.3) with a $m = 1$-dimensional trait vector and a 1×1 nonsingular matrix M with entry $\sigma^2 > 0$. From the formula

$$\mathcal{R}_0(\mathbf{x}, \mathbf{u}, \mathbf{v}) = b(v)\frac{1 - \mu_l(v)}{\mu_a(v)}\exp(-c_{el}x_1 - c_{ea}x_3)\exp(-c_{pa}x_3)$$

for the reproduction number, a straightforward differentiation gives

$$\partial_v \mathcal{R}_0(\mathbf{0}_3, \mathbf{u}, \mathbf{v})|_{\mathbf{v}=\mathbf{u}} = \frac{3}{11}\beta\frac{6633 - 8866u - 5675u^2 + 6488u^3 - 3820u^4 + 1200u^5}{(10v^2 - 13v + 11)^2}$$

which has three real roots only one of which lies in the interval $0 \leq u \leq 1$, namely

$$u^* = 0.62415.$$

Then

$$\mathcal{R}_0^* = \mathcal{R}_0(\mathbf{0}_3, u^*, v)|_{v=u^*} = 12.196\beta.$$

From the negativity of

$$\partial_{vv}\mathcal{R}_0(\mathbf{0}_3, \mathbf{u}^*, \mathbf{v})|_{\mathbf{v}=\mathbf{u}^*} = -66.260\beta$$

and the fact that κ is positive (because the density-dependent terms in the projection matrix are all decreasing exponentials of density), we arrive at the following conclusions from Corollary 11 and Theorem 73 for the Darwinian LPA model with projection matrix (8.25) and components (8.26):

> for σ^2 small, the extinction equilibrium $(\mathbf{x}, \mathbf{u})^T = (\mathbf{0}_3, u^*)^T$ loses stability as $\mathcal{R}_0^* = 12.196\beta$ increases through 1 with the result that a forward-stable bifurcation of positive equilibria $(\mathbf{x}, u)^T$ occurs.

8.3. EXAMPLES

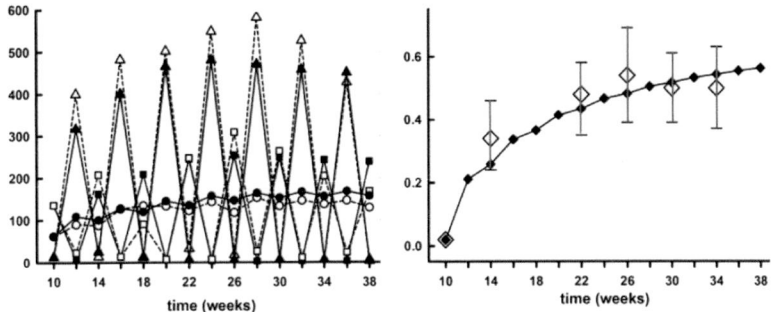

Figure 8.4: The black symbols in the left and right graphs show the time series of $\mathbf{x}(t) = (x_1, x_2, x_3)^T$ and $u(t)$ for the solution of the Darwinian LPA model with projection matrix (8.25) and trait-dependent components (8.26) with parameter values $c_{pa} = 0.015$, $c_{el} = 0.0093$, $c_{ea} = 0.011$ and $\sigma^2 = 0.40$. The initial condition is $\mathbf{x}(0) = (13, 136, 62)^T$ and $u(0) = 0.02$. In the left graph, the triangles, squares, and circles are larvae $x_1(t)$, pupae $x_2(t)$, and adults $x_2(t)$, respectively. The open symbols in the left graph are data from one replicate of the experiment reported in [255] and the open diamonds in the right graph are the averaged +allele frequencies over all replicates (with error bars).

Recall that this result is restricted to a neighborhood of the bifurcation point. The (replicated & controlled) experiments reported in [255] were carried out for a population with $\beta = 11$ or $\mathcal{R}_0^* = 134.16$, which is not close to 1. Indeed, numerical simulations of the Darwinian model indicate that the positive equilibrium is not stable for this value of $\mathcal{R}_0^*$, but instead there is an attracting 2-cycle. See Figure 8.4. This suggests that a period-doubling bifurcation occurs as $\mathcal{R}_0^*$ increases, a rigorous justification for which remains to be provided.

8.3.3 A Multiple Trait Discrete Logistic (Beverton–Holt) Model

The equations

$$\begin{aligned}
x(t+1) &= b_0 \exp\left(-\frac{u_1(t)}{2w_1^2} - \frac{u_2(t)}{2w_2^2}\right) \frac{x(t)}{1+c_0 x(t)} \\
u_1(t+1) &= u_1(t) + \sigma_1^2 \left(-\frac{u_1(t)}{w_1^2} + c_1 c_0 \frac{x(t)}{1+c_0 x(t)}\right) \\
u_2(t+1) &= u_2(t) + \sigma_2^2 \left(-\frac{u_2(t)}{w_1^2} + c_2 c_0 \frac{x(t)}{1+c_0 x(t)}\right).
\end{aligned} \quad (8.27)$$

describe a Darwinian logistic (Beverton–Holt) model obtained from Equation (8.7) with $s_a = 0$ and $b_0 := bs_n > 0$ and with reproduction allocation fraction f and competition coefficient c that are dependent on $m = 2$ traits, as given by

$$f(\mathbf{v}) = \exp\left(-\frac{v_1^2}{2w_1^2} - \frac{v_2^2}{2w_2^2}\right),$$

$$c(\mathbf{v} - \mathbf{u}) = c_0 \exp\left(-c_1(v_1 - u_1) - c_2(v_2 - u_2)\right)$$

with

$$w_i > 0, \quad c_0 > 0, \quad c_1, c_2 \geq 0.$$

In this model, the maximum birth rate $b_0 f(\mathbf{v})$ is attained by individuals with trait pair $\mathbf{v} = \mathbf{0}_2$. In this example, the competition coefficient $c(\mathbf{v} - \mathbf{u})$ expresses a of birth rate density dependence. Individuals with larger traits v_i suffer less density-related loss in fertility. The coefficient c_0 is the density coefficient for individuals with traits equal to the population mean.

Finally, we take the variance/covariance matrix to be

$$\mathbf{V} = \begin{pmatrix} \sigma_1^2 & 0 \\ 0 & \sigma_2^2 \end{pmatrix}, \sigma_i^2 > 0, \tag{8.28}$$

i.e., we assume there is no covariance between the two traits.

From

$$r(x, \mathbf{u}, \mathbf{v}) = b_0 \exp\left(-\frac{v_1^2}{2w_1^2} - \frac{v_2^2}{2w_2^2}\right) \frac{1}{1 + c(\mathbf{v} - \mathbf{u})x}, \tag{8.29}$$

a calculation shows

$$\nabla_{\mathbf{v}} r(0, \mathbf{u}, \mathbf{v})|_{\mathbf{v}=\mathbf{u}} = -b_0 \exp\left(-\frac{u_1^2}{2w_1^2} - \frac{u_2^2}{2w_2^2}\right) \begin{pmatrix} \frac{u_1}{w_1^2} \\ \frac{u_2}{w_2^2} \end{pmatrix}$$

and hence the only critical trait $\mathbf{u}^* \in \Upsilon = \mathbb{R}^2$ is $\mathbf{u}^* = \mathbf{0}_2$. The matrix

$$H^* = \begin{pmatrix} -\frac{1}{w_1^2} & 0 \\ 0 & -\frac{1}{w_2^2} \end{pmatrix},$$

is invertible and negative definite. Since the only density factor in the the discrete logistic (Beverton–Holt) equation is a decreasing function of x, we know from formula (8.18) that $\kappa > 0$ (or one can calculate $\kappa = c_0 > 0$ using the formula).

Corollary 11 implies the extinction equilibrium $(x, \mathbf{u})^T = (0, \mathbf{0}_2)^T$ destabilizes as $r_0 = R_0 = b_0$ increases through 1 and Theorem 73 implies, for sufficiently small $\sigma_i^2 > 0$, that the resulting bifurcation of positive equilibria is forward and stable.

Sample simulations in Figure 8.5 illustrate this bifurcation.

8.3.4 Non-equilibrium Dynamics: A Darwinian Ricker Model

The focus in the previous examples in this section was on the existence and stability of equilibria, both extinction and positive equilibria, and on the utilization of the general bifurcation results in Section 8.2. We know of course that outside a neighborhood of the bifurcation point (in particular, for large values of $\mathcal{R}_0$) difference equation models sometimes exhibit non-equilibrium dynamics. The iconic example of this is the famous Ricker equation and its period-doubling route to chaotic dynamics. We expect, of course, no less from evolutionary versions of difference equation models.

As an illustration, consider a Darwinian version of the Ricker equation obtained from the model in Section 8.1.2 with $s_a = 0$ (semelparity) and the discrete logistic density factor $1/(1 + c_0 x)$ replaced by $\exp(-cx)$

$$\begin{aligned} x(t+1) &= b_0 \exp\left(-w_0 u^2(t)\right) \exp\left(-c_0 x(t)\right) x(t) \\ u(t+1) &= \left(1 - 2\sigma^2 w_0\right) u(t), \end{aligned} \tag{8.30}$$

8.3. EXAMPLES

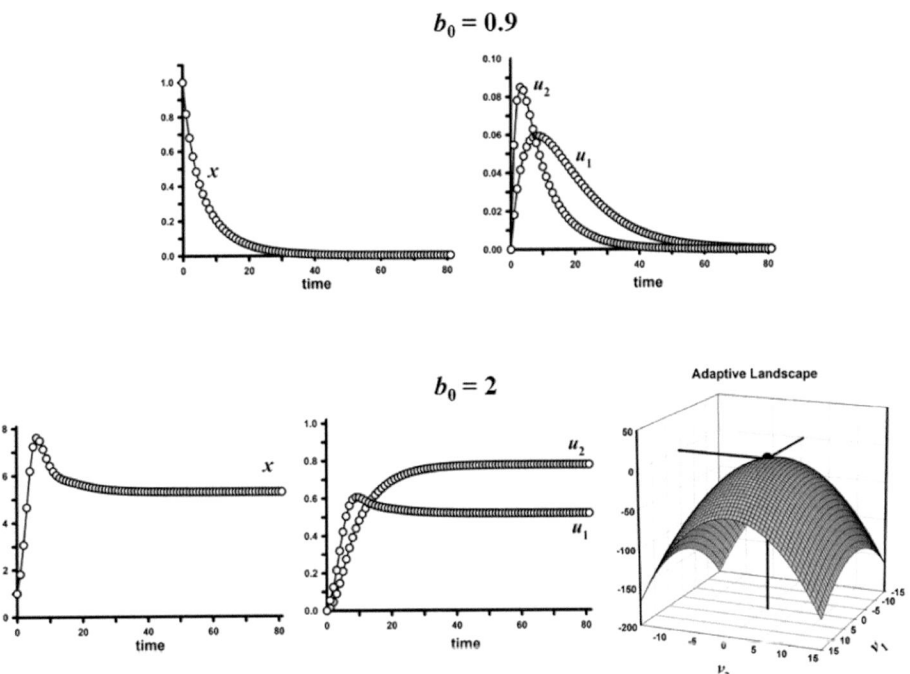

Figure 8.5: Two sample solutions of the system (8.27) are shown with parameter values $w_1 = 1.5$, $w_2 = 1$, $c_0 = 0.1$, $c_1 = 1$, $c_2 = 1.5$, $\sigma_1^2 = 0.2$ and $\sigma_2^2 = 0.4$ and with initial condition are $(x, \mathbf{u})^T = (1, \mathbf{0}_2)^T$. The top row of time-series graphs shows the solution approaching the extinction equilibrium $(x, \mathbf{u})^T = (0, \mathbf{0}_2)^T$ when $b_0 = 0.9 < 1$. The bottom row shows the solution approaching a positive equilibrium $(x, \mathbf{u})^T = (5.284, \mathbf{u})^T$, with trait component $\mathbf{u} \approx (0.779, 0.519)^T$, when $b_0 = 2 > 1$. Also shown is the adaptive landscape at the equilibrium with the equilibrium trait component located at its global maximum, indicating that the trait pair is an ESS.

where $b_0 := bs_n$. The results in Section 8.2 imply that the extinction equilibrium $(x, u)^T = (0, 0)^T$ destabilizes as b_0 increases through 1 with the result that a forward and stable bifurcation of positive equilibria results. It is not difficult to solve the equilibrium equations for the equilibria

$$\begin{pmatrix} x \\ u \end{pmatrix} = \begin{pmatrix} \frac{1}{c_o} \ln b_0 \\ 0 \end{pmatrix} \tag{8.31}$$

which exist and are positive for all $b_0 > 1$. The Jacobian evaluation at this equilibrium has eigenvalues

$$\lambda_1 = 1 - \ln b_0 < 1$$
$$\lambda_2 = 1 - 2\sigma^2 w_0 < 1$$

and, for $b_0 > 1$. Applying the linearization principle, we obtain part (a) of the following theorem.

Theorem 74 *(a) Suppose $\sigma^2 < 1/w_0$. The extinction equilibrium $(x, u)^T = (0, 0^T)$ of the Darwinian Ricker model (8.30) is stable if $b_0 < 1$ and unstable if $b_0 > 1$. For $b_0 > 1$, the positive equilibrium (8.31) is stable if $b_0 < e^2$ and unstable if $b_0 > e^2$.*

(b) Suppose $\sigma^2 > 1/w_0$. Then all solutions $(x(t), u(t))^T$ of (8.30) with $x(0) \geq 0$ satisfy $\lim_{t \to \infty} u^2(t) = +\infty$ and $\lim_{t \to \infty} x(t) = 0$.

Proof. To prove part (b), we note that $y(t) := u^2(t)$ satisfies the linear difference equation $y(t+1) = (1 - 2\sigma^2 w_0)^2 y(t)$. Since $\sigma^2 > 1/w_0$ implies $(1 - 2\sigma^2 w_0)^2 > 1$, it follows that $\lim_{t \to \infty} y(t) = +\infty$. Choose $t^* > 0$ so large that $t \geq t^*$ implies $b_0 \exp(-w_0 u^2(t)) < 1/2$. Then, for any solution $(x(t), u(t))^T$ of (8.30) with $x(0) \geq 0$, we have $0 \leq x(t+1) < x(t)/2$ for $t \geq t^*$, which implies $\lim_{t \to \infty} x(t) = 0$. ∎

For slow evolution ($\sigma^2 < 1/w_0$), the Darwinian Ricker model (8.31) displays a similar equilibrium scenario as the non-evolutionary Ricker model. This is illustrated in Figure 8.6 by the virtually identical bifurcation diagrams for the Ricker equation and this Darwinian Ricker equation. We conclude that this evolutionary Ricker model implies when compared to the non-evolutionary Ricker model, that evolution has no effect on the onset of complex dynamics (i.e., non-equilibrium dynamics). This conclusion is not robust, however; it is highly dependent on the assumptions made on how the Ricker equation coefficients depend on the evolving trait. Other evolutionary versions of the Ricker model shows destabilization of the positive equilibrium at a critical value of b_0 which, depending on model parameter values, can be greater than or smaller e^2. See [80] and Problem 6.

Part (b) of Theorem 74 shows that the Darwinian Ricker model (8.31) predicts global extinction occurs if evolution proceeds at too fast a rate (a kind of evolutionary suicide). How robust this conclusion is for other Darwinian models remains an interesting open question. In the remainder of this chapter, we will focus on slow evolution, i.e., small values of σ^2.

8.3.5 A Darwinian Leslie–Gower Competition Model

In previous sections and examples, we studied Darwinian models of a single species. The methodology applies to multi-species models as well [298]. In this section, we give an example

8.3. EXAMPLES

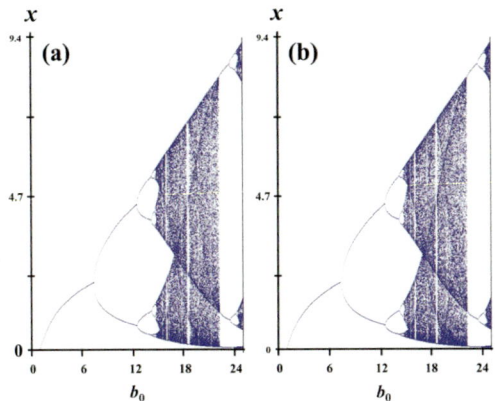

Figure 8.6: (a) The familiar bifurcation diagram for the Ricker equation $x(t+1) = b_0 \exp(-c_0 x_t) x_t$ with $c_0 = 1$ (b) The bifurcation diagram showing the x component of the Darwinian Ricker equation (8.30) with $c_0 = 1$ and $\sigma^2 = 1$.

based on the Leslie–Gower competition model studied in Chapter 3. That model is described by the system of two difference equations:

$$x_1(t+1) = \frac{b_1}{1 + c_{11}x_1(t) + c_{12}x_2(t)} x_1(t)$$

$$x_2(t+1) = \frac{b_2}{1 + c_{21}x_1(t) + c_{22}x_2(t)} x_2(t)$$

with $b_i > 1$. In Section 3.4, we found that there are four generic asymptotic outcomes predicted by this classic model. These outcomes are determined by the location on the map shown in Figure 8.7 of the point with coordinates $(c_{12} - c_{22}, c_{21} - c_{11})$. For example, if this point lies in the quadrant marked SE, then the equilibrium point $(x_1^*, x_2^*)^T = (0, (b_2 - 1)/c_{22})^T$ is globally attracting on $\mathring{\mathbb{R}}_+^2$ and species x_1 is competitively eliminated.

A Darwinian version of the Leslie–Gower model (with $b_1 = b_2 = b$) in which the competition coefficients evolve is studied in [256]. The model equations constitute a system of four difference equations:

$$x_1(t+1) = x_1 \frac{b}{1 + c(u_1(t), u_1(t)) x_1(t) + c(u_1(t), u_2(t)) x_2(t)}$$

$$x_2(t+1) = x_2 \frac{b}{1 + c(u_2(t), u_1(t)) x_1(t) + c(u_2(t), u_2(t)) x_2(t)}$$

$$u_1(t+1) = u_1(t) - \sigma_1^2 \frac{d(u_1(t), u_1(t)) x_1(t) + d(u_1(t), u_2(t)) x_2(t)}{1 + c(u_1(t), u_1(t)) x_1(t) + c(u_1(t), u_2(t)) x_2(t)}$$

$$u_2(t+1) = u_2(t) - \sigma_2^2 \frac{d(u_2(t), u_1(t)) x_1(t) + d(u_2(t), u_2(t)) x_2(t)}{1 + c(u_2(t), u_1(t)) x_1(t) + c(u_2(t), u_2(t)) x_2(t)},$$

where

$$d(v, u) := \partial_v c(v, u)$$

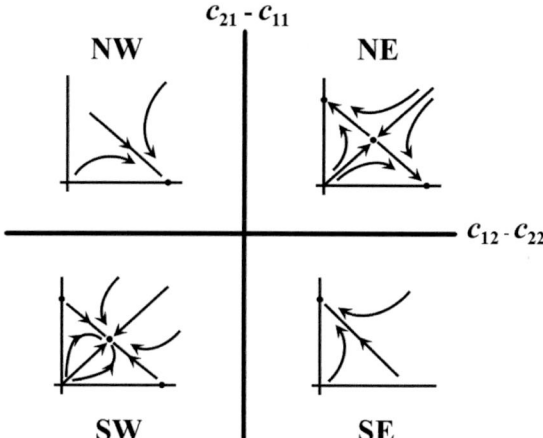

Figure 8.7: The phase portraits of the Leslie–Gower two-species competition model as they depend on the location of the point $P = (c_{12} - c_{22}, c_{21} - c_{11})$.

and u_i is the mean trait of species x_i. One question of interest concerns the effect of evolution on the competitive outcome. Specifically, we are interested in whether evolutionary adaptation can change the outcome from what it would be if evolution did not occur. If no evolution occurs ($\sigma_i^2 = 0$), then both mean traits u_i remain constant over time and then the competitive outcome is determined by the location of the point

$$P(u_1, u_2) = (c(u_1, u_2) - c(u_2, u_2), c(u_2, u_1) - c(u_1, u_1))$$

on the map in Figure 8.7. If evolution occurs (at least one $\sigma_i^2 > 0$), then the point $P(u_1(t), u_2(t))$ moves along a path on the map. Supposing for the moment that the Darwinian model system is a stable equilibrium $(x_1, x_2, u_1, u_2)^T = (x_1^*, x_2^*, u_1^*, u_2^*)^T$, we ask whether or not there are conditions under which the initial and final points $P(u_1(0), u_2(0))$ and $P(u_1^*, u_2^*)$ lie in different quadrants on the map, indicating that evolution has produced a different competitive outcome. For example, if $P(u_1(0), u_2)$ and $P(u_1^*, u_2)$ lie in the SE and NW quadrants, respectively, then the competitive outcome has reversed from x_2 being the winning species to x_1 being the winning species. We focus on this particular outcome reversal.

One case considered in [256], which is the only case we consider here, is when only one of the species evolves. If we take that species to be x_1, then $\sigma_2^2 = 0$ and the mean trait u_2 of species x_2 remains constant for all time. The model then reduces to a system of three difference equations:

$$\begin{aligned}
x_1(t+1) &= x_1 \frac{b}{1 + c(u_1(t), u_1(t)) x_1(t) + c(u_1(t), u_2) x_2(t)} \\
x_2(t+1) &= x_2 \frac{b}{1 + c(u_2, u_1(t)) x_1(t) + c(u_2, u_2) x_2(t)} \\
u_1(t+1) &= u_1(t) - \sigma_1^2 \frac{d(u_1(t), u_1(t)) x_1(t) + d(u_1(t), u_2) x_2(t)}{1 + c(u_1(t), u_1(t)) x_1(t) + c(u_1(t), u_2) x_2(t)}.
\end{aligned} \qquad (8.32)$$

8.3. EXAMPLES

As in [256], we model the trait dependence of the competition coefficients by

$$c(v,u) = a \exp\left(\frac{v^2}{2\sigma_k^2} - \frac{(v-u)^2}{2\sigma_\alpha^2}\right) \qquad (8.33)$$

with coefficients $a, \sigma_k^2, \sigma_\alpha^2 > 0$ and $\sigma_i^2 \geq 0$. In this case,

$$d(v,u) = \left(\frac{\sigma_\alpha^2 - \sigma_k^2}{\sigma_k^2 \sigma_\alpha^2} v + \frac{1}{\sigma_\alpha^2} u\right) c(v,u) \qquad (8.34)$$

in the trait equation for $u_1(t)$ in (8.32).

The equilibrium equations associated with (8.32) are, if $\sigma_1^2 > 0$,

$$x_1 = x_1 \frac{b}{1 + c(u_1,u_1)x_1 + c(u_1,u_2)x_2}$$

$$x_2 = x_2 \frac{b}{1 + c(u_2,u_1)x_1 + c(u_2,u_2)x_2}$$

$$0 = d(u_1,u_1)x_1 + d(u_1,u_2)x_2,$$

in which u_2 is the fixed mean trait of species x_2. Since we are interested in conditions under which x_1 competitively eliminates x_2 we begin by investigating the existence and stability properties of the exclusion equilibrium $(x_1, x_2, u_1)^T = (x_1^*, 0, u_1^*)^T$ whose components $x_1 = x_1^* > 0$ and $u_1 = u_1^*$ satisfy the equations:

$$1 = \frac{b}{1 + c(u_1, u_1)x_1}$$

$$0 = d(u_1, u_1).$$

From (8.34), we get that the second equation is

$$0 = a \frac{u_1}{\sigma_k^2} \exp\left(\frac{u_1^2}{2\sigma_k^2}\right),$$

and hence $u_1^* = 0$. The first equation then implies

$$x_1^* = \frac{b-1}{c(u_1^*, u_1^*)} = \frac{b-1}{a}.$$

Thus, there exists an x_2-exclusion equilibrium if and only if $b > 1$ (in which case it is unique), namely,

$$E_1 : \begin{pmatrix} x_1 \\ x_2 \\ u_1 \end{pmatrix} = \begin{pmatrix} \frac{b-1}{a} \\ 0 \\ 0 \end{pmatrix}, \quad b > 1.$$

The Jacobian associated with (8.32), when evaluated at E_1, turns out to be a block triangular matrix of the form

$$\begin{pmatrix} \lambda_1 & * & 0 \\ 0 & \lambda_2 & 0 \\ 0 & * & \lambda_3 \end{pmatrix}$$

(where the asterisks denoted unneeded terms) whose eigenvalues lie on the diagonal

$$\lambda_1 = \frac{1}{b} > 0$$

$$\lambda_2 = \frac{b}{1 + (b-1)\exp\left(\frac{\sigma_\alpha^2 - \sigma_k^2}{2\sigma_k^2 \sigma_\alpha^2} u_2^2\right)} > 0$$

$$\lambda_3 = 1 - \frac{b-1}{b}\frac{\sigma_1^2}{\sigma_k^2} < 1.$$

Since $b > 1$, we have $0 < \lambda_1 < 1$. Furthermore, some algebra shows that $0 < \lambda_2 < 1$ if $\sigma_k^2 < \sigma_a^2$ and that $-1 < \lambda_3 < 1$ if $\sigma_1^2 < 2\sigma_k^2 \frac{b}{b-1}$.

Theorem 75 *The Darwinian Leslie–Gower model (8.32) with (8.33) and (8.34) has a unique x_2-exclusion equilibrium E_1 if $b > 1$ and it is (locally asymptotically) stable if $\sigma_k^2 < \sigma_a^2$ and $\sigma_1^2 < 2\sigma_k^2 \frac{b}{b-1}$. The equilibrium E_1 is unstable if $\sigma_k^2 > \sigma_a^2$.*

Theorem 75 gives conditions when x_2 is competitively eliminated by x_1. The stability result in Theorem 75 is local so we might wonder whether, under these same conditions, there is also a stable x_1-exclusion equilibrium $(x_1, x_2, u)^T = (0, x_2^*, u_1^*)^T$, $x_2^* > 0$. The components $x_2 = x_2^*$ and $u = u_1^*$ of such an equilibrium must satisfy the equilibrium equations

$$1 = \frac{b}{1 + c(u_2, u_2) x_2}$$

$$0 = d(u_1, u_2).$$

Using the definitions (8.33) and (8.34), we solve the second equation for $u_1 = u_1^*$ and use the answer in the first equation to solve for $x_2 = x_2^*$. The result is the x_1-exclusion equilibrium

$$E_2: \begin{pmatrix} x_1 \\ x_2 \\ u_1 \end{pmatrix} = \begin{pmatrix} 0 \\ \frac{b-1}{a}\exp\left(-\frac{u_2^2}{2\sigma_k^2}\right) \\ \frac{\sigma_k^2}{\sigma_k^2 - \sigma_\alpha^2} u_2 \end{pmatrix} \text{ if } b > 1.$$

The Jacobian associated with (8.32), when evaluated at E_2, is a triangular matrix of the form

$$\begin{pmatrix} \lambda_1 & 0 & 0 \\ * & \lambda_2 & 0 \\ * & * & \lambda_3 \end{pmatrix}$$

(where the asterisks denoted unneeded terms) whose eigenvalues lie on the diagonal

$$\lambda_1 = \frac{b}{1 + (b-1)\exp\left(\frac{\sigma_\alpha^2}{2\sigma_k^2}\frac{u_2^2}{\sigma_k^2 - \sigma_\alpha^2}\right)} > 0$$

$$\lambda_2 = \frac{1}{b} > 0$$

$$\lambda_3 = \frac{1 + (b-1)\exp\left(\frac{\sigma_\alpha^2}{2\sigma_k^2}\frac{u_2^2}{\sigma_k^2 - \sigma_\alpha^2}\right)\left(1 + \frac{\sigma_k^2 - \sigma_\alpha^2}{\sigma_k^2 \sigma_\alpha^2}\sigma_1^2\right)}{1 + (b-1)\exp\left(\frac{\sigma_\alpha^2}{2\sigma_k^2}\frac{u_2^2}{\sigma_k^2 - \sigma_\alpha^2}\right)} > 0.$$

Some algebra shows $\lambda_1 > 1$ if $\sigma_k^2 < \sigma_a^2$ and $\lambda_3 > 1$ if $\sigma_k^2 > \sigma_a^2$ so that E_2 is unstable if $\sigma_k^2 \neq \sigma_a^2$.

Theorem 76 *The Darwinian Leslie–Gower model (8.32) with (8.33) and (8.34) has a unique x_1-exclusion equilibrium E_2 if $b > 1$ and it is unstable if $\sigma_k^2 \neq \sigma_a^2$.*

Although we have not established global dynamics, the local stability results in Theorems 75 and 76 suggest a competitive outcome analogous to that in the NW quadrant on the map in Figure 8.7, namely that x_1 competitively excludes x_2 for all initial conditions, when $\sigma_k^2 < \sigma_a^2$. Sample simulations of the model equations shown in Figure 8.8 support this conclusion. Furthermore, if the mean trait u_2 of species x_2 and the initial mean trait $u_1(0)$ of species x_1 are such that the point $P(u_1(0), u_2)$ lies in the SE quadrant of the map in Figure 8.7, then, if evolution does not occur, x_1 is eliminated in favor of x_2, while, if evolution does occur, the reverse outcome results, i.e., x_2 is eliminated in favor of x_1. For an example, see Figure 8.8. A laboratory experiment with two species of *Tribolium* that exhibited this phenomenon is reported in [103] and is discussed using the Darwinian Leslie–Gower model in [256].

There are obviously ecological scenarios of interest other than the one considered here, as well as unresolved mathematical questions, such as the existence and stability of coexistence equilibria (x_1^*, x_2^*, u_1^*), $x_i^* > 0$, the global stability of equilibria, and the evolution of both species $\sigma_i^2 > 0$. For a study of these and other scenarios and issues concerning the Darwinian Leslie–Gower model, see [256].

In the next section, we will investigate some of these questions, with a focus on the global stability of coexistence equilibria.

Exercises 8.3

1. Consider model equations (8.9)–(8.10) with $s_a = 0$ (semelparity) and inequality (8.11). We showed in Section 8.1.2 that the positive equilibrium (8.14) is globally asymptotically stable when it exists, i.e., when $bs_n > 1$, if evolution is not too fast in the sense that $\sigma^2 < \sigma_0^2$. We also showed that, if evolution is too fast, in the sense that $\sigma^2 > \sigma_0^2$, then the positive equilibrium is unstable. Show in this case $\lim_{t \to \infty} u^2(t) = +\infty$ and $\lim_{t \to \infty} x(t) = 0$ for all solutions with $x(0) \geq 0$. Thus, this model implies that fast evolution leads to extinction (or *evolutionary suicide*).

2. Consider the Darwinian Ricker equation $x(t+1) = [bfs_n + s_a(1-f)]\exp(-cx(t))x(t)$ with trait-dependent f and c given by (8.8). (a) Repeat the extinction and positive equilibrium analysis carried out for the Darwinian logistic equation in Example 8.1.2. (b) Apply Theorems 72 and 73.

3. Write down the Darwinian model equations associated with
$$r(x, u, v) = bf(v)s_n\beta(c(v-u)x) + s_a(1-f(v))\sigma(c(v-u)x),$$
$$b > 0, \quad 0 < s_n \leq 1, \quad 0 \leq s_a < 1,$$
$$f(\cdot), c(\cdot) \in C^2(\mathbb{R} \to \mathbb{R}_+), \quad \beta(\cdot), \sigma(\cdot) \in C^2(\nleq \to \mathbb{R}_+),$$
$$0 \leq f(\cdot) \leq 1, \quad f(0) = 1, \quad f'(0) = 0,$$
$$0 \leq \sigma(\cdot) \leq 1, \quad \beta(0) = \sigma(0) = 1.$$

Show that $(x, u)^T = (0, 0)^T$ is an extinction equilibrium and apply Theorems 72 and 73. Under what conditions is the bifurcation forward and under what conditions is it backward?

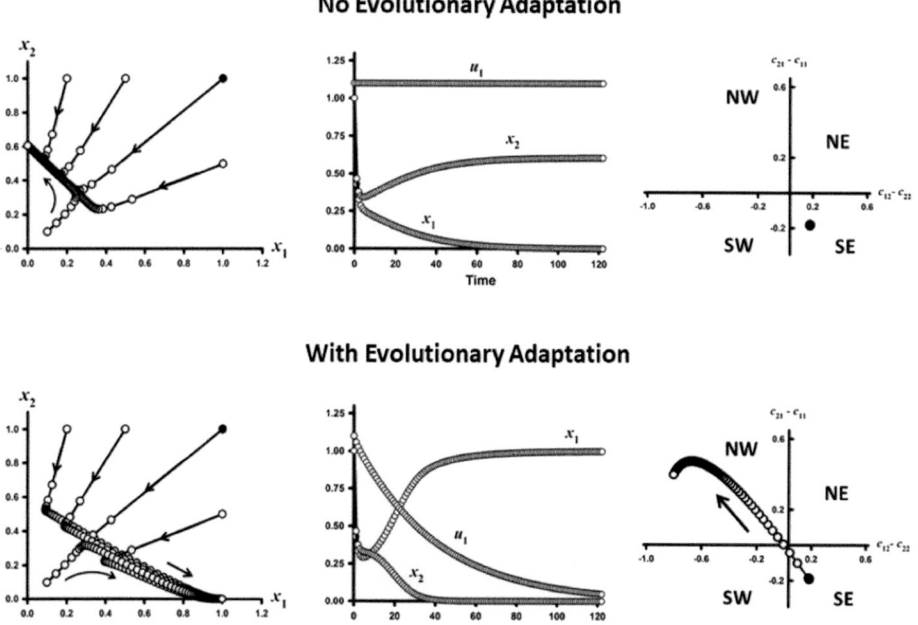

Figure 8.8: Shown are the results of simulations of the Darwinian Leslie–Gower model (8.32) with coefficients (8.33) and (8.34) and with parameter values $b = 2$, $a = 1$, $u_2 = 1$, $\sigma_k^2 = 1$, and $\sigma_\alpha^2 = 3$. **Top row:** In these plots, $\sigma_1^2 = 0$ and there is no adaptive evolution. The graph of the left shows the $(x_1(t), x_2(t))^T$ components, in the (x_1, x_2)-plane, from five solutions $(x_1(t), x_2(t), u_1(t))^T$ with the initial conditions $(x_1(0), x_2(0), u_1(0))^T = (1, 1, , 1.1)^T$, $(0.2, 1, , 1.1)^T$, $(0.5, 1, , 1.1)^T$, $(1, 0.5, , 1.1)^T$, and $(0.1, 0.1, , 1.1)^T$. Note that species $x_1(t)$ goes extinct and $\lim_{t \to \infty} x_2(t) = 0.6065$. The time series for the solution with the first initial condition (shown by the black dot) is shown in the middle graph. The right-hand graph shows that the point $P(1, 1, 1.1)$ lies in the SE quadrant. **Bottom row:** In these plots, evolutionary adaptation of species x_1 occurs with $\sigma_1^2 = 0.05$. Now species $x_2(t)$ goes extinct while $\lim_{t \to \infty} x_1(t) = 1$ and the mean trait of $x_1(t)$ evolves to 0. In this case, the point $P(x_1(t), x_2(t), u_1(t))$ migrates from the SE quadrant to the NW quadrant, demonstrating the reversal of the competitive outcome.

4. In what ways do the analyses and conclusions in the multi-trait Example 8.3.3 change if the population equation is changed from a discrete logistic (Beverton–Holt) to a Ricker equation?

5. Let $\varphi(x_2) = 1/(1+cx)$ and $\alpha(x_2) = 1$ in the model equations (8.21)–(8.22). Find formulas for all equilibria and show, using the linearization principle, that positive equilibria exist and are, if σ^2 is not too large, locally asymptotically stable for all $\mathcal{R}_0^* > 1$.

6. (a) Derive the equations for $x(t)$ and $u(t)$ for a Darwinian Ricker model using
$$r(x, u, v) = b_0 f(v) \exp(-c(v-u)x)$$
with $f(v) = \exp(-w_0 v^2)$, $w_0 > 0$ and $c_0 = c(0) > 0$, $c_1 = c'(0) \neq 0$.

   (b) For the model in (a) show that $(x, u)^T = (0, 0)^T$ is the only extinction equilibrium and that $r^* = b_0$. Apply the bifurcation Theorems 72 and 73 in Section 8.2.

   (c) Find formulas for the bifurcating positive equilibria and use the Jury conditions to study their stability (by linearization).

   (d) Use your results in (c) to find conditions under which the positive equilibria lose stability as b_0 increases through a critical value $b_0^* > e^2$. In this case, we could say evolution selects against nonequilibrium and complex dynamics. Also, find conditions under which the positive equilibria lose stability as b_0 increases through a critical value $b_0^* < e^2$. In this case, evolution promotes the onset of nonequilibrium and complex dynamics.

7. Rework Example 8.3.1 with $b(v)$ and $s_1(v)$ replaced by
$$b(v) = b(1 - \exp(-v))$$
$$s_1(v) = s_1 \exp(-dv), \quad d > 0.$$

8. Rework Example 8.3.5 with the Leslie–Gower competition model replaced by the competition model
$$x_1(t+1) = b_1 \exp(-c_{11}x_1(t) - c_{12}x_2(t))x_1(t)$$
$$x_2(t+1) = b_2 \exp(-c_{21}x_1(t) - c_{22}x_2(t))x_2(t).$$

8.4 Global Stability

In this section, we will investigate the global stability of coexistence equilibria of a Darwinian version of the discrete logistic (Beverton-Holt) model. The main tool we are going to use is the theory of maps that were introduced by Hal Smith [280].

Definition 33 *Let X be an ordered metric space. A continuous map $F : X \to X$ is mixed monotone if there exists a map (not necessarily continuous) $f : X \times X \to X$ satisfying*

(i) $F(\mathbf{x}) = f(\mathbf{x}, \mathbf{x})$ for all $\mathbf{x} \in X$;

(ii) for $\mathbf{y} \in X$ and $\mathbf{x}_1 \leq \mathbf{x}_2$, we have $f(\mathbf{x}_1, \mathbf{y}) \leq f(\mathbf{x}_2, \mathbf{y})$;

(iii) for $\mathbf{x} \in X$ and $\mathbf{y}_1 \leq \mathbf{y}_2$, we have $f(\mathbf{x}, \mathbf{y}_2) \leq f(\mathbf{x}, \mathbf{y}_1)$.

In [123], the authors developed a method to define the map f mentioned above. We now give the first example to illustrate this method.

Example 106 *Let us consider the discrete logistic (Beverton–Holt) model*
$$x(t+1) = \frac{bx(t)}{1+cx(t)}.$$

The corresponding evolution model with one trait may be developed by letting $b(v) = b_0 e^{-\frac{v^2}{2}}$ *and* $c(v,u) = c(v-u)$, *with* $c(0) = c_0$. *Hence, we have the following evolutionary model* [242]:

$$x(t+1) = \left.\frac{b_0 e^{-\frac{v^2}{2}}}{1+c(v-u(t))x(t)}\right|_{v=u(t)}$$

$$u(t+1) = u(t) + \sigma^2 \left.\left(\frac{\partial}{\partial v}\left(\frac{b_0 e^{-\frac{v^2}{2}}}{1+c(v-u(t))x(t)}\right)\right)\right|_{v=u(t)}.$$

This leads to the model

$$\begin{aligned} x(t+1) &= \frac{b_0 e^{-\frac{u^2}{2}} x(t)}{1+c_0 x(t)} \\ u(t+1) &= (1-\sigma^2)u(t) - \frac{c_1 \sigma^2 x(t)}{1+c_0 x(t)} \end{aligned}, \qquad (8.35)$$

where $c_1 = \frac{d}{dv}(c(v-u))|_{v=u}$.

Next we will show that the map F representing model (8.35) is mixed monotone.

Lemma 14 *Assume that* $c_1 < 0$. *The map F of model (8.35) is mixed monotone, if* $b_0 > \sigma^2$.

Proof. Define the map f as follows $f((x_1, u_1), (x_2, u_2)) = F(x_1, u_2)$. We are going to use the southeast order, i.e., if $\mathbf{x} = (x_1, u_1)$, $\mathbf{y} = (x_2, u_2)$, then $\mathbf{x} \leq \mathbf{y}$ if $x_1 \leq x_2$ and $u_1 \geq u_2$. First, assume that $c_1 < 0$ and $\mathbf{x} \leq \mathbf{y}$. Then

$$f((x_1, u_1), (x, u)) = F(x_1, u) = \left(\frac{b_0 e^{-\frac{u^2}{2}} x_1}{1+c_0 x_1}, (1-\sigma^2)u - \frac{\sigma^2 c_1 x_1}{1+c_0 x_1}\right)$$

and

$$F(x_2, u) = \left(\frac{b_0 e^{-\frac{u^2}{2}} x_2}{1+c_0 x_2}, (1-\sigma^2)u - \frac{\sigma^2 c_1 x_2}{1+c_0 x_2}\right).$$

It is clear that $F(x_1, u) \leq F(x_2, u)$. Hence F satisfies (ii) in Definition 33. Note that item (i) in Definition 33 is trivially satisfied.

Next, we prove item (iii).

$$f((x, u), (x_1, u_1)) = F(x, u_1) = \left(\frac{b_0 e^{-\frac{u^2}{2}} x}{1+c_0 x}, (1-\sigma^2)u_1 - \frac{\sigma^2 c_1 x}{1+c_0 x}\right)$$

$$f((x, u), (x_2, u_2)) = F(x, u_2) = \left(\frac{b_0 e^{-\frac{u^2}{2}} x}{1+c_0 x}, (1-\sigma^2)u_2 - \frac{\sigma^2 c_1 x}{1+c_0 x}\right)$$

8.4. GLOBAL STABILITY

Thus $F(x, u_2) \leq F(x, u_1)$ and (iii) is proved. ∎

System (8.35) has an extinction equilibrium $E_0^* = (0, 0)$. Now the Jacobian matrix at E^* is given by

$$JF(E_0^*) = \begin{pmatrix} b_0 & 0 \\ -\sigma^2 c_1 & 1 - \sigma^2 \end{pmatrix}$$

with eigenvalues $\lambda_1 = b_0 > 1$ and $|\lambda_2| = |1 - \sigma^2| < 1$ if $0 < \sigma^2 < 2$. Hence E_0^* is a saddle.

To find the coexistence equilibrium we solve the isocline equations.

$$b_0 e^{-\frac{u^2}{2}} = 1 + c_0 x \tag{8.36}$$

$$u = \frac{-c_1 x}{1 + c_0 x}. \tag{8.37}$$

From (8.36), we have

$$u = \Psi_1(x) = \sqrt{2 \ln \left(\frac{b_0}{1 + c_0 x} \right)} \tag{8.38}$$

$$u = \Psi_2(x) = \frac{-c_1 x}{1 + c_0 x}. \tag{8.39}$$

The map $\Psi_1(x)$ is defined on $x \in \left(0, \frac{b_0 - 1}{c_0}\right)$, $b_0 > 1$, with $\Psi_1(0) = \sqrt{2 \ln(b_0)} > 0$ and $\Psi_1' = \frac{-c_0}{(1+c_0)\sqrt{2 \ln\left(\frac{b_0}{1+c_0 x}\right)}} < 0$. Hence $\Psi_1(x)$ is decreasing on $\left(0, \frac{b_0 - 1}{c_0}\right)$. Now the $\Psi_2(x)$ is defined on $(0, \infty)$ with $c_1 < 0$. Note that $\Psi_2'(x) = \frac{-c_1}{(1+c_0 x)^2} > 0$ which implies that $\Psi^2(x)$ is strictly increasing. Let $\Psi(x) = \Psi_1(x) - \Psi_2(x)$ which is defined on $\left(0, \frac{b_0 - 1}{c_0}\right)$. Now

$$\Psi'(x) = \frac{-c_0}{(1 + c_0 x)\sqrt{2 \ln\left(\frac{b_0}{1+c_0 x}\right)}} + \frac{c_1}{(1 + c_0 x)^2} < 0$$

and thus $\Psi(x)$ is decreasing on $\left(0, \frac{b_0 - 1}{c_0}\right)$. Moreover, $\Psi(0) = \sqrt{2 \ln(b_0)} > 0$ and $\Psi\left(\frac{b_0 - 1}{c_0}\right) = \frac{c_1}{c_0}\left(\frac{b_0 - 1}{b_0}\right) < 0$. By the intermediate value theorem, there exists a unique positive fixed point $E^* = (x^*, u^*)^T$ (Figure 8.9.)

The Jacobian matrix at E^* is given by

$$JF(E^*) = \begin{pmatrix} \frac{1}{1+c_0 x^*} & \frac{c_1 x^{*2}}{1+c_0 x^*} \\ \frac{-c_1 \sigma^2}{(1+c_0 x^*)^2} & 1 - \sigma^2 \end{pmatrix}.$$

Hence E^* is locally asymptotically stable if

$$\det JF = \frac{1 - \sigma^2}{1 + c_0 x^*} + \frac{c_1^2 \sigma^2 x^{*2}}{(1 + c_0 x^*)^3} < 1, \tag{8.40}$$

$$\det JF > -\operatorname{tr} JF - 1 \text{ or } 2 - \sigma^2 + \frac{2 - \sigma^2}{1 + c_0 x^*} + \frac{c_1^2 \sigma^2 x^{*2}}{(1 + c_0 x^*)^3} > 0 \tag{8.41}$$

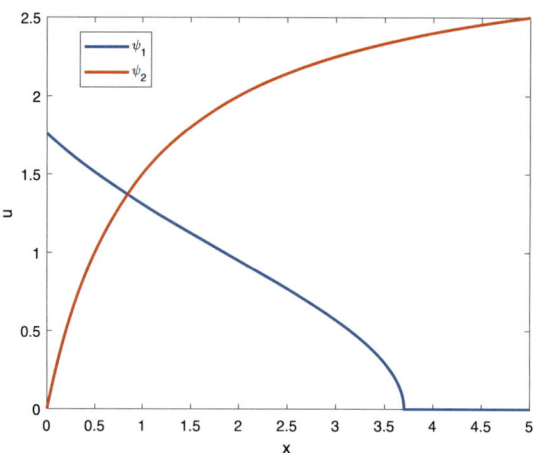

Figure 8.9: Existence of a unique positive fixed point for $b_0 = 4.7$ and $c_1 = 3$.

which is always true if $0 < \sigma^2 < 2$ (Figure 8.9).

$$\det JF > \operatorname{tr} JF - 1 \text{ or } 1 - \frac{1}{1 + c_0 x^*} + \frac{c_1^2 \sigma^2 x^{*2}}{(1 + c_0 x^*)^3} > 0 \tag{8.42}$$

which is always true.

This yields the following result.

Theorem 77 E^* *is locally asymptotically stable if condition* (8.40) *holds and* $0 < \sigma^2 < 2$.

Definition 34 *A region D in the domain of a map F is called an absorbing region if all the orbits of the points in the domain of F are in D after a number of iterations.*

Lemma 15 (Absorbing region) *Let $b_0 > 1$, $1 < \sigma^2 < 2$ and $c_1 \neq 0$. Then the map F in model* (8.35) *has a compact invariant absorbing region D. Moreover, D is a subset of the fourth quadrant when $c_1 > 0$ and it is a subset of the first quadrant when $c_1 < 0$.*

Proof. Assume that $c_1 > 0$. Now on the x-axis, $0 \leq x \leq b_0/c_0$, and $0 \leq u \leq c_1 \sigma^2/c_0$. Hence the set $D = \{(x, u) : 0 \leq x \leq b_0/c_0, 0 \leq u \leq c_1 \sigma^2/c_0\}$ is an invariant absorbing region of the map F that is a subset of the first quadrant. Analogously, there is an invariant absorbing region $D = \{(x, u) : 0 \leq x \leq b_0/c_0, c_1 \sigma^2/c_0 \leq u \leq 0\}$ in the fourth quadrant when $c_1 < 0$. ∎

As a consequence of this result, we have the following theorem.

Theorem 78 (Global Stability) *[123] Let $(c_1 \neq 0)$ and $\left(1 < \sigma^2 < 2 - b_0 e^{-1/2}\right)$ with $(b_0 > 1)$. Then the equilibrium point $((x^*, u^*)^T)$ of the model* (8.35) *is globally asymptotically stable provided it is locally asymptotically stable.*

Proof. Assume that $c_1 < 0$. Let D be the absorbing area as defined in Lemma 15, and let $B = D - \{(x, u) : x = 0\}$ Let $\mathbf{a} = (x_1, u_1), \mathbf{b} = (x_2, u_2) \in B$. We claim that, if $\mathbf{a} \leq_{se} \mathbf{b}$, $f(\mathbf{a}, \mathbf{b}) \leq_{se} \mathbf{a}$, and

$f(\mathbf{b}, \mathbf{a}) \geq_{se} \mathbf{b}$, then $\mathbf{a} = \mathbf{b}$. Let us assume by contradiction that $\mathbf{a} \neq \mathbf{b}$. So either $\mathbf{a} <_{se} \mathbf{b}$ or $\mathbf{a} >_{se} \mathbf{b}$. Note that $\mathbf{a} <_{se} \mathbf{b}$ implies that $x_1 < x_2$ and $u_1 > u_2$. Now $f(\mathbf{a}, \mathbf{b}) \leq_{se} \mathbf{a}$ and $f(\mathbf{b}, \mathbf{a}) \geq_{se} \mathbf{b}$ implies that $F(x_1, u_2) \leq_{se} (x_1, u_1)$ and $F(x_2, u_1) \geq_{se} (x_2, u_2)$, respectively. Consequently, $\alpha b_0 e^{-u_2^2/2} - 1 - c_0 x_1 \leq 0$ and $\alpha - b_0 e^{-u_1^2/2} + 1 + c_0 x_2 \leq 0$. Hence $b_0(e^{-u_2^2/2} - e^{u_1^2/2} \leq c_0(x_1 - x_2)$ which is false, and, consequently, $\mathbf{a} = \mathbf{b}$. A similar conclusion may be obtained if $\mathbf{b} <_{se} \mathbf{a}$. Now let $(x, u) \in B$. Then its orbit closure $\overline{O(x, u)} \subset D$ and thus compact. But since the origin is a saddle, it follows that $\mathbf{c} = \inf \omega(x, u) \in B$ and $\mathbf{d} = \sup \omega(x, u) \in B$, where $\omega(x, u)$ is the omega limit set of (x, u). Note that $\omega(x, u)$ is invariant. This implies that both $\mathbf{c}$ and $\mathbf{d}$ are in $\omega(x, u)$. If $(y_1, v_1) \in \omega(x, u)$, then there exists $(y_2, v_2) \in \omega(x, u)$ such that $F(y_2, v_2) = (y_1, v_1)$. Hence, $\mathbf{c} \leq_{se} (y_2, v_2) \leq_{se} \mathbf{d}$. Let $\mathbf{c} = (z, w)$ and $\mathbf{d} = (k, s)$. Then one may show that $f(\mathbf{c}, \mathbf{d}) \leq_{se} (y_1, v_1)$, and $f(\mathbf{d}, \mathbf{c}) \geq_{se} (y_1, v_1)$. Since (y_1, v_1) was arbitrary chosen from $\omega(x, u)$, it follows that $f(\mathbf{c}, \mathbf{d}) \leq_{se} \mathbf{c}$ and $f(\mathbf{d}, \mathbf{c}) \geq_{se} \mathbf{d}$, which implies that $\mathbf{c} = \mathbf{d}$. Therefore the omega limit set of the point (x, u) is a fixed point in D. Since the origin is a saddle and the interior fixed point (x^*, u^*) is locally asymptotically stable, it follows for all points $(x, u) \in D$, $\omega(x, u) = (x^*, u^*)$. Hence (x^*, u^*) is globally asymptotically stable. Now for $c_1 > 0$, the proof follows by letting $c_1 = -\bar{c}_1$, where $\bar{c}_1 < 0$. ∎

8.5 Secondary Bifurcations

Recall that, for the nonevolutionary discrete logistic (Beverton–Holt), the extinction equilibrium is globally asymptotically stable if $0 < b_0 < 1$ and loses its stability at the bifurcation parameter $b_0 = 1$. And when $b_0 > 1$, the positive equilibrium is globally asymptotically stable and no bifurcation occurs afterward. For the evolutionary model (8.35) with $b_0 > 1$, however, the only possible bifurcation is a Neimark-Sacker bifurcation; see ([242]).

Consider the evolutionary Ricker model

$$\begin{aligned} x(t+1) &= b_0 x(t) e^{\alpha - \frac{u^2}{2} - c_0 x(t)} \\ u(t+1) &= u(t) + \sigma^2 \left(-u(t) - \frac{\partial c(z)}{\partial z}\bigg|_{z=0} \right) x(t) \end{aligned} \qquad (8.43)$$

which we can re-write as

$$\begin{aligned} x(t+1) &= x(t) e^{\alpha - \frac{u^2}{2} - c_0 x(t)} \\ u(t+1) &= (1 - \sigma^2) u(t) - c_1 \sigma^2 x(t), \end{aligned} \qquad (8.44)$$

where $c_1 := \frac{d}{dz} c(z)|_{z=0}$, and $\alpha = \ln b_0$.

There are three coefficients in Equation (8.44). The coefficient $b_0 = e^\alpha$ is the maximal possible fertility rate, as a function of the trait v, and the coefficient σ^2 is the speed of evolution. The coefficient c_1 is the sensitivity of the competition $c(z)$ to changes in the difference $z = v - u$ at when $v = u$. If $c_1 \neq 0$, then c_1 measures the difference between the competition intensities experienced by individuals that have the population mean trait and those whose traits are slightly different from the mean. For example, if $c_1 > 0$, then an individual that inherits a trait slightly larger (smaller) than the mean u will experience increased (decreased) intraspecific competition. These interpretations can also hold, of course, if $c_1 < 0$, that is an individual that inherits a trait slightly smaller (larger) than the mean u will experience increased (decreased) intraspecific competition. Now c_1 maybe equal 0 and the ecological reason for this assumption is that it is sometimes assumed in evolutionary game theory models that individual experiences maximum competition when its trait equals the population mean.

The non-evolutionary Ricker model destabilizes in period-doubling bifurcation at the critical value $\alpha = 2$. For the Darwinian versions of the Ricker equation considered here, we arrive at several general conclusions. If $c_1 = 0$ in the trait-dependent density coefficient $c(v-u)$, then there is no change in the destabilization point for the fertility rate e^α. As shown in the figures, both models destabilize in period-doubling bifurcations at the same critical value $\alpha = 2$. In this sense, we conclude that evolution has no effect on the onset of non-equilibrium and complex dynamics. The opposite is true in the case of hierarchical trait-dependent competition coefficients, i.e., when $c_1 \neq 0$. In this case, the onset of non-equilibrium and complex dynamics is delayed to a larger critical value of α when evolution proceeds slowly (i.e., σ^2 is small). In this case, we say that slow evolution selects against non-equilibrium and complex dynamics. If, on the other hand, evolution proceeds at a faster speed, then there are two differences with the non-evolutionary Ricker equation, depending on the magnitude of the density effects, i.e., the size of c_1. First, the onset of non-equilibrium and complex dynamics can lead not to a period-doubling bifurcation, but to a Neimark-Sacker bifurcation. Secondly, in the latter case, the bifurcation point can be either later or earlier than 2. In the latter case (and only in this case), which occurs for larger σ^2 and c_1 values, we can conclude that evolution promotes non-equilibrium and complex dynamics.

Consider the following set of parameters

$$c_0 = 1, \sigma^2 = 0.8, \ c_1 = 2.$$

c_1 and σ^2 are fixed and we change α accordingly. The Figures (8.12(a),(b),(c),(d)) show the stable dynamics of the population and their trait, converging to the positive equilibrium $(0.51, 1.02)$ for the initial condition $(0.4, 0.5)$. Beyond the value $\alpha = 1.085$, the evolutionary system loses its stability, and an attracting closed invariant curve appears (Figures 8.10, 8.11, and 8.12).

Hence the system undergoes a Neimark Sacker bifurcation. For instance, if $\alpha > 1.085$, then the eigenvalues of the Jacobian matrix of $JF(0.51, 1.02)$ at the survival equilibrium point are

$$\lambda_{1,2} = 0.345 \pm 0.9007191571\, i. \tag{8.45}$$

Exercises 8.4

1. Prove that inequalities (8.41) and (8.42) hold for the evolutionary discrete logistic (Beverton–Holt) model (8.35).

2. Consider the evolutionary Ricker model (8.44).

   (i) Show that the map is mixed monotone if $1 < \alpha < 1$, $1 < \sigma^2 < 2$, and $c_1 \neq 0$.

   (ii) Show that the interior equilibrium is globally asymptotically stable under the conditions of (i).

8.5. SECONDARY BIFURCATIONS

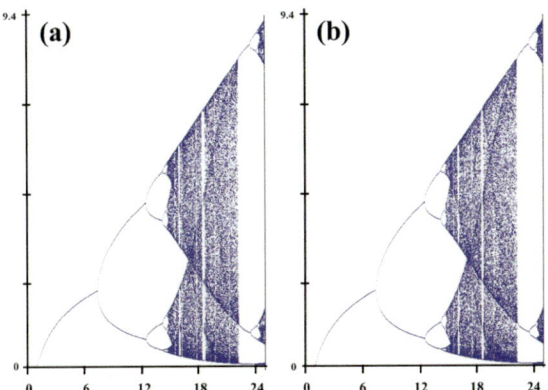

Figure 8.10: (a) The familiar bifurcation diagram for the Ricker equation (3.55)a with $c_0 = 1$, (b) The bifurcation diagram showing the x component of the Darwinian Ricker equation with $c_0 = 1$, $c_1 = 0$ and $\sigma^2 = 1$.

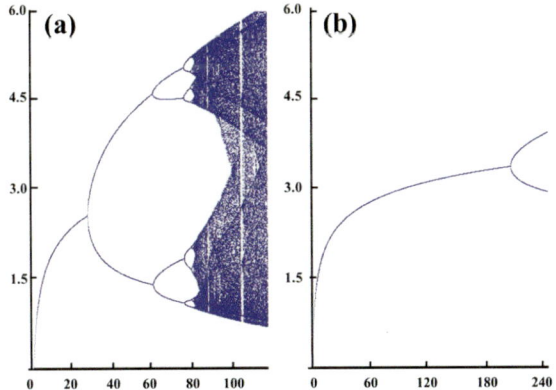

Figure 8.11: The bifurcation diagram showing the x component of the Darwinian Ricker equation with $c_0 = 1$, and (a) $c_1 = 0.5$ and $\sigma^2 = 0.5$ and (b) $c_1 = 0.6$ and $\sigma^2 = 0.5$.

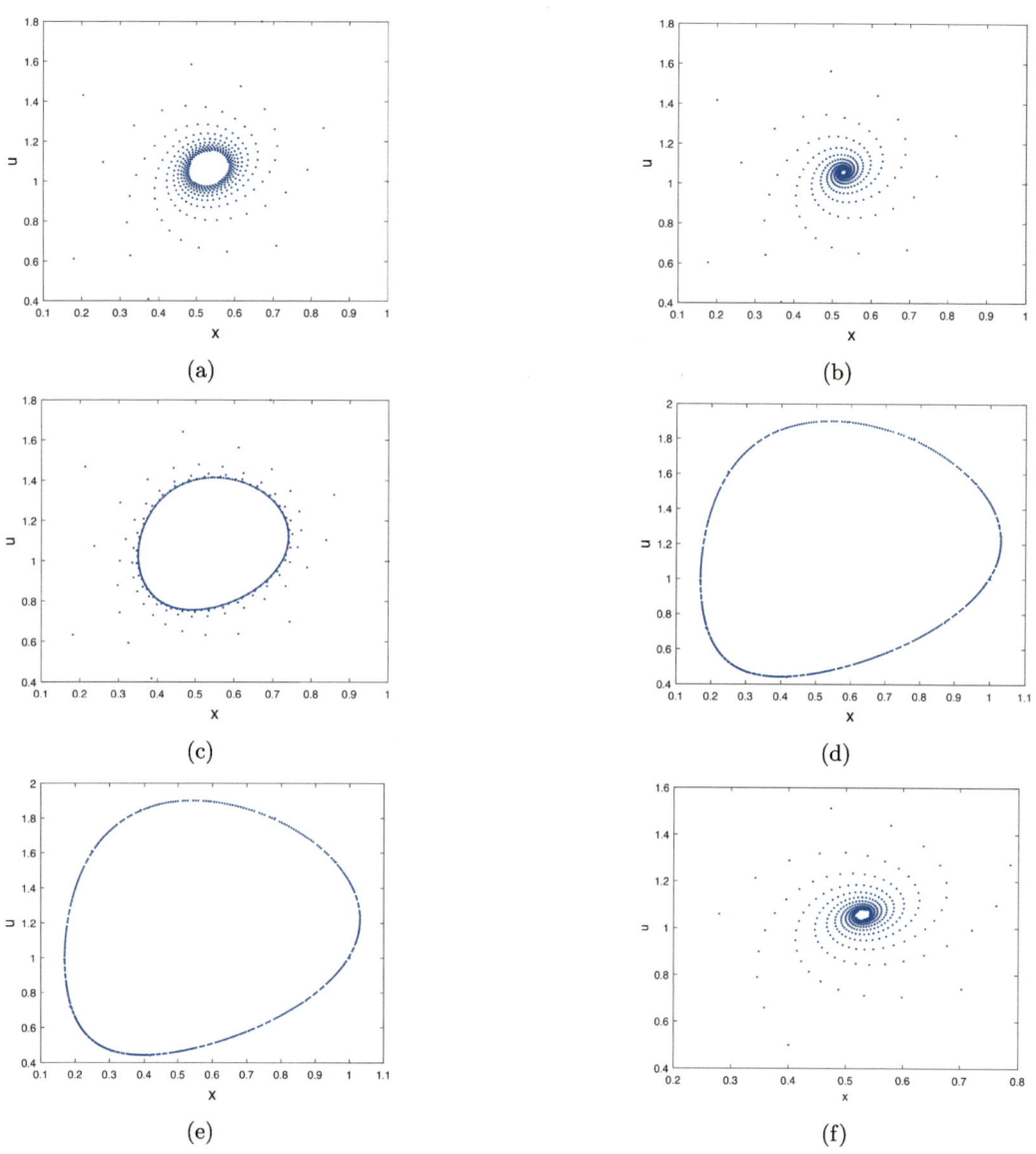

Figure 8.12: Supercritical Neimark-Sacker bifurcation diagram of the evolutionary Ricker model, under the assumption that $c_0 = 1$.

3. Consider the 2-species Leslie–Gower evolutionary model

$$x_1(t+1) = \frac{b_1 e^{-\frac{u_1^2}{2}} x_1(t)}{1 + c_{01} x_1(t) + c_{12} x_2(t)}$$

$$x_2(t+1) = \frac{b_2 e^{-\frac{u_2^2}{2}} x_2(t)}{1 + c_{21} x_1(t) + c_{02} x_2(t)} \tag{8.46}$$

$$u_1(t+1) = (1 - \sigma_1^2) u_1(t) - \frac{c_1 \sigma^2 x_1(t)}{1 + c_{01} x_1(t)}$$

$$u_2(t+1) = (1 - \sigma_2^2) u_2(t) - \frac{c_2 \sigma^2 x_2(t)}{1 + c_{02} x_2(t)}$$

where u_1, u_2 are the mean traits of species x_1 and x_2, and σ_1^2, σ_2^2 are the speed of evolution of species x_1 and x_2, respectively. Assume that $0 < \sigma_1^2, \sigma_2^2 < 2$, and $b_1, b_2 > 1$.

  (i) Show that there exists a unique interior equilibrium.
  (ii) Show that the extinction equilibrium is a saddle.
  (iii) Give conditions under which the interior equilibrium is locally asymptotically stable.
  (iv) Prove that the map representing the model is mixed monotone.
  (v) Prove that, if the interior equilibrium is locally asymptotically stable, then it is globally asymptotically stable.

4. Consider the 2-species evolutionary Ricker model

$$x_1(t+1) = x_1(t) e^{\alpha - \frac{u_1^2}{2} - c_{01} x(t) - c_{12} y(t)}$$

$$x_2(t+1) = x_2(t) e^{\beta - \frac{u_2^2}{2} - c_{12} x(t) - c_{02} y(t)} \tag{8.47}$$

$$u_1(t+1) = (1 - \sigma_1^2) u_1(t) - c_1 \sigma_1^2 x_1(t)$$

$$u_2(t+1) = (1 - \sigma_2^2) u_2(t) - c_0 \sigma_2^2 x_2(t).$$

  (i) Under the conditions $1 < \sigma_1^2, \sigma_2^2 < 2$, $0 < \alpha, \beta < 1$, the map representing this model is mixed monotone.
  (ii) Find conditions under which the interior equilibrium point is locally asymptotically stable.
  (iii) Show that, if the interior equilibrium is locally asymptotically stable, then it is globally asymptotically stable.

5. Consider the following evolutionary discrete logistic (Beverton–Holt) model with immigration

$$x(t+1) = \frac{b_0 e^{-\frac{u^2}{2}} x(t)}{1 + c_0 x(t)} + H$$

$$u(t+1) = (1 - \sigma^2) u(t) - \frac{c_1 \sigma^2 x(t)}{1 + c_0 x(t)} \tag{8.48}$$

(i) Investigate the local stability of the equilibrium points.

(ii) Determine if the map representing the model is mixed monotone.

(iii) Determine if local stability of the interior equilibrium implies global stability.

(iv) Does the system undergo a period-doubling bifurcation and/or Neimark-Sacker bifurcation.

6. Conjecture: The conclusions in Problem 2 concerning the evolutionary single-species model are valid for the case $0 < \sigma^2 < 1$. Prove the conjecture.

7. (Research Project) Develop an evolutionary Host-Parasitoid model (see Bedding model (3.43)). Then investigate the local and global stability of the equilibrium points.

8. (Research Project) Investigate in details the Neimark-Sacker bifurcation of the evolutionary discrete logistic (Beverton–Holt) model (8.35).

9. (Research Project) Investigate in detail the period-doubling and the Neimark-Sacker bifurcation of the evolutionary single-species Ricker model (8.44)

10. (Research Project) Investigate the bifurcation of the evolutionary Leslie–Gower model (8.46).

11. (Research Project) Investigate the bifurcation of the evolutionary 2-species Ricker model (8.47).

12. (Research Project) Consider the evolutionary discrete logistic (Beverton–Holt) model with immigration.

$$\begin{aligned} x(t+1) &= \frac{b_0 e^{-\frac{u^2}{2}} x(t)}{1 + c_0 x(t)} + H \\ u(t+1) &= (1-\sigma^2) u(t) - \sigma^2 \frac{c_1 x(t)}{1 + c_0 x(t)}. \end{aligned} \quad (8.49)$$

Investigate the stability and the bifurcation of this model.

13. (Research Project) Repeat Problem 12 for the evolutionary Ricker model with immigration.

$$\begin{aligned} x(t+1) &= x(t) e^{\alpha - \frac{u^2}{2} - c_0 x(t)} + H \\ u(t+1) &= (1-\sigma^2) u(t) - c_1 \sigma^2 x(t). \end{aligned} \quad (8.50)$$

Chapter 9
Nonautonomous Models

9.1 Introduction

In the previous chapters, we used *autonomous* population models with constant model parameters (coefficients) to study fundamental principles in theoretical population biology. These parameters account for a variety of vital rates, mechanisms, and processes that affect a population's growth or decay, including fertility rates, survival probabilities, competition coefficients, and so on. More realistic models would not assume these parameters remain constant over time, particularly for natural populations.

There are any number of reasons why a model's parameters might change over time, such as changes in the population's physical environment, life history cycles, metabolic processes, etc. For example, a specific parameter, say a fertility rate, might fluctuate randomly due to stochastic environmental fluctuations, which would lead us to a stochastic difference equation, or it might suffer regular environmental fluctuations that proceed in a deterministic manner. An example of the latter case occurs if an environmental shift occurs from one state to another, which would lead to an asymptotically autonomous difference equation. Another example occurs if environmental fluctuations are regular and recurrent, which leads to a difference equation with periodic parameters—a so-called *periodically forced difference equation*. These fluctuations lead to models that are represented by nonautonomous difference equations [92, 122, 132, 133, 220] We begin our investigation of nonautonomous models with this case.

9.2 Examples of Nonautonomous Models

1. Recall the basic linear equation models from Chapter 1:
$$x(t+1) = rx(t)$$
$$x(t+1) = rx(t) + \Lambda,$$
where
$$r = b + \sigma$$
$$= \text{fertility} + \text{survival}.$$

Nonconstant parameters give

$$x(t+1) = r(t)x(t)$$
$$x(t+1) = r(t)x(t) + \Lambda(t),$$

where

$$r(t) = b(t) + \sigma(t).$$

A special case is when b, σ and Λ undergo periodic oscillations with a common period $p > 1$. That is,

$$r(t+p) = r(t) \text{ and } \Lambda(t+p) = \Lambda(t).$$

2. Nonlinear examples when $\sigma = 0$.

$$\text{discrete logistic (Beverton–Holt): } x(t+1) = b\frac{x(t)}{1+cx(t)},$$

$$\text{Ricker: } x(t+1) = bx(t)\exp(-cx(t)).$$

A special case is when b and c undergo periodic oscillations with a common period $p > 1$. That is,

$$b(t+p) = b(t) \text{ and } c(t+p) = c(t) \text{ for all } t = 0, 1, \ldots.$$

9.3 Linear Periodic Difference Equations

Consider the general nonautonomous linear equation

$$x(t+1) = r(t)x(t) + \Lambda(t), \text{ with } x(0) = x_0. \qquad (9.1)$$

Of interest is the existence of a solution to this equation. By the variation of constants formula from [120], the unique solution of Equation (9.1) is

$$x(t) = \begin{cases} \left(\Pi_{i=0}^{t-1} r(i)\right) x_0 + \sum_{j=0}^{t-1} \left(\Pi_{i=j+1}^{t-1} r(i)\right) \Lambda(j), & \text{for } t = 1, 2, \cdots \\ x_0 & \text{for } t = 0. \end{cases} \qquad (9.2)$$

The derivation of Equation (9.2) is left as an exercise.

When $\Lambda(t) = 0$, then Equation (9.1) reduces to the following nonautonomous *homogeneous linear equation*,

$$x(t+1) = r(t)x(t), \text{ with } x(0) = x_0. \qquad (9.3)$$

Using the variation of constants formula, Equation (9.2), we obtain the following solution of model (9.3):

$$x(t) = \begin{cases} \Pi_{i=0}^{t-1} r(i) x_0 & \text{for } t = 1, 2, \cdots \\ x(0) = x_0 & for t = 0. \end{cases}$$

As a result, we obtain the following result for the periodically forced homogenous equation.

9.3. LINEAR PERIODIC DIFFERENCE EQUATIONS

Theorem 79 *In model (9.3), let*
$$r(t+p) = r(t),$$
where the period $p > 1$. Then the population goes extinct if
$$r_0 := \Pi_{i=0}^{p-1} r(i) < 1 \text{ or } x(0) = 0,$$
while it persists if
$$r_0 := \Pi_{i=0}^{p-1} r(i) > 1 \text{ and } x(0) > 0.$$
Furthermore, there exists a p-periodic solution if and only if $r_0 = 1$.

9.3.1 Attenuance Versus Resonance in Linear Periodic Models

One ecological question that has been studied by means of periodic models concerns the effect of a periodic environment on a population. Is the average of the resulting population oscillations in the periodically forced model less than the average of the carrying capacity in the corresponding model with constant model parameters [54], [55], [96] and [265]? For example, periodically forced continuous-time and discrete-time logistic equations have been used to show that a periodic environment is always deleterious. Others have obtained similar results for populations governed by periodically forced single-species population models such as the discrete logistic (Beverton–Holt) and Ricker models [96], [128], [129], [130], [141], [142], [192] and [193].

In Chapter 1, for the following autonomous linear population model without structure:
$$x(t+1) = \Lambda + \sigma x(t), \text{ where } \sigma \in (0,1) \text{ and } \Lambda \in [0, \infty), \tag{9.4}$$
we obtained that
$$x(t) = \left(x(0) - \frac{\Lambda}{1-\sigma}\right)\sigma^t + \frac{\Lambda}{1-\sigma},$$
and the population model has a globally asymptotically stable fixed point at the carrying capacity
$$x^* := K = \frac{\Lambda}{1-\sigma}.$$
When $\Lambda = 0$, then $K = 0$ and the population goes extinct. However, when $\Lambda > 0$ then the population persists on the positive fixed point $K = \frac{\Lambda}{1-\sigma}$.

To introduce K explicitly in model (9.4), we replace Λ by $K(1-\sigma)$ and obtain
$$x(t+1) = K(1-\sigma) + \sigma x(t), \text{ where } K > 0.$$
When this model is subjected to $p - periodic$ forcing in either the carrying capacity or survival probability, it becomes
$$x(t+1) = K(t)(1 - \sigma(t)) + \sigma(t) x(t), \tag{9.5}$$
where for all $t \in \mathbb{Z}_+$, $\sigma(t) \in (0,1)$, $\sigma(t) = \sigma(t+p)$, $K(t) = K(t+p) > 0$ and $p > 1$. Model (9.5) is an example of a single-species nonautonomous population model with a periodically forced constant recruitment function, where the carrying capacity is positive.

In [141], the authors obtained that the population governed by model (9.5) persists on a globally asymptotically stable cycle,
$$\{\overline{x}(0), \overline{x}(1), \overline{x}(2), ..., \overline{x}(p-1)\}.$$

The proof uses general results of [129] and [134]. To introduce the general results, we consider the following nonautonomous periodic system of period p:

$$x(t+1) = f_t(x(t)), \qquad (9.6)$$

where if $t \geq p$, then

$$f_t = f_{t \mod p},$$

where $\mod k$ denotes modulo k. Let

$$\{f_0, f_1, ..., f_{p-1}\}$$

be the set of maps that define Equation (9.6), where $f_p = f_0$, $f_{p+1} = f_1$, and so on.

Theorem 80 *[134] Let*

$$f : X \to X$$

be a continuous map on a connected metric space X. If a r − cycle, c_r, is globally asymptotically stable, then c_r must be a fixed point.

Theorem 81 *[129] Assume that X is a connected metric space and each map f_i of (9.6) is a continuous map on X. Let*

$$c_r = \{x(0), x(1), ..., x(r-1)\}$$

be a r − cycle of the p − periodic Equation (9.6). If c_r is globally asymptotically stable, then r divides p.

Proof. Let

$$\Phi_p = f_{p-1} \circ f_{p-2} \circ \cdots \circ f_1 \circ f_0$$

be the composition map of the maps of Equation (9.6). Then the map Φ_p is continuous. We have the following two possibilities.

(1) If

$$\Phi_p(x(0)) = x(0),$$

then r must divide p.

(2) If

$$\Phi_p(x(0)) \neq x(0), \text{ then } \Phi_p^m(x(0)) = x(0),$$

where $m = \frac{s}{p}$, where s is the least common multiple of r and p. Since $x(0)$ is globally asymptotically stable, it follows from Theorem 80 that $m = 1$. Hence, r divides p. ∎

Now, we prove that model (9.5) has a globally asymptotically stable cycle,

$$\{\overline{x}(0), \overline{x}(1), \overline{x}(2), ..., \overline{x}(p-1)\}.$$

Theorem 82 *[141] Model (9.5) has a globally asymptotically stable r-periodic cycle that starts at*

$$\overline{x}(0) = \frac{K(p-1)\widehat{\sigma}(p-1) + \sigma(p-1)K(p-2)\widehat{\sigma}(p-2) + \cdots + K(0)\widehat{\sigma}(0)\Pi_{t=1}^{p-1}\sigma(t)}{1 - \Pi_{t=0}^{p-1}\sigma(t)}, \qquad (9.7)$$

where r divides p, and for each $t \in \{0, 1, 2, \cdots, p-1\}$

$$\widehat{\sigma}(t) = (1 - \sigma(t)).$$

9.3. LINEAR PERIODIC DIFFERENCE EQUATIONS

Proof. For each $t \in \{0, 1, 2, ..., p-1\}$, define
$$f_t : \mathbb{R}_+ \to \mathbb{R}_+$$
by
$$f_t(x) = K(t)(1 - \sigma(t)) + \sigma(t) x.$$
The set of iterates of the $p - periodic$ dynamical system $\{f_0, f_1, f_3, ..., f_{p-1}\}$ is equivalent to the set of population sequences generated by model (9.5). Moreover,
$$f_0(x(0)) = K(0)(1 - \sigma(0)) + \sigma(0) x(0),$$

$$\begin{aligned}f_1 \circ f_0(x(0)) &= K(1)(1 - \sigma(1)) + \sigma(1)(\sigma(0) x(0) + K(0)(1 - \sigma(0))) \\ &= K(1)(1 - \sigma(1)) + \sigma(1) K(0)(1 - \sigma(0)) + \sigma(1) \sigma(0) x(0).\end{aligned}$$

By mathematical induction, for $p > 1$
$$\begin{aligned}f_{p-1} \circ \cdots \circ f_1 \circ f_0(x(0)) &= K(p-1)(1 - \sigma(p-1)) + \sigma(p-1) K(p-2)(1 - \sigma(p-2)) + \cdots \\ &\quad + K(0)(1 - \sigma(0)) \Pi_{t=1}^{p-1} \sigma(t) + x(0) \Pi_{t=0}^{p-1} \sigma(t).\end{aligned}$$

The fixed point of $f_{p-1} \circ \cdots \circ f_1 \circ f_0$, which is also the initial point of a periodic cycle of model (9.5) is given by Equation (9.7).

Since $f_{p-1} \circ \cdots \circ f_1 \circ f_0$ is an affine map with slope $0 < \Pi_{t=0}^{p-1} \sigma(t) < 1$, its fixed point is globally asymptotically stable. Hence, the cycle is globally asymptotically stable. Using Theorem 81, we obtain that the period of the globally asymptotically stable must be a divisor of p. ∎

The length of the globally asymptotically stable cycle predicted by Theorem 82 can be any divisor of p. In the following example, we illustrate a globally asymptotically $2 - cycle$ in a $4 - periodic$ linear model.

Example 107 *Set the following parameter values in model (9.5), where $p = 4$.*
$$K(0) = 1, \quad K(1) = 2, \quad K(2) = \tfrac{359}{364}, \quad K(3) = 2.$$
$$\sigma(0) = \tfrac{1}{10}, \quad \sigma(1) = \tfrac{9}{10}, \quad \sigma(2) = \tfrac{2}{10}, \quad \sigma(3) = \tfrac{9}{10}.$$

Using Equation (9.7), we obtain that the $4 - periodic$ linear model has a globally asymptotically stable $2 - cycle$
$$\left\{\frac{101}{91}, \frac{92}{91}\right\}.$$

Periodically forced models such as model (9.5) usually generate cycles [128], [129], [130], [131], [141] and [142]. To study the effects of periodic fluctuations on populations governed by periodically forced models, we use the following definition and compare the average of the cycles to the average of the carrying capacity [96].

Definition 35 *[96] Consider the $p-$ periodic model*

$$x(t+1) = f_t(x(t)),$$

where there exists $p \in \mathbb{Z}_+$ such that

$$f_{t+p}(x(t)) = f_t(x(t)).$$

A periodic orbit of the $p-$periodic model is attenuant (resonant) if its average value is less (greater) than the average of the carrying capacities of the map $f_t(x)$.

We use the following bounds on the average of the periodic orbit to study the impact of periodic forcing on the population governed by model (9.5).

Theorem 83 *[141] In model (9.5),*

$$\frac{1 - \max\{\sigma(t)\}}{1 - \min\{\sigma(t)\}} \sum_{t=0}^{p-1} K(t) \leq \sum_{t=0}^{p-1} \overline{x}(t) \leq \frac{1 - \min\{\sigma(t)\}}{1 - \max\{\sigma(t)\}} \sum_{t=0}^{p-1} K(t).$$

Proof. Let $K(t) = K(t \mod p)$ and $\sigma(t) = \sigma(t \mod p)$. Then

$$\overline{x}(t+1) = K(t)(1 - \sigma(t)) + \sigma(t)\overline{x}(t) \text{ for all } t.$$

Since $\overline{x}(t)$ is $p-$ *periodic*,

$$\sum_{t=0}^{p-1} \overline{x}(t) = \sum_{t=0}^{p-1} (K(t)(1-\sigma(t)) + \sigma(t)\overline{x}(t)) = \sum_{t=0}^{p-1} K(t)(1-\sigma(t)) + \sum_{t=0}^{p-1} \sigma(t)\overline{x}(t).$$

Hence,

$$(1 - \max\{\sigma(t)\}) \sum_{t=0}^{p-1} K(t) + \min\{\sigma(t)\} \sum_{t=0}^{p-1} \overline{x}(t)$$

$$\leq \sum_{t=0}^{p-1} \overline{x}(t) \leq (1 - \min\{\sigma(t)\}) \sum_{t=0}^{p-1} K(t) + \max\{\sigma(t)\} \sum_{t=0}^{p-1} \overline{x}(t).$$

Hence,

$$\frac{1 - \max\{\sigma(t)\}}{1 - \min\{\sigma(t)\}} \sum_{t=0}^{p-1} K(t) \leq \sum_{t=0}^{p-1} \overline{x}(t) \leq \frac{1 - \min\{\sigma(t)\}}{1 - \max\{\sigma(t)\}} \sum_{J=0}^{p-1} K(t).$$

∎

Corollary 12 *[141] In model (9.5), if $\min\{\sigma(t)\} = \max\{\sigma(t)\}$, then*

$$\sum_{t=0}^{p-1} \overline{x}(t) = \sum_{t=0}^{p-1} K(t)$$

and the $p-$ periodic orbit is neither attenuant or resonant.

9.3. LINEAR PERIODIC DIFFERENCE EQUATIONS

In Corollary 12, the average total population size remains the same as the average carrying capacity whenever the survival probability is constant while the carrying capacity is periodic. Also, in Theorem 82, when the survival probability is periodic while the carrying capacity is constant, by starting with the initial population size $x_0 = K$, we see that the globally attracting periodic orbit is the constant carrying capacity. Consequently, the total population size remains constant and there is neither attenuance nor resonance in model (9.5) when either the carrying capacity or survival probability is constant.

To illustrate the combined impact of periodic carrying capacity and periodic survival probability on the average biomass, we restrict ourselves to a $period-2$ forcing of both $\sigma(J)$ and $K(J)$, where $J = 0, 1$. Now, let

$$\mathcal{R}_d = \frac{K(0)\sigma(1) + K(1)\sigma(0)}{K(0)\sigma(0) + K(1)\sigma(1)} \text{ and } Sgn = (K(0) - K(1))(-\sigma(0) + \sigma(1)).$$

Recall that, for each $t \in \{0, 1\}$,

$$f_t(x) = K(t)(1 - \sigma(t)) + \sigma(t)x.$$

Thus, when the population is at the carrying capacity and $x = K(t)$, then $\sigma(t)K(t)$ is the population that survives the unit interval described by f_t. Similarly, $\sigma(t+1)K(t)$ is the population that survives the unit interval described by f_{t+1}. Thus, the threshold parameter, $\mathcal{R}_d$, gives the effective ratio of the average survivors in the periodic versus constant environments.

$\mathcal{R}_d < 1$ is equivalent to $Sgn < 0$. Consequently, $\mathcal{R}_d < 1$ when the carrying capacity and the survival probability are oscillating in synchrony and $\mathcal{R}_d > 1$ when they are asynchronous. In the next theorem, we show that the globally attracting $2-cycle$ predicted by Theorem 82 is attenuant and the $period-2$ forcing is detrimental to the population when $\mathcal{R}_d < 1$. However, if $\mathcal{R}_d > 1$ then the $2-cycle$ is resonant and the $period-2$ forcing is beneficial to the population.

Theorem 84 *[141] Model (9.5) with $p = 2$ has a globally attracting attenuant $2-cycle$ if $Sgn < 0$ and a globally asymptotically stable resonant $2-cycle$ if $Sgn > 0$. That is, model (9.5) with $p = 2$ has a globally asymptotically stable attenuant $2-cycle$ when the carrying capacity and the survival probability are oscillating in synchrony, and a globally asymptotically stable $2-cycle$ when they are asynchronous.*

Proof.

$$\bar{x}(0) + \bar{x}(1) = \frac{K(1)(1-\sigma(1)) + K(0)(1-\sigma(0))\sigma(1)}{1 - \sigma(0)\sigma(1)}$$

$$+ \frac{K(0)(1-\sigma(0)) + K(1)(1-\sigma(1))\sigma(0)}{1 - \sigma_0\sigma_1}$$

$$= \frac{K(0)(1-\sigma(0))(1+\sigma(1)) + K(1)(1-\sigma(1))(1+\sigma(0))}{1 - \sigma(0)\sigma(1)}.$$

To investigate the occurrence of attenuance and resonance, we consider

$$\bar{x}(0) + \bar{x}(1) - (K(0) + K(1)) = \frac{K(0)(1-\sigma(0))(1+\sigma(1)) + K(1)(1-\sigma(1))(1+\sigma(0))}{1 - \sigma(0)\sigma(1)}$$

$$- (K(0) + K(1))$$

$$= \frac{K(0)(-\sigma(0) + \sigma(1)) + K(1)(-\sigma(1) + \sigma(0))}{1 - \sigma(0)\sigma(1)}.$$

Since $1 - \sigma(0)\sigma(1) > 0$, the occurrence of resonance is equivalent to

$$K(0)(-\sigma(0) + \sigma(1)) + K(1)(-\sigma(1) + \sigma(0)) > 0$$
$$\iff K(0)\sigma(1) + K(1)\sigma(0)$$
$$> K(0)\sigma(0) + K(1)\sigma(1)$$
$$\iff Sgn > 0.$$

Proceeding using similar arguments establishes that the occurrence of attenuance is equivalent to $Sgn < 0$. ∎

9.4 Periodically Forced Discrete Logistic (Beverton–Holt) Model

In this section, we consider the discrete logistic (Beverton–Holt) model,

$$x(t+1) = b\frac{x(t)}{1+cx(t)}, \text{ where } b, c > 0.$$

If $b < 1$, then the population governed by the discrete logistic (Beverton–Holt) model goes extinct. Consequently, to study a persistent population in a seasonal environment and with a seasonal birth rate, we assume that $b > 1$. As a proxy for the environment, we use the "carrying capacity", that is, the positive equilibrium of the equation, which we denote by K:

$$x^* = K := \frac{b-1}{c},$$

and place it explicitly into the equation by the substitution $c = \frac{(b-1)}{K}$ to get

$$x(t+1) = b\frac{K}{K+(b-1)x(t)}x(t)..$$

Consequently, the discrete logistic (Beverton–Holt) model with periodically forced growth rate, $b(t)$, and carrying capacity, is

$$x(t+1) = \frac{b(t)K(t)}{K(t)+(b(t)-1)x(t)}x(t), \tag{9.8}$$

where $b(t+p) = b(t) > 1$, and $K(t+p) = K(t)$ for all $t \in \mathbb{Z}_+$.

9.4.1 The Cushing–Henson Conjecture: Discrete Logistic (Beverton–Holt) Recruitment

For the $p-periodic$ discrete logistic (Beverton–Holt) model,

$$x(t+1) = b\frac{K(t)}{K(t)+(b-1)x(t)}x(t), \text{ with } b > 1, K(t+p) = K(t) > 0 \text{ and } p \geq 2,$$

in [97], Cushing and Henson conjectured the following:

9.4. PERIODICALLY FORCED DISCRETE LOGISTIC (BEVERTON–HOLT) MODEL

1. There is a positive $p-periodic$ solution
$$\{\overline{x}(0), \overline{x}(1), \cdots, \overline{x}(p-1)\}$$
that globally attracts all positive solutions.

2. The average
$$av(\overline{x}(t)) = \frac{1}{p}\sum_{i=0}^{p-1}\overline{x}(i)$$
satisfies the inequality
$$av(\overline{x}(t)) < av(K(t)).$$

That is, a periodically forced carrying capacity has a deleterious effect on the average of the asymptotically stable population cycle.

Cushing and Henson proved both conjectures for $p = 2$ in [58]. To study extensions of the conjectures, we write model (9.8) in the form $x(t+1) = f_t(x(t))$ for each $t \in \mathbb{Z}_+$, where
$$f_t(x) = \frac{b(t)K(t)}{K(t)+(b(t)-1)x}x.$$

Hence, the composite map
$$\Phi_p(x) = f_{p-1} \circ f_{p-2} \circ \cdots \circ f_0(x)$$
defines a discrete semidynamical system on $\mathbb{R}^+$ given by
$$x(t+1) = \Phi_p(x(t)).$$

Therefore,
$$\Phi_p(x) = \frac{Q(p-1)L(p-1)x}{L(p-1)+E(p-1)x},$$
where
$$L(p-1) = K(p-1)...K(0), \ Q(p-1) = b(p-1)...b(0),$$
and
$$E(p-1) = K(p-1)E(p-2)+(b(p-1)-1)b(p-2)b(p-3)...\mu(0)K(p-2)K(p-3)...K(0).$$
On letting
$$E(0) = b(0)-1,$$
we obtain
$$E(p-1) = \left(\Pi_{j=0}^{p-2}K(j+1)\right)(b(0)-1)+\sum_{s=0}^{p-2}\left[(b(s+1)-1)\Pi_{j=s+1}^{p-2}K(j+1)\Pi_{j=0}^{s}b(j)\Pi_{j=0}^{s}K(j)\right].$$

Notice that
$$\Pi_{j=1}^{0}K(j+1) = 1.$$

Furthermore,
$$\Phi_p(x) = \frac{Q(p-1)L(p-1)x}{L(p-1) + E(p-1)x}$$
is an autonomous discrete logistic (Beverton–Holt) map. Consequently, the unique positive fixed point of Φ_p is globally asymptotically stable in $(0, \infty)$. Hence, the p-periodic equation has a unique globally asymptotically stable p-periodic cycle.

Model (9.8) has a p-periodic solution if and only if
$$\Phi_p(\overline{x}(0)) = \overline{x}(0).$$
That is,
$$\overline{x}(0) = \frac{Q(p-1)L(p-1)\overline{x}(0)}{L(p-1) + E(p-1)\overline{x}(0)}.$$
Hence,
$$\overline{x}(0) = \frac{L(p-1)(Q(p-1)-1)}{E(p-1)}$$
is a point in the unique periodic orbit of model (9.8). Since H is the autonomous discrete logistic (Beverton–Holt) map (or belong to the class $\mathcal{K}$ defined in [129]), the unique periodic orbit is globally asymptotically stable. This establishes the first of the Cushing–Henson conjectures for $p \geq 2$ with $b(t+p) = b(t) > 1$ and $K(t+p) = K(t)$ for all $t \in \mathbb{Z}_+$.

9.4.2 Deleterious Effects of Periodic Forcing

Now, we use a proof of [130] that is similar to that of [192] to establish the second conjecture,
$$av(\overline{x}(t)) < av(K(t)).$$
Let
$$c(p) = \{\overline{x}(0), \overline{x}(1), \cdots, \overline{x}(p-1)\}$$
be the unique periodic orbit of model (9.8). Then
$$\begin{aligned}
av(\overline{x}(t)) &= \frac{1}{p}\sum_{i=0}^{p-1}\overline{x}(i) = \frac{1}{p}\sum_{i=0}^{p-1}\overline{x}(i+1) \\
&= \frac{1}{p}\sum_{i=0}^{p-1}\frac{b(i)K(i)}{K(i)+(b(i)-1)\overline{x}(i)}\overline{x}(i) \\
&= \frac{1}{p}\sum_{i=0}^{p-1}\frac{\left(\frac{b(i)K(i)}{b(i)-1}\right)\left(\frac{(b(i)-1)}{K(i)}\overline{x}(i)\right)}{1+\left(\frac{(b(i)\mu(i)-1)}{K(i)}\overline{x}(i)\right)} \\
&= \frac{1}{p}\frac{\left[\sum_{i=0}^{p-1}\left(\frac{b(i)K(i)}{b(i)-1}\right)f\left(\frac{(b(i)-1)}{K(i)}\overline{x}(i)\right)\right]}{\sum_{i=0}^{p-1}\left(\frac{b(i)K(i)}{b(i)-1}\right)}\sum_{i=0}^{p-1}\left(\frac{b(i)K(i)}{b(i)-1}\right),
\end{aligned}$$
where
$$f(x) = \frac{x}{1+x}.$$

Clearly, $f'(x) > 0$ and $f''(x) < 0$ for all x. Thus, the concave function f satisfies the Jensen's inequality

$$f\left(\frac{\sum_{i=0}^{p-1} w(i) u(i)}{\sum_{i=0}^{p-1} w(i)}\right) > \frac{\sum_{i=0}^{p-1} w(i) f(u(i))}{\sum_{i=0}^{p-1} w(i)}.$$

Letting

$$w(i) = \frac{b(i) K(i)}{b(i) - 1}, u(i) = \frac{(b(i) - 1)}{K(i)} \overline{x}(i)$$

and applying Jensen's inequality we find that

$$av(\overline{x}(t)) < \frac{1}{p} \sum_{i=0}^{p-1} \frac{b(i) K(i)}{b(i) - 1} f\left(\frac{\sum_{i=0}^{p-1} \left(\frac{b(i)K(i)}{b(i)-1}\right) \frac{(b(i)-1)}{K(i)} \overline{x}(i)}{\sum_{i=0}^{p-1} \frac{b(i)K(i)}{(b(i)-1)}}\right)$$

$$= \frac{1}{p} \sum_{i=0}^{p-1} \frac{b(i) K(i)}{b(i) - 1} \frac{\sum_{i=0}^{p-1} b(i) \overline{x}(i) / \sum_{i=0}^{p-1} \frac{b(i)K(i)}{(b(i)-1)}}{1 + \frac{\sum_{i=0}^{p-1} b(i)\overline{x}(i)}{\sum_{i=0}^{p-1} \frac{b(i)K(i)}{(b(i)-1)}}}$$

$$= \frac{1}{p} \sum_{i=0}^{p-1} \frac{b(i) K(i)}{b(i) - 1} \left[\frac{\frac{1}{p}\sum_{i=0}^{p-1} b(i) \overline{x}(i)}{\frac{1}{p}\sum_{i=0}^{p-1} \frac{b(i)K(i)}{(b(i)-1)} + \frac{1}{p}\sum_{i=0}^{p-1} b(i)\overline{x}(i)}\right].$$

Let

$$T_p = \frac{1}{p} \sum_{i=0}^{p-1} \frac{b(i) K(i)}{(b(i) - 1)}.$$

Then

$$av(\overline{x}(t)) < T_p \frac{b^* av(\overline{x}(t))}{T_p + b_* av(\overline{x}(t))},$$

where

$$b^* = \max\{b(i)\} \text{ and } b_* = \min\{b(i)\}.$$

Hence,

$$av(\overline{x}(t)) < T_p \frac{(b^* - 1)}{b_*}$$

and

$$av(\overline{x}(t)) < \frac{b^* (b^* - 1)}{b_* (b_* - 1)} av(K(t)). \tag{9.9}$$

While Equation (9.9) gives an upper bound on the constant relating the two averages, it suffers from the deficiency that if one of the μ values is close to unity, the constant becomes very large. If $\mu(t)$ is constant, then

$$b_* = b^*$$

and

$$av(\overline{x}(t)) < av(K(t)).$$

A periodically forced carrying capacity has a deleterious effect on the average of the asymptotically stable population cycle whenever the growth rate, $\mu(t)$, is constant. In the next section, we refine the estimate

$$av(\overline{x}(t)) < \frac{b^*(b^*-1)}{b_*(b_*-1)} av(K(t))$$

for period $p = 2$.

9.4.3 Discrete Logistic (Beverton–Holt) Recruitment with $p = 2$

In this section, we consider model (9.8), where $p = 2$. To refine the estimate, Equation (9.9), we make use of the following two auxiliary results.

Lemma 16 *[130] Assume $\alpha, \beta, x, y \in (0,1)$, where*

$$\alpha + \beta = 1.$$

Then

$$\frac{xy}{\alpha x + \beta y} - \beta x - \alpha y = \frac{-\alpha\beta(x-y)^2}{\alpha x + \beta y}. \tag{9.10}$$

Proof. Let

$$g(x,y) = \frac{xy}{\alpha x + \beta y} - \beta x - \alpha y.$$

Then

$$(\alpha x + \beta y) g(x,y) = \left\{ (1 - \alpha^2 - \beta^2) xy - \alpha\beta(x^2 + y^2) \right\} = -\alpha\beta(x-y)^2.$$

∎

The next auxiliary result follows from elementary calculus.

Lemma 17 *[130] For $x, a > 1$, define*

$$u(x,a) = \frac{|a-x|}{ax-1}.$$

Then

$$u(a,a) = 0 \leq u(x,a) < 1.$$

As in [130], consider the two discrete logistic (Beverton–Holt) functions,

$$f(x) = \frac{b_f x_f}{x_f + (b_f - 1)x} x \text{ and } g(x) = \frac{b_g x_g}{x_g + (b_g - 1)x} x.$$

Thus, x_f and b_f (respectively, x_g and b_g) denote, respectively, the stable fixed point and growth rate of f (respectively, g). Next, we derive formulas for the fixed points, $x_{f \circ g}$ and $x_{g \circ f}$ of the composition functions $f \circ g$ and $g \circ f$, respectively.

$$(g \circ f)(x) = \frac{b_f b_g x_f x_g}{x_f x_g + [(b_f - 1)x_g + (b_g - 1)b_f x_f]x} x = \frac{b_f b_g x_{g \circ f} x}{x_{g \circ f} + (b_f b_g - 1)x}, \tag{9.11}$$

9.4. PERIODICALLY FORCED DISCRETE LOGISTIC (BEVERTON–HOLT) MODEL

where
$$x_{g \circ f} = \frac{x_g x_f}{r x_g + s x_f}, r = \frac{b_f - 1}{b_f b_g - 1} \text{ and } s = \frac{b_g - 1}{b_f b_g - 1} b_f.$$

Notice that
$$r + s = 1.$$

From Equation (9.11), we see that the composition of two discrete logistic (Beverton–Holt) models f and g is again a discrete logistic (Beverton–Holt) model where $x_{g \circ f}$ given explicitly in the equation. As a result, the composition model has a globally asymptotically stable fixed point.

Let
$$f_0(x) = \frac{b(0) x_0}{K(0) + (b(0) - 1) x} x \text{ and } f_1(x) = \frac{b(0) x_1}{K(1) + (b(0) - 1) x} \quad (9.12)$$

and let x_0 be the fixed point of $f_1 \circ f_0$. Then we have
$$x_0 = x_{f_1 \circ f_0} = \frac{K(0) K(1)}{r K(1) + s K(0)} = r K(0) + s K(1) - \frac{r s (K(1) - K(0))^2}{r K(1) + s K(0)}. \quad (9.13)$$

By letting
$$\lambda = b(0) b(1) - 1$$
and substituting from Equation (9.11), we obtain
$$\lambda x_0 = (b(0) - 1) K(0) + (b(1) - 1) b(0) K(1) - \frac{(b\mu(0) - 1)(b(1) - 1) b(0) (K(1) - K(0))^2}{(b(0) - 1) K(1) + (b(1) - 1) b(0) K(0)}.$$

A similar expression for x_1 is obtained by interchanging all subscripts. Adding the two and letting
$$\bar{x} = \frac{x_0 + x_1}{2}, \; \overline{K} = \frac{K(0) + K(1)}{2} \text{ and } \widehat{K} = K(0) - K(1),$$
we obtain
$$\bar{x} = \overline{K} + \sigma \frac{\widehat{K}}{2} - \Delta \frac{(b(0) - 1)(b(1) - 1) b(0) (\widehat{K})^2}{2 b(0) b(1) - 1} [128], \quad (9.14)$$

where
$$\Delta = \frac{b(0)((b(1))^2 - 1) K(0) + b(1)((b(0))^2 - 1) K(1)}{b(0)(b(1) - 1)^2 (K(0))^2 + (b(0) - 1)(b(1) - 1)(b(0) b(1) + 1) K(0) K(1) + b(1)(b(0) - 1)^2 (K(1))^2} > 0 \quad (9.15)$$

and
$$\sigma = \frac{b(0) - b(1)}{b(0) b(1) - 1}.$$

From Lemma 17,
$$0 \le |\sigma| < 1.$$

When $p = 2$, this equality is the final chapter in the saga of the Cushing–Henson conjecture.

Remark 23 *[130] In the case*
$$b(0) = b(1) = b,$$
the expression of Equation (9.14) reduces to
$$\bar{x} = \overline{K} - \frac{1}{2} \frac{b(K(0) + K(1))}{b(K(0))^2 + (b^2 + 1)K(0)K(1) + b(K(1))^2} (K(1) - K(0))^2,$$
an exact expression for the difference in the averages.

Remark 24 *[130] For certain values of the b's and K's, the state average $\bar{x}$ can exceed $\overline{K}$. For example, when*
$$b(0) = 4, b(1) = 2, K(0) = 11 \text{ and } K(1) = 7,$$
then
$$\bar{x} = 9.23 > \overline{K} = 9.$$

9.5 Nonautonomous Ricker Model

First, we consider the autonomous Ricker model,
$$x(t+1) = bx(t)\exp(-cx(t)), \text{ where } b, c > 0.$$
As with the discrete logistic (Beverton–Holt) model, if $b < 1$, then the population governed by the Ricker model goes extinct. Consequently, to study a persistent population in a seasonal environment and with a seasonal birth rate, we assume that $b > 1$. As a proxy for the environment, we use the "carrying capacity", that is, the positive equilibrium of the equation, which we denote by K:
$$x^* = K := \frac{\ln b}{c},$$
and place it explicitly into the equation by the substitution $c = \frac{\ln b}{K}$ to get
$$x(t+1) = x(t)\exp\left[r\left(1 - \frac{x(t)}{K}\right)\right],$$
where $r = \ln b > 0$. Consequently, we obtain the nonautonomous Ricker model
$$x(t+1) = x(t)\exp\left[r(t)\left(1 - \frac{x(t)}{K(t)}\right)\right], \qquad (9.16)$$
where $x(0) \geq 0$, $\{r(t)\}$ and $\{K(t)\}$ are strictly positive sequences of real numbers defined for all $t \in \{0, 1, 2, ...\}$. Next, we establish the following persistence result for model (9.16).

Theorem 85 *[303] If there exist positive constants r_*, r^*, K_* and K^* such that*
$$0 < r_* \leq r(t) \leq r^*, \quad 0 < K_* \leq K(t) \leq K^*, \quad t \in \{0, 1, 2, ...\},$$
then any positive solution of
$$x(t+1) = x(t)\exp\left[r(t)\left(1 - \frac{x(t)}{K(t)}\right)\right]$$

satisfies
$$u_* \leq \liminf_{t\to\infty} x(t) \leq \limsup_{t\to\infty} x(t) \leq u^*, \qquad (9.17)$$
where
$$u^* = \frac{K^*}{r^*}\exp(r^*-1) \text{ and } u_* = K_*\exp\left[r^*\left(1-\frac{u^*}{K_*}\right)\right].$$

Proof. We use two cases to show that
$$\limsup_{t\to\infty} x(t) \leq u^*.$$

CASE 1. There exists a positive integer t_0 such that
$$x(t_0) < x(t_0+1).$$

Now
$$1 - \left(\frac{x(t_0)}{K(t_0)}\right) > 0$$
implies that
$$x(t_0) < K(t_0) \leq K^*.$$

Using the fact that
$$\max_{x\in\mathbf{R}} x\exp[r(1-x)] = \frac{1}{r}\exp(r-1), \text{ where } r>0,$$
we have
$$\begin{aligned} x(t_0+1) &= x(t_0)\exp\left[r(t_0)\left(1-\frac{x(t_0)}{K(t_0)}\right)\right] \\ &\leq K(t_0)\frac{x(t_0)}{K(t_0)}\exp\left[r^*\left(1-\frac{x(t_0)}{K(t_0)}\right)\right] \\ &\leq \frac{K^*}{r^*}\exp(r^*-1) = u^*. \end{aligned}$$

Next, we establish that $x(t) \leq u^*$ for $t \geq t_0$.

In fact, if there exists an integer $s \geq t_0+2$ such that $x(s) > u^*$ by letting s^* be the least integer between t_0 and s such that
$$x(s^*) = \max_{t_0 \leq t \leq s} x(t),$$
we obtain $s \geq t_0+2$ and $x(s^*) > x(s^*-1)$, which implies $x(s^*) \leq u^* < x(s)$. This is impossible.

CASE 2.
$$x(t) \geq x(t+1) \text{ for } t \in \{0,1,2,...\}.$$

Now
$$1 - \frac{x(t)}{K(t)} \leq 0 \text{ for } t \in \{0,1,2,...\}.$$

Thus,
$$x(t) \geq K(t) \geq K^* \text{ for } t \in \{0,1,2,...\}.$$

Since $\{x(t)\}$ is a nonincreasing sequence that is bounded below by K_*, we know that
$$\lim_{t\to\infty} x(t) = \overline{x} \geq K^*.$$
Letting $t \to \infty$ in model (9.16), we obtain
$$\overline{x} = \lim_{t\to\infty} K(t) \leq K^* \leq u^*.$$
Hence,
$$\limsup_{t\to\infty} x(t) \leq u^*.$$
Next, we show that
$$\liminf_{t\to\infty} x(t) \geq u_*. \tag{9.18}$$
Since $\lim_{t\to\infty} \sup x(t) \leq u^*$, for each $\varepsilon > 0$, there exists a large integer t^* such that
$$x(t) \leq u^* + \varepsilon \text{ for } t \geq t^*.$$
We consider two cases.

CASE (i) There exists a positive integer $\bar{t}_0 \geq t^*$ such that $x(\bar{t}_0 + 1) < x(\bar{t}_0)$. Similar to the proof of CASE 1, we obtain
$$x(t) \geq K_* \exp\left[r^*\left(1 - \frac{u^* + \varepsilon}{K_*}\right)\right] \text{ for } t \geq t^*.$$

CASE (ii) $x(t+1) \geq x(t)$ for $t \geq t^*$.

Since $x(t) \leq u^* + \varepsilon$ for $t \geq t^*$, we know that $\lim_{t\to\infty} x(t) = L$. Letting $t \to \infty$ in model (9.16), we obtain
$$\lim_{t\to\infty} K(t) = L.$$
Hence,
$$L = \lim_{t\to\infty} x(t) = \lim_{t\to\infty} K(t) \geq K_* \geq K_* \exp\left[r^*\left(1 - \frac{u^* + \varepsilon}{K_*}\right)\right].$$
Combining CASE (i) and CASE (ii), we see that
$$\liminf_{t\to\infty} x(t) \geq K_* \exp\left[r^*\left(1 - \frac{u^* + \varepsilon}{K_*}\right)\right].$$
Since ε is arbitrary, we know that (9.18) holds. Combining
$$\limsup_{t\to\infty} x(t) \leq u^*$$
with (9.18) completes the proof. ∎

Remark 25 *From the proof of Theorem 85, we see that, if either $\lim_{t\to\infty} K(t)$ does not exist or $r^* \neq 1$, then $u_* \leq x(t) \leq u^*$ eventually holds.*

9.5.1 Periodically Forced Ricker Model

In this section, we consider model (9.16), where $\{r(t)\}$ and $\{K(t)\}$ are positive *periodic* sequences of real numbers defined for all $t \in \{0, 1, 2, ...\}$. We first establish the following existence result.

Theorem 86 *[303]* *If $\{r(t)\}$ and $\{K(t)\}$ are positive periodic sequences with a common period p, that is,*

$$r(t+p) = r(t) \text{ and } K(t+p) = K(t) \text{ for all } t \in \{0, 1, 2, ...\},$$

then there exists a $p-$periodic solution for model (9.16).

Proof. If $K(t) = K$, a constant, then $x(t) = K$ is a solution of model (9.16), and Theorem 86 holds. Now, we assume that $\{K(t)\}$ is not constant so that $\lim_{t \to \infty} K(t)$ does not exist. By the assumptions, there exist positive constants r_*, r^*, K_* and K^* such that

$$0 < r_* \leq r(t) \leq r^*, \quad 0 < K_* \leq K(t) \leq K^*, \quad t \in \{0, 1, 2, ...\},$$

where $r_* = \min_t \{r(t)\}$, $r^* = \max_t \{r(t)\}$, $K_* = \min_t \{K(t)\}$, and $K^* = \max_t \{K(t)\}$. Using the proof of Theorem 85, it is easy to see that

$$x(0) \in [u_*, u^*] \text{ implies } x(t) \in [u_*, u^*] \text{ for } t \in \{0, 1, 2, ...\}. \qquad (9.19)$$

Now, we define the map

$$F : [u_*, u^*] \to [u_*, u^*] \text{ by } F(x(0)) = x(p).$$

From model (9.16), we see that $x(p)$ depends continuously on $x(0)$. Thus, F is continuous and maps the interval $[u_*, u^*]$ into itself. Therefore, F has a fixed point x^*. Let $x(0) = x^*$, then the corresponding solution $\{\widetilde{x}(t)\}$ of model (9.16) is a $p-$periodic solution to model (9.16) in $[u_*, u^*]$. ∎

With an additional condition, in the next theorem, we establish the global asymptotic stability of the periodic solution obtained in Theorem 86.

Theorem 87 *[303]* *Suppose*

$$r(t+p) = r(t) \text{ and } K(t+p) = K(t) \text{ for all } t \in \{0, 1, 2, ...\},$$

and

$$\frac{K^*}{K_*} \exp(r^* - 1) \leq 2,$$

where $r_* = \min_t \{r(t)\}$, $r^* = \max_t \{r(t)\}$, $K_* = \min_t \{K(t)\}$, and $K^* = \max_t \{K(t)\}$. If $\{\widetilde{x}(t)\}$ is a periodic solution of model (9.16), then for every positive solution $\{x(t)\}$ of model (9.16),

$$\lim_{t \to \infty} (x(t) - \widetilde{x}(t)) = 0.$$

Proof. Note that if $K(t) = K$, a constant, then $\frac{K^*}{K_*}\exp(r^* - 1) \leq 2$ implies that $r^* \leq 1 + \ln 2 < 2$. Using results from [304], we know that

$$\lim_{t \to \infty} x(t) = K.$$

This implies that
$$\lim_{t\to\infty} (x(t) - \widetilde{x}(t)) = 0,$$
where $\widetilde{x}(t) = K$.

Now, assume that $\{K(t)\}$ is not constant. Let $x(t) = \widetilde{x}(t)\exp(y(t))$. Then model (9.16) is transformed to the equation
$$y(t+1) = y(t) - \frac{r(t)}{K(t)}\widetilde{x}(t)(\exp(y(t)) - 1).$$

Define $V(t) = y^2(t)$. Then
$$\begin{aligned}
\Delta V(t) &= V(t+1) - V(t) \\
&= (y(t+1) - y(t))(y(t+1) + y(t)) \\
&= -\frac{r(t)}{K(t)}\widetilde{x}(t)(\exp(y(t)) - 1)\left(2y(t) - \frac{r(t)}{K(t)}\widetilde{x}(t)(\exp(y(t)) - 1)\right) \\
&= -\frac{r(t)}{K(t)}\widetilde{x}(t)(\exp(\theta y(t)))\left(2 - \frac{r(t)}{K(t)}\widetilde{x}(t)(\exp(\theta y(t)))\right)y^2(t),
\end{aligned}$$
for some $\theta \in (0,1)$. Since $\widetilde{x}(t)(\exp(\theta y(t)))$ lies between $\widetilde{x}(t)$ and $x(t)$, by Theorem 85 and Remark 25, we know that there exists a positive integer t_1 such that
$$2 - \frac{r(t)}{K(t)}(\exp(\theta y(t))) \geq 2 - \frac{r^*u^*}{K_*} = 2 - \frac{K^*}{K_*}\exp(r^* - 1) \geq 0, \text{ for } t \geq t_1.$$

This implies that $\{V(t)\}$ is nonincreasing for $t \geq t_1$. As a result,
$$\lim_{t\to\infty} V(t) = v^* \in [0, \infty).$$

Next, we establish that $v^* = 0$.

Proof of Claim: If $v^* > 0$, then $y(t) \geq \sqrt{v^*}$ for $t \geq t_1$. Since $\{K(t)\}$ is not constant, there exists an integer q with $0 \leq q < p$ such that $K(q) > K_*$. Recall that
$$\Delta V(t) = -\frac{r(t)}{K(t)}\widetilde{x}(t)(\exp(y(t)) - 1)\left(2y(t) - \frac{r(t)}{K(t)}\widetilde{x}(t)(\exp(y(t)) - 1)\right).$$

Consequently, we have
$$\Delta V(q + tp) \leq -\frac{r_*}{K^*}\left(2 - \frac{r^*}{K(q)}u^*\right)v^* < 0, \text{ for } t \geq t_1.$$

This implies that $\sum_{t=0}^{\infty} \Delta V(t)$ diverges to $-\infty$. But from $\lim_{t\to\infty} V(t) = v^* \in [0, \infty)$,
$$\sum_{t=0}^{\infty} \Delta V(t) = v^* - V(0),$$
a contradiction. Therefore, $v^* = 0$. Thus, $\lim_{t\to\infty} y(t) = 0$, and the equation
$$\lim_{t\to\infty}(x(t) - \widetilde{x}(t)) = 0$$
holds. ∎

9.5. NONAUTONOMOUS RICKER MODEL

Remark 26 By Theorem 86, $\{\widetilde{x}(t)\}$ is the global attractor for all positive solutions of model (9.16). Hence, $\{\widetilde{x}(t)\}$ is the unique $p-periodic$ solution of model (9.16).

Remark 27 When $r(t) = r$, a constant, then the $p-periodic$ model (9.16) reduces to

$$x(t+1) = x(t)\exp\left[r\left(1 - \frac{x(t)}{K(t)}\right)\right], \ K(t+p) = K(t), \ t \in \{0,1,2,...\}. \tag{9.20}$$

In [271], it was shown that, when

$$\frac{K^*}{K_*}\exp(r^* - 1) \leq 2,$$

then the unique $p-periodic$ solution of model (9.20) is attenuant.

9.5.2 Neither Attenuance nor Resonance when $r(t) = K(t)$

When $r(t) = K(t)$, then the $p-periodic$ model (9.16) reduces to

$$x(t+1) = x(t)\exp(r(t) - x(t)), \ r(t+p) = r(t), \ t \in \{0,1,2,...\}. \tag{9.21}$$

In [271], it was shown that model (9.21) exhibits neither attenuance nor resonance. To illustrate this, we note that if $0 < r(t) < 2$, model (9.21) has a globally asymptotically stable $k-cycle$ [271]. Let

$$C_k = \{\overline{x}_0, \overline{x}_1, ..., \overline{x}_{k-1}\}$$

be the unique $k-periodic$ cycle. Then

$$\begin{aligned}\overline{x}_0 &= \overline{x}_k = \overline{x}_{k-1}e^{r(k-1)-\overline{x}_{k-1}} \\ &= \overline{x}_{k-2}e^{r(k-2)-\overline{x}_{k-2}}e^{r(k-1)-\overline{x}_{k-1}},\end{aligned}$$

and by iteration, we get

$$\overline{x}_0 = \overline{x}_0 e^{\sum_{t=0}^{k-1} r(t) - \sum_{t=0}^{k-1} \overline{x}_t}.$$

Hence,

$$\frac{1}{k}\sum_{t=0}^{k-1} r(t) = \frac{1}{k}\sum_{t=0}^{k-1} \overline{x}_t.$$

That is, model (9.21) exhibits neither attenuance nor resonance [271].

Exercises 9.2-9.5

1. Consider the Ricker (exponential) 2-periodic model

    $$x(t+1) = b(t)x(t)e^{-cx(t)}, \ b(t+2) = b(t), \ 0 < b(t) < e^2, \ t = 0,1,2,....$$

    (i) Find the 2-periodic cycle c_2.
    (ii) Show that c_2 is globally asymptotically stable on $(0, \infty)$.
    (iii) Show that the system is attenuant.

2. Consider the discrete logistic (Beverton–Holt) 2-periodic model
$$x(t) = \frac{bK(t)x(t)}{K(t) + (b-1)x(t)}, \quad K(t+2) = K(t), \ b > 0.$$

   (i) Find the 2-periodic cycle c_2.

   (ii) Show that c_2 is globally asymptotically stable.

   (iii) Determine if the system is attenuant or resonant.

3. Let
$$f(x) = \frac{x}{1+x}.$$

   Clearly, $f'(x) > 0$ and $f''(x) < 0$ for all x. Prove that the concave function f satisfies the Jensen's inequality
$$f\left(\frac{\sum_{i=0}^{p-1} w(i) u(i)}{\sum_{i=0}^{p-1} w(i)}\right) > \frac{\sum_{i=0}^{p-1} w(i) f(u(i))}{\sum_{i=0}^{p-1} w(i)}.$$

4. Give an example of a globally attracting 3-cycle in a 6-periodic linear model of the form
$$x(t+1) = K(t)(1 - \sigma(t)) + \sigma(t) x(t),$$
   where
$$\sigma(t) = \sigma(t+6) \text{ and } K(t) = K(t+6) \text{ for all } t \in \mathbb{Z}_+.$$

5. Consider the 4-periodic discrete logistic (Beverton–Holt) model
$$x(t+1) = \frac{b(t)K(t)x(t)}{K(t) + (b(t)-1)x(t)}$$
   with $b(t+4) = b(t)$, $K(t+4) = K(t)$, $b(0) = 3$, $b(1) = 4$, $b(2) = 2$, $b(3) = 5$, $K(0) = 1$, $K(1) = \frac{6}{17}$, $K(2) = 2$, and $K(3) = \frac{4}{17}$.

   (i) Show that there are no 4-periodic cycles.

   (ii) Show that there is a unique 2-periodic cycle c_2, then compute it.

   (iii) Using the linearization principle determine the stability (local, global, unstable) of c_2.

6. Consider the p-periodic discrete logistic (Beverton–Holt) model
$$x(t+1) = \frac{bK(t)x(t)}{K(t) + (b(t)-1)x(t)} - f_t(x(t))$$
   with $K(t+p) = K(t) > 0$, $b > 1$.

   (i) Find a formula for $\Phi_p(x) = f_{p-1} \circ f_{p-2} \circ \cdots \circ f_0(x)$.

   (ii) Find a formula for the p-periodic cycle c_p of the system.

   (iii) Show that the p-periodic cycle is globally asymptotically stable.

7. Consider the following 2-periodic Ricker model with the Allee effect (1.46)

$$x(t+1) = x(t)\exp\{\alpha(t) - x(t)\}\exp\left\{\frac{m}{1+sx(t)}\right\}$$
$$= f_t(x(t)),$$

where $\alpha(t+2) = \alpha(t)$, $t = 0, 1, 2, \ldots$. Let A_{f_0}, A_{f_1} be the threshold Allee of f_0 and f_1, and K_{f_0}, K_{f_1} be the carrying capacities of f_0 and f_1. Assume that $\alpha_1 > \alpha_0$.

(i) Show that $f_0 \circ f_1(x) < f_1 \circ f_0(x)$ for all $x > 0$.

(ii) Show that there exists a fixed point $K_{f_0 f_1}$ of $f_0 \circ f_1$ and a fixed point $K_{f_1 f_1}$ of $f_1 \circ f_0$ such that $K_{f_0} < K_{f_0 f_1} < K_{f_1 f_0} < K_{f_1}$.

(iii) Show that there exists a fixed point $A_{f_0 f_1}$ of $f_0 \circ f_1$ and a fixed point $A_{f_1 f_0}$ of $f_1 \circ f_0$ such that $f_1^{-1}(A_{f_1}) < A_{f_1 f_0} < A_{f_0}$, and $A_{f_1} < A_{f_0 f_1} < f_0^{-1}(A_0)$.

8. A map f on an interval $[a,b] \subset \mathbb{R}$, b may be ∞, is called unimodal if

(i) $f(b) = 0$ when b is finite, or $\lim_{x \to \infty} f(x) = 0$,

(ii) There is a critical point c_f of f such that f is strictly increasing on $[0, c_f)$ and strictly decreasing on $(c_f, b]$ or (c_f, ∞).

Let f, g be two unimodal maps on $[a, b]$, with $f(x) > g(x)$ on $[a, b]$, x_f^*, and x_g^* are positive fixed points of f and g, respectively.

(i) Show that the composition map $f \circ g$ has a fixed point $x_{f \circ g}^*$ such that $x_g^* < x_{f \circ g}^*$.

(ii) If $c_b, c_f > x_f^*$,

(a) show that $f \circ g(x) > g \circ f(x)$ for all $x \in [x_g^*, x_f^*]$,

(b) show that $x_g^* < x_{g \circ f}^* < x_{f \circ g}^* < x_f^*$.

9. [129] Consider the p-periodic discrete logistic (Beverton–Holt) model

$$x(t+1) = \frac{bK(t)x(t)}{K(t) + (b-1)x(t)}, \quad b > 1,$$

where $K(t+p) = K(t)$ for all $t \in \mathbb{Z}^+$.

(i) Show that if the equation has a r-periodic cycle, then r must divide p.

(ii) Show that the equation has a r-periodic cycle c_r, $r \neq p$, if

$$\frac{L_{p-1}}{M_{p-1}} = \frac{L_{r-1}}{M_{r-1}},$$

where

$$L_s = K_s K_{s-1} \ldots K_0$$

$$M_s = \prod_{j=0}^{s-1} K_{j+1} + \sum_{m=0}^{s-1}\left(\prod_{i=m+1}^{s-1} K_{j+1}\right) b^{m+1} K_m K_{m-1} \ldots K_0$$

and $\bar{x} \in c_r$ is given by
$$\bar{x} = \left(\frac{b^r - 1}{b - 1}\right) \frac{L_{r-1}}{M_{r-1}}.$$

10. [129] Let $\mathcal{F}$ be a class of continuous functions $f : \mathbb{R}^+ \to \mathbb{R}^+$ with the following properties.

    (i) Every $f \in \mathcal{F}$ is concave, i.e., $f(\alpha x + \beta y) \geq \alpha f(x) + \beta f(y)$ for all $x, y \in \mathbb{R}^+$, where $\alpha, \beta \geq 0$, $\alpha + \beta = 1$.

    (ii) There exist x_1, x_2 such that $f(x_1) > x_1$ and $f(x_2) < x_2$, i.e., the graph of f crosses the diagonal.

      (a) Prove that if f and g are concave and f is increasing, then $f \circ g$ is concave.
      (b) $\mathcal{F}$ is closed under the operation of composition, i.e., $f, g \in \mathcal{F}$ implies $f \circ g \in \mathcal{F}$.
      (c) Each $f \in \mathcal{F}$ has a unique globally asymptotically stable fixed point $x_f^* > 0$.
      (d) If $f, g \in \mathcal{F}$ with $x_f^* < x_g^*$, then $x_f^* < x_{f \circ g}^* < x_g^*$ and $x_f^* < x_{g \circ f}^* < x_g^*$.

11. [134] Let $f : \mathbb{R} \to \mathbb{R}$ be a continuous map. Prove that if f has a r-cycle c_r which is globally asymptotically stable, then c_r must be a fixed point.

12. [129] Let $f : \mathbb{R} \to \mathbb{R}$ be continuous maps, with $f_{t+p} = f_t$, $t \in \mathbb{Z}^+$. Assume that the equation $x(t+1) = f_t(x(t))$ has r-periodic cycle c_r which is globally asymptotically stable, then r must divide p.

9.6 Nonlinear Periodic Systems

This section will investigate the dynamics of nonlinear periodically forced systems of difference equations. The study will be divided into two sections. We will begin with monotone periodic systems.

9.6.1 Monotone Periodic Systems

In Chapter 3, we introduced an autonomous competitive system (monotone maps) and investigated the global stability of the coexistence equilibrium (Theorem 33). The theory was applied to Leslie–Gower and Ricker's competition models. In this section, we extend the results to the general map (9.22). Our exposition is based on the work of Balreira and Luis [35].

Let $F : \Omega \subset \mathbb{R}^2_+ \to \mathbb{R}^2_+$ be a competitive map of class C^1. Recall that a map is competitive if whenever $\mathbf{x} \leq_K \mathbf{z}$, then $F(\mathbf{x}) \leq F(\mathbf{z})$, where $\leq_K$ is the southeast order, i.e., $(x_1, y_1) \leq (x_2, y_2)$ if $x_1 \leq x_2$ and $y_1 \geq y_2$. This can be checked easily noting the sign structure of the Jacobian matrix

$$JF(\mathbf{x}) = \begin{pmatrix} + & - \\ - & + \end{pmatrix}.$$

We also assume that Ω is K-convex, that it contains every line segment joining two points in Ω that are ordered with respect to two points in Ω that are ordered with respect to $\leq_K$. In addition, we make the following four assumptions.

H_1: The map F is competitive.

9.6. NONLINEAR PERIODIC SYSTEMS

H$_2$: The extinction fixed point $(0,0)^T$ is a repeller.

H$_3$: The map F has two exclusion fixed points $E_0^* = (x_0^*, 0)$ and $E_1^* = (0, y_1^*)$ that are globally asymptotically stable when restricted to each axis, but a saddle in Ω.

H$_4$: F is a Kolmogorov map, i.e., $F(x,y) = (xf(x,y), yg(x,y))$. Moreover, the reduced map $\tilde{F}(x,y) = (f(x,y), g(x,y))$ is injective.

We are now ready to state the main result.

Theorem 88 *[35] Consider the p-periodic system*

$$\mathbf{x}(t+1) = F_t(\mathbf{x}(t)), \qquad (9.22)$$

where $F_{t+p} = F_t$, $F_t : \Omega \subset \mathbb{R}_+^2 \to \mathbb{R}_+^2$, Ω is a competitive region. Assume that each map F_t satisfies ($\mathbf{H_1}$)-($\mathbf{H_4}$). If

$$\Phi_p = F_{p-1} \circ F_{p-2} \circ \cdots \circ F_1 \circ F_0$$

satisfies ($\mathbf{H_3}$)-($\mathbf{H_4}$), then the periodic system has a globally asymptotically stable periodic cycle of period p or a divisor of p in the interior of Ω.

Example 108 *Consider the 2-periodic Leslie–Gower model where*

$$F_0(x,y) = \begin{pmatrix} \frac{a_0 x}{1+c_{11}x+c_{12}y} \\ \frac{b_0 y}{1+c_{21}x+c_{22}y} \end{pmatrix},$$

$$F_2(x,y) = \begin{pmatrix} \frac{a_1 x}{1+c_{11}x+c_{12}y} \\ \frac{b_1 y}{1+c_{21}x+c_{22}y} \end{pmatrix}.$$

Each one of these maps was introduced as Example 65. We assume that $a_i, b_i > 1$, $i = 0, 1$, and $c_{21} < c_{11}$, $c_{12} < c_{22}$. It was shown in Example 65 that each of the maps F_0 and F_1 satisfies the assumptions ($\mathbf{H_1}$)-($\mathbf{H_4}$). So it remains to show that the composition map $\Phi_2 = F_1 \circ F_0$ satisfies ($\mathbf{H_3}$)-($\mathbf{H_4}$). To show that Φ_2 satisfies ($\mathbf{H_3}$). Putting $\Phi_2(x,0) = (x,0)$, we find the 2-periodic cycle

$$\overline{\mathbf{x}}_1 = \begin{pmatrix} \frac{a_0 a_1 - 1}{c_{11}(1+a_0)} \\ 0 \end{pmatrix}, \quad \overline{\mathbf{x}}_2 = \begin{pmatrix} \frac{a_0 a_1 - 1}{c_{11}(1+a_1)} \\ 0 \end{pmatrix}.$$

It was shown in Section 3.1 that this 2-periodic cycle is globally asymptotically stable on the x-axis.

Now the Jacobian matrix of the map F_0 is given by

$$JF_0(\overline{\mathbf{x}}_1) = \begin{pmatrix} \dfrac{a_0}{\left(1+c_{11}\left(\frac{a_0 a_1 - 1}{c_{11}(1+a_0)}\right)\right)^2} & \dfrac{-a_0 c_{12}\left(\frac{a_0 a_1 - 1}{c_{11}(1+a_0)}\right)}{\left(1+c_{11}\left(\frac{a_0 a_1 - 1}{c_{11}(1+a_0)}\right)\right)^2} \\ 0 & \dfrac{1+c_{21}\left(\frac{a_0 a_1 - 1}{c_{11}(1+a_0)}\right)}{\left(1+c_{21}\left(\frac{a_0 a_1 - 1}{c_{11}(1+a_0)}\right)\right)^2} \end{pmatrix}$$

$$= \begin{pmatrix} \dfrac{(1+a_0)^2}{a_0(1+a_1)^2} & * \\ 0 & \dfrac{a_1}{1+\frac{c_{21}}{c_{11}}\left(\frac{a_0 a_1 - 1}{1+a_0}\right)} \end{pmatrix}$$

$$JF_0(\overline{\mathbf{x}}_2) = \begin{pmatrix} \dfrac{(1+a_1)^2}{a_1(1+a_0)^2} & ** \\ 0 & \dfrac{a_0}{1+\frac{c_{21}}{c_{11}}\left(\frac{a_0 a_1 - 1}{1+a_1}\right)} \end{pmatrix}$$

$$\text{Now } JF(x_1)JF(x_0) = \begin{pmatrix} \dfrac{1}{a_0 a_1} & *** \\ 0 & \dfrac{a_0 a_1}{\left(1+\frac{c_{21}}{c_{11}}\left(\frac{a_0 a_1 - 1}{1+a_0}\right)\right)\left(1+\frac{c_{21}}{c_{11}}\left(\frac{a_0 a_1 - 1}{1+a_1}\right)\right)} \end{pmatrix}.$$

Hence the eigenvalues are

$$\lambda_1 = \frac{1}{a_0 a_1} < 1, \qquad \lambda_2 = \frac{a_0 a_1}{A_1 A_2}, A_1 = 1 + \frac{c_{21}}{c_{11}}\left(\frac{a_0 a_1 - 1}{1+a_0}\right),$$

where

$$A_2 = 1 + \frac{c_{21}}{c_{11}}\left(\frac{a_0 a_1 - 1}{1+a_1}\right)$$

$$A_1 A_2 = 1 + \frac{c_{21}}{c_{11}}(a_0 a_1 - 1)^2 \left[\frac{1}{1+a_1} + \frac{1}{1+a_0} + \frac{c_{21}}{c_{11}}\frac{(a_0 a_1 - 1)}{(1+a_0)(1+a_1)}\right].$$

Since $\frac{c_{21}}{c_{11}} < 1$, one can show that $A_1 A_2 < a_0 a_1$ (Problem 3) and, consequently, $\lambda_2 > 1$. Hence the 2-periodic cycle on the x-axis is a saddle. Similarly, one can show that the 2-periodic cycle on the y-axis is a saddle (Problem 1). Assumption (**H$_4$**) follows easily by the fact that the composition of two injective maps is injective (Problem 2). Therefore, by Theorem 88 the 2-periodic cycle is globally asymptotically stable.

In the next section, we investigate a special class of non-monotone periodic maps.

9.6.2 Mixed Monotone Periodic Systems

In Chapter 8, Section 33, we introduced the notion of mixed monotone maps and develop a global stability theory that was applied to the Leslie–Gower model (and the Ricker competition model). In this subsection, we will extend the global stability Theorem 78 to periodic systems.

Theorem 89 *The composition of p mixed monotone maps is a mixed monotone map.*

Proof. First, we prove that the composition of two mixed maps is a monotone map. Then we use mathematical induction to prove that this is true for the composition of p-maps. The details are left to the reader as Problem 6. ∎

9.6. NONLINEAR PERIODIC SYSTEMS

Next, we focus on the stability analysis of the 2-periodic cycles of mixed monotone periodic systems when $p = 2$. We begin our analysis by stating a perturbation theorem that will be crucial in our analysis.

Consider the difference equation

$$\mathbf{x}(t+1) = F(\mathbf{x}(t), \alpha_t), \qquad (9.23)$$

where $F: U \times G \to U$ is continuous, $U \subset \mathbb{R}_+^2$, $G \subset \mathbb{R}_+$.

We start our analysis with a perturbation result that is crucial in our investigation of the global stability of periodic systems.

Theorem 90 [124] Let $\mathbf{x}_0^*$ be the interior equilibrium point of $F_0(\mathbf{x})$, i.e., $F(\mathbf{x}_0^*, \alpha_0) = \mathbf{x}_0^*$. Assume that $\mathbf{x}_0^*$ is globally asymptotically stable hyperbolic equilibrium. Then there exists $\delta > 0$ and a unique $\mathbf{x}^*(\alpha) \in U$ for $\alpha \in \mathbf{B}(\alpha_0, \delta)$ such that $F(\mathbf{x}^*(\alpha), \alpha) = \mathbf{x}^*(\alpha)$ and $F^t(\mathbf{z}) \to \mathbf{x}^*(\alpha)$ as $t \to \infty$ for all $\mathbf{z} \in U$.

Applying this theorem to a 2-periodic system yields the following result.

Theorem 91 (Global stability of the 2-periodic system) Assume that $\mathbf{x}_0^*$ is globally asymptotically stable hyperbolic equilibrium of the map F_0 in which $\alpha = \alpha_0$. Then for sufficiently small $\delta > 0$ and $\alpha_1 = \alpha_0 \pm \delta$, there is a 2-periodic cycle of the composition map $F_1 \circ F_0$ that is globally asymptotically stable.

Proof. By Theorem 89, the second iteration $F^2 = F \circ F$ of the mixed map F is a mixed monotone map. Now we perturb F^2 and write it as the composition of two maps $G = F_1 \circ F_0$, where $F_0 = F$, in which $\alpha = \alpha_0$, and F_1 where $\alpha_1 = \alpha_0 \pm \delta$. Theorem 90 implies there exists an interior 2-periodic cycle that is globally asymptotically stable. ∎

We now give an example to illustrate this theorem.

Example 109 *Consider the evolutionary Leslie Gower model*

$$\begin{cases} x(t+1) = b_t x(t) \exp\left(-\frac{u^2(t)}{2}\right) \frac{1}{1+c_0 x(t)} \\ u(t+1) = (1-\sigma^2)u(t) - \frac{c_1 \sigma^2 x(t)}{1+c_0 x(t)}, \end{cases} \qquad (9.24)$$

where $b_{t+2} = b_t$. Now by Theorem 78 for $b_0 > 1$, the interior equilibrium of the map

$$F_0(x, u) = \begin{pmatrix} \frac{b_0 x(e^{-u^2})}{1+c_0 x} \\ (1-\sigma^2)u \frac{c_1 \sigma^2 x}{1+c_0 x} \end{pmatrix}$$

is globally asymptotically stable.

Then by Theorem 91, for sufficiently small δ, there exists a globally asymptotically interior 2-periodic cycle of system (9.24), with $b_1 = b_0 \pm \delta$.

Here is an example of numerical simulations. This is the periodic version of Equation (8.44).

Example 110

$$\begin{aligned} x(t+1) &= x(t) e^{\alpha(t) - \frac{u^2}{2} - c_0 x(t)} \\ u(t+1) &= (1-\sigma^2)u(t) - c_1 \sigma^2 x(t), \end{aligned} \qquad (9.25)$$

with $\alpha(t+2) = \alpha(t)$. Let $c_0 = 1$, $\sigma^2 = 1.5$, $c_1 = 2$ and $\alpha_0 = 0.3$. Then the fixed point $(0.210977, -0.421954)^T$ is a globally asymptotically stable fixed point of the map F_0, where

$$F_0(x,u) = \begin{pmatrix} x\exp\left(0.3 - \frac{u^2}{2} - x\right) \\ -0.5u - 2\sigma^2 x \end{pmatrix}.$$

Now, with $\alpha_1 = \alpha_0 + 0.2 = 0.5$, it follows from Theorem 90 that

$$C_2 = \left\{ \begin{pmatrix} \bar{x}(0) \\ \bar{u}(0) \end{pmatrix}, \begin{pmatrix} \bar{x}(1) \\ \bar{u}(1) \end{pmatrix} \right\} \approx \left\{ \begin{pmatrix} 0.271644 \\ -0.461582 \end{pmatrix}, \begin{pmatrix} 0.251217 \\ -0.58414 \end{pmatrix} \right\}$$

is a globally asymptotically stable 2−periodic cycle of System (110)), where the composition map is $G = F_1 \circ F_0$ with

$$F_1(x,u) = \begin{pmatrix} x\exp\left(0.5 - \frac{u^2}{2} - x\right) \\ -0.5u - 2\sigma^2 x \end{pmatrix}.$$

One may extend Theorem 91 to a p-periodic system using the same argument in the proof of Theorem 91.

Theorem 92 *(global stability of the p-periodic system)*

Assume that the period of the system p is even, and $\mathbf{x}_0^*$ is globally asymptotically stable hyperbolic equilibrium of the map F_0 in which $\alpha = \alpha_0$. Then for sufficiently small $\delta_i > 0$ and $\alpha_{i+1} = \alpha_i \pm \delta_i$, $i = 0, 1, \ldots, p-2$ such that $0 < \alpha_0 \pm \sum_{i=0}^{p-1} \delta_i < 1$, there is a p−periodic cycle which is globally asymptotically stable

Proof. The proof is similar to the proof of the case $p = 2$ and is left to the reader as Problem 7. ∎

9.6.3 Asymptotically Periodic Systems

In this subsection, we investigate general nonautonomous models that are asymptotically periodic. The following result generalizes Theorem (46) on asymptotically autonomous systems. Let $\mathbb{R}^n_+$ denote the cone of nonnegative vectors in $\mathbb{R}^n$ and let $int(\mathbb{R}^n_+)$ and $\partial(\mathbb{R}^n_+)$ denote the interior and the boundary of $\mathbb{R}^n_+$, respectively. Let G_t, $F_t : \mathbb{R}^n_+ \longrightarrow \mathbb{R}^n_+$ be continuous functions for all $t \in \mathbb{Z}_+$ and $t = 0, 1, \ldots, p$, such that $G_{t+p} = G_t$, for some $p \geq 1$. Assume that

$\mathbf{A_1}$: F_t converges uniformly to G_t as $t \to \infty$.

Then $\mathbf{x}(0) \in \mathbb{R}^n_+$ implies that the solutions of the nonautonomous difference equation

$$\mathbf{x}(t+1) = F_t(\mathbf{x}(t)), \tag{9.26}$$

satisfies $\mathbf{x}(t) \in \mathbb{R}^n_+$, for all $t \in \mathbb{Z}_+$, where $\mathbf{x} = (x_1, x_2, \ldots, x_n) \in \mathbb{R}^n_+$.

The same is true for solutions of the limiting nonautonomous periodic equation

$$\mathbf{x}(t+1) = G_t(\mathbf{x}(t)), \tag{9.27}$$

where we assume

$\mathbf{A_2}$:

(i) $F_t : int(\mathbb{R}^n_+) \longrightarrow int(\mathbb{R}^n_+)$.

(ii) $G_t : int(\mathbb{R}_+^n) \longrightarrow int(\mathbb{R}_+^n)$.

Theorem 93 *Assume $\mathbf{A_1}$ and $\mathbf{A_2}$ and the limiting periodic equation (9.27) has periodic cycle C_p of period p or a divisor of p in $\mathbb{R}_+^n$. Then*

(i) *if $c_p \in int(\mathbb{R}_+^n)$, and if it is globally asymptotically stable on $int(\mathbb{R}_+^n)$, then all solutions of the nonautonomous difference equation with $\mathbf{x}(0) \in int(\mathbb{R}_+^n)$ tend to c_p.*

(ii) *if $c_p \in \partial(\mathbb{R}_+^n)$, and if it is globally asymptotically stable on $int(\mathbb{R}_+^n)$, then all solutions of the nonautonomous difference equation with $\mathbf{x}(0) \in int(\mathbf{R}_+^n)$ tend to c_p.*

The next example illustrates the effectiveness of this theorem

Example 111 *Consider the nonautonomous Leslie–Gower model*
$$\mathbf{x}(t+1) = F_t(\mathbf{x})(t)),$$
where $\mathbf{x} = (x,y)^T$ and
$$F_t(x,y) = \begin{pmatrix} \frac{a_t x}{1+c_{11}x+c_{12}y} \\ \frac{b_t y}{1+c_{21}x+c_{22}y} \end{pmatrix}$$
with, $a_t, b_t > 1, t \in \mathbb{Z}^+$. Assume that the sequence $\{F_t\}$ converges uniformly to the periodic map $\{G_0, G_1\}$, in which $(a_t, b_t) \to ((r_0, s_0), (r_1, s_1))$ as $t \to \infty$, where
$$G_i(x,y) = \begin{pmatrix} \frac{r_i x}{1+c_{11}x+c_{12}y} \\ \frac{s_i y}{1+c_{21}x+c_{22}y} \end{pmatrix}, \quad i = 0, 1.$$

From Example 108, the periodic system $(G_0, G_1)^T$ has a globally asymptotically stable 2-periodic cycle c_2. Now by Theorem 88, it follows by Theorem 93 that all solutions of this nonautonomous system tend to c_2.

Exercises 9.6.1-9.6.3

1. Consider the 2-periodic Leslie–Gower model 108 with $a_i, b_i > 1$, $i = 0, 1$.

    (i) Show that the 2-periodic cycle on the x-axis is a saddle.

    (ii) Show that the 2-periodic cycle on the y-axis is a saddle.

2. Show that the composition of two injective maps $F_1, F_0 : \mathbb{R}^2 \to \mathbb{R}^2$ is also injective.

3. In Example 108, show that $A_1 A_2 < a_0$, where
$$A_1 = 1 + \frac{c_{21}}{c_{11}}\left(\frac{a_0 a_1 - 1}{1+a_0}\right) \text{ and } A_2 = 1 + \frac{c_{21}}{c_{11}}\left(\frac{a_0 a_1 - 1}{1+a_1}\right).$$

4. Consider the 2-periodic Ricker competition model
$$x(t+1) = x(t)\exp\left(\alpha_t - c_{11}x(t) - c_{12}y(t)\right)$$
$$y(t+1) = r(t)\exp\left(\beta_t - c_{21}x(t) - c_{22}y(t)\right)$$
$\alpha_{t+2} = \alpha_t$, $\beta_{t+2} = \beta_t$, $0 < \alpha_t, \beta_t < 1$.

(i) Show that the system has a 2-periodic cycle in the interior of $\mathbb{R}_+^2$.

(ii) Prove that this interior 2-periodic cycle is globally asymptotically stable.

(iii) Show that on the x-axis, there is a 2-periodic cycle which is a saddle.

(iv) Show that on the y-axis there is a 2-periodic cycle which is a saddle.

(v) Prove that there is a 2-periodic interior cycle that is globally asymptotically stable.

5. Consider the nonautonomous single-species Ricker model

$$x(t+1) = x(t)\exp\left(\alpha_t - c_0 x(t)\right),$$

where $0 < \alpha_t < 2$. Assume that $\alpha_t \to \{r_0, r_1\}$ as $t \to \infty$. Show that this equation has a globally asymptotically stable 2-periodic cycle.

6. Prove that the composition of p-mixed monotone maps is a mixed monotone map when $p > 1$ is a positive interior.

7. Prove Theorem (92).

8. Consider the evolutionary discrete logistic (Beverton–Holt) model [75]

$$x(t+1) = \frac{b(u(t))x(t)}{1 + c_0 x(t)} \tag{9.28}$$

$$u(t+1) = u(t) + \alpha^2 \frac{b'(u(t))}{b(u(t))}. \tag{9.29}$$

(i) Find the survival equilibrium $E^* = (x^*, u^*)^T$.

(ii) Show that E^* is globally asymptotically stable on $\mathbb{R}^+ \times U$, for some open neighborhood U of u^* if $b(u^*) > 1$ and

$$\left|1 + \sigma^2 \frac{b''(u^*)}{b(u^*)}\right| < 1.$$

(iii) Assume that the trait equation (9.29) has a locally asymptotically hyperbolic p-periodic solution $v(t)$ and $\prod_{t=0}^{1} b(v(t)) > 1$. Show that there exists a survival p-periodic solution $(z(t), v(t))^T$ of Equation (9.28) which is globally asymptotically stable on $\mathbb{R}^+ \times U$ for some neighborhood U of v_0.

9. Consider the 2-species periodic Leslie–Gower evolutionary model (see the autonomous evolutionary Leslie–Gower model 8.46)

$$x_1(t+1) = \frac{b_1(t)e^{-\frac{u_1^2}{2}}x_1(t)}{1 + c_{01}x_1(t) + c_{12}x_2(t)}$$

$$x_2(t+1) = \frac{b_2(t)e^{-\frac{u_2^2}{2}}x_2(t)}{1 + c_{21}x_1(t) + c_{02}x_2(t)} \tag{9.30}$$

$$u_1(t+1) = (1 - \sigma_1^2)u_1(t) - \frac{c_1\sigma^2 x_1(t)}{1 + c_{01}x_1(t)}$$

$$u_2(t+1) = (1 - \sigma_2^2)u_2(t) - \frac{c_2\sigma^2 x_2(t)}{1 + c_{02}x_2(t)},$$

where $0 < \alpha(t), \beta(t) < 1$ are of period 2,

   (i) Show that the system is mixed monotone.
   (ii) Show there exists a 2-periodic cycle which is globally asymptotically stable.

10. Consider now the 2-species nonautonomous periodic evolutionary Ricker model given by

$$\begin{cases} x(t+1) = x(t)\exp\left(\alpha_t - \frac{u_1^2(t)}{2} - c_{11}(0)x(t) - c_{12}y(t)\right) \\ y(t+1) = y(t)\exp\left(\beta_t - \frac{u_2^2(t)}{2} - c_{21}x(t) - c_{22}(0)y(t)\right) \\ u_1(t+1) = (1-\sigma_1^2)u_1(t) - \sigma_1^2 c_1 x(t) \\ u_2(t+1) = (1-\sigma_2^2)u_2(t) - \sigma_2^2 c_2 y(t) \end{cases}, \qquad (9.31)$$

where we assume that $\alpha_{t+2} = \alpha$ and $\beta_{t+2} = \beta_t$ for all $t = 0, 1, 2, \ldots$, $1 < \sigma^2 < 2$.

   (i) Show that the system is mixed monotone.
   (ii) Show there exists a 2-periodic cycle which is globally asymptotically stable.

11. (Research Project) Consider again system (9.31) with $0 < \sigma^2 < 2$

   (i) Show that the system is mixed monotone.
   (ii) Show there exists a 2-periodic cycle which is globally asymptotically stable.

9.6.4 Small Amplitude Perturbations

Consider the periodically forced Ricker equation

$$x(t+1) = x(t)\exp\left[r(t)\left(1 - \frac{x(t)}{K(t)}\right)\right] \qquad (9.32)$$

in which the oscillations in the coefficients $r(t)$ and $K(t)$ are p-periodic and of small amplitude $\alpha > 0$ around their averages $av(r(t)) = r_0 > 0$ and $av(K(t)) = K_0 > 0$. Specifically,

$$r(t) = r_0 + \alpha r_1(t), \quad K(t) = K_0 + \alpha K_1(t), \qquad (9.33)$$

where $r_1(t)$ and $K_1(t)$ are p-periodic and

$$av(r_1(t)) = av(K_1(t)) = 0.$$

For $|\alpha|$ sufficiently small, both $r(t)$ and $K(t)$ are positive valued for all t. This is a motivating example for the periodically forced equations we consider in this section, which have the form

$$\mathbf{x}(t+1) = F_t(\mathbf{x}(t), \alpha). \qquad (9.34)$$

We assume

A3: (a) for each t, $F_t(\cdot, \cdot) \in C^2(\Omega \times \Lambda \to \Omega)$, where Ω and Λ are open sets in $\mathbb{R}^m$ and $\mathbb{R}$, respectively, with $0 \in \Lambda$;
   (b) $F_{t+p}(\cdot, \cdot) = F_t(\cdot, \cdot)$ for all t ;
   (c) $F_t(x, 0) \equiv F(x)$ for all $x \in \Omega$ and t.

We view Equation (9.34) as a perturbation of the autonomous equation

$$\mathbf{x}(t+1) = F(\mathbf{x}(t)). \tag{9.35}$$

If $\bar{\mathbf{x}}(t)$ is a q-periodic solution of (9.35), then each point in the cycle $\bar{\mathbf{x}}(0), \bar{\mathbf{x}}(1), \ldots, \bar{\mathbf{x}}(q-1)$ is a fixed point of the qth composite map

$$C_q(\mathbf{x}) := F(\mathbf{x}) \circ F(\mathbf{x}) \cdots \circ F(\mathbf{x}). \tag{9.36}$$

Let $J_q(\mathbf{x})$ denote the Jacobian of the composite $C_q(\mathbf{x})$. The q-periodic solution $\bar{\mathbf{x}}(t)$ has q phase shifts all of which are also q-periodic solutions of the equation. Let $\bar{\mathbf{x}}_i(t)$ denote the ith phase shift, i.e., the solution of (9.35) with initial condition $\bar{\mathbf{x}}_i(0) = \bar{\mathbf{x}}(i)$.

Theorem 94 *Assume A3 and that the autonomous equation (9.35) has a q-periodic solution $\bar{\mathbf{x}}(t)$. Let m be the least common multiple of p and q and assume no eigenvalue of the Jacobian $J_q(\bar{\mathbf{x}}(i))$ is an $(m/q)^{th}$ root of unity. Then for sufficiently small α there exists an m-periodic solution $\bar{\mathbf{x}}_i(t,\alpha)$ of (9.34), for which $\bar{\mathbf{x}}_i(\cdot,\alpha)$ is twice continuously differentiable in α and $\bar{\mathbf{x}}_i(t,0) = \bar{\mathbf{x}}_i(t)$.*

Proof. Let

$$C_j(\mathbf{x},\alpha) := F_{j-1}(\mathbf{x},\alpha) \circ F_{j-2}(\mathbf{x},\alpha) \circ \cdots \circ F_0(\mathbf{x},\alpha)$$

denote the jth composite of $F_t(\mathbf{x},\alpha)$ and note that the mth composite of $F_t(\mathbf{x},\alpha)$

$$C_m(\mathbf{x},\alpha) := F_{m-1}(\mathbf{x},\alpha) \circ F_{m-2}(\mathbf{x},\alpha) \circ \cdots \circ F_{p-1}(\mathbf{x},\alpha) \circ \cdots \circ F_0(\mathbf{x},\alpha)$$

can be written (because $F_t(\mathbf{x},\alpha)$ is p-periodic in t)

$$C_m(\mathbf{x},\alpha) = C_p(\mathbf{x},\alpha) \circ C_p(\mathbf{x},\alpha) \circ \cdots \circ C_p(\mathbf{x},\alpha),$$

where $C_p(\mathbf{x},\alpha)$ appears m/q times. A solution $\bar{\mathbf{x}}_i(t,\alpha)$ of (9.34) is an m-periodic solution if and only if $\bar{\mathbf{x}}_i(0,\alpha)$ is a fixed point of the map defined by the mth composite, i.e., is a solution of

$$\mathbf{x} = C_m(\mathbf{x},\alpha), \tag{9.37}$$

i.e., of the equation $\mathbf{x} - C_m(\mathbf{x},a) = 0$. By assumption, this equation has a solution $\mathbf{x} = \bar{\mathbf{x}}_i(0)$ when $\alpha = 0$. To prove the theorem we apply the Implicit Function Theorem to solve the equation for $x = x(\alpha)$, $x(0) = \bar{\mathbf{x}}_i(0)$, which is then the initial condition for a m-periodic solution $\bar{\mathbf{x}}_i(t,\alpha)$. The Implicit Function Theorem which requires that the Jacobian of $\mathbf{x} - C_m(\mathbf{x},a)$ with respect to $\mathbf{x}$, i.e., $I - J_m(\mathbf{x},\alpha)$, be nonsingular at $\mathbf{x} = \bar{\mathbf{x}}_i(0)$ and $\alpha = 0$. By the chain rule $J_m(\bar{\mathbf{x}}_i(0),0) = (J_a(\bar{\mathbf{x}}_i(0)))^{m/q}$ whose eigenvalues are the m/q powers of the eigenvalues of $J_a\bar{\mathbf{x}}_i(0)$, none of which equals 1 by assumption. Thus, $I - J_m(\bar{\mathbf{x}}_i(0))$ is nonsingular. ∎

Theorem 94 as well as the following theorem and other results concerning periodically forced cycles, can be found in [68] and [165].

Theorem 95 *Assume A3 and that the autonomous equation (9.35) has a q-periodic solution $\bar{\mathbf{x}}(t)$.*

(a) *If $\rho(J_q(\bar{\mathbf{x}}(i))) < 1$ (i.e., $\bar{\mathbf{x}}(t)$ is stable by linearization), then for small α there exists a (locally asymptotically) stable m-periodic solution $\bar{\mathbf{x}}_i(t,\alpha)$ of (9.34) that is twice continuously differentiable in α and that satisfies $\bar{\mathbf{x}}_i(t,0) = \bar{\mathbf{x}}(t)$*

(b) *If no eigenvalue of the Jacobian $J_q(\bar{\mathbf{x}}(i))$ is an $(m/q)^{th}$ root of unity and if $\rho(J_q(\bar{\mathbf{x}}(i))) > 1$, then the m-periodic solution $\bar{\mathbf{x}}_i(t,\alpha)$ of (9.34) is unstable for small α.*

9.6. NONLINEAR PERIODIC SYSTEMS

Proof.

(a) $\rho(J_q(\bar{\mathbf{x}}(i))) < 1$ implies no eigenvalue is a root of unity and therefore by Theorem 94 an m-periodic solution exists as described. By continuity, $\rho(J_q(\bar{\mathbf{x}}_i(t,\alpha))) < 1$ for α sufficiently small and therefore $\bar{\mathbf{x}}_i(t,\alpha)$ is (locally asymptotically) stable by the linearization principle.

(b) By Theorem 94 an m-periodic solution exists as described. By continuity, $\rho(J_q(\bar{\mathbf{x}}_i(t,\alpha))) > 1$ for α sufficiently small and therefore $\bar{\mathbf{x}}_i(t,\alpha)$ is unstable by the linearization principle.

∎

Remark 28 *The spectrum (set of eigenvalues) of $J_q(\bar{\mathbf{x}}(i))$ is the same for all phases i and therefore to apply Theorems 94 and 95 we need consider only one Jacobian, say $J_q(\bar{\mathbf{x}}(0))$. Moreover, it follows, when these theorems apply, that there exist q solutions $\bar{\mathbf{x}}_i(t,\alpha)$ of period m and either all are stable or all are unstable.*

9.6.5 Examples

The following example, which utilizes the periodically forced Ricker equation (9.32), illustrates the use of Theorems 94 and 95 in several selected cases. In part (a), the unperturbed Ricker equation ($\alpha = 0$) has a stable equilibrium and is subjected to 2-periodic oscillations in its coefficients. In parts (b) and (c), the unperturbed Ricker equation has an unstable equilibrium and a stable 2-periodic solution. In part (b) the coefficients of this equation are subjected to 2-periodic oscillations and in part (c) to 3-periodic oscillations.

Example 112 *In this example we consider selected examples of the p-periodically forced Ricker equation (9.32) with coefficients (9.33).*

(a) *If $0 < r_0 < 2$ then when $\alpha = 0$ (and the equation becomes the standard Ricker equation) there exists a (globally) asymptotically stable $q = 1$ cycle, i.e., an equilibrium. Suppose the coefficients (9.33) in this Ricker equation have period $p = 2$. Then Theorem 95(A) implies there exists a stable, $m = 2$ periodic solution of the periodically forced Ricker (9.32) for all small amplitudes α. A numerical example illustrating this result is shown in Figure 9.1(A) using the p-periodic coefficients*

$$K(t) = 1 + \alpha \cos\left(\frac{2\pi}{p}t\right), \quad r(t) = r_0 + \alpha\alpha_r \cos\left(\frac{2\pi}{p}t\right), \tag{9.38}$$

where $|\alpha| < 1$ and $|\alpha\alpha_r| < 1$. Here α_r determines the relative amplitude between $K(t)$ and $r(t)$.

(b) *If we change r_0 so that $2 < r_0 < 2.52546$, then when $\alpha = 0$ the Ricker equation has a hyperbolic, stable $q = 2$ periodic solution and an unstable equilibrium (the eigenvalue of the Jacobian is greater than 1). Theorem 95 implies, for small amplitudes $\alpha \neq 0$, that there exist three $m = 2$ periodic solutions: one is unstable and collapses to the equilibrium at $\alpha = 0$ and the other two are stable and collapse to the two phases of the 2-cycle at $\alpha = 0$. See Figure 9.1(B) for example.*

(c) Finally, suppose $2 < r_0 < 2.52546$ as in (b) but the coefficients (9.33) have period $p = 3$. Then in a similar application of Theorem 95 we find, for small amplitudes α, that there exist three $m = 6$ periodic solutions: one is unstable and collapses to the equilibrium at $\alpha = 0$ and the other two are stable and collapse to the two phases of the 2-cycle at $\alpha = 0$. See Figure 9.1(B) for an example.

Theorems 94 and 95 concern the existence and stability of periodic solutions of small amplitude, p-periodic perturbations of an autonomous equation which has a q-periodic cycle. It is also often of interest to know other properties of these periodic solutions, such as their amplitude, phases and averages in relationships with those model equation's coefficients and/or with the unperturbed cycle. A classic way to investigate such properties, at least for small amplitudes α, is to study the lower order terms in a Taylor expansion of $\bar{\mathbf{x}}_i(t, \alpha)$ in α centered at $\alpha = 0$. Each point $\bar{\mathbf{x}}_i(t, \alpha)$, $t = 0, 1, ..., q-1$, solves the composite equation (9.37)

$$\bar{\mathbf{x}}_i(t, \alpha) = C_m(\bar{\mathbf{x}}_i(t, \alpha), \alpha) \tag{9.39}$$

and the coefficients in its Taylor expansion

$$\bar{\mathbf{x}}_i(t, \alpha) = \bar{\mathbf{x}}_i(t) + \partial_\alpha \bar{\mathbf{x}}_i(t, \alpha)|_{\alpha=0} \alpha + \frac{1}{2} \partial_\alpha^2 \bar{\mathbf{x}}_i(t, \alpha)|_{\alpha=0} \alpha^2 + O(\alpha^2)$$

can be found by implicit differentiation of Equation (9.39) with respect to α followed by evaluation at $\alpha = 0$. Typical applications of this perturbation methodology utilize the first order (linear) Taylor polynomial approximation to the cycle.

Example 113 *Consider the periodically forced Ricker equation in Example 112(a). In that example, the autonomous (unperturbed) Ricker equation has a stable equilibrium which is perturbed to a stable 2-cycle when the coefficients (9.38) have period 2 and a small amplitude α. In this example we illustrate the use of first order perturbation approximation to study this 2-periodic solution for the numerical example in Figure 9.1(A) for which the periodic Ricker equation is*

$$x(t+1) = x(t) \exp\left[\left(1.9 - 50\alpha(-1)^t\right)\left(1 - \frac{x(t)}{1 + \alpha(-1)^t}\right)\right].$$

The unperturbed equation when $\alpha = 0$

$$x(t+1) = x(t) \exp[1.9(1 - x(t))]$$

has a stable equilibrium $\bar{x}_1(t) = 1$. The initial condition $\bar{x}(0, \alpha)$ for the perturbed 2-periodic solution $\bar{x}_1(t, \alpha)$ guaranteed by Theorem 94 is a fixed point of the composite map, i.e.

$$x = x \exp\left[r(0)\left(1 - \frac{x}{K(0)}\right)\right] \exp\left[r(1)\left(1 - \frac{x \exp\left[r(0)\left(1 - \frac{x}{K(0)}\right)\right]}{K(1)}\right)\right] \tag{9.40}$$

$$r(t) = 1.9 - 50\alpha(-1)^t, \quad K(t) = 1 + \alpha(-1)^t$$

which, after the cancellation of x and the taking of a logarithm of both sides, reduces to the equation $0 = g(x, \alpha)$ where

$$g(x, \alpha) := (1.9 - 50\alpha)\left(1 - \frac{x}{1+\alpha}\right) + (1.9 + 50\alpha)\left(1 - \frac{x \exp\left[(1.9 - 50\alpha)\left(1 - \frac{x}{1+\alpha}\right)\right]}{1 - \alpha}\right). \tag{9.41}$$

9.6. NONLINEAR PERIODIC SYSTEMS

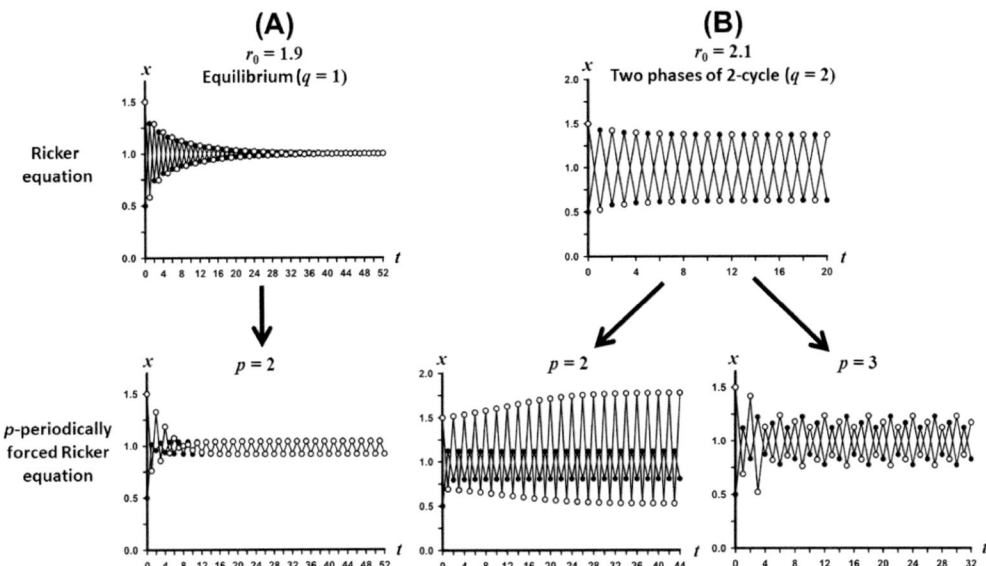

Figure 9.1: In each plot there is shown two times-series solutions, with initial conditions $x(0) = 0.5$ and 1.5, of the Ricker p-periodically forced Ricker (9.32) with coefficients (9.38) and $K_0 = 1$ and $\alpha_r = -50$. The two row plots are without periodic forcing, i.e., $\alpha = 0$ and the bottom row are plots with small amplitude forcing $\alpha = 0.01$. **(A)** With $r_0 = 1.9$ the unperturbed Ricker equation has a hyperbolic, stable equilibrium (i.e., a $q = 1$ periodic solution) which, when $p = 2$ periodically forced, results in a stable 2-periodic solution. **(B)** With $r_0 = 2.1$ the unperturbed Ricker equation has a hyperbolic, stable $q = 2$ periodic solution which has two phase shifts. The upper plot shows one of the computed solutions ($x(0) = 0.5$) tending to one phase of the 2-cycle and the other ($x(0) = 1.5$) tending to the other phase of the 2-cycle. When forced with period $p = 2$ (lower left plot) we see the two initial conditions tending to two 2-periodic solutions which are not phase shifts of each other. When forced with period $p = 3$ (lower right plot) we see the two initial conditions tending to two 6-periodic solutions ($m = 6$ is the least common multiple of 2 and 3) which are not phase shifts of each other.

The solution x of the equation $0 = g(x,\alpha)$ that equals 1 when $\alpha = 0$ provides the initial conditions $x = \bar{x}(0,\alpha)$ for the 2-periodic solution. We can approximate $\bar{x}(0,\alpha)$ for small α using the Taylor expansion

$$\bar{x}(0,\alpha) = 1 + c_1\alpha + c_2\alpha^2 + O(\alpha^3)$$

$$c_1 = \left.\frac{d\bar{x}(0,\alpha)}{d\alpha}\right|_{\alpha=0}, \quad c_2 = \frac{1}{2}\left.\frac{d^2\bar{x}(0,\alpha)}{d\alpha^2}\right|_{\alpha=0}.$$

One way to calculate c_1 and c_2 is to use implicit differentiation of the equation $0 = g(\bar{x}(0,\alpha),\alpha)$ with respect to α followed by an evaluation at $\alpha = 0$. Another way is to use a computer algebra program to calculate the Taylor expansion of $g(\bar{x}(0,\alpha),\alpha)$ with respect to α centered at $\alpha = 0$ and set the resulting coefficients of α and α^2 equal to 0. This leads to simple equations to solve for c_1 and c_2. Specifically, setting the linear and quadratic coefficients of

$$g(1 + c_1\alpha + c_2\alpha^2 + O(\alpha^3), \alpha) = -(0.19c_1 + 3.61)\alpha$$
$$+ (0.1805c_1^2 + 3.249c_1 - 0.19c_2 - 107.23)\alpha^2 + O(\alpha^3)$$

equal to 0 we get the equations

$$0 = 0.19c_1 + 3.61$$
$$0 = 0.1805c_1^2 + 3.249c_1 - 0.19c_2 - 107.23$$

that we can solve sequentially for $c_1 = -19.0$ and $c_2 = -546.32$ to obtain

$$\bar{x}(0,\alpha) = 1 - 19\alpha - 546.32\alpha^2 + O(\alpha^3).$$

The second term $\bar{x}(1,\alpha)$ in the 2-periodic solution can be calculated by applying the Ricker equation (9.42) to this initial condition:

$$\bar{x}(1,\alpha) = \bar{x}_1(0,\alpha)\exp\left[(1.9 - 50\alpha)\left(1 - \frac{\bar{x}_1(0,\alpha)}{1+\alpha}\right)\right]$$

$$= \left(1 - 19\alpha - 546.32\alpha^2 + O(\alpha^3)\right)\exp\left[(1.9 - 50\alpha)\left(1 - \frac{1 - 19\alpha - 546.32\alpha^2 + O(\alpha^3)}{1+\alpha}\right)\right]$$

$$= 1 + 19\alpha - 546.31\alpha^2 + O(\alpha^3).$$

In summary, we have Taylor expansions for both terms in the 2-periodic solution

$$\bar{x}(t,\alpha): \quad \{1 - 19\alpha - 546.32\alpha^2 + O(\alpha^3),\ 1 + 19\alpha - 546.31\alpha^2 + O(\alpha^3)\}.$$

From these approximations we can obtain some features of the 2-periodic solution of the $p = 2$ periodic forced Ricker equation (9.42).

For small $\alpha > 0$ the stable 2-periodic solution $\bar{x}(t,\alpha)$
· oscillates in-phase with $r(t)$ and out-of-phase with $K(t)$;
· has an average $1 - 546.32\alpha^2 + O(\alpha^3) < 1$ and therefore is attenuant.
These features can be seen for $\alpha = 0.01$ upon close inspection of the plots in Figure 9.1(A).

9.6. NONLINEAR PERIODIC SYSTEMS

Example 114 *Consider the periodically forced Ricker equation in Example 112(b). In that example, the autonomous (unperturbed) Ricker equation has a stable 2-cycle each of whose two phases is perturbed to a stable 2-cycle when the coefficients (9.38) have period 2 and a small amplitude α. In this example we illustrate the use of first order perturbation approximation to study both of these 2-periodic solutions for the numerical example in Figure 9.1(B) for which the periodic Ricker equation is*

$$x(t+1) = x(t)\exp\left[\left(2.1 - 50\alpha(-1)^t\right)\left(1 - \frac{x(t)}{1+\alpha(-1)^t}\right)\right]. \tag{9.42}$$

(Note that $\cos(\pi t) = (-1)^t$.) The unperturbed equation when $\alpha = 0$

$$x(t+1) = x(t)\exp\left[2.1\left(1 - x(t)\right)\right]$$

has a stable 2-periodic solution with the two phases

$$\bar{x}_1(t) \; : \; \{0.62929, 1.3707\}$$
$$\bar{x}_2(t) \; : \; \{1.3707, 0.62929\}$$

whose initial conditions are fixed points of the composite map $x = F_1(F_0(x,\alpha),\alpha)$ where

$$F_t(x,\alpha) = x\exp\left[\left(2.1 - 50\alpha(-1)^t\right)\left(1 - \frac{x}{1+\alpha(-1)^t}\right)\right].$$

The initial conditions $\bar{x}_1(0,\alpha)$ and $\bar{x}_2(0,\alpha)$ for the perturbed 2-periodic solutions $\bar{x}_1(t,\alpha)$ and $\bar{x}_2(t,\alpha)$ guaranteed by Theorem 94 are fixed points of the composite equation (9.40) with

$$r(t) = 2.1 - 50\alpha(-1)^t, \quad K(t) = 1 + \alpha(-1)^t$$

which, after the cancellation of x and the taking of a logarithm of both sides, reduces to the equation $0 = g(x,\alpha)$ where

$$g(x,\alpha) := (2.1 - 50\alpha)\left(1 - \frac{x}{1+\alpha}\right) + (2.1 + 50\alpha)\left(1 - \frac{x\exp\left[(2.1-50\alpha)\left(1-\frac{x}{1+\alpha}\right)\right]}{1-\alpha}\right). \tag{9.43}$$

Note that the two initial conditions 0.62929 and 1.3707 for the two phases of the 2-periodic solution of the unperturbed Ricker are solutions of the equation $0 = g(x,0)$. The solutions of $0 = g(x,\alpha)$ provide the initial conditions $\bar{x}_1(0,\alpha)$ and $\bar{x}_2(0,\alpha)$ for the 2-periodic $\bar{x}_1(t,\alpha)$ and $\bar{x}_2(t,\alpha)$ solutions. We can approximate $\bar{x}_1(0,\alpha)$ and $\bar{x}_2(0,\alpha)$ for small α using the Taylor expansions

$$\bar{x}_1(0,\alpha) = 0.62929 + \left.\frac{d\bar{x}_1(0,\alpha)}{d\alpha}\right|_{\alpha=0} \alpha + O(\alpha^2)$$

$$\bar{x}_2(0,\alpha) = 1.3707 + \left.\frac{d\bar{x}_2(0,\alpha)}{d\alpha}\right|_{\alpha=0} \alpha + O(\alpha^2),$$

where we calculate the derivatives $d\bar{x}_i(0,\alpha)/d\alpha|_{\alpha=0}$ by implicit differentiation of the equation $0 = g(\bar{x}_i(0,\alpha),\alpha)$ with respect to α followed by an evaluation at $\alpha = 0$:

$$0 = \left.\frac{dg(\bar{x}_i(0,\alpha),\alpha)}{dx}\right|_{\alpha=0} \left.\frac{d\bar{x}_i(0,\alpha)}{d\alpha}\right|_{\alpha=0} + \left.\frac{dg(\bar{x}_i(0,\alpha),\alpha)}{d\alpha}\right|_{\alpha=0}$$

$$\left.\frac{d\bar{x}_i(0,\alpha)}{d\alpha}\right|_{\alpha=0} = -\left.\frac{dg(\bar{x}_i(0,\alpha),\alpha)}{d\alpha}\right|_{\alpha=0} \left(\left.\frac{dg(\bar{x}_i(0,\alpha),\alpha)}{dx}\right|_{\alpha=0}\right)^{-1}.$$

Using the definition (9.43) of $g(x,\alpha)$, we find that

$$\left.\frac{d\bar{x}_1(0,\alpha)}{d\alpha}\right|_{\alpha=0} = 17.355$$

$$\left.\frac{d\bar{x}_2(0,\alpha)}{d\alpha}\right|_{\alpha=0} = 35.749$$

and hence we obtain the first order (linear) approximations to the initial conditions for the two 2-periodic solutions

$$\bar{x}_1(0,\alpha) \approx 0.62929 + 17.354\alpha$$

$$\bar{x}_2(0,\alpha) \approx 1.3707 + 35.748\alpha.$$

The second terms $\bar{x}_1(1,\alpha)$ and $\bar{x}_2(1,\alpha)$ in each 2-periodic solution can be calculated by applying the Ricker equation (9.42) to these initial conditions to get

$$\bar{x}_1(1,\alpha) = \bar{x}_1(0,\alpha)\exp\left[(2.1 - 50\alpha)\left(1 - \frac{\bar{x}_1(0,\alpha)}{1+\alpha}\right)\right]$$

$$\approx (0.62929 + 17.354\alpha)\exp\left[(2.1 - 50\alpha)\left(1 - \frac{0.62929 + 17.354\alpha}{1+\alpha}\right)\right]$$

$$= 1.3707 - 35.748\alpha + O(\alpha^2)$$

and

$$\bar{x}_2(1,\alpha) = \bar{x}_2(0,\alpha)\exp\left[(2.1 - 50\alpha)\left(1 - \frac{\bar{x}_2(0,\alpha)}{1+\alpha}\right)\right]$$

$$\approx (1.3707 + 35.749\alpha)\exp\left[(2.1 - 50\alpha)\left(1 - \frac{1.3707 + 35.749\alpha}{1+\alpha}\right)\right]$$

$$= 0.62929 - 17.354\alpha + O(\alpha^2).$$

Finally, we arrive at the first order perturbation approximation of the two 2-periodic cycles

$$\bar{x}_1(t,\alpha) \;:\; \{0.62929 + 17.354\alpha,\; 1.3707 - 35.748\alpha\}$$

$$\bar{x}_2(t,\alpha) \;:\; \{1.3707 + 35.748\alpha,\; 0.62929 - 17.354\alpha\}.$$

From these approximations we can obtain some features of the two 2-periodic solutions of the $p=2$ periodic forced Ricker equation (9.42).

For small $\alpha > 0$ the two stable 2-periodic solutions $\bar{x}_1(t,\alpha)$ and $\bar{x}_2(t,\alpha)$
· are out-of-phase, but are not phase shifts of each other;
· have different averages:

$$\text{av}(\bar{x}_1(t,\alpha)) = 1 - 9.197\alpha$$

$$\text{av}(\bar{x}_2(t,\alpha)) = 1 + 9.197\alpha$$

· $\bar{x}_1(t,\alpha)$ is attenuant while $\bar{x}_2(t,\alpha)$ is resonant
   (since the average of the 2-periodic solution when $\alpha = 0$ equals 1);
· $\bar{x}_2(t,\alpha)$ has a larger amplitude than $\bar{x}_1(t,\alpha)$ and oscillates in-phase with $K(t)$, whereas $\bar{x}_1(t,\alpha)$ oscillates in-phase with $r(t)$.

These features can be seen for $\alpha = 0.01$ upon close inspection of the plots in Figure 9.1.

9.6.6 A Case Study

In a series of papers [58], [160], [161], [162], [164] the nonlinear LPA model introduced in Section 6.6 was used to study the periodic responses of laboratory cultures of flour beetles (*T. castaneum*) to periodic oscillations in their habitats reported in the seminal paper by D. A. Jillson [176] and in subsequent experiments reported in [162]. The astonishing resonance phenomena observed by Jillson, under certain conditions, and the subsequent LPA model studies were the stimulus for the Cushing–Henson conjectures studied in Section 9.4.1. Another central goal in these studies was to verify the existence, by means of controlled, replicated experiments, the model predicted multiple periodic attractors.

When grown in a constant volume of nutrient medium (that occupied by 20 grams) in the laboratory setting reported in [162], the flour beetle population dynamics was that of a 2-periodic cycle. Experiments (with replicates and controls) were conducted in which the volume was periodically oscillated with period 2. The time unit in the LPA model was the census time of 2weeks (which corresponded to the time spent in the L-stage and the P-stage).

The LPA model used in these studies was introduced in Section 6.6. It is a three-dimensional, structured population matrix model $\mathbf{x}(t+1) = \mathcal{L}(\mathbf{x}(t))\mathbf{x}(t)$ with an (extended) Leslie matrix projection matrix (6.53)

$$\mathcal{L}(\mathbf{x}) = \begin{pmatrix} 0 & 0 & b\exp(-c_{el}x_1 - c_{ea}x_3) \\ 1-\mu_l & 0 & 0 \\ 0 & \exp(-c_{pa}x_3) & 1-\mu_a \end{pmatrix} \quad (9.44)$$

and state variable

$$\mathbf{x} = \begin{pmatrix} x_1 \\ x_2 \\ x_3 \end{pmatrix} = \begin{pmatrix} L \\ P \\ A \end{pmatrix}$$

consisting of larval, pupal (and callow), and fertile adult life cycle stages L, P, and A. A modification that includes the volume V of a cultures habitat was derived in [58] and [162] and has population projection matrix

$$\mathcal{L}(\mathbf{x}) = \begin{pmatrix} 0 & 0 & b\exp\left(\frac{-c_{el}x_1 - c_{ea}x_3}{V}\right) \\ 1-\mu_l & 0 & 0 \\ 0 & \exp\left(\frac{-c_{pa}x_3}{V}\right) & 1-\mu_a \end{pmatrix}.$$

If the habitat volume changes over time, then V is replaced by $V(t)$. In [162], inspired by one of Jillson's experiments, V is given a period 2 oscillation with amplitude α

$$V(t) = 1 + \alpha(-1)^t, \quad 0 \leq \alpha < 1$$

around the average volume of 1. (The volume in the experiments, namely that occupied by 20 grams of medium, is taken as the unit of volume.) This results in a 2-periodically forced LPA model

$$\mathbf{x}(t+1) = \mathcal{L}_t(\mathbf{x}(t), \alpha)\mathbf{x}(t), \quad (9.45)$$

where

$$\mathcal{L}_t(\mathbf{x}, \alpha) = \begin{pmatrix} 0 & 0 & b\exp\left(\frac{-c_{el}x_1 - c_{ea}x_3}{1+\alpha(-1)^t}\right) \\ 1-\mu_l & 0 & 0 \\ 0 & \exp\left(\frac{-c_{pa}x_3}{1+\alpha(-1)^t}\right) & 1-\mu_a \end{pmatrix} \quad (9.46)$$

is a 2-periodically forced Leslie projection matrix.

In a constant habitat ($\alpha = 0$) the model reduces to $\mathbf{x}(t+1) = \mathcal{L}(\mathbf{x}(t))\mathbf{x}(t)$ with projection matrix (9.44). Parameterization of this model, together with experimental manipulations utilized in experiments, give the coefficient values [162]

$$b = 6.598, \quad \mu_l = 0.2055, \quad \mu_a = 0.1 \\ c_{el} = 0.1, \quad c_{ea} = 0.01, \quad c_{pa} = 0.0047. \tag{9.47}$$

and the projection matrix

$$\mathcal{L}(\mathbf{x}) = \begin{pmatrix} 0 & 0 & 6.598\exp(-0.1x_1 - 0.01x_3) \\ 0.7945 & 0 & 0 \\ 0 & \exp(-0.0047x_3) & 0.9 \end{pmatrix}. \tag{9.48}$$

Solving the equilibrium equation $\mathbf{x} = \mathcal{L}(\mathbf{x})\mathbf{x}$ numerically, we obtain the positive equilibrium

$$\mathbf{x}_e = \begin{pmatrix} 23.36 \\ 18.56 \\ 110.44 \end{pmatrix}.$$

The Jacobian of $\mathcal{L}(\mathbf{x})\mathbf{x}$, when evaluated at $\mathbf{x}_e$, has eigenvalues

$$\lambda_1 = -2.337, \quad \lambda_2 = 8.442 \times 10^{-1}, \quad \lambda_3 = 5.291 \times 10^{-3} \tag{9.49}$$

and hence the equilibrium $\mathbf{x}_e$ is unstable ($|\lambda_1| > 1$).

Solving the composite equation $\mathbf{x} = \mathcal{L}(\mathcal{L}(\mathbf{x})\mathbf{x})\mathcal{L}(\mathbf{x})\mathbf{x}$ numerically for a nonzero fix point

$$\mathbf{x} = \begin{pmatrix} 162 \\ 0 \\ 243 \end{pmatrix}$$

we obtain from it a 2-periodic solution

$$\mathbf{x}_1(t): \begin{pmatrix} 162 \\ 0 \\ 243 \end{pmatrix} \to \begin{pmatrix} 0 \\ 129 \\ 219 \end{pmatrix} \to \begin{pmatrix} 162 \\ 0 \\ 243 \end{pmatrix} \to \begin{pmatrix} 0 \\ 129 \\ 219 \end{pmatrix} \to \ldots \tag{9.50}$$

and its phase shift

$$\mathbf{x}_2(t): \begin{pmatrix} 0 \\ 129 \\ 219 \end{pmatrix} \to \begin{pmatrix} 162 \\ 0 \\ 243 \end{pmatrix} \to \begin{pmatrix} 0 \\ 129 \\ 219 \end{pmatrix} \to \begin{pmatrix} 162 \\ 0 \\ 243 \end{pmatrix} \to \ldots$$

An application of Theorems 94 and 95 implies, for small amplitudes α, the existence of two stable 2-periodic solutions $\mathbf{x}_1(t,\alpha)$ and $\mathbf{x}_2(t,\alpha)$ (near $\mathbf{x}_1$ and $\mathbf{x}_2$) and an unstable 2-periodic solution $\mathbf{x}_3(t,\alpha)$ (near $\mathbf{x}_e$).

Laboratory experiments reported in [162] were designed to corroborate the occurrence of these stable cycles in cultures of flour beetles for several amplitudes α. For $\alpha = 0.4$ (which oscillates the habitat between 28 and 12 grams) the 2-periodic solutions can be numerically calculated from the composite equation. These turn out to be the stable 2-periodic solutions

$$\mathbf{x}_1(t, 0.4): \begin{pmatrix} 92 \\ 0 \\ 142 \end{pmatrix} \to \begin{pmatrix} 0 \\ 73 \\ 128 \end{pmatrix} \to \begin{pmatrix} 92 \\ 0 \\ 142 \end{pmatrix} \to \begin{pmatrix} 0 \\ 73 \\ 128 \end{pmatrix} \to \cdots$$

$$\mathbf{x}_2(t, 0.4): \begin{pmatrix} 0 \\ 180 \\ 306 \end{pmatrix} \to \begin{pmatrix} 227 \\ 0 \\ 340 \end{pmatrix} \to \begin{pmatrix} 0 \\ 180 \\ 306 \end{pmatrix} \to \begin{pmatrix} 227 \\ 0 \\ 340 \end{pmatrix} \to \cdots$$

and the unstable (saddle) 2-periodic solution

$$\mathbf{x}_3(t, 0.4): \begin{pmatrix} 64 \\ 3 \\ 121 \end{pmatrix} \to \begin{pmatrix} 4 \\ 51 \\ 111 \end{pmatrix} \to \begin{pmatrix} 64 \\ 3 \\ 121 \end{pmatrix} \to \begin{pmatrix} 4 \\ 51 \\ 111 \end{pmatrix} \to \cdots .$$

Note that these model predicted cycles imply some oscillatory characteristics that are easily observed in experimental data:

- $\mathbf{x}_2(t, 0.4)$ is resonant in each of the three stages $x_1 = L$, $x_2 = P$, and $x_3 = A$ and the larval stage x_1 oscillates in-phase with the habitat volume $V(t) = 1 + 0.4(-1)^t$;

- $\mathbf{x}_1(t, 0.4)$ is attenuant in each of the three stages and the larval stage x_1 oscillates out–of–phase with the habitat volume.

Controlled and replicated experiments carried out for over two years verified the occurrence of these model predicted cycles; see [162] for a full account of the project. In Figure 9.2 the larval components from three of numerous replicates are displayed. Plot (A) in Figure 9.2 shows an example of a population grown in a constant habitat, which the model predicts has the stable 2-periodic solution (9.50) that oscillates between 0 and a high of 162 larvae. Typical of real data, there are stochastic deviations from any successfully predicted deterministic dynamic. In these experiments there was virtually no census errors, but environmental and demographic stochastic are present [59], [88]. Plots (B) and (C) in Figure 9.2 show the larval components of two replicates grown in a 2-periodically oscillating habitat with $\alpha = 0.4$ and initiated at initial conditions predicted to approach the resonant and the attenuant 2-periodic solutions, respectively (namely $\mathbf{x}(0) = (150, 200, 150)^T$ and $(150, 0, 150)^T$, respectively). Notably, the replicate in (C) shows the characteristics of low amplitude oscillations out-of-phase with the oscillating habitat during the first 60 weeks of the experiment, but at week 62 a stochastic event moved the culture to the basin of attraction of the resonant 2-periodic solution, as seen by the increased amplitude and the change in phase. This shift in phase and amplitude was observed in virtually all replicates and is explained by stochasticity and the fact that the basin of attraction of the attenuant 2-periodic solution is relatively small. For more study of this model and the experiments and of a stochastic version of the model see [162].

Exercises 9.6.4-9.6.6

1. Prove that the spectrum (set of eigenvalues) of the Jacobian $J_q(\bar{x}(i))$ of the composite $C_q(x)$ is the same for all i phases of the q-periodic solution $x(t)$.

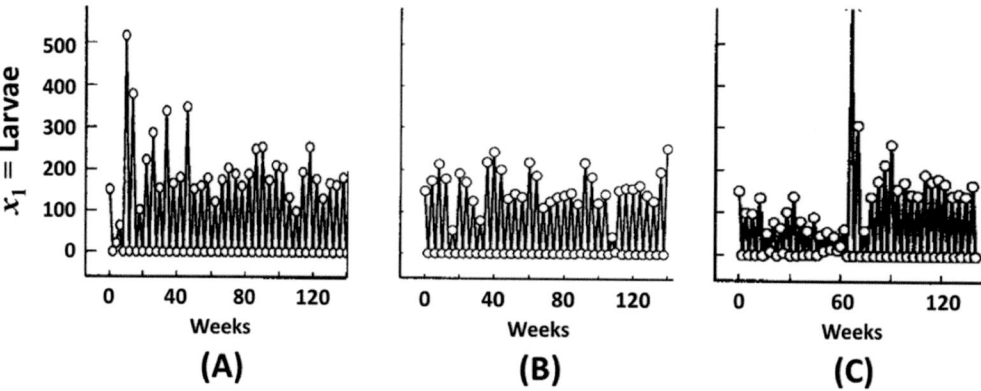

Figure 9.2: Shown are plots of the larval component of some sample replicates taken from data gathered in replicated and controlled experiments with flour beetles that were designed to corroborate the model predictions of the periodically forced LPA model. In (A) the habitat is held constant and the model predicts a stable 2-periodic solution (alternating between 0 and a high of 162). When the habitat oscillates with amplitude $\alpha = 0.4$ the model predicts a resonant 2-periodic solution in-phase with the habitat and an attenuant 2-periodic solution out-of-phase with the habitat. The initial conditions in (B) and (C) predict approach to the former and latter 2-periodic solutions, respectively. Notice that the replicated in (C) switches from the attenuant to the resonant solution due to a stochastic event that moves it to the resonant solution's basin of attraction.

2. Apply Theorems 94 and 95 to the p-periodically forced discrete logistic (Beverton–Holt) equation

$$x(t+1) = b(t) \frac{1}{1 + c(t) x(t)} x(t)$$

with $b(t) = b_0 + \alpha b_1(t)$ and $c(t) = c_0 + \alpha c_1(t)$ where $b_0, c_0 > 1$ and where $b_1(t)$ and $c_1(t)$ are p-periodic with $av(b_1(t)) = av(c_1(t)) = 0$.

3. Consider the 2-periodically forced discrete logistic (Beverton–Holt) equation in Problem 2 with a constant inherent birth rate $b(t) = b_0$ and a 2-periodic intra-specific competition coefficient $c(t) = c_0 + \alpha (-1)^t$, $0 \le \alpha < c_0$. Use the composite map to show that there exists a positive 2-periodic solution if and only if $b_0 > 1$ and find a formula for it. Use the formula to show it is (locally asymptotically) stable and attenuant and it oscillates in-phase with $c(t)$.

4. Consider the 2-periodically forced discrete logistic (Beverton–Holt) equation in Problem 2 with a 2-periodic inherent birth rate $b(t) = b_0 + \alpha (-1)^t$, $0 \le \alpha < b_0$, and a constant intra-specific competition coefficient $c(t) = c_0$. Use the composite map to show that there exists a positive 2-periodic solution if $b_0 > 1$ and α is sufficiently small and find a formula for it. Use the formula to show it is (locally asymptotically) stable and attenuant and it oscillates out-of-phase with $b(t)$.

5. Consider the 2-periodically forced discrete logistic (Beverton–Holt) equation in Problem 2 with a 2-periodic inherent birth rate $b(t) = b_0 + \alpha(-1)^t$, $0 \le \alpha < b_0$, and a 2-periodic intra-specific competition coefficient $c(t) = c_0 - \alpha(-1)^t$, $0 \le \alpha < c_0$ that oscillate out-of-phase. Apply Theorems 94 and 95 to show that there exists a positive 2-periodic solution if $b_0 > 1$ and α is sufficiently small. Find a formula for the 2-periodic solutions and use it to show that it oscillates out-of-phase with $b(t)$ and in-phase with $c(t)$. Determine conditions on c_0 under which the 2-periodic solution is attenuant and conditions on c_0 under which it is resonant.

6. (a) Calculate the Jacobian of the LPA model $\mathbf{x}(t+1) = \mathcal{L}(\mathbf{x}(t))\mathbf{x}(t)$ with projection matrix (9.48) and use it to calculate the eigenvalues (9.49).

   (b) Use a computer program to calculate the three fixed points of the composite map of the LPA model (9.45)-(9.46), with parameter values (9.47) and amplitude $\alpha = 0.4$, that provide the initial conditions for the 2-periodic solutions $\mathbf{x}_1(t, 0.4)$, $\mathbf{x}_1(t, 0.4)$, and $\mathbf{x}_1(t, 0.4)$.

   (c) Use a computer program to calculate the Jacobian of the composite evaluated at each of the fixed points in (b) to show that the eigenvalues associated with the perturbed 2-periodic solutions are

$$\mathbf{x}_1(t, 0.4): \quad \lambda = 1.23 \times 10^{-5},\ 0.57 \pm 0.44i$$
$$\mathbf{x}_2(t, 0.4): \quad \lambda = 0,\ 0.31 \pm 0.35i$$
$$\mathbf{x}_3(t, 0.4): \quad \lambda = 2.66,\ 0.70,\ -1.87 \times 10^{-4}.$$

Bibliography

[1] J.D. Aber, Why don't we believe the models? Bull. Ecol. Soc. Am. **78**(3), 232–3 (1997)

[2] P.A. Abrams, Modelling the adaptive dynamics of traits involved in inter- and intraspecific interactions: an assessment of three methods. Ecol. Lett. **4**, 166–175 (2001)

[3] P.A. Abrams, 'Adaptive Dynamics' and 'adaptive dynamics'. J. Evol. Biol. **18**(5), 1162–1165 (2005)

[4] A.S. Ackleh, H. Caswell, R.A. Chiquet, T. Tang, A. Veprauskas, Sensitivity analysis of the recovery time for a population under the impact of an environmental disturbance. Nat. Resour. Model. **32**(1), e12166 (2019)

[5] J.C. Alexander, J.A. Yorke, Z. You, I. Kan, Riddled basins. Int. J. Bifurc. Chaos **2**, 795–813 (1992)

[6] W.C. Allee, *Animal Aggregations, a Study in General Sociology* (University of Chicago Press, Chicago, 1931)

[7] W.C. Allee, *The Social Life of Animals*, 3rd edn. (William Heineman Ltd, London and Toronto, 1941)

[8] W.C. Allee, O. Park, T. Park, K. Schmidt, *Principles of Animal Ecology* (W. B. Saunders Company, Philadelphia, 1949)

[9] L. Allen, Some discrete-time SI, SIR and SIS epidemic models. Math. Biosc. **124**, 83–105 (1994)

[10] L.J.S. Allen, F.G. Fagan, G. Hognas, H. Fagerholm, Population extinction in discrete-time stochastic population models with an Allee effect. J. Differ. Equ. Appl. **11**, 273–293 (2006)

[11] L. Allen, P. van den Driessche, The basic reproduction number in some discrete-time epidemic models. J. Differ. Equ. Appl. **14**(10–11), 1127–1147 (2008)

[12] D.J. Allwright, Hypergraphic functions and bifurcations in recurrence relations. SIAM J. Appl. Math. **34**, 687–691 (1978)

[13] R. Anderson, R.M. May, Population biology of infectious disease: Part I. Nature **280**, 361–367 (1979)

[14] L. Assas, B. Dennis, S. Elaydi, E. Kwessi, G. Livadiotis, A stochastic modified Beverton-Holt model with Allee effects. J. Differ. Equ. Appl. **22**(1), 37–54 (2016)

[15] L. Assas, S. Elaydi, E. Kwessi, G. Livadiotis, D. Ribble, Hierarchical competition models with Allee effect. J. Biol. Dyn. **9**(1), 34–51 (2014)

[16] L. Assas, S. Elaydi, E. Kwessi, G. Livadiotis, B. Dennis, Hierarchical competition models with Allee effect II: the case of immigration. J. Biol. Dyn. **9**, 288–316 (2015)

[17] A.S. Ackleh, M.I. Hossain, A. Veprauskas, A. Zhang, Persistence and stability analysis of discrete-time predator-prey models: a study of population and evolutionary dynamics. J. Differ. Equ. Appl. **25**(11), 2019,1568–1603 (2019)

[18] S. Baigent, Z. Hou, S. Elaydi, E.C. Balreira, R. Luis, A global picture for the planar Ricker map: convergence to fixed points and identification of the stable/unstable manifolds. J. Differ. Equ. Appl. (2023). https://doi.org/10.1080/10236198.2023.2222855

[19] N.T.J. Bailey, *The Mathematical Theory of Infectious Diseases and Its Applications* (Charles Griffin, London, 1975)

[20] E.C. Balreira, S. Elaydi, R. Luis, Local stability implies global stability for the planar Ricker competition model. Discrete, Contin. Dyn. Sys. Ser. B **19**(2), 323–351 (2014)

[21] E.C. Balreira, S. Elaydi, R. Luis, Global dynamics of triangular maps. Nonlinear Anal. Theory, Methods, Appl. Ser. A **104**, 75–83 (2014)

[22] E.C. Balreira, S. Elaydi, R. Luis, Global stability of higher dimensional monotone maps. J. Differ. Equ. Appl. **23**(12), 2037–2071 (2017)

[23] E.C. Balreira, S. Elaydi, R. Luís, Local stability implies global stability for the planar Ricker competition model. Discrete Contin. Dyn. Syst. B **19**(2), 323–351 (2014)

[24] M. Barfield, R.D. Holt, R. Gomulkiewicz, Evolution in stage-structured populations. Am. Nat. **177**(4), 397–409 (2011)

[25] F. Bartha, A. Garab, T. Krisztin, Local stability implies global stability for the 2-dimensional Ricker map. J. Differ. Equ. Appl. **19**(12), 2043–2078 (2013)

[26] J.R. Beddington, C.A. Free, J.H. Lawson, Dynamic complexity in predator-prey models framed in difference equations. Nature **(255)**, 58–60 (1975)

[27] H. Behncke, Periodical Cicadas. J. Math. Biol. **40**, 413–431 (2000)

[28] L. Berec, E. Angulo, F. Courchamp, Multiple Allee effects and population management. Trends Ecol. Evol. **22**, 185–191 (2007)

[29] L. Berec, Models of Allee effects and their implications for population and community dynamics. Biophys. Rev. Lett. **03**(1), 157–181 (2008)

BIBLIOGRAPHY

[30] A. Berman, R.J. Plemmons, *Nonnegative Matrices in the Mathematical Sciences*, Classics in Applied Mathematics 9, Society for Industrial and Applied Mathematics (Philadelphia, 1994)

[31] H. Bernadelli, Population waves. J. Burma Res. Soc. **31**, 1–18 (1941)

[32] D. Bernoulli, Essai d'une nouvelle analyse de la mortalité caus ée par la petite vérole. Mém. Math. Phys. Acad. R. Sci. Paris **1**, 1–45 (1766)

[33] J. Best, C. Castillo-Chavez, A. Yakubu, Hierarchical competition in discrete-time models with dispersal. Fields Inst. Commun. **36**, 59–72 (2003)

[34] R.J.H. Beverton, S.J. Holt, On the dynamics of exploited fish population. Fish. Invest. Ser. II **19**, 1–533 (1957)

[35] K.W. Blayneh, Hierarchical size-structured population model **9**, 527–539 (2009)

[36] J.S. Brashares, J.R. Werner, A.R. Sinclair, Social "meltdown" in the demise of an island endemic: Allee effects and the Vancouver Island marmot. J. Anim. Ecol. **79**(5), 965–973 (2010)

[37] F. Brauer, C. Castillo-Chavez, *Mathematical Models in Population Biology and Epidemiology*, Texts in Applied Mathematics 40 (Springer, New York, 2001)

[38] F. Brauer, Z. Feng, C. Castillo-Chavez, Discrete epidemic models. Math. Biosc. Eng. **7**(1), 1–15 (2010)

[39] J.S. Brown, Y. Cohen, T.L. Vincent, Adaptive dynamics with vector-valued strategies. Evol. Ecol. Res. **9**, 719–756 (2007)

[40] M.B. Bulmer, Periodical insects. Am. Nat. **111**, 1099–1117 (1982)

[41] B.P. Brooks, Linear stability conditions for a first order 4-dimensional discrete dynamic. J. Appl. Computat. Math. **3**, 2014-5

[42] C. Castillo-Chavez, A.-A. Yakubu, Dispersal, disease, and life history evolution. Math. Biosc. **173**, 35–53 (2001)

[43] J. Carter, A.S. Ackleh, B.P, Leonard, H. Wang, Giant panda population dynamics and bamboo life history: a structured population approach to examining carrying capacity when the prey in semelparous. Ecol. Model. **123**, 207–223 (1999)

[44] C. Castillo-Chavez, A.-A. Yakubu, Dispersal, disease and life-history evolution. Math. Biosc. **173**, 35–53 (2001)

[45] H. Caswell, *Matrix Population Models: Construction, Analysis and Interpretation*, 2nd edn. (Sinauer Associates, Inc., Publishers, Sunderland, Massachusetts, 2001)

[46] H. Caswell, *Sensitivity Analysis: Matrix Methods in Demography and Ecology*, Demographic Research Monographs. Springer Open (2019). https://doi.org/10.1007/978-3-030-10534-1

[47] P. Chesson, R.R. Warner, Environmental variable promotes coexistence in lottery competitive systems. Am. Nat. **117**, 923–943 (1981)

[48] R.A. Chiquet, B. Ma, A.S. Ackleh, N. Pal, N. Sidorovskaia, Demographic analysis of sperm whales using matrix population models. Ecol. Model. **248**, 71–79 (2013)

[49] N. Chitnis, J.M. Cushing, J.M. Hyman, Bifurcation analysis of a mathematical model for malaria transmission. SIAM J. Appl. Math. **67**(1), 24–45 (2006)

[50] N. Chitnis, J.M. Hyman, J.M. Cushing, Determining important parameters in the spread of malaria through the sensitivity analysis of a mathematical model. Bull. Math. Biol. **70**, 1272–1296 (2008)

[51] S.N. Chow, J.K. Hale, *Methods of Bifurcation Theory*, Grundlehren der Mathematischen Wissenschaften [Fundamental Principles of Mathematical Science], vol. 251 (Springer, 1982)

[52] G. Chowell, C. Castillo, P.W. Fenimore, C.M. Kribs-Zaleta, L. Arriola, J.M. Hyman, Model parameters and outbreak control for SARS. Emerg. Infect. Dis. **10**(7), 1258–1263 (2004)

[53] L.C. Cole, The population consequences of life history phenomena. Quar. Rev. Biol. **29**, 103–137 (1954)

[54] B.D. Coleman, On the growth of populations with a narrow spread in reproductive age. 1. General Theory and examples. J. Math. Biol. **6**, 1–19 (1978)

[55] C.S. Coleman, J.C. Frauenthal, Satiable egg eating predators. Math. Biosci. **63**, 99–119 (1983)

[56] W.A. Coppel, The solution of equations by iteration. Math. Proc. Camb. Philos. Soc. **51**(01), 41–43 (1955)

[57] R.F. Costantino, R.A. Desharnais, J.M. Cushing, B. Dennis, Chaotic dynamics in an insect population. Science **275**, 389–391 (1997)

[58] R.F. Costantino, J.M. Cushing, B. Dennis, R.A. Desharnais, S.M. Henson, Resonant population cycles in temporally fluctuating habitats. Bull. Math. Biol. **60**(2), 247–275 (1998)

[59] R.F. Costantino, R.A. Desharnais, J.M. Cushing, B. Dennis, S.M. Henson, A.A. King, The flour beetle Tribolium as an effective tool of discovery. Adv. Ecol. Res. **37**, 101–141 (2005)

[60] F.C. Courchamp, T.H. Clutton-Brock, B.T. Grenfell, Inverse density dependence and the Allee effect. Trends Ecol. Evol. **14**, 405–410 (1999)

[61] F. Courchamp, L. Berec, J. Gascoigne, *Allee Effects in Ecology and Conservation* (Oxford University Press, Oxford, Great Britain, 2008)

[62] K.M. Crowe, A nonlinear ergodic theorem for discrete systems. J. Math. Biol. **32**, 179–191 (1994)

[63] P. Cull, Stability of discrete one-dimensional population models. Bull. Math. Biol. **50**(1), 67–75 (1988)

[64] J.M. Cushing, A strong ergodic theorem for some nonlinear matrix models for structured population growth. Nat. Resour. Model. **3**(3), 331–357 (1989)

[65] J.M. Cushing, A strong Ergodic theorem for some nonlinear matrix models. Nat. Resour. Model. **3**(3), 331–357 (1989)

[66] J.M. Cushing, The dynamics of hierarchical age-structured populations. J. Math. Biol. **12**, 705–729 (1994)

[67] J.M. Cushing, *An Introduction to Structured Population Dynamics, Conference Series in Applied Mathematics*, vol. 71 (SIAM, Philadelphia, 1998). ISBN 0-89871-417-6

[68] J.M. Cushing, Periodically forced nonlinear systems of difference equations. J. Differ. Equ. Appl. **3**, 547–561 (1998)

[69] J.M. Cushing, Nonlinear semelparous Leslie models. Math. Biosci. Eng. **3**(1), 17–36 (2006)

[70] J.M. Cushing, *Matrix Models and Population Dynamics*, a chapter in Mathematical Biology, ed. by M. Lewis, A.J. Chaplain, J.P. Keener, P.K. Maini, IAS/Park City Mathematics Series, vol. 14 (American Mathematical Society, Providence, RI, 2009), pp. 47–150

[71] J.M. Cushing, Three stage semelparous Leslie models. J. Math. Biol. **59**, 75–104 (2009)

[72] J.M. Cushing, A bifurcation theorem for Darwinian matrix models. Nonlinear Stud. **17**(1), 1–13 (2010)

[73] J.M. Cushing, On the relationship between r and $R0$ and its role in the bifurcation of equilibria of Darwinian matrix models. J. Biol. Dyn. **5**, 277–297 (2011)

[74] J.M. Cushing, A dynamic dichotomy for a system of hierarchical difference equations. J. Differ. Equ. Appl. **18**(1), 1–26 (2012)

[75] An Evolutionary Beverton-Holt model, Theory and Application of Difference Equations and Discrete Dynamical Systems, ed. by Z. Alsharawi, J.M. Cushing, S. Elaydi, Springer Proceedings in Mathematics & Statistics (102) (2014), pp. 127–141

[76] J.M. Cushing, Backward bifurcations and strong Allee effects in matrix models for the dynamics of structured populations. J. Biol. Dyn. **8**, 57–73 (2014)

[77] J.M. Cushing, On the fundamental bifurcation theorem for semelparous Leslie models, in *Mathematics of Planet Earth: Dynamics, Games and Science*, ed. by J.P. Bourguignon, R. Jeltsch, A. Pinto, M. Viana, CIM Mathematical Sciences Series (Springer, Berlin, 2015), pp. 215–251

[78] J.M. Cushing, One dimensional maps as population and evolutionary dynamic models, Applied Analysis in Biological and Physical Sciences, ed. by J.M. Cushing, M. Saleem, H.M. Srivastava, M.A. Khan, M. Merajuddin, Springer Proceedings in Mathematics & Statistics, vol. 186 (Springer, India, 2016), pp. 41–62

[79] J.M. Cushing, Difference equations as models of evolutionary dynamics. J. Biol. Dyn. **13**(1), 103–127 (2019)

[80] J.M. Cushing, A Darwinian Ricker equation, in *Progress on Difference Equations and Discrete Dynamical Systems*, ed. by S. Baigent, M. Bhoner, S. Elaydi (Springer Nature, Switzerland AG, 2020), pp. 231–243

[81] J.M. Cushing, A bifurcation theorem for Darwinian matrix models and an application to the evolution of reproductive life-history strategies. J. Biol. Dyn. **15**(sup1), S190–S213 (2021). https://doi.org/10.1080/17513758.2020.1858196

[82] J. M. Cushing, *Matrix Models for Population, Disease, and Evolutionary Dynamics, Student Mathematical Library*, vol. 106 (American Mathematical Society, 2024) ISBN 978-1-4704-7334-1

[83] J.M. Cushing, J. Li, On Ebenman's model for the dynamics of a population with competing juveniles and adults. Bull. Math. Biol. **51**(6), 687–713 (1989)

[84] J.M. Cushing, Z. Yicang, The net reproductive value and stability in structured population models. Nat. Resour. Model. **8**(4), 1–37 (1994)

[85] J.M. Cushing, Z. Yicang, The net reproductive value and stability in matrix population models. Nat. Resour. Model. **8**, 297–333 (1994)

[86] J.M. Cushing, B. Dennis, R.A. Desharnais, R.F. Costantino, An interdisciplinary approach to understanding nonlinear ecological dynamics. Ecol. Model. **92**, 111–119 (1996)

[87] J.M. Cushing, B. Dennis, R.A. Desharnais, R.F. Costantino, Moving toward an unstable equilibrium: saddle nodes in population systems. J. Anim. Ecol. **67**(1), 298–306 (1998)

[88] J. Cushing, R.F. Costantino, B. Dennis, R.A. Desharnais, S.M. Henson, *Chaos in Ecology: Experimental Nonlinear Dynamics, Theoretical Ecology Series*, vol. 1 (Academic Press (Elsevier Science), New York, 2003). ISBN: 0-12-1988767

[89] J.M. Cushing, S. Levarge, N. Chitnis, S.M. Henson, Some discrete competition models and the competitive exclusion principle. J. Differ. Equ. Appl. **10**(13–15), 1139–1151 (2004)

[90] J.M. Cushing, S.M. Henson, C.C. Blakburn, Multiple mixed-type attractors in a competition model. J. Biol. Model. **1**(4), 347–362 (2007)

[91] J.M. Cushing, Shandelle M. Henson, Chantel C. Blackburn, Multiple mixed-type attractors in a competition model. J. Biol. Dyn. **1**(4), 347–362 (2007)

[92] J.M. Cushing, S.M. Henson, Stable bifurcations in nonlinear semelparous Leslie models. J. Biolo. Dyn. **6**, 80–102 (2012)

[93] J.M. Cushing, F. Martins, A.A. Pinto, A. Veprauskas, A bifurcation theorem for evolutionary matrix models with multiple traits. J. Math. Biol. **75**(1), 491–520 (2017). https://doi.org/10.1007/s00285-016-1091-4

[94] J.M. Cushing, A.P. Farrell, A bifurcation theorem for nonlinear matrix models of population dynamics. J. Differ. Equ. Appl. **26**(1), 25–44 (2019)

[95] J.M. Cushing, A.P. Farrell, A bifurcation theorem for nonlinear matrix models of population dynamics. J. Differ. Equ. Appl. **26**(1), 25–44 (2019). https://doi.org/10.1080/10236198.2019.1699916

[96] J.M. Cushing, S.M. Henson, Global dynamics of some periodically forced, monotone difference equations. J. Differ. Equ. Appl. **7**, 859–872 (2001)

[97] J.M. Cushing, S.M. Henson, A periodically forced Beverton-Holt equation. J. Differ. Equ. Appl. **8**, 1119–1120 (2002)

[98] E. D'Aniello, S. Elaydi, The structure of ω-limit sets of asymptotically non-autonomous discrete dynamical systems. Discrete Contin. Dyn. Syst. Ser. B **25**(3), 903–915 (2020)

[99] C. Darwin, *The Origin of Species* (Avenel Books, London, 1859)

[100] N.V. Davydova, O. Diekmann, S.A. van Gils, Year class coexistence or competitive exclusion for strict biennials. J. Math. Biol. **46**, 95–131 (2003)

[101] O. Diekmann, N.V. Davydova, S.A. van Gils, On a boom or bust year class cycle. J. Differ. Equ. Appl. **11**, 327–335 (2005)

[102] N.V. Davydova, O. Diekmann, S.A. van Gils, On circulant populations. I. The algebra of semelparity. Linear Algebra Appl. **398**, 185–243 (2005)

[103] P.S. Dawson, A conflict between Darwinian fitness and population fitness in Tribolium "competition" experiments. Genetics **62**, 413–419 (1969)

[104] B. Dennis, Allee effects: population growth, critical density, and the chance of extinction. Nat. Resour. Model. **3**, 481–538 (1989)

[105] B. Dennis, R.A. Desharnais, J.M. Cushing, S.M. Henson, R.F. Costantino, Estimating chaos and complex dynamics in an insect population. Ecol. Monogr. **71**(2), 277–303 (2001)

[106] B. Dennis, R.A. Desharnais, J.M. Cushing, R.F. Costantino, Nonlinear demographic dynamics: mathematical, models, statistical methods, and biological experiments. Ecol. Monogr. **65**(3), 261–281 (1995)

[107] F. Dercole, S. Rinaldi, *Analysis of Evolutionary Processes: The Adaptive Dynamics Approach and Its Applications* (Princeton University Press, New Jersey, 2008)

[108] R.A. Desharnais, R.F. Costantino, J.M. Cushing, S.M. Henson, B. Dennis, Chaos and population control of insect outbreaks. Ecol. Lett. **4**(3), 229–235 (2001)

[109] R.A. Desharnais, R.F. Costantino, J.M. Cushing, S.M. Henson, B. Dennis, A.A. King, Experimental support of the scaling rule for demographic stochasticity. Ecol. Lett. **9**, 537–547 (2006)

[110] R.L. Devaney, *An Introduction to Dynamical Systems*, 2nd edn. (Addison-Wesley, Reading, Massachusetts, 1989)

[111] U. Dieckmann, R. Law, The dynamical theory of coevolution: a derivation from stochastic ecological processes. J. Math. Biol. **34**, 569–612 (1996)

[112] O. Diekmann, Y. Wang, P. Yan, Carrying simplices in discrete competitive systems and age-structured semelparous populations. Discrete Contin. Dyn. Syst. **20**(1), 37–52 (2008)

[113] O. Diekmann, H. Hessterbeek, T. Britton, *Mathematical Tools for Understanding Infectious Disease Dynamics* (Princeton University Press, Princeton, New Jersey, 2013)

[114] O. Diekmann, N. Davydova, S. van Gils, On a boom and bust year class cycle. J. Differ. Equ. Appl. **11**, 4-5, 327-335

[115] B. Ebenman, Niche differences between age classes and intraspecific competition in age-structured populations. J. Theor. Biol. **124**, 25–33 (1987)

[116] B. Ebenman, Competition between age-classes and population dynamics. J. Theor. Biol. **131**, 389–400 (1988)

[117] J. Edmunds, J.M. Cushing, R.F. Costantino, S.M. Henson, B. Dennis, R.A. Desharnais, Park's Tribolium competition experiments: a non-equilibrium species coexistence hypothesis. J. Anim. Ecol. **72**, 703–712 (2003)

[118] F.E. Egler, W.A. Niering, Mostly a misunderstanding, I believe. Bull. Ecol. Soc. Am. **79**(4), 256–257 (1998)

[119] S.N. Elaydi, *Discrete Chaos: With Applications in Science and Engineering*, 2nd edn. (Chapman & Hall/CRC, 2008)

[120] S.N. Elaydi, *An Introduction to Difference Equations*, 3rd edn. (Springer, New York, 2005)

[121] S. Elaydi, W. Harris, On the computation of A^n. SIAM Rev. **40**(4), 965–971 (1998)

[122] S. Elaydi, Nonautonomous difference equations: open problems and conjectures, difference, and differential equations, in *The Fields Institute of Mathematical Sciences*, ed. by S. Elaydi et al. (2004), pp. 423–429

[123] S. Elaydi, Y. Kang, R. Luis, Global asymptotic stability of evolutionary periodic Ricker competition model. J. Differ. Equ. Appl. (2023)

[124] S. Elaydi, Y. Kang, R. Luís, The effects of evolution on the stability of competing species. J. Biol. Dyn. **16**(1), 816–839 (2022)

[125] S. Elaydi, E. Kwessi, G. Livadiotis, Hierarchical competition models with the Allee effect III: multispecies. J. Biol. Dyn. **12**(1), 271–287 (2018)

[126] S.N. Elaydi, R. Sacker, Basin of attraction of periodic orbits of maps on the real line. J. Differ. Equ. Appl. **10**, 881–888 (2004)

[127] S.N. Elaydi, R.J. Sacker, Population models with Allee effect: a new model. J. Biol. Dyn. **4**, 397–408 (2010)

[128] S.N. Elaydi, R.J. Sacker, Global stability of periodic orbits of nonautonomous difference equations and population biology. J. Differ. Equ. **208**(1), 258–273 (2005)

[129] S.N. Elaydi, R.J. Sacker, Global stability of periodic orbits of nonautonomous difference equations and population biology on Cushing-Henson conjectures, in *Proceedings of ICDEAS8, Brno* (2003)

[130] S.N. Elaydi, R.J. Sacker, Nonautonomous Beverton-Holt equation and the Cushing-Henson conjectures. J. Differ. Equ. Appl. **11**(4–5), 337–346 (2005)

BIBLIOGRAPHY

[131] S.N. Elaydi, R.J. Sacker, Periodic difference equations, population biology, and the Cushing-Henson conjectures. Math. Biosci. **201**(1–2), 195–207 (2006)

[132] S. Elaydi, R.J. Sacker, Nonautonomous Beverton-Holt equations the Cushing-Henson conjectures. J. Differ. Equ. Appl. **11**, 337–347 (2005)

[133] S. Elaydi, R.J. Sacker, Skew-product dynamical systems: applications to difference equations, NOVA 2006, *Proceeding of the UAE Math Day*

[134] S.N. Elaydi, A.-A. Yakubu, Global stability of cycles: Lotka-Volterra competition model with stocking. J. Differ. Equ. Appl. **8**(6), 537–549 (2002)

[135] P.L. Errington, Some contributions of a fifteen-year local study of the northern bob-white to a knowledge of population phenomena. Ecol. Monog. **15**, 1–34 (1945)

[136] N.H. Fefferman, J.M. Reed, A vital rate sensitivity analysis of notable age distributions and short-term planning. J. Wildl. Manag. **70**(3), 649–656 (2006). https://doi.org/10.2193/0022-541X(2006)70[649:AVRSAF]2.0.CO;2

[137] M. Feigenbaum, Quantitative universality for a class of nonlinear transformations. J. Stat. Phy. **19**, 25–52 (1978)

[138] R.A. Fisher, *The Genetical Theory of Natural Selection* (Clarendon Press, Oxford, 1930)

[139] J.E. Franke, A.-A. Yakubu, Extinction in systems of Bobwhite quail populations. Can. Appl. Math. Q. **3**(2), 173–201 (1995)

[140] J.E. Franke, A. Yakubu, Mutual exclusion versus coexistence for discrete competitive systems. J. Math. Biol. **30**, 161–168 (1991)

[141] J.E. Franke, A.-A. Yakubu, Population models with periodic recruitment functions and survival rates. J. Differ. Equ. Appl. **11**(14), 1169–1184 (2005)

[142] J. Franke, A.-A. Yakubu, Signature function for predicting resonant and attenuant population cycles. Bull. Math. Biol. **68**, 2069–2104 (2006)

[143] J. Gani, Some problems of epidemic theory. J. R. Stat. Soc. Ser. A (General) **141**(3), 323–347 (1978)

[144] J.C. Gascoigne, R.N. Lipcius, Allee effects driven by predation. J. Appl. Ecol. (2004)

[145] G.F. Gause, *The Struggle for Existence* (Hafner, New York, 1934)

[146] G.F. Gause, Experimental demonstration of Volterra's periodic oscillation in the numbers of animals. J. Exp. Biol. **12**, 44–48 (1935)

[147] G.F. Gause, La théorie mathématique de la lutte pour la vie. Actualités Scientifiques et Industrielles **227**, 1–63 (1935)

[148] B.T. Grenfell, A.P. Dobson, *Ecology of Infectious Diseases in Natural Populations* (University Press, Cambridge, 1995)

[149] A.B. Gumel, S. Lenhart (eds.), *Modeling Paradigms and Analysis of Disease Transmission Models* (American Mathematical Society, Providence, RI)

[150] M. Guzowska, R. Luis, S. Elaydi, Bifurcation and invariant manifolds of the logistic competition model. J. Differ. Equ. Appl. **17**(12), 1581–1872 (2011)

[151] M. Gyllenberg, A.V. Osipov, G. Soderbacka, Bifurcation analysis of a metapopulation model with sources and sinks. J. Nonlinear Sci. **6**, 329–366 (1996)

[152] J.W. James, *Modeling Biological Systems: Principles and Applications* (Chapman & Hall, New York, 1996)

[153] W.H. Hamer, Epidemic disease in England. Lancet **1**, 733–739 (1906)

[154] G. Hardin, The competitive exclusion principle. Science **131**, 1292–1297 (1960)

[155] A.J. Harry, C.M. Kent, V.L. Kocic, Global behavior of solutions of a periodically forced Sigmoid Beverton-Holt model. J. Biol. Dyn. **6**(2), 212–234 (2102)

[156] M.P. Hassell, Density dependence in single-species populations. J. Anim. Ecol. **44**, 283–295 (1975)

[157] M. Hassell, N. Comins, Discrete-time models for two-species competition. Theor. Popul. Biol. **9**, 202–221 (1976)

[158] A. Hastings, Complex interactions between dispersal and dynamics: lessons from coupled logistic equations. Ecology **75**, 1362–1372 (1993)

[159] S. Henson, J.M. Cushing, Hierarchical models of interspecific competition: scramble versus contact. J. Math. Biol. **34**, 755–772 (1996)

[160] S.M. Henson, J.M. Cushing, The effect of periodic habitat fluctuations on a nonlinear insect population model. J. Math. Biol. **36**, 201–226 (1997)

[161] S.M. Henson, J.M. Cushing, R.F. Costantino, B. Dennis, R.A. Desharnais, Phase switching in population cycles. Proc. R. Soc. Lond. B **265**, 2229–2234 (1998)

[162] S.M. Henson, R.F. Costantino, J.M. Cushing, B. Dennis, R.F. Costantino, Multiple attractors, saddles, and population dynamics in periodic habitats. Bull. Math. Biol. **61**, 1121–1149 (1999)

[163] S.M. Henson, R.F. Costantino, J.M. Cushing, R.A. Desharnais, B. Dennis, A.A. King, Lattice effects observed in chaotic dynamics of experimental populations. Science **294**, 602–605 (2001)

[164] S.M. Henson, R.F. Costantino, J.M. Cushing, B. Dennis, R.A. Desharnais, Basins of attraction: population dynamics with two locally stable 4-cycles. Oikos **98**, 17–24 (2002)

[165] S.M. Henson, Multiple attractors and resonance in periodically-forced population models. Physica D: Nonlinear Phenom. **140**, 33–49 (2000)

[166] S.M. Henson, A.A. King, R.F. Costantino, J.M. Cushing, B. Dennis, R.A. Desharnais, Explaining and predicting patterns in stochastic population systems. Proc. R. Soc. Lond. B **270**, 1549–1553 (2003)

[167] H.W. Hethcote, The mathematics of infectious diseases. SIAM Rev. **42**(4), 599–653 (2000)

[168] R.V. Hogg, E.A. Tanis, D.L. Zimmerman, *Probability and Statistical Inference* (Pearson, 2015)

[169] F. Hoppenstaedt, *Mathematical Theories of Populations: Demographics, Genetics and Epidemics, CBMS-NSF Regional Conference Series in Applied Mathematics*, vol. 20 (SIAM, Philadelphia, 1975)

[170] J. Hofbauer, J.W.-H. So, Uniform persistence and repellors for maps. Proc. Am. Math. Soc. **107**(4), 1137–1142 (1989)

[171] R.A. Horn, C.R. Johnson, *Matrix Analysis* (Cambridge University Press, Cambridge, 1985)

[172] J. Impagliazzo, *Deterministic Aspects of Mathematical Demography, Biomathematics*, vol. 13 (Springer, Berlin, 1985)

[173] W.T. Jamieson, On the global behavior of the LPA model when $c_{pa} = 0$. J. Differ. Equ. Appl. **26**(3), 353–361 (2020)

[174] S.R.-J. Jang, Allee effects in a discrete-time host-parasitoid model. J. Differ. Equ. Appl. **12**(2), 165–181 (2006)

[175] Sophia Jang, Sandra Diamond, A host-parasitoid interaction with Allee effects on the host. J. Comput. Math. Appl. **53**, 89–103 (2007)

[176] D.A. Jillson, Insect populations respond to fluctuating environments. Nature **288**, 699–700 (1980)

[177] D.M. Johnson, A.M. Liebhold, P.C. Tobin, O.N. Bjornstad, Allee effects and pulsed invasion by the gypsy moth. Nature **444**(7117), 361–363 (2006)

[178] F. Jones, J. Perry, Modeling populations of cyst-nematodes. J. Appl. Ecol. **15**, 349–371 (1978)

[179] E. Jury, *Theory and Applications of the Z-Transform* (Wiley, 1964)

[180] Y. Kang, D. Ambruster, Y. Kuang, Dynamics of a plant-herbivore model. J. Biol. Dyn. **2**, 101–89 (2008)

[181] Y. Kang, P. Chesson, Relative nonlinearity and permanence. Theor. Popul. Biol. **78**(1), 26–35 (2010)

[182] Y. Kang, S.K. Sasmal, A.R. Bhowmick, J. Chattopadhyay, A host-parasitoid system with predation-driven component Allee effects in host population. J. Biol. Dyn. **9**, 213–232 (2015)

[183] Y. Kang, H. Smith, Global dynamics of a discrete two-species Lottery-Ricker competition model. J. Biol. Dyn. **6**(2), 358–376 (2012)

[184] Y. Kang, A.-A. Yakubu, Weak Allee effects and species coexistence. Nonlinear Anal. Real World Appl. **12**(6), 3329–3345

[185] S. Kapcak, S. Elaydi, A. Ufektepe, Stability of a predator-prey model with refuge effect. J. Differ. Equ. Appl. **22**(7), 989–1004 (2016)

[186] S. Kapcak, V. Ufuktube, S. Elaydi, Stability and invariant manifolds of a generalized Beddington host-parasitoid model. J. Biol. Dyn. **7**, 233–253 (2013)

[187] W.O. Kermack, A.G. McKendrick, A contribution to the mathematical theory of epidemics. Proc. R. Soc. Lond. **115**, 700–721 (1927)

[188] H. Kestelman, Mappings with non-vanishing Jacobian. Am. Math. Mon. **78**(6), 662–663 (1971)

[189] H. Kielhöfer, *Bifurcation Theory: An Introduction with Applications to Partial Differential Equations*, vol. 156 (Springer, New York, 2011)

[190] A.A. King, R.F. Costantino, J.M. Cushing, S.M. Henson, R.A. Desharnais, B. Dennis, Anatomy of a chaotic attractor: subtle model predicted patterns revealed in population data. Proc. Natl. Acad. Sci. **101**(1), 408–413 (2003)

[191] B. Kipchumba, Construction and analysis of a Leslie matrix population model for Amboseli elephants, Masters thesis, Applied Mathematics, University of Nairobi (2013)

[192] V.L. Kocic, A note on the Cushing-Henson conjecture. J. Differ. Equ. Appl. **10**(8), 791–79 (2004)

[193] R. Kon, A note on attenuant cycles of population models with periodic carrying capacity. J. Differ. Equ. Appl. **10**(8), 791–79 (2004)

[194] R. Kon, Invisibility of missing year-classes in Leslie matrix models for a semelparous biennial population, in *Proceedings of the Czech-Japanese Seminar in Applied Mathematics*, COE Lecture Note, vol. 3. Faculty of Mathematics, Kyushu University (2005), pp. 64–75

[195] R. Kon, Competitive exclusion between year-classes in a semelparous biennial population, in *Mathematical Modeling of Biological Systems*, vol. II, ed. by A. Deutsch, R. Bravo de la Parra, R. de Boer, O. Diekmann, P. Jagers, E. Kisdi, M. Kretzschmar, P. Lansky, H. Metz (Birkhäuser, Boston, 2007), pp. 79–90

[196] R. Kon, Bifurcations of cycles in nonlinear semelparous Leslie matrix models. J. Math. Biol. **80**(4), 1187–1207 (2020)

[197] R. Kon, Permanence induced by life-cycle resonances: the periodical cicada problem. J. Biol. Dyn. **6**(2), 855–890 (2012)

[198] R. Kon, Y. Saito, Y. Takeuchi, Permanence of single-species stage-structured models. J. Math. Biol. **48**, 515–528 (2004)

[199] R. Kon, Y. Saito, Y. Takeuchi, Permanence of single-species stage-structured models. J. Math. Biol. **48**, 515–528 (2005)

[200] R. Kon, Y. Iwasa, Single-class orbits in nonlinear Leslie matrix models for semelparous populations. J. Math. Biol. **55**(5–6), 781–802 (2007)

[201] Y. Kuang, J.M. Cushing, Global stability in a nonlinear difference-delay equation model of flour beetle population growth. J. Differ. Equ. Appl. **2**, 31–37 (1995)

BIBLIOGRAPHY

[202] Yu.A. Kuznetsov, *Elements of Applied Bifurcation Theory*, 3rd edn. (Springer, 2004)

[203] J.H. Lambert, Toedlichkeit der Kinderblattern Beytrage. Buchhandlung der Realschule 1772, 3:568

[204] R. Lande, Natural selection and random genetic drift in phenotypic evolution. Evolution **30**, 314–334 (1976)

[205] R. Lande, A quantitative genetic theory of life history evolution. Ecology **33**, 607–615 (1982)

[206] S.D. Lane, N.J. Mills, Intraspecific competition and density dependence in an Ephestia knehniella. Oikos **101**, 578–590 (2003)

[207] J.P. LaSalle, *The Stability of Dynamical Systems, CBMS-NSF Regional Conference Series in Applied Mathematics*, vol. 25 (SIAM, Philadelphia, 1976). ISBN 978-0-89871-022-9

[208] J.P. LaSalle, The stability and control of discrete processes. *Applied Mathematical Sciences*, vol. 82 (Springer, New York, 1986)

[209] P.H. Leslie, On the use of matrices in certain population mathematics. Biometrika **33**, 183–212 (1945)

[210] P.H. Leslie, Some further notes on the use of matrices in population mathematics. Biometrika **35**, 213–245 (1948)

[211] P.H. Leslie, T. Park, D.M. Mertz, The effect of varying the initial numbers on the outcome of competition between two Tribolium species. J. Anim. Biol. **37**, 9–23 (1957)

[212] S.A. Levin, C.P. Goodyear, Analysis of an age-structured fishery model. J. Math. Biol. **9**, 245–274 (1980)

[213] D.R. Levitan, M.A. Sewell, F.-S. Chia, How distribution and abundance influence fertilization success in the sea Urchin Strongylocentotus Franciscanus. Ecology **73**(1), 248–254 (1992)

[214] E.G. Lewis, On the generation and growth of a population. Sanky **6**, 93–96 (1943)

[215] S.J. Lewis, A note on the strong ergodic theorem of some discrete models. J. Differ. Equ. Appl. **3**(1), 55–63 (1997)

[216] R.C. Lewontin, Evolution and the theory of games. J. Theor. Biol. **1**(3), 382–403 (1961)

[217] W. Lidicker, The Allee effect: its history and future importance. Open Ecol. J. **3**(1), 71–82 (2010)

[218] G. Livadiotis, S. Elaydi, General Allee effect in two-species population biology. J. Biol. Dyn. **6**(2), 959–973 (2012)

[219] R. Luís, S. Elaydi, H. Oliveira, Stability of a Ricker-type competition model and the competitive exclusion principle. J. Biol. Dyn. **5**(6), 636–660 (2011)

[220] G. Livadiotis, L. Assas, S. Elaydi, E. Kwessi, D. Ribble, Competition models with Allee effects. J. Differ. Equ. Appl. **20**(8), 1127–1151. https://doi.org/10.1080/10236198.2014.897341

[221] R. Luis, S. Elaydi, H. Oliveira, Towards a theory of periodic difference equation and population biology, dynamics, games and science I, *Springer Proceedings in Mathematics*, ed. by peixoto et al. (2011), pp. 287–322

[222] C.-K. Li, H. Schneider, Applications of Perron-Frobenius theory to population dynamics. J. Math. Biol. **44**, 450–462 (2002)

[223] C.-K. Li, H. Schneider, Applications of Perron-Frobenius theory to population dynamics. J. Math. Biol. **44**, 450–462 (2002)

[224] T.Y. Li, J.A. Yorke, Period three implies chaos. Am. Math. Month. **82**, 985–992 (1975)

[225] J. Li, B. Song, X. Wang, An extended discrete Ricker population model with Allee effects. J. Differ. Equ. Appl. **13**(4), 309–321 (2007)

[226] A. Liapunov, Probleme general de la stabilite du movement, 1907 translation of Russian original reprinted in Ann. Math. Stud. 17 (Princeton, 1949)

[227] P. Liu, S. Elaydi, Discrete competitive and cooperative models of Lotka Volterra type. J. Comput. Anal. Appl. **3**, 53–73 (2001)

[228] I.M. Longini, Jr., The generalized discrete-time epidemic model with immunity: a synthesis. Math. Biosci. **82**, 19–41 (1986)

[229] J.L. Lush, *Animal Breeding Plans* (Iowa State Press, Ames, Iowa, 1937)

[230] M. Martcheva, *An Introduction to Mathematical Epidemiology, Texts in Applied Mathematics 61* (Springer, 2015)

[231] R. May, Host-parasitoid systems in patchy environments: a phenomenological model. J. Anim. Ecol. **47**(3), 833–844 (1978)

[232] R.M. May, Simple mathematical models with very complicated dynamics. Nature **261**, 459–469 (1977)

[233] R. May, Simple mathematical models with very complicated dynamics. Nature **261**, 559–467 (1976)

[234] R. May, G. Oster, Bifurcation and dynamic complexity in simple ecological models. Am. Nat. **110**, 573–599 (1976)

[235] J. Maynard Smith, *Models in Ecology* (Cambridge University Press, Cambridge, 1974)

[236] M.A. McCarthy, The Allee effect, finding mates and and theoretical models. Ecol. Model. **103**, 99–102 (1997)

[237] B.J. McGill, J.S. Brown, Evolutionary game theory and adaptive dynamics of continuous traits. Annu. Rev. Ecol. Evol. Syst. **38**, 403–435 (2007)

[238] E.P. Meissen, K.R. Salau, J.M. Cushing, A global bifurcation theorem for Darwinian matrix models. J. Differ. Equ. Appl. **22**(8), 1114–1136 (2016)

[239] J.G. Miltonand J. Belair, Chaos, noise, and extinction in models of population growth. Theor. Popul. Biol. **37**, 273–290 (1990)

[240] E. Mjølhus, A. Wikan, T. Solberg, On synchronization in semelparous populations. J. Math. Biol. **50**, 1–21 (2005)

[241] K. Mokni, S. Elaydi, M. Ch-Chaoui, A. Eladdadi, Discrete evolutionary population models: a new approach. J. Biol. Dyn. **14**(1), 454–478 (2020)

[242] M. Ch-Chaoui, K. Mokni, A discrete evolutionary Beverton-Holt population model. Int. J. Dyn. Control **11** (2023)

[243] C.J. Moss, The demography of an African elephant (*Loxodonta africana*) population in Amboseli, Kenya. J. Zool. **255**(2), 145–156. https://doi.org/10.1017/S0952836901001212

[244] J.I. Neimark, On some cases of periodic motions depending on parameters. Dokl. Akad. Nauk **SSSR 129**, 736–739 (1964) (in Russian)

[245] A.J. Nicholson, V.A. Bailey, The balance of animal populations, Part I. Proc. Zool. Soc. Lond. **(3)**, 551–598 (1935)

[246] S.O. Ojung'a, J.O. Nyakinda, E. Okuto, J.A. Mullah, Invasive species population status modeling using stage based matrix: Mount Elgon ecosystem. Math. Theory Model. **9**(2) (2019). https://doi.org/10.7176/MTM

[247] J.M. Ortega, *Matrix Theory, A Second Course* (Plenium, New York, 1987)

[248] T. Park, Experimental studies of interspecific competition. II. Temperature, humidity, and competition in two species of Tribolium. Physiol. Zool. **27**, 177–238 (1954)

[249] T. Park, D.B. Mertz, W. Grodinski, T. Prus, Cannibalistic predation in populations of flour beetles. Physiol. Zool. **38**, 289–321 (1965)

[250] P. van den Driessche, Reproduction numbers of infectious disease models. Infect. Dis. model. 1–16 (2017)

[251] P. van den Driessche, A.A. Yakubu, Disease extinction versus persistence in discrete-time epidemic models, Bull. Math. Biol. (2018). https://doi.org/10.1007/s11538-018-0426-2

[252] C. Pfister, A. Bradbury, Harvesting red sea urchins: recent effects and future predictions. Ecol. Appl. **6**(298), 298–310 (1996)

[253] P. Rabinowitz, Some global results for nonlinear eigenvalue problems. J. Funct. Anal. **7**, 487–513 (1971)

[254] R.C. Rael, T.L. Vincent, R.F. Costantino, J.M. Cushing, Evolution of corn oil sensitivity in the flour beetle, in *Advances in Dynamic Game Theory*, ed. by Jørgensen, Quincampoix and Vincent (Birkhäuser, Berlin, 2007), pp. 367–376

[255] R.C. Rael, R.F. Costantino, J.M. Cushing, T.L. Vincent, Using stage-structured evolutionary game theory to model the experimentally observed evolution of genetic polymorphism. Evol. Ecol. Res. **11**, 141–151 (2009)

[256] R.C. Rael, T.L. Vincent, J.M. Cushing, Competitive outcomes changed by evolution. J. Biol. Dyn. **5**(3), 227–252 (2011)

[257] J.D. Reeve, D.J. Rhoses, P. Turchin, Scramble competition in the Souther Pine beetle Dendroctonus frontalis. Ecol. Entomol. **23**(4), 433–443 (1998)

[258] W.E. Ricker, Stock and recruitment. J. Fish. Res. Board Can. **11**(5), 559–623 (1954)

[259] M. Rietkerk, F.V.D. Bosch, J. Ven de Koppel, Site-specific properties and irreversible vegetation, changes in semi-arid grazing systems. Oikos **80**, 2441–252 (1997)

[260] S.L. Robertson, J.M. Cushing, Spatial segregation in stage-structured populations with an application to Tribolium. J. Biol. Dyn. **5**(5), 398–409 (2011)

[261] S.L. Robertson, J.M. Cushing, R.F. Costantino, Life stage: interactions and spatial patterns. Bull. Math. Biol. **74**, 491–508 (2012)

[262] S.L. Robertson, J.M. Cushing, A bifurcation analysis of stage-structured density dependent integrodifference equations. J. Math. Anal. Appl. **388**, 490–499 (2012)

[263] D. Roff, *Evolution of Life Histories: Theory and Analysis* (Chapman and Hall, New York, USA, 1992)

[264] D.A. Roff, *Life History Evolution* (Oxford University Press, Oxford, England, 2001)

[265] S. Rosenblat, Population models in a periodically fluctuating environment. J. Math. Biol. **9**, 23–36 (1980)

[266] R. Ross, *The Prevention of Malaria*, 2nd edn. (John Murray, London, 1911)

[267] B. Ryals, R.J. Sacker, Global stability in the 2D Ricker equation. J. Differ. Equ. Appl. **21**(11), 1068–1081 (2015)

[268] B. Ryals, R.J. Sacker, Global stability in the 2D Ricker equation-revisited. Discrete Contin. Dyn. Syst. Ser. B **22**(2), 585–604 (2017)

[269] B. Ryals, A note on a parameter bound for global stability in the 2D coupled Ricker equation. J. Differ. Equ. Appl. **24**(2), 240–244 (2018)

[270] R. Sacker, On invariant surfaces and bifurcation of periodic solutions of ordinary differential equations, Report IMM-NYU **333** (1964), New York University

[271] R. Sacker, A note on periodic Ricker Maps. J. Differ. Equ. Appl. **13**(1), 89–92 (2007)

[272] I. Scheuring, Allee effect increases the dynamical stability of populations. J. Theor. Biol. **199**, 407–414 (1999)

[273] S. Schreiber, Allee effects, extinctions, and chaotic transients in simple population models. Theor. Popul. Biol. **64**, 201–209 (2003)

[274] S.J. Schreiber, Chaos and sudden extinction in simple ecological models. J. Math. Biol. 239–260 (2001)

[275] F.M. Scudo, J.R. Ziegler, *The Golden Age of Theoretical Ecology: 1923-1940*, Lecture Notes in Biomathematics (Springer, Berlin, 1978)

[276] Shandelle M. Henson, Robert A. Desharnais, R.F. Costantino, J.M. Cushing, Brian Dennis, *Nonlinear Patterns in Population Dynamics: Theory and Experiments* (Chapman & Hall/CRC, to appear 2025)

[277] A.N. Sharkovsky, Co-existence of cycles of a continuous mapping of a line into itself. Ukrainian Math. Z. **16**, 61–71 (1964)

[278] D. Singer, Stable orbits and bifurcation of maps of the interval. SIAM J. Appl. Math. **2**(35), 260–267 (1978)

[279] H. Smith, Planar competitive and cooperative difference equations. J. Differ. Equ. Appl. **3**(5–6), 335–357 (1998)

[280] H.L. Smith, Global stability for mixed monotone systems. J. Differ. Equ. Appl. **14**(10–11), 1159–1164 (2008)

[281] H. Smith, *Monotone Dynamical Systems: An Introduction to the Theory of Competitive and Cooperative Systems* (Mathematical Society of Japan, 2009)

[282] H. Smith, Periodic competitive differential equations and the discrete dynamics of competitive maps. J. Differ. Equ. **64**(2), 165–194 (1986)

[283] H. Smith, Planar competitive and cooperative difference equations. J. Differ. Equ. Appl. **3**(5–6), 335–357 (1998)

[284] H. Smith, P. Waltman, Perturbation of a globally stable steady state. Proc. Am. Math Soc. **127**(2), 447–453 (1999)

[285] E. Sober, *The Nature of Selection: Evolutionary Theory in Philosophical Focus* (University of Chicago Press, 1984)

[286] S.C. Stearns, *The Evolution of Life Histories* (Oxford University Press, Oxford, UK, 1992)

[287] A. Stephens, W.J. Sutherland, R.P. Freckleton, What is the Allee effect? OKOS **87**, 185–190 (1999)

[288] P. Turchin, Rarity of density dependence or population regulation with lags. Nature **344**, 660–663 (1990)

[289] M.B. Usher, A matrix approach to the management of renewable resources, with special reference to selection forests. J. Appl. Ecol. **3**, 355–367 (1966)

[290] M.B. Usher, A matrix approach to the management of renewable resources, with special reference to selection forests-two extensions. J. Appl. Ecol. **6**, 347–348 (1969)

[291] M.B. Usher, A matrix model for forest management. Biometrics **25**, 309–315 (1969)

[292] E. van der Jeijen, M.J. Crawley, R.M. Nisbet, *The Dynamics of a Herbivore-plant Interaction, Insect Populations: In Theory and Practice*, ed. by J.P. Dempster, I.F.G. McLean (Chapman and Hall, London, 1998)

[293] R.R. Veit, M.A. Lewis, Dispersal, population growth, and the Allee effect: dynamics of the House Finch invasion of Eastern North America. Am. Nat. **148**(2), 255–274 (1996)

[294] A. Veprauskas, *On the dynamic dichotomy between positive equilibria and synchronous 2-cycles in matrix population models*, Ph.D. dissertation, University of Arizona (2016)

[295] A. Veprauskas, Synchrony and the dynamic dichotomy in a class of matrix population models. SIAM J. Appl. Math. **78**(5), 2491–2510 (2018)

[296] A. Veprauskas, J.M. Cushing, Evolutionary dynamics of a multi-trait semelparous model. Discrete Contin. Dyn. Syst. Ser. B **21**(2), 655–676 (2016)

[297] A. Veprauskas, J.M. Cushing, A juvenile-adult population model: climate change, cannibalism, reproductive synchrony, and strong Allee effects. J. Biol. Dyn. **11**, Sup1, 1–24 (2017)

[298] T.L. Vincent, J.S. Brown, *Evolutionary Game Theory, Natural Selection, and Darwinian Dynamics* (Cambridge University Press, Cambridge, UK, 2005)

[299] E.O. Wilson, *The Future of Life* (Alfred Knopf, New York, 2002)

[300] A.-A. Yakubu, Introduction to discrete-time epidemic models. DIMACS Ser. Discrete Math. Theor. Comput. Sci. **75**, 83–109 (2010)

[301] A.-A. Yakubu, C. Castillo-Chavez, Interplay between local dynamics and dispersal in discrete-time metapopulation models. J. Theor. Biol. **218**, 273–288 (2002)

[302] S.-R. Zhou, D.-Y. Zhang, Allee effects and the neutral theory of biodiversity. Funct. Ecol. **20**(3), 509–513 (2006)

[303] Z. Zhou, X. Zou, Stable periodic solutions in a discrete periodic logistic equation. Appl. Math. Lett. **16**, 165–171 (2003)

[304] Q.Q. Zhang, Z. Zhou, Global attractivity of a nonautonomous discrete logistic model. Hokkaido Math. J. **29**, 37–44 (2000)

[305] N. Ziyadi, A.-A. Yakubu, Predator-induced and mating limitation-induced Allee effects in a discrete-time SIS epidemic model. Comput. Math. Appl. **66**(11), 2196–2210 (2013)

Index

A

Absorbing set, 170
Adaptive (or fitness) landscape, 384
Adaptive landscape, 395
African elephants, 86
Age structured model, 282
Allee effect, 33, 235, 246, 253, 297
Allwright-Singer, 32
Asymptotically periodic systems, 442
Attenuance, 419
Attracting, 19, 121
Attractor, 312
Autonomous equation, 5

B

Backward bifurcation, 292, 300
Backward-unstable bifurcation, 389, 393
Basic juvenile-adult model, 285
Basin of attraction, 19
Beddington et al. model, 131, 178
Beverton-Holt equation, 12
Beverton–Holt recruitment, 428
Bifurcation, 174, 411
Bifurcation analysis, 37
Bifurcation diagram, 38, 48
Bifurcation of endemic equilibria, 338
Bifurcation point, 37, 282, 390
Boundary equilibrium pair, 291

C

Cannibalism, 321
Center, 107
Chaos, 45
Characteristic polynomial, 63, 65
Cobweb analysis, 20

Comparison method, 161
Competition exclusion principle, 115
Competition models, 118
Competition models with the Allee effect, 253
Competitive, 158, 438
Competitive exclusion principle, 154
Competitive region, 439
Competitive systems, 158
Complex eigenvalues, 106
Component Allee effect, 242
Contest models, 15
Convex function, 265
Cooperative system, 164
Critical curve, 160
Critical number, 265
Critical trait, 390
Cushing-Henson conjecture, 424

D

Darwinian dynamic models, 379
Darwinian juvenile-adult model, 391
Darwinian Leslie-Gower competition model, 400
Darwinian Leslie-Gower model, 404
Darwinian logistic (Beverton-Holt) model, 381
Darwinian LPA model, 396
Darwinian Ricker equation, 413
Darwinian Ricker model, 398
Definition of sensitivity, 49
Deleterious effects of periodic forcing, 426
Demographic Allee effect, 33
Direction of bifurcation, 292
Discrete logistic (Beverton-Holt) equation, 11
Discrete-time SIR model, 190
Discrete-time SIRS model, 201

Disease acquired herd immunity, 227
Disease-free equilibrium point, 202
Dissipative, 171, 288
Distinct real eigenvalues, 105
Dominant eigenvalue, 63, 69, 91
Dynamic dichotomy, 304, 308, 313

E

Ebenman's model, 307
Eigenvalue, 62
Eigenvector, 62
Elaydi and Yakubu, 142
Endemic equilibria, 336
English Boarding School Influenza Outbreak, 200
Epidemic model, 216
Equilibrium pair, 291, 389
Error magnification, 51
Escape function, 192
Evolutionarily stable strategy (ESS), 380
Evolutionary Leslie-Gower model, 441
Evolutionary models, 379
Evolutionary stable strategies, 384
Evolutionary suicide, 405
Example Levin-Goodyear model, 290
Exclusion region, 272
Existence and stability of periodic solutions, 446
Exponential model equation, 13
Extinction equilibrium, 122, 287, 291
Extinction equilibrium pair, 291, 389

F

Feigenbaum number, 46
Fertility matrix, 287
Fertility rate, 2
Fiber, 275
Final size equation, 198
Final size of the epidemic, 194
Flour beetle, 319
Focus, 107
Forward bifurcation, 251, 292
Forward invariant, 288
Forward-stable bifurcation, 389, 393
Fredholm Alternative, 308
Fundamental Bifurcation Theorem, 286
Fundamental Bifurcation Theorem for Darwinian Models, 385
Fundamental Theorem of Demography, 75, 76, 281

G

Global bifurcation, 295
Global stability, 149, 287, 407
Globally asymptotically stable, 19, 29

H

Hadamard product, 92
Hartman-Grobman Theorem, 146
Herbivore-plan model, 263
Herd immunity, 205
Hierarchical models, 155, 269, 278
Hierarchical system, 156
Hierarchical type, 398
Host-parasitoid model, 267
Hyperbolic equilibrium (fixed) points, 37
Hyperbolic fixed point, 122

I

Immediate basin of attraction, 272
Imprimitive Leslie model, 301
Infected, 190
Influenza, 200
Inherent, 286
Inherent population growth rate, 286, 287
Inherent population projection matrix, 287, 291
Inherent reproduction number, 286, 287
Interspecific competition, 115
Intraspecific competition, 115
Invariant loops, 177, 316
Irreducible, 67, 68
Isoclines, 118

J

Jacobian matrix, 121
Jury's test, 108
Juvenile-adult Leslie matrix, 92
Juvenile-adult Leslie model, 87
Juvenile-adult model, 285
Juvenile-adult model with Allee effect, 297

INDEX

K
k-cycle, 40
Kermack-McKendrick continuous-time model, 191
Kestelman, 159
Kolmogorov, 156
Kolmogorov map, 439
Kolmogorov-type, 172, 380

L
Lagged Beverton-Holt model, 148
LaSalle Invariance Principle, 151
LaSalle's Invariance Principle, 187
Leslie age-structured model, 56
Leslie logistic model, 283
Leslie matrix, 58
Leslie matrix model, 60
Leslie model, 282
Leslie projection matrix, 63
Leslie-Gower model, 118
Levin-Goodyear model, 290, 296, 297
Liapunov exponent λ, 50
Liapunov exponents, 51
Liapunov functions, 150, 220
Liapunov Stability Theorem, 150
Life cycle graph, 60
Linear (density-independent) models, 3
Linear Leslie matrix model, 282
Linear matrix equation, 281
Linear periodic difference equations, 418
Linear systems, 101
Linearization principle, 25
Linearization principle of cycles, 138
Local bifurcation, 292
Local stability of competition systems, 122
Locally asymptotically stable (LAS), 19, 121
LPA model, 319, 321

M
Magnification of error, 51
Malaria model, 367
Matrix equation, 58
May model, 135
Mean infectious period, 199
Mediterranean flour moth *Ephestia kuehniella*, 15

Mixed monotone, 407, 438
Mixed monotone periodic systems, 440
Model of gonorrhea infection, 152
Modeling density effects, 10
Models with Constant or Asymptotically Constant Population Size, 222
Models with disease-free population cycles, 359
Models with migration/immigration, 6
Models with multiple attractors with no Allee effects, 236
Models with multiple infected classes, 345
Models with the Allee effects, 242
Models without migration, 3
Monotone, 30, 438
Monotone periodic systems, 438
Multiple trait, 397
Mutant species, 380

N
N-species Leslie-Gower competition model, 253
Negative density effect, 285
Neimark-Sacker bifurcation, 122, 174, 177, 319, 323
Neimark-Sacker Bifurcation Theorem, 177
Neither attenuance nor resonance, 435
Neubert-Caswell, 171
Next generation matrix, 79, 287
Nicholson-Bailey model, 129, 135
Non-extinction equilibrium, 291
Non-hyperbolic equilibrium points, 37
Nonautonomous, 5
Nonautonomous linear, 104
Nonautonomous models, 417
Nonautonomous Ricker model, 430
Nonhyperbolic fixed point, 122
Nonlinear (density-dependent) competition models, 115
Nonlinear autonomous (density-dependent) models, 9
Nonlinear Leslie model, 290
Nonlinear matrix equation, 283, 286
Nonnegative equilibrium pair, 291
Norm, 120
Normalized age distribution, 284
Numerical scheme to compute Liapunov exponents, 51

O
Operator norm, 289

P
$p - periodic$ dynamical system, 421
P. Aurelia, 115
P. Bursaria, 115
P. Caudatum, 115
Paramecium, 115
Period-doubling bifurcation, 37, 45, 122, 174, 175, 323, 362
Periodic mixed monotone maps, 440
Periodic orbits, 40, 137
Periodic point of period k, 40
Periodically forced Beverton-Holt model, 424
Periodically forced Ricker model, 433
Permanent, 263, 288
Perron-Frobenius Theorem, 66, 69, 287
Persistence, 44, 169
Phase space analysis, 105
Pine beetle *Dendroctonus frontalis*, 15
Poisson processes, 192
Population cycles, 40, 137
Population growth rate, 75, 281
Population projection matrix, 281
Population's demographic vector $\mathbf{x}(t)$, 58
Positive density effects, 10
Positive equilibrium pair, 291
Predator-prey models, 129
Predator–prey models with Allee effects, 262
Primitive, 70
Probability of survival, 2
Projection matrix, 58, 75, 285

Q
Quadratic model equation, 14

R
Red blood cell production, 112
Reducible, 67
Repeated real eigenvalues, 105
Reproduction number $\mathcal{R}_0$, 4, 76, 212, 282
Resonance, 419
Ricker Competition Model, 118
Ricker equation, 13
Ricker hierarchical competition model, 269
Ricker model, 24

S
Saddle, 123
Saddle-node bifurcation, 174, 182
SARS model, 373
Schwarzian derivative, 32, 38
Scramble competition model, 16
Secondary bifurcations, 316
SEIR, 355
SEIR model, 357
Semelparous, 12, 316
Semelparous Leslie model, 301, 304
Semi-stable, 275
Sensitive dependence, 49
Sensitivity, 49
Sensitivity analysis, 87
Sensitivity matrix, 90
Sharkovsky, 49
Single class 2-cycle, 303
Single class cycle, 303
Single-class 3-cycle, 316
Single-class k-cycle, 304
SIR, 355
SIR model, 193
SIS Discrete-Time Epidemic Models With Birth and Death, 209
SIS model, 221
Small amplitude perturbations, 445
Solution of the projection equation, 75
Source, 107
Spectral radius, 121
Stability analysis, 120
Stability and the direction of bifurcation, 294
Stability of endemic equilibria, 340
Stability of equilibrium points, 19
Stable, 19, 121
Stable bifurcation, 293, 300, 389
Stable Manifold Theorem, 144, 145
Strange attractor, 346
Strategy, 380
Strong Allee effect, 33, 235, 242, 253, 264, 300
Strong Allee effect with three attractors, 310
Strong ergodic property, 76, 285
Strongly connected, 67
Structured population model, 56, 281

INDEX

Susceptible, 190
Synchronous, 303
Synchronous cycle, 303, 316

T
Theorem Direction of Bifurcation, 292
Trait, 380
Transcritical bifurcation, 37, 281, 291
Transition matrix, 287
Transmission coefficient, 192

U
Uniformly persistent, 288
Unimodal, 46

Unstable bifurcation, 293
Usher matrix, 86

V
Vancouver Island marmot, 245
Variation of constants formula, 104

W
Weak Allee effect, 242, 253

GPSR Compliance

The European Union's (EU) General Product Safety Regulation (GPSR) is a set of rules that requires consumer products to be safe and our obligations to ensure this.

If you have any concerns about our products, you can contact us on ProductSafety@springernature.com

In case Publisher is established outside the EU, the EU authorized representative is:

Springer Nature Customer Service Center GmbH
Europaplatz 3
69115 Heidelberg, Germany

Batch number: 08152540

Printed by Printforce, the Netherlands